A practical introduction to electronic circuits

A practical introduction to electronic circuits

Third edition

MARTIN HARTLEY JONES
Managing Director, Engineering
Industrial Group
Smiths Industries plc

Published by the Press Syndicate of the University of Cambridge
The Pitt Building, Trumpington Street, Cambridge CB2 1RP
40 West 20th Street, New York, NY 10011-4211, USA
10 Stamford Road, Oakleigh, Melbourne 3166, Australia

First published 1977
Reprinted 1978, 1979, 1980, 1981 (with corrections), 1982, 1983
Second edition 1985
Reprinted 1987, 1988, 1990, 1992, 1993
Third edition 1995

Printed in Great Britain at the University Press, Cambridge

A Catalogue record for this book is available from the British Library

Library of Congress cataloguing in publication data

Jones, Martin Hartley, 1942–
A practical introduction to electronic circuits / Martin Hartley Jones. — 3rd ed.
p. cm.
Includes bibliographical references and index.
ISBN 0–521–47286–5. — ISBN 0–521–47879–0 (pbk.)
1. Electronic circuits. I. Title.
TK7867.J62 1995
621.3815—dc20 95–4035 CIP

ISBN 0 521 47286 5 hardback
ISBN 0 521 47879 0 paperback

Cover illustration from radar processing circuit.
Courtesy Kelvin Hughes Ltd.

TO CHRISTOPHER, TIMOTHY AND ELIZABETH

CONTENTS

PREFACE TO THE THIRD EDITION

The popularity of the first and second editions has been very gratifying. This third edition uses the same practical approach to provide an update relevant to today's exciting electronics scene whilst retaining the character of the basic text.

It is a sign of the relentless advance of electronics that key applications today such as the CD player, cellular telephone and fax machine were nowhere on the scene at the original publication date of 1977. The second edition in 1985 then recognised the burgeoning microcomputer industry, though some home computers popular at that time have now disappeared without trace.

Today, a decade later, the industry has seen remarkable progress, especially in computer processing speed, memory capacity and component packing density. Surface-mount technology is the norm and more and more signal processing and recording is carried out digitally. Fortunately for the student and the experimenter, the basic component and circuit elements explained in the book continue to be just as relevant as ever. Components are still widely manufactured in the wire-ended packages best suited to experimental work; the favourite BC107 transistor remains commonplace. In those circuits where components have become obsolete, suitable updates have been introduced.

In some areas, developing technology has changed the relative importance of particular topics. For example switch-mode power supplies are now deserving of greater coverage, as is the whole theme of analogue to digital conversion. Some other topics no longer as relevant have been quietly dropped. The ubiquitous PC now features in some characteristically practical interfacing experiments in addition to the programming examples using the Acorn computers popular in education.

Thank you to my colleagues at Kelvin Hughes for their advice and assistance, to Maureen Brown for her typing and to my wife Sylvia for her continuing secretarial help and motivation. I am also grateful to my sons, Chris and Tim, for their useful suggestions on the updates.

Martin Hartley Jones
July 1994

PREFACE TO THE SECOND EDITION

There have of course been many advances in electronics since the preparation of the first edition and this has been a welcome opportunity to bring the text up to date. The decade leap is reflected particularly in the two new chapters (13 and 14) on digital techniques and computers. Here I have aimed to present the important relationship of the microcomputer chip to other circuits, both digital and analogue, in a digestible form with plenty of experiments.

In the remainder of the book, many detailed changes have updated it without destroying the logical structure which has by now become familiar to students on many college and university courses. The approach throughout remains a practical one and is still based on my experience of teaching electronics in the Department of Pure and Applied Physics at UMIST. In recent years my work has been in industrial electronic design, further experience which has helped me to keep the text relevant to real applications.

In acknowledging gratefully the help I have received in preparing this second edition, I must first thank those readers of the first edition who wrote in with encouraging suggestions and amendments. Thank you too to my eldest son Christopher for his help in devising and testing the computer-based experiments. Alan Jubb was a great help in checking the manuscript of the new material.

Finally thank you again to my wife Sylvia not only for her typing skill but also for her encouragement and patience.

Martin Hartley Jones
February 1985

PREFACE TO THE FIRST EDITION

Anyone who is interested in technology is aware of the importance of electronics. Despite its all-pervading influence, however, electronics retains a strong element of mystery for many people who are otherwise well informed on technical matters. In this book, I aim to provide a practically-based explanation of the subject to try to dispel the mystery.

In the study of electronics, I have always found practical experience to be an invaluable stimulant and confidence-builder. There is enormous satisfaction in 'lashing up' a new circuit on the test bench and seeing it work for the first time. As an experimental physicist, I have never regarded electronic design as an end in itself, but rather as a valuable tool in research and development. The practical approach is therefore of primary importance in the electronics content of the UMIST Pure and Applied Physics degree course. Experience of teaching electronics to people of various backgrounds has encouraged me in the belief that a book based on a practical viewpoint can be valuable at all levels from school, through college or university, to the industrial laboratory.

This book aims to cover a wide range of circuit building bricks, analogue and digital, discrete and integrated. It is from these bricks that an elaborate system is constructed, whether it be a colour television set or a computer. Sufficient practical information is given to enable the reader to construct the circuits and test them in the laboratory. There are no formal exercises: I hope that the reader will be stimulated to test for himself whether the circuits work as I describe.

Very few of the circuits are critical regarding component layout. Most can be quickly soldered together on a tagboard or printed-circuit stripboard. A 6 V or 9 V radio battery makes a perfectly adequate power supply in most instances. A simple audio generator, an oscilloscope and a multimeter will facilitate testing.

The only mathematics required for a satisfactory understanding of most circuits is Ohm's law:

$$\text{current} = \frac{\text{voltage}}{\text{resistance}}.$$

Differential calculus is occasionally used as a convenient way to measure the slope at a particular point on a curve. For the sections of the book which deal with frequency or phase response, some knowledge of reactive circuits, using j-notation, will be helpful.

The bibliography at the end of the book lists some introductory texts which will fill in the background for the reader with little experience of electrical circuits. The bibliography also lists books which discuss detailed topics to a level which is inappropriate here.

I would commend every reader to make maximum use of the many data sheets and application notes published by all semiconductor manufacturers. The various monthly electronics magazines, readily available from news-agents, are also invaluable sources of practical information. Electronics is a rapidly-developing subject. Whilst this book aims to provide a foundation of knowledge and experience that will be relevant for many years, it is only by familiarity with the latest literature that the electronic engineer can remain abreast of the current state of the art.

It is impossible in a reasonable space to identify and thank individually all those who have contributed to this book. I must thank all my staff colleagues and students, both past and present, whose astute questions have so often initiated a broadening of my own understanding. My special thanks must go to my wife, Sylvia. Not only has she skilfully typed the manuscript, but she has also been a constant source of encouragement, and has accepted with equanimity the domestic disorder inevitable in this type of enterprise.

Martin Hartley Jones
November 1975

1

Amplification and the transistor

1.1 Amplification

The single most important function in electronics can be expressed in one word: *amplification*. This is the process whereby the power of a signal is increased in magnitude. A simple mechanical example of amplification is provided by the power steering system on cars and commercial vehicles, where a small force applied to the steering wheel by the driver is amplified hydraulically to produce the force required to move the front wheels of the vehicle. Here is the basic feature of an amplifier: a small input signal is used to control a more powerful output signal. The extra power is drawn from some external energy source, the latter being the vehicle engine in this instance.

The earliest example of electrical amplification is the electromagnetic relay, invented by Joseph Henry in 1835, and used by Samuel Morse to increase the power of weak telegraph signals. It was the relay which made possible the first long-distance telegraph line, from Baltimore to Washington, which was opened in 1844. As can be seen in fig. 1.1, the weak incoming signal is used to operate an electromagnet which attracts an armature and closes electrical contacts; these contacts then switch a powerful outgoing signal which is transmitted to the next leg of the line. The dots and dashes of the strong output signal are thus a faithful replica of the weak input. Relays are still used extensively in power switching systems, but are generally being superseded by electronic methods.

Electronic amplifying devices are known generally as *active* components to distinguish them from non-amplifying circuit elements such as resistors, capacitors and inductors, which are grouped under the heading of *passive* components.

The most everyday application of electronic amplification is the ordinary radio, which receives a tiny input signal at its antenna (typically less than one

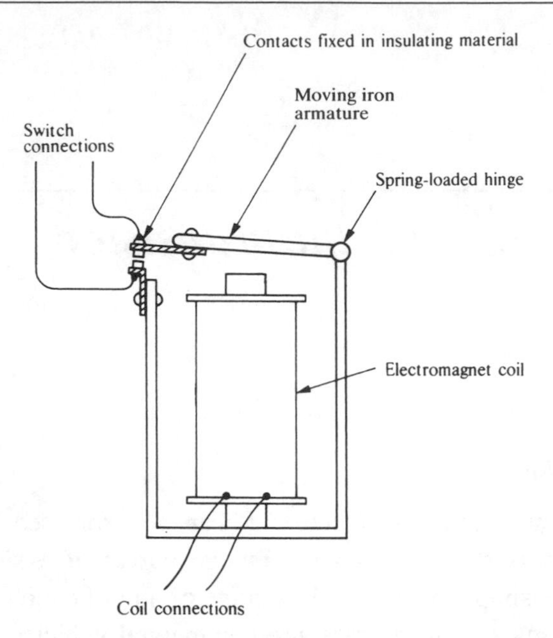

Fig. 1.1. The electromagnetic relay – an example of electrical amplification.

microwatt) and yet can turn out a power of several watts to the loudspeaker. The extra power involved is drawn from a battery or the a.c. mains.

1.2 The transistor as an amplifying device

The *bipolar junction transistor*, better known simply as the transistor, is the most common active device in electronics. Before discussing how a transistor works, it is useful to look at what it can do. For this purpose, we shall regard the transistor as a 'black box' whose circuit symbol is shown in fig. 1.2.

The transistor is a current-controlled amplifying device: if a small current flows between the base and emitter, it gives rise to a much larger current between collector and emitter. The name *transistor* is in fact derived from the two words *transfer-resistor*: a small base current is transferred to the collector circuit in greatly magnified form.

Some simple experiments will demonstrate this high current gain of the transistor. In the simple lamp circuit of fig. 1.3(*a*), it is clear that the bulb should light when the two free wires (numbered 1 and 2) are connected together. It is equally obvious that, if you try to complete the circuit through your body by holding one wire in each hand, insufficient current will flow to

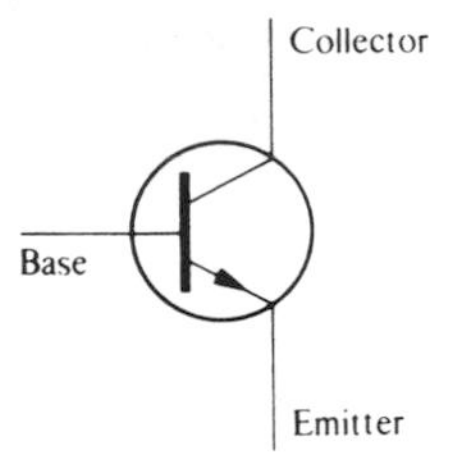

Fig. 1.2. Circuit symbol for the bipolar transistor (npn-type).

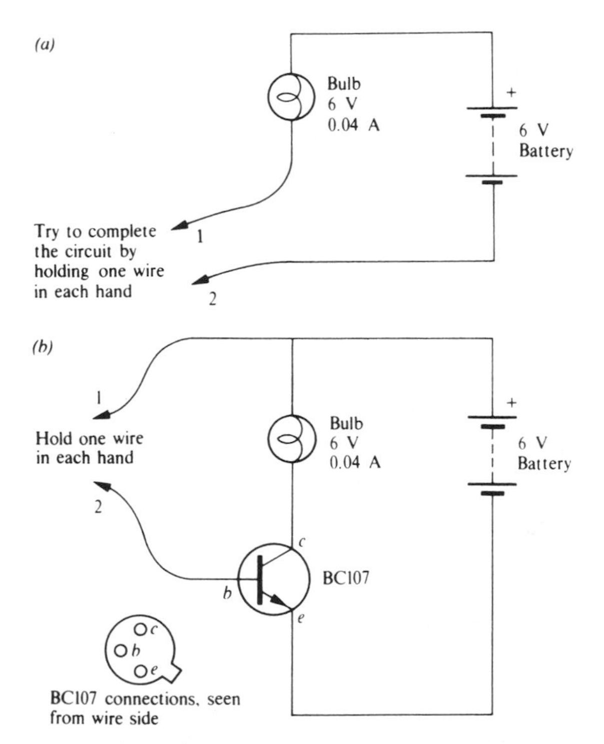

Fig. 1.3. Experimental bulb circuit to illustrate current amplification by transistor. (*a*) Tiny current in body will not light bulb. (*b*) Transistor amplifies body current to light bulb.

light the bulb. In fact the current is limited to less than 1 mA by body resistance, whilst the bulb specified needs 40 mA to light fully.

Now look at the circuit of fig. 1.3(*b*), where the bulb is in the collector–emitter circuit of a transistor. Hold the free wires once again; this time, the small current flows via your body resistance, from the battery into the

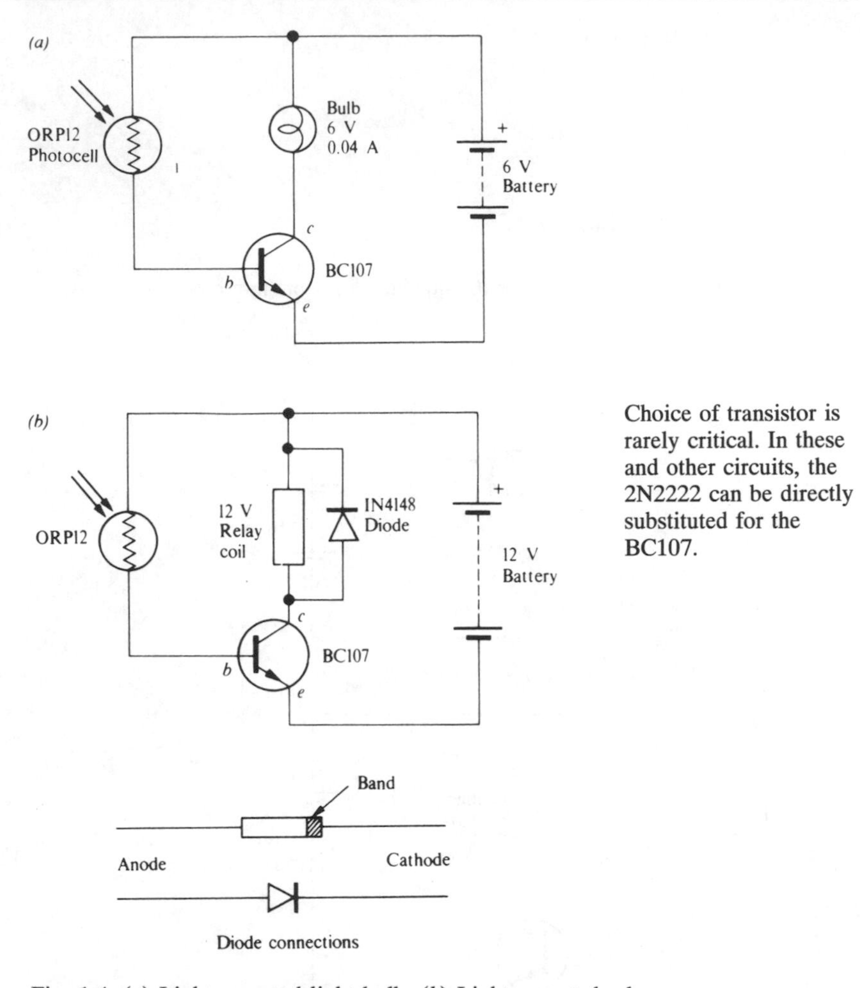

Choice of transistor is rarely critical. In these and other circuits, the 2N2222 can be directly substituted for the BC107.

Fig. 1.4. (*a*) Light-operated light bulb. (*b*) Light-operated relay.

base–emitter circuit of the transistor. The transistor acts as a current amplifier and will light the bulb, albeit dimly. Moistening the hands helps to reduce skin resistance and gives a better result. The tiny current in your body is controlling a current some hundred times larger in the bulb.

The circuit of fig. 1.4(*a*) takes the experiment a stage further by using the transistor to make a simple light-operated switch. The transistor base circuit is completed here by the ORP12 cadmium-sulphide photocell which behaves as a light-dependent resistor. When the cell is in the dark, its resistance is several megohms and negligible base current flows in the transistor. In reasonably bright light, the cell resistance falls to a few kilohms and the base current

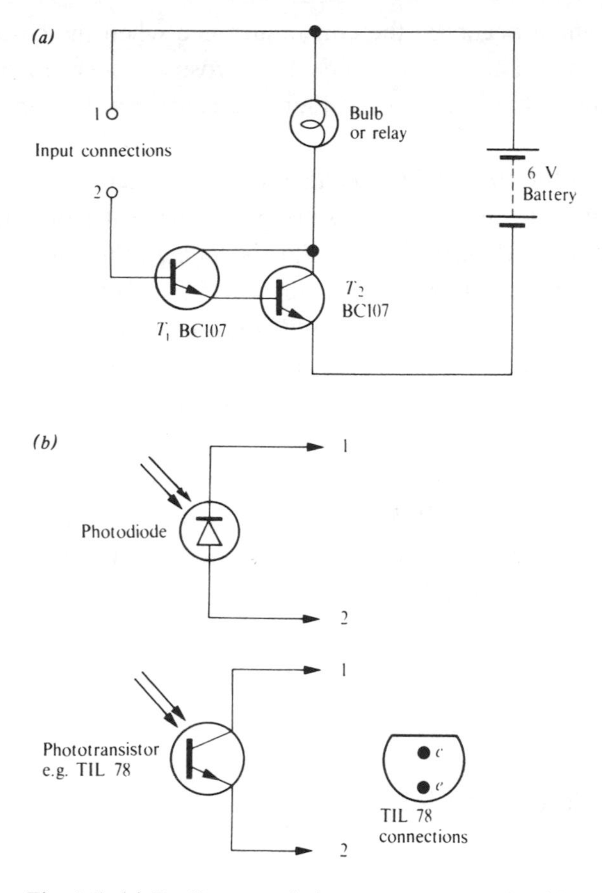

Fig. 1.5. (*a*) Darlington pair increases current gain. (*b*) Connection of photodiode and phototransistor to Darlington pair.

of about a milliamp lights the lamp, thanks to the amplification of the transistor.

Turning the cell to face the bulb makes an 'electronic candle': in the dark, the bulb is out, but if light momentarily falls on the cell, the bulb lights and remains lit, the photocell current being sustained by its light. To extinguish the 'candle', it is only necessary to interrupt the light path between bulb and photocell.

In fig. 1.4(*b*), the collector current is used to operate the coil of a relay; the relay contacts can then be used to switch on or off any required device, such as a motor to open a garage door when the car headlamps illuminate the ORP12 cell. The IN4148 diode connected across the relay coil serves to clip off the voltage surge which occurs in the coil inductance when the current is switched off. A diode should always be connected across a solenoid which is transistor-controlled; the voltage surge can otherwise cause breakdown in the transistor.

Fig. 1.5 shows a way of further increasing the current gain of a circuit. Known as a *Darlington pair*, the two transistors give a current gain equal to the products of their individual current gains. This is because the base–emitter current of T_2 is equal to the collector–emitter current of T_1. If the connections 1 and 2 are held, one in each hand, the bulb will light brightly: the circuit is much more sensitive than fig. 1.4(*a*). The extra current gain means that a photodiode or phototransistor can be used as a light sensor if connected as shown in fig. 1.5(*b*).

The current gain of a transistor is normally given the symbol h_{FE}, and its value may be anything from 10 to 1000 depending on the type of transistor. The current gain of the BC107 transistor usually lies in the range 100 to 400; current gain is not a closely controlled parameter.

In the Darlington pair,

$$h_{FE\text{total}} = h_{FE1} \times h_{FE2}.$$

1.3 Introduction to solid-state devices

1.3.1 General

The transistor is made up of two different types of semiconductor material. In order to understand how a transistor works it is necessary to look at some of the properties of that unusual class of materials called semiconductors.

1.3.2 Semiconductors

Solid materials may be divided into three classes as far as electrical properties are concerned: conductors, insulators and semiconductors. The class into which a material falls depends on the behaviour of the electrons in the outermost orbit of the atoms. In the case of an insulator, such as polythene, these *valence* electrons are tightly bound to the nucleus; very few are able to break free from their atoms to conduct an electric current. A conductor, however, such as copper, has a great cloud of free electrons present at all temperatures above absolute zero; the valence electrons are very weakly bonded indeed to their parent atoms so that they drift freely.

Semiconductors are exceptional materials. The one most commonly used in transistors is silicon; germanium is also in use. Both these elements are tetravalent, i.e. there are four electrons in the outer orbit of the atoms. Silicon and germanium crystals have a very neat system by which the atoms are held together in stable form: this is known as covalent bonding. It is a fact that when an atom has eight valence electrons it turns out to be in a very stable state (the inert gases are in such a state). Atoms in a silicon or germanium crystal have a mutual sharing arrangement whereby each nucleus has a 'half share' in eight valence electrons instead of the exclusive possession of four valence electrons that an isolated atom would have. Such an arrangement of silicon atoms is shown diagrammatically in fig. 1.6(*a*); each bond drawn between the atoms represents a shared valence electron. It is interesting to note at this stage that the extreme hardness of diamond is due to the tetravalent carbon atoms adopting this type of ordered covalent crystal structure. Diamond is actually classed as a semiconductor, though the tight bonding which gives rise to its physical hardness renders it a very poor electrical conductor indeed. It is fortunate that we have much better and cheaper alternative materials available for transistors!

1.3.3 Electrons and holes

The perfect array of silicon atoms in fig. 1.6(*a*) is only found at temperatures near absolute zero. At room temperature, thermal vibration of the atoms causes a few bonds to fracture; electrons break away from the atoms and are free to wander through the crystal. Where an electron breaks free it leaves behind a *hole* or absence of negative charge, which can also appear to move if it is filled by an electron from an adjacent atom. Fig. 1.6(*b*) represents a section of silicon crystal structure at room temperature with a free electron and the resulting hole.

This availability of free electrons makes the silicon a conductor of electricity, albeit a very poor one. If the silicon is connected into circuit, for instance across a battery, then the applied field draws the free electrons towards the positive terminal, whilst further free electrons are made available at the negative terminal and can travel through the semiconductor by hopping from hole to hole. Thus a current flow is established. As the temperature of the semiconductor is raised, more bonds are broken, more electrons and holes become available and the conductivity increases. It is interesting to note that this temperature effect is directly opposite to the effect observed in conducting metals: in a conductor there is already such a cloud of free electrons available,

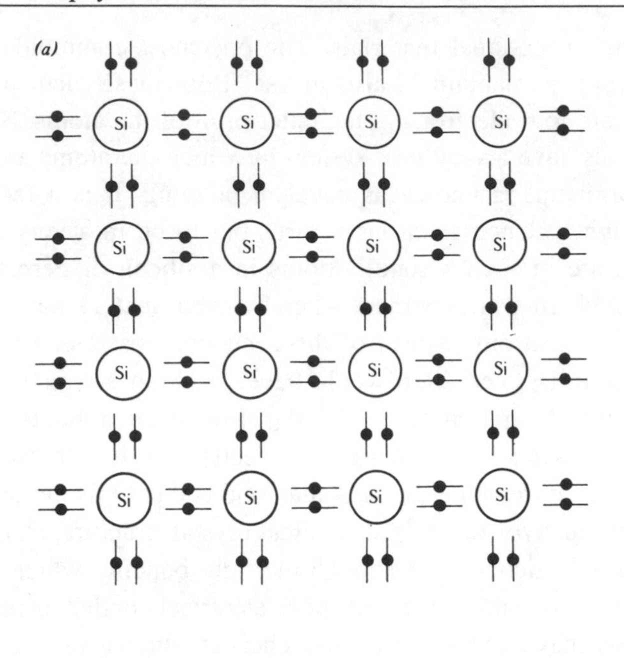

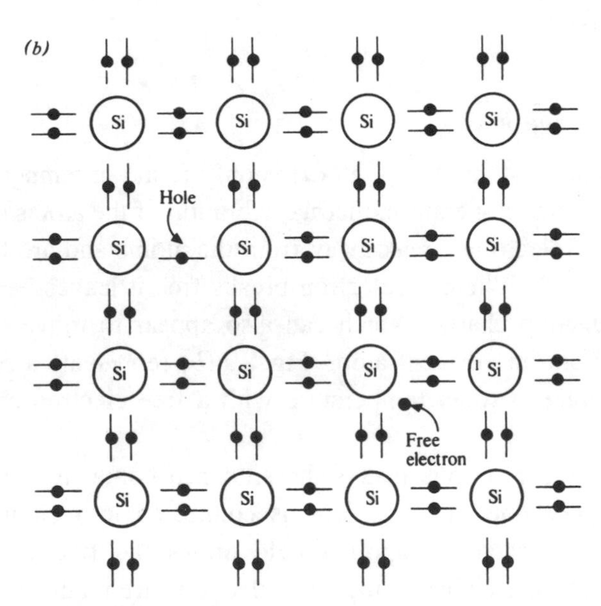

Fig. 1.6. (*a*) Atoms in a silicon crystal showing covalent electron bonds. (*b*) Crystal lattice as (*a*), but with free electron and hole produced by thermal agitation.

even at low temperatures, that the limiting factor as far as conduction is concerned is not the availability of free electrons, but the ability of those electrons to flow past the metal atoms. As the temperature increases in a conductor, the amplitude of vibration of the atoms is increased so that they offer a greater obstruction to the free electrons. Thus, in a conductor, resistance increases with increasing temperature whilst, in a semiconductor, resistance decreases with increasing temperature. The very weak conductivity exhibited by semiconductors in their pure form is known as *intrinsic* conductivity.

1.3.4 Extrinsic conductivity

The addition of impurities to a semiconductor can produce interesting results. Certain impurity atoms are able to fit into the crystal lattice without introducing undue strain and, if the valency of these atoms is different from that of the semiconductor, then conductivity is greatly increased. Fig. 1.7 shows the effect of introducing pentavalent phosphorus atoms into a silicon crystal. Four of the five valence electrons are involved in bonding with the silicon lattice, but the remaining electron is so weakly bound that it is free to move about the crystal and thus to conduct current. The introduction of impurities to semiconductors is known as *doping* and the resulting conductivity known as

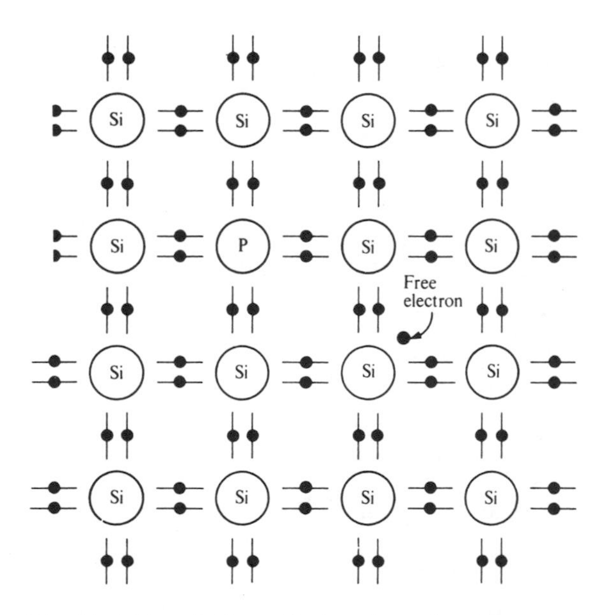

Fig. 1.7. Silicon lattice with phosphorus impurity atom. A free electron is now available for conduction (n-type semiconductor).

extrinsic conductivity. A pentavalent impurity such as phosphorus is known as a *donor* impurity, since it gives free electrons to the lattice. Because the extrinsic conductivity is due to free *negative* charges (electrons), this type of doped semiconductor is known as *n-type*.

Fig. 1.8 shows the effect of introducing tervalent boron atoms into a crystal of silicon. Although it has only three valence electrons, the boron atom accepts an extra electron from one of the adjacent silicon atoms to complete its covalent bonds. This leaves a *hole* or absence of a valence electron in the lattice and this hole is free to move and thus act as a means of conduction. Of course, it is the valence electrons that do the shifting, but the result is that the hole is shuffled from atom to atom. A tervalent impurity such as boron is known as an *acceptor* impurity because of its ability to accept an electron when it enters the lattice. Because conduction is now due to the *positive* holes, the doped semiconductor is known as *p-type*.

It is important to realize that neither p-type nor n-type material possesses an *overall* electric charge. In each case, the total number of electrons is balanced by an identical number of protons in the atomic nuclei. The p and n designations simply refer to the type of charge responsible for conduction within the crystal.

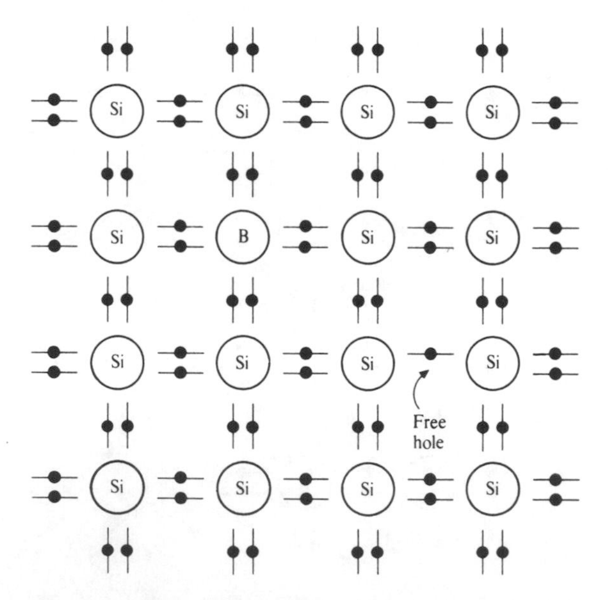

Fig. 1.8. Silicon lattice with boron impurity atom. A free hole is now available for conduction (p-type semiconductor).

1.3.5 Majority and minority carriers

In doped semiconductors, although it is the impurities that are chiefly responsible for conduction, the thermal breaking of bonds, which causes the intrinsic conductivity of pure semiconductors, is still continuing. Thus, in an n-type material, as well as the free electrons from the donor impurity, there are a few holes present because of thermal generation of electron–hole pairs. Likewise in p-type material there are a few thermally generated electrons. The current carriers deliberately introduced by doping are known as majority carriers, whilst the thermally generated ones are called minority carriers.

1.3.6 Compensation

It is possible to convert n-type material to p-type or vice versa simply by adding sufficient impurity to outnumber the majority carriers. The free electrons, present when the material was n-type, fill holes in the p-type material and are thus immobilized. This principle, which is known as compensation, is used to good effect in the manufacture of transistors where different impurities are diffused into the crystal at various stages of manufacture.

1.3.7 The pn junction

The operation of a semiconductor device such as the transistor depends on the effects occurring at the junction between p- and n-type materials. It is essential at this stage to understand that a semiconductor junction is a change from p- to n-type material within the same continuous crystal lattice. Simply placing a piece of p-type material in contact with a piece of n-type will not normally result in a pn junction.

Fig. 1.9 shows a pn junction *diode* with a metallic contact on each side. Underneath the junction diagram is a simple graph showing how the potential varies through the junction. As soon as the junction is formed, some of the free electrons near the junction in the n-type material cross over to fill some holes in the p-type; in doing so, they leave behind a net positive charge, whilst at the same time giving the p-type material a negative charge. These charges form a potential barrier which opposes further migration of electrons across the junction so that a state of equilibrium is reached. In this state, the region near the junction is relatively clear of holes and free electrons as a result of the initial migration. This is called a *depletion layer* and is typically less than one micron wide.

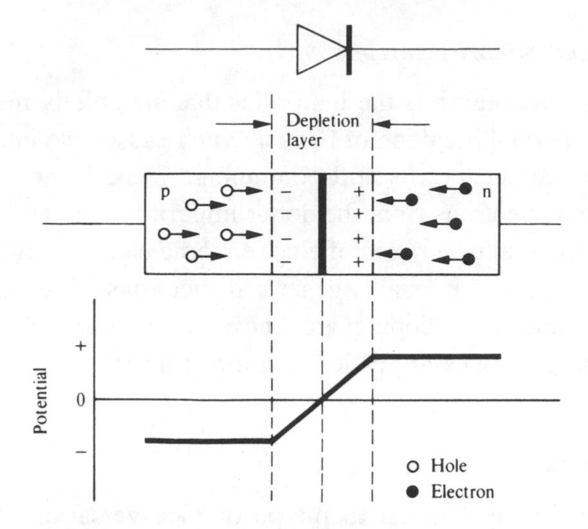

Fig. 1.9. Diode circuit symbol with pn junction, showing depletion layer and variation of potential.

1.3.8 Biased pn junction

If an external d.c. supply is connected to a pn junction, the potential barrier at the depletion layer is either increased or decreased depending on the polarity of the external supply or *bias*. Fig. 1.10 shows the two conditions of (*a*) reverse bias, where the potential barrier is reinforced and the depletion layer widened, and (*b*) forward bias, where the effect of the barrier is decreased and the depletion layer narrowed. Under reverse-biased conditions, the only current flowing across the junction is the tiny one due to thermal breaking of bonds in both p- and n-type materials. The minority carriers are of the appropriate polarity to be drawn across the junction. At room temperature this reverse current is, however, so small in a silicon junction (typically 1 nA) as to be negligible for most practical purposes. When the junction is forward-biased, however, the potential barrier is decreased in height, equilibrium is upset and some electrons in the n-region and holes in the p-region are able to cross the junction. The greater the forward-bias voltage, the lower the potential barrier becomes and the more electrons and holes cross the depletion layer. Hence a net current flow is established across the junction.

It is important to note that, as the forward e.m.f. applied across the junction is increased, so the effective resistance of the junction is decreased owing to the reduction of the potential barrier. The result is that a very small increase in applied voltage in the forward direction results in a large increase in current.

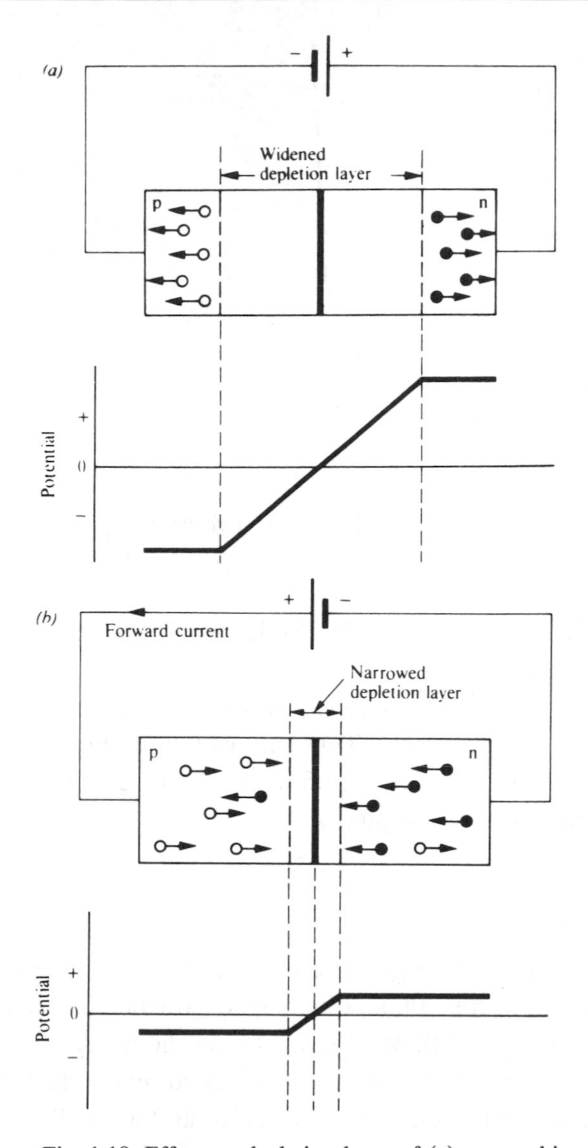

Fig. 1.10. Effect on depletion layer of (*a*) reverse bias, (*b*) forward bias.

Typically, in a small silicon diode, a forward voltage of 0.6 V produces a current of 1 mA, and a forward voltage of 0.8 V a current of 100 mA. The forward and reverse characteristics of a typical small silicon diode are shown on a graph of current against applied e.m.f. in fig. 1.11. It is clear from the graph that a silicon junction does not begin to conduct significantly until a forward e.m.f. in the region of 0.5 V is present. Germanium junctions exhibit

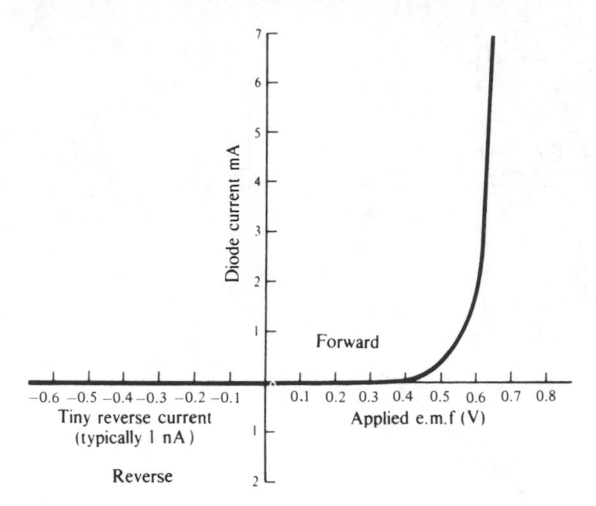

Fig. 1.11. Forward and reverse current–voltage characteristics of a pn silicon diode.

a smaller potential drop, in the region of 0.2 V. Diode characteristics are examined in more detail in chapter 6.

In brief, a diode allows current to flow freely in one direction but presents a nearly infinite resistance the other way. This *one-way* characteristic points to an important use of diodes: rectification, the conversion of a.c. to d.c. Applications in this field are discussed in chapter 9.

1.3.9 Avalanche breakdown

Under reverse-biased conditions, although the diode appears as an insulator, if the applied voltage is steadily increased, a point is reached where the junction suddenly begins to conduct (fig. 1.12). This is due to the thermally generated electrons acquiring sufficient energy as they are accelerated by the field in the depletion layer to generate further electron–hole pairs as they collide with silicon atoms. These newly-liberated carriers then generate still further carriers and the avalanche sets in. The process is non-destructive as long as the current is externally limited to avoid overheating the junction. Avalanche breakdown, which can occur at voltages from 5 V to 1000 V or more, depending on the construction of the diode and degree of doping of the silicon, sets a limit on the permissible peak inverse voltage in rectifier circuits.

Special diodes with low reverse breakdown voltages are manufactured: these are usually called Zener diodes after Carl Zener who suggested, in 1934, a mechanism for electrical breakdown. Zener diodes are made using very

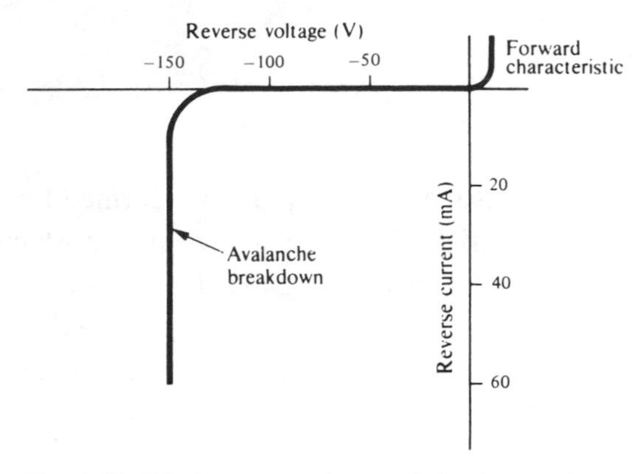

Fig. 1.12. Diode reverse characteristic showing typical avalanche breakdown.

heavy doping, creating a high concentration of majority carriers and forcing the depletion layer to remain thin, even under reverse-biased conditions; the resulting high potential gradient can result in breakdown under a reverse e.m.f. as low as 3 V. The very sharp breakdown characteristic shown in fig. 1.12 indicates that the p.d. across a diode under breakdown conditions is almost constant over a wide range of reverse current. For this reason, Zener diodes are used to stabilize the output voltage of power supplies (chapter 9). Strictly speaking, Zener breakdown, which involves tunnelling through the potential barrier, only applies to diodes breaking down at less than 5 V; above 5 V they should be called avalanche diodes.

1.3.10 Junction capacitance and varactor diodes

A reverse-biased diode behaves like a small capacitor, a typical capacitance value of a small silicon diode being 2 pF. The depletion layer acts as the insulating dielectric between the n and p conducting ‘plates’. Furthermore, the capacitance falls slightly as the reverse-bias voltage is increased because the depletion layer becomes wider. By appropriate choice of doping, special varactor diodes are manufactured which typically exhibit a capacitance change from 10 pF to 2 pF as the reverse bias is increased from 2 V to 30 V. Such diodes are extensively used in tuning VHF radio and UHF TV sets, the facility for electrical control of the tuning capacitance often being exploited in special circuits which enable the set to lock onto the desired station automatically (automatic frequency control).

1.4 The transistor

1.4.1 Introduction

The bipolar junction transistor consists of two pn junctions formed by a sandwich of doped semiconductor material. Fig. 1.13 shows the most common arrangement, the npn transistor. A thin layer of lightly doped p-type material (the base) is sandwiched between two thicker layers of n-type material (emitter and collector); the p-type base layer may be thinner than one micron.

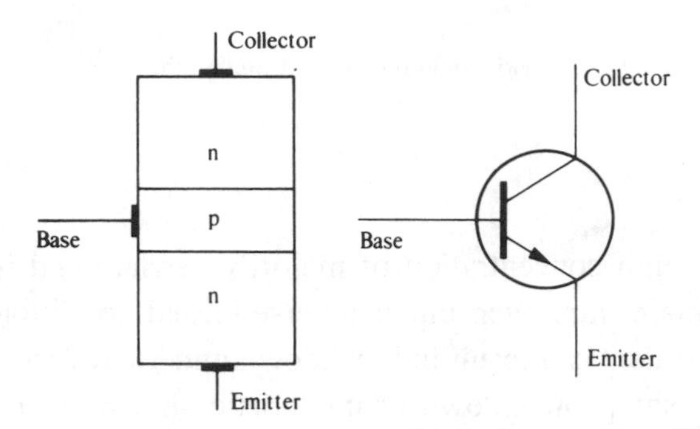

Fig. 1.13. The npn transistor – construction and circuit symbol.

1.4.2 Transistor action

Fig. 1.14 shows a transistor connected into a simple *common-emitter* circuit. In fig. 1.14(*a*) no base current is flowing whilst in fig. 1.14(*b*) the switch S is closed, allowing a current to flow from battery B_1 into the base of the transistor. Consider fig. 1.14(*a*) first of all; the important point to note is that the collector–base junction is reverse-biased, the resulting potential barrier preventing any flow of majority carriers. Thus, neglecting leakage, the current in the collector circuit is effectively zero with switch S open. Now consider what happens when S is closed (fig. 1.14(*b*)). The base–emitter junction becomes forward-biased whilst the collector–base junction remains reverse-biased. Owing to the forward bias on the base–emitter junction, electrons from the n-type emitter cross into the p-type base, where they diffuse across towards the depletion layer at the base–collector junction. Once the electrons reach the depletion layer, being minority carriers in the base region, they have a 'downhill run' through the potential barrier and are rapidly swept into the collector, thus establishing a collector current in the transistor. The action of

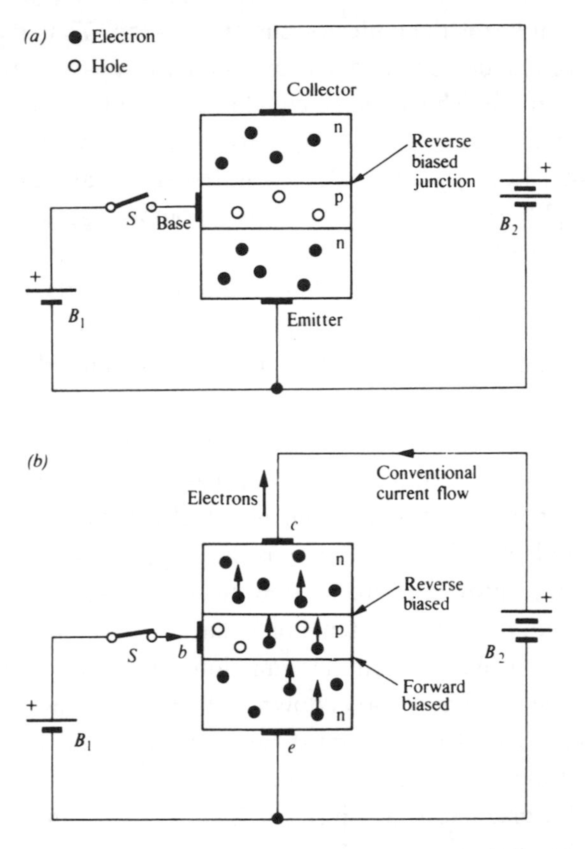

Fig. 1.14. Diagrams showing transistor action: (*a*) no base current, (*b*) base current flowing.

forward-biasing the base–emitter junction is like opening a gate and allowing a current to flow in the collector–emitter circuit. This is transistor action.

A further point requires explanation. Why do the electrons not recombine with holes in the p-type base region as they diffuse to the collector? The answer is that, by making the base of very lightly doped p-type material, that is with a low concentration of holes, and at the same time using a very thin base, there is only a small chance of an electron being waylaid by a hole and recombining. When an electron does recombine in the base region, it upsets for a moment the equilibrium, because the base has captured a negative charge. This is corrected by a hole supplied by base battery B_1. It is the supply of holes to compensate for recombination in the base which constitutes the base current of the transistor. It is therefore the base current which avoids a build-up of negative charge on the base and keeps the base–emitter junction forward-biased; this in turn keeps the collector current flowing. Thus the transistor is a current-controlled device. The current gain (h_{FE}) is the ratio of collector current to base current. This must be equal to the number of electrons per second making a successful trip from emitter to collector, divided by the number which recombine. In a typical small silicon transistor, an electron in the base region has roughly a 1 in 100 chance of recombining, so that the current gain is of the order of 100.

It is the fact that both electron and hole currents are involved in its operation which gives the transistor its full title of bipolar junction transistor. This term distinguishes it from the unipolar or field-effect transistor which will be considered in the next chapter.

It was mentioned earlier that, in a forward-biased junction, both electrons and holes are involved in conduction; in the forward-biased base–emitter junction of the transistor we have so far only considered electrons crossing the junction. This assumption is justified in practice as the n-type emitter is deliberately very heavily doped, providing many free electrons, whilst the base region is very lightly doped, providing so few holes that they can be neglected in considering conduction in the base–emitter junction. So heavily doped is the emitter region that the avalanche breakdown voltage of a base–emitter junction is typically only 6 V. This fact must be borne in mind in certain switching circuits, where care must be taken to avoid excessive reverse bias, but it can also be a useful characteristic, since the base–emitter junction of a small silicon transistor behaves as a 6 V Zener diode and is sometimes used that way.

1.4.3 Second-order effects

Fig. 1.15 shows a graph of collector current plotted against base current for a small silicon transistor: there is clearly a linear relationship between I_C and I_B over most of the collector current range. At low values of base current, however, the current gain is somewhat reduced. This is explained by considering the behaviour of electrons in the base: at very low base currents the electrons which cross from the emitter to the base region do not have any strong encouragement to reach the collector. It is only when they reach the collector–base depletion layer that they are swept up by the field; before this they are simply diffusing across the base in random fashion, prey to recombination with any hole they may encounter on the way. At higher values of base current, conditions are healthier for the electrons. The holes injected by the base current build up a slight electric field in the base which helps to draw the electrons into the depletion layer. Thus,

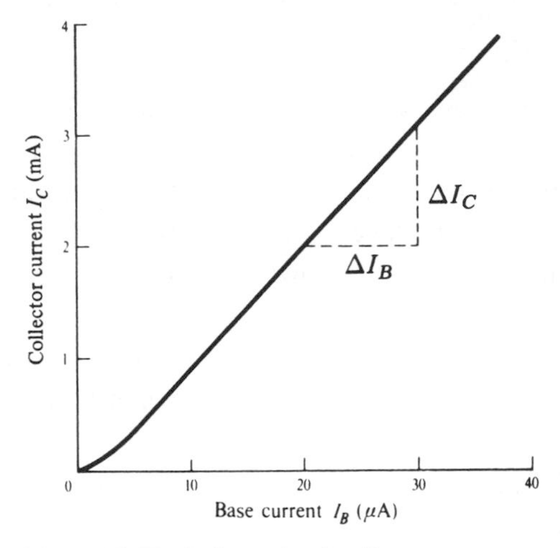

Fig. 1.15. Typical graph of collector current against base current in a silicon transistor.

at moderate collector currents of about 1 mA, a transistor will exhibit higher current gain than at low collector currents around 10 μA.

At very high collector currents, when the hole population in the base is very high, gain begins to fall. The base behaves as if it were more heavily doped than it really is, so that a significant fraction of the current across the base–emitter junction consists of holes going from base to emitter as well as the useful electrons going the other way towards the collector. Thus, more and more of the base current is 'wasted' and the current gain falls. This effect is of importance in power amplifiers, where it can lead to waveform distortions at high collector currents.

Because of the slightly non-linear relation between collector current and base current, there are two ways of specifying the current gain of a transistor in the common-emitter circuit. The d.c. current gain is simply obtained by dividing collector current by base current; this is given the symbol h_{FE}, $\bar{\beta}$ or α' and is important in switching circuits. For most amplification purposes, however, we are only concerned with small variations in collector current, and a more appropriate method of specifying current gain is to divide the *change* in collector current by the *change* in base current and obtain the small signal current gain h_{fe} or β where, from fig. 1.15,

$$h_{fe} = \frac{\Delta I_C}{\Delta I_B}.$$

For most practical purposes, h_{FE} and h_{fe} may be considered identical.

1.4.4 Collector–base leakage current

Since the collector–base junction is reverse-biased, there exists a tiny leakage current into the base from the collector, termed I_{CBO}, because it is measured with the emitter open circuit. In a silicon transistor at room temperature, I_{CBO} is small, usually less than 0.01 μA. However, when the transistor is in the common–emitter connection, the fact that I_{CBO} flows into the base means that, if the base is open-circuit as in fig. 1.14(a), I_{CBO} must flow into the emitter, at which point it is indistinguishable from an external base current. Thus I_{CBO} is amplified by the transistor and gives rise to a leakage between collector and emitter $I_{CEO} = h_{FE}I_{CBO}$, which can be as high as 1 μA. Since I_{CBO} is largely a result of thermal breaking of bonds, it increases with temperature, roughly doubling in value for every 18 deg C rise in temperature. When I_{CBO} becomes comparable with the normal circuit collector current, a transistor is usually considered too hot. Silicon junctions can operate up to 200 °C, but germanium junctions, which have a much higher leakage, can only go to about 85 °C.

When silicon transistors operate at room temperature I_{CBO} and I_{CEO} can usually be completely neglected. In a germanium transistor, I_{CBO} is typically 2 μA at room temperature (20 °C), so that if h_{FE} is 100, then I_{CEO} will be 200 μA. This relatively high leakage current is one reason why germanium devices are obsolete except for special purposes, where the low forward p.d. of the germanium junction is required.

1.4.5 npn and pnp transistors

The description of transistor action given above refers to npn devices, which are the most common; pnp devices are also readily available and are very useful in a whole range of *complementary* circuits since they offer characteristics identical to npn transistors but with the opposite polarity of supply voltage. Whereas in the npn transistor the collector current consists of electrons, in the pnp transistor it consists of holes. Likewise, the base current is an electron-flow instead of a hole-flow. Fig. 1.16 shows a pnp structure and its circuit symbol.

1.4.6 Optoelectronic devices

Reverse leakage current in a pn junction is, as we have seen, due to minority carriers.

The energy which liberates these hole–electron pairs is normally purely

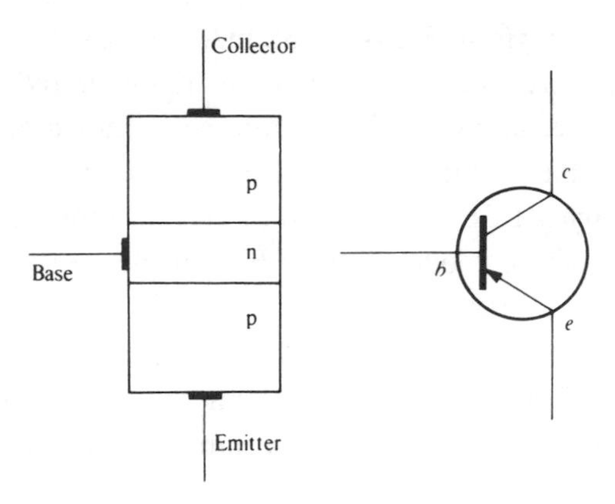

Fig. 1.16. The pnp transistor – construction and circuit symbol.

thermal. If, however, light is allowed to fall on the junction, a great increase in minority carrier population results. Electrons and holes are liberated by the energy of the incident photons, causing a great increase in reverse leakage current.

A *photodiode* is a normal pn junction mounted in a case with a transparent window. Normally operated in reverse-bias, typical diode current is 1 nA in the dark, increasing to 1 μA for illumination intensity of 1 mW/cm^2. This intensity is typical of that given by a 60 W bulb at a distance of about 30 cm (200 lux).

A *phototransistor* is simply a normal transistor with a transparent window in the case. Some phototransistors, like the TIL78, are moulded in transparent plastic; the top is normally convex to act as a lens, focusing the light onto the transistor, thus increasing the effective sensitivity of the device and making it directional.

When light falls on a transistor, minority carriers are liberated at both junctions, but it is those at the reverse-biased collector–base junction which give rise to the photocurrent. Just as the thermal collector–base leakage current I_{CBO} is amplified by transistor action to give the larger collector–emitter leakage I_{CEO}, so the collector–base photocurrent is similarly increased. The sensitivity of a phototransistor is typically a hundred times that of a photodiode. The base connection is normally unused; in fact cheaper phototransistors such as the TIL78 only have collector and emitter connections brought out.

No section on optoelectronics is complete without mentioning the *light-emitting diode* (LED). Junctions of certain semiconducting compounds, notably gallium phosphide and gallium arsenide, emit light when forward-biased. Forward currents from 5 mA to 80 mA are usual, a series resistor being used to limit the current drawn. LEDs are available in red, green, yellow and a rather dim blue, give ample brilliance for use as indicator lamps and have an indefinite life. In the circuits of figs. 1.3, 1.4 and 1.5, the 6 V, 0.04 A bulb can be replaced by a LED with a 100 Ω current-limiting series resistor. Remember to connect the cathode of the LED on the negative side of the circuit so that it is *forward*-biased.

The combination of a LED and a phototransistor has produced the valuable device called an *opto-isolator*. Consisting simply of a phototransistor looking at a LED, it allows the transfer of signals from one circuit to another with complete electrical isolation.

Extension of the optical coupling has produced *optical fibre* signal transmission technology.

1.5 Testing transistors

In experimental electronics, it is useful to have a simple method of testing transistors. The two parameters which best indicate the health of a transistor are collector–emitter leakage current, I_{CEO}, and d.c. current gain, h_{FE}. These can both be measured with the circuit in fig. 1.17. With the switch S open,

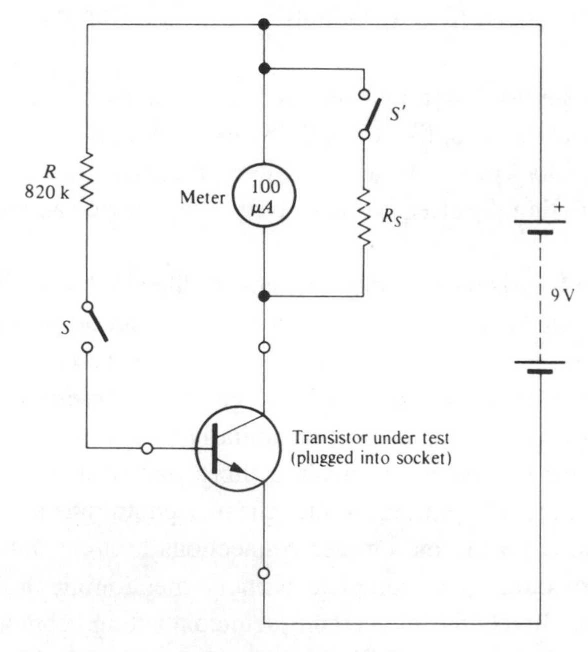

Fig. 1.17. Transistor test circuit.

no base current flows and the milliammeter in the collector circuit reads the leakage current I_{CEO}. When S is closed, a base current of approximately 10 μA flows through R (about 0.6 V is dropped across the base–emitter junction so that $I_B = (9 - 0.6)/820\,000 \approx 10\ \mu\text{A}$). Thus, the increase in collector current in milliamps is equal to $h_{FE}/100$.

To simplify measurement, the meter may be a 0–100 μA instrument with a shunt R_S switched into circuit by S' as S is closed. Thus, small leakage currents are readily measured on the 100 μA range, then the meter is shunted to read 10 mA full scale for the measurement of h_{FE}. To test pnp transistors, the battery and meter connections are reversed.

1.6 Voltage amplification

1.6.1 Introduction

Just as Morse's telegraph used a sequence of voltage pulses as its signals so, in electronic circuits, signals usually take the form of a.c. or d.c. voltages. Devices such as a phono cartridge or microphone produce a.c. voltages which must be amplified before they are useful. Some signal sources, such as the phototransistor and some nuclear radiation detectors, may produce an output current signal, but this is normally converted into a voltage signal before being amplified. Voltage amplification is therefore of great importance, and the bipolar transistor, despite its basic function as a current-amplifying device, finds its chief application in voltage amplifiers.

1.6.2 Load resistor

Fig. 1.18(*a*) shows a very simple voltage amplifier; the output voltage signal, V_{out}, is developed by virtue of the collector current flowing in the load resistor, R_L. This illustrates one of the most important applications of resistors in electronic circuits: the conversion of a current into a voltage. The input voltage, V_{in}, is applied to the base–emitter junction and gives rise to a base current i_b dependent on the base–emitter junction resistance. This base current causes a much larger collector current i_c to flow, producing a voltage drop i_cR_L across R_L. This p.d. is proportional to V_{in} but much larger in magnitude.

An important feature of this circuit is the earth rail, also called ground, 0 V or common, marked with the symbol shown. The earth rail is common to the input signal, output signal and the d.c. supply, and is normally the reference point for all circuit voltages.

1.6.3 Working point and bias

The circuit of fig. 1.18(*a*) is, as might be anticipated, a somewhat over-simplified voltage amplifier. It will only respond to positive input voltages and, furthermore, only to voltages greater than 0.5 V, this latter being the e.m.f. necessary to forward-bias the base–emitter junction. It is clear that, if the circuit is to amplify small signals faithfully, then the base–emitter junction must be forward-biased even when there is no signal. Normal alternating signal voltages have both positive and negative swings, so that the output at the collector must be able to move upwards towards the d.c. supply rail (negative

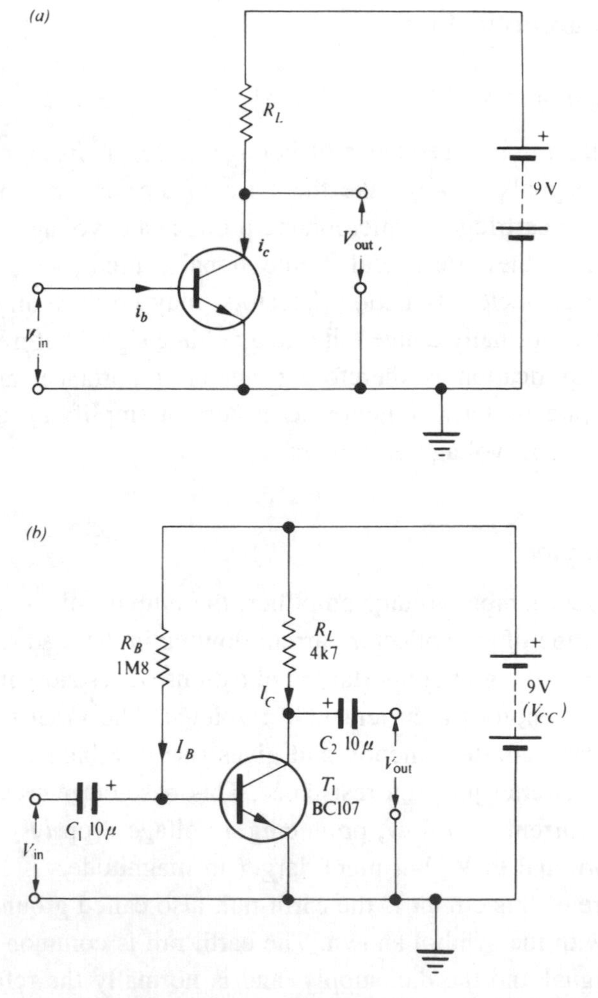

Fig. 1.18. The transistor used as a voltage amplifier: (*a*) primitive form, (*b*) with bias.

input) or downwards towards the earth rail (positive input). The implication here is that, when the input signal is zero (normally called the *quiescent* condition) the transistor should draw sufficient collector current so that the collector voltage sits midway between earth and supply level, ready to swing in either direction according to the polarity of the input signal.

Fig. 1.18(*b*) shows a circuit which will achieve the required result. A small silicon transistor such as the BC107 will operate very well with a quiescent collector current of 1 mA. Under this condition, a good working (operating) point requires that the collector sit midway between 0 V and +9 V, i.e. 4.5 V must be dropped in R_L. Thus, by Ohm's law, $R_L = 4.5\ \text{V}/1\ \text{mA} = 4500\ \Omega$. The nearest preferred resistor value is 4.7 kΩ (see appendix 1). Algebraically,

$$\begin{aligned} V_{CE} &= V_{CC} - I_C R_L \\ &= V_{CC} - h_{FE} I_B R_L, \end{aligned}$$

where V_{CC} is supply voltage.

If we assume a typical value of 200 for the d.c. current gain (h_{FE}) of the BC107, a 1 mA collector current requires a base current, I_B of 1/200 mA, or 5 μA. R_B, the base bias resistor, supplies this current, the value being decided once again by Ohm's law,

$$\begin{aligned} R_B &= \frac{V_{CC}}{I_B} = \frac{9}{5 \times 10^{-6}} \\ &= 1.8\ \text{M}\Omega. \end{aligned}$$

The base–emitter voltage V_{BE} (approximately 0.6 V) is here neglected by comparison with the much larger supply voltage V_{CC}.

1.6.4 Coupling capacitors

Coupling capacitors C_1 and C_2 are used to isolate external circuitry from the d.c. voltages present on base and collector under quiescent conditions. This ability of a capacitor to block d.c. voltage whilst transmitting a.c. is very valuable in electronics and arises from the tendency of a capacitor to hold its charge, and therefore the p.d. between its plates, constant. An increase in potential on one plate therefore causes a corresponding increase in potential on the other. An a.c. signal, which changes its potential many times a second, is thus transmitted faithfully from one plate to the other. A d.c. voltage, however, allows the capacitor time to adjust its charge to accommodate a new p.d. between its plates and is therefore not transmitted. This time taken to adjust to a new p.d. is dependent on the *time constant* of the circuit, which should be long compared with one period of the lowest a.c. frequency to be transmitted. This concept is considered in more detail in chapter 8. In the simple voltage amplifier being considered, 10 μF coupling capacitors provide a time constant adequate for the faithful transmission of a.c. down to a frequency of 10 Hz.

The positive sign at one side of the capacitor symbol is for guidance in the connection of electrolytic capacitors, where the insulating dielectric layer is an extremely thin film of electrolytically-deposited aluminium oxide. Such capacitors provide high capacitance values in compact form at low cost, but, except in the case of special non-polarized types, must be connected in circuit with the correct polarity.

1.6.5 Stabilizing the operating point

A serious disadvantage of fig. 1.18(*b*) is that the quiescent collector voltage is completely dependent on transistor h_{FE}. Now h_{FE} is subject to large variations between different specimens of a given type of transistor. For instance, although a typical value for the h_{FE} of a BC107 is 200, the manufacturers specify that it may be as low as 90 or as high as 450. These variations will upset the d.c. operating point. For instance, an h_{FE} of 100 instead of 200 will cause the transistor to draw only 0.5 mA instead of 1 mA, thus dropping only 2.35 V instead of 4.7 V across R_L; this increased quiescent voltage means that the circuit will only be able to handle positive output voltage swings of about 2 V instead of 4 V (the negative output swing capability will be increased to about 6 V, but this is of little use if the positive swing is restricted).

The consequences of using a transistor with $h_{FE} = 400$ are even more serious. Here the collector current will be doubled to 2 mA. A quick calculation shows that the full 9 V supply voltage will then be dropped across R_L. The transistor is said to be saturated, or bottomed. In practice, a small collector–emitter voltage (typically 0.2 V) will be present. Any further increase in base current has little effect: it is, of course, impossible to drop more than V_{CC} across R_L. Since the collector of a saturated transistor is virtually at earth potential, the circuit is now useless for linear amplification, no negative-going output being available. The saturated condition is discussed further in section 1.7.

To return to the linear amplifier of fig. 1.18(*b*), some improvement in design is necessary to increase its tolerance of variations in h_{FE}. Even if we were able to select transistors with an h_{FE} of 200, and this is very expensive for circuit manufacturers, h_{FE} increases with temperature, so that the circuit would still not be reliable. Fig. 1.19 shows a very simple but effective improvement.

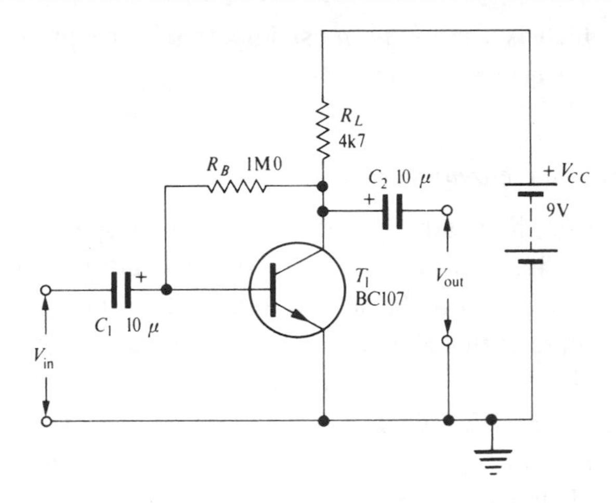

Fig. 1.19. Voltage amplifier with stabilization of operating point.

Instead of connecting R_B directly to V_{CC}, we halve its value and connect it to the collector ($V_{CE} \approx V_{CC}/2$). In this way the quiescent base bias current is now dependent upon the quiescent collector voltage. Even with high values of h_{FE}, the transistor cannot saturate because, if the collector voltage drops then so does the base bias current, 'throttling back' the collector current. Conversely, if h_{FE} is low, the quiescent collector voltage will be high, giving a higher value of I_B.

Base bias current is now given by

$$I_B = \frac{V_{CE}}{R_B}$$

and, as before,

$$V_{CE} = V_{CC} - h_{FE} I_B R_L .$$

Combining these equations gives

$$V_{CE} = \frac{V_{CC}}{1 + h_{FE} R_L / R_B} .$$

With R_L and R_B as in fig. 1.19, if $h_{FE} = 100$, $V_{CE} \approx 6$ V, and if $h_{FE} = 400$, $V_{CE} \approx 3$ V. Although there is still a variation in operating point, it is not serious unless large signals demand the absolute maximum output voltage swing capability. The circuit of fig. 1.19 will work with a very wide range of transistors and is a useful general purpose voltage amplifier. The principle of designing the circuit to be self-compensating for variations in h_{FE} is just one example of *negative feedback*, which is one of the most important concepts in electronics, and is explored further in chapter 4.

1.6.6 Stabilized voltage amplifier

For some applications, even the relatively small variations in operating point produced by the circuit of fig. 1.19 are excessive. If d.c. conditions must be virtually independent of transistor h_{FE} the stabilized circuit shown in fig. 1.20 may be used. The first feature of this circuit is resistor R_3 in the emitter circuit, which means that the emitter no longer sits at earth potential, but at a small potential above earth, given by $I_E R_3$, where I_E is the emitter current. A second feature is that, instead of using a single resistor to set the base *current* at a given value, the potential divider R_1, R_2 fixes the base *potential* relative to earth. The current through the potential divider is of the order of ten times the base current, so the latter has little influence on base potential. Since the base–emitter junction is forward-biased, it only has a small p.d. across it (approximately 0.6 V in a silicon transistor) so that the emitter potential is equal to the base potential less 0.6 V.

Thus, if the base potential relative to earth is V_B and the emitter potential relative to earth is V_E,

$$V_E = V_B - 0.6,$$

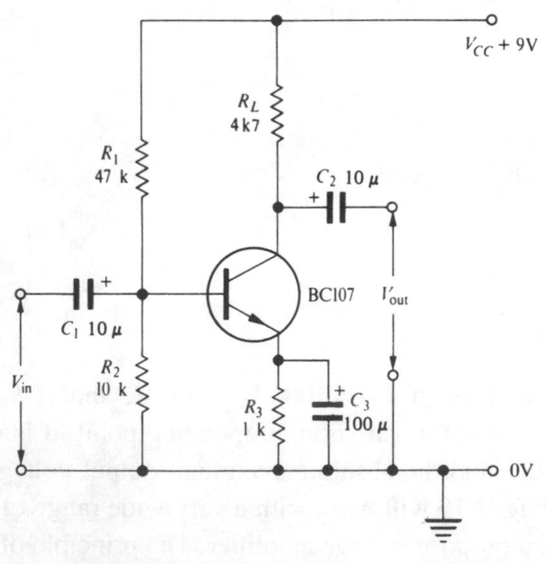

Fig. 1.20. Stabilized amplifier with emitter resistor.

but

$$V_E = I_E R_3$$

therefore

$$I_E = \frac{V_B - 0.6}{R_3}.$$

Hence the emitter current I_E is fixed by the chosen values of V_B and R_3.

With the component values of fig. 1.20, the base potential is fixed at 1.6 V by R_1 and R_2; the emitter potential is therefore approximately 1.0 V, giving the required emitter current of 1 mA in the 1 kΩ emitter resistor.

Now

$$I_E = I_C + I_B$$

and

$$I_B \ll I_C,$$

therefore

$$I_E \approx I_C.$$

Hence the collector current is also approximately 1 mA.

It is interesting to note that the above analysis of the circuit did not involve the transistor h_{FE}. In fact the only parameter of any significance in the circuit is V_{BE} which is assumed to be 0.6 V and varies very little (<0.1 V) from one transistor to another. In designing a stabilized circuit, the voltage drop in the emitter resistor should be large in comparison with possible variations in V_{BE}, but not so large that the available output voltage swing is seriously reduced (the collector can now only swing between V_{CC} and V_E instead of between V_{CC} and earth). A drop of 1 V is usually suitable. The large capacitor C_3 shunts the emitter resistor in order that no a.c. signal voltage appears on the emitter. Without C_3, the voltage gain would be drastically reduced by negative feedback because any signal voltage across R_3 opposes the input signal.

1.6.7 Measurement of voltage gain

The voltage gain of an amplifier may be conveniently measured by injecting a sine-wave signal from a signal generator into the input and then using an oscilloscope to measure the output signal V_{out} and compare it with the input signal V_{in}.

Voltage gain

$$A_V = \frac{V_{\text{out}}}{V_{\text{in}}}.$$

For the circuits shown in this chapter, voltage gain is of the order of 150 to 200. The theoretical calculation of voltage gain is outlined in chapter 6.

1.7 Saturated operation

There is a clear difference between the simple switching circuits at the beginning of this chapter (section 1.2), and the linear amplifier circuits just discussed. In normal operation of the linear amplifier, the collector current is at all times directly proportional to the base current. In a switching circuit, however, such as the one in fig. 1.21, the collector current is chiefly determined by the supply voltage V_{CC} and by the load resistance R_L. We briefly encountered saturation as an undesirable condition in the voltage amplifier, but this state is of sufficient importance to merit further discussion.

Consider what happens to the collector current in fig, 1.21 as the base current is gradually increased from zero. With switch S_1 open, no base current flows and the collector current is negligible. Closing S_1 gives rise to a finite base current $I_B = V_{CC}/R_B$, neglecting the p.d. across the base–emitter junction. Collector current through the load R_L will thus be given by $I_C = h_{FE}V_{CC}/R_B$. In

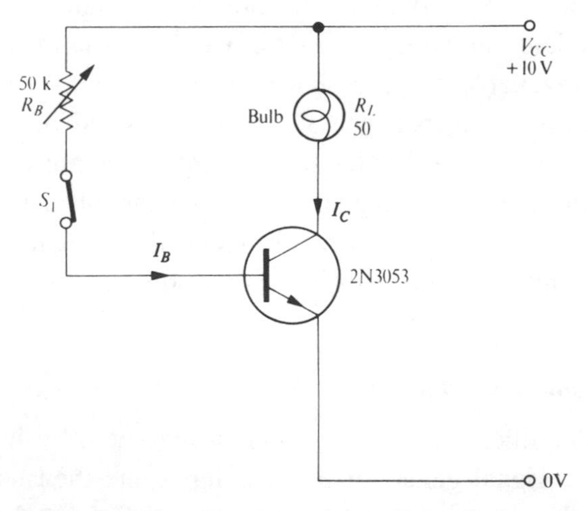

Fig. 1.21. Illustration of saturation. Transistor is acting as a switch to turn on the bulb.

practical terms, with the component values shown, if the transistor $h_{FE} = 100$, and R_B is set to maximum (50 kΩ),

$$I_C = \frac{100 \times 10}{50\,000} \text{A}$$

$$= 20 \text{ mA.}$$

The potential drop in R_L is given by $R_L I_C$ and in this case is equal to $50 \times 0.02 = 1$ V. The transistor is in its linear mode of operation; further decrease of R_B results in increased base current, increased collector current and therefore increased potential drop in R_L. The circuit could be used as a voltage amplifier in this condition.

Now consider the case where

$$R_B = h_{FE} R_L,$$

therefore base current

$$I_B = \frac{V_{CC}}{R_B} = \frac{V_{CC}}{h_{FE} R_L}.$$

Hence, collector current

$$I_C = \frac{h_{FE} V_{CC}}{h_{FE} R_L} = \frac{V_{CC}}{R_L}.$$

The transistor is here behaving just like a pair of switch contacts as far as the load is concerned. It is clear from Ohm's law that the load current cannot in any circumstances exceed the value V_{CC}/R_L. Further increase in transistor base current cannot therefore increase collector current, which is now determined only by the load resistance and supply voltage. The transistor is *saturated* or *bottomed.*

In practice, there is always a small voltage drop between the collector and emitter of a saturated transistor, usually denoted $V_{CE(\text{sat})}$. It is normally less than 1 V and may be as little as 0.1 V in transistors specially designed for switching. Generally, $V_{CE(\text{sat})}$ decreases as the base–emitter junction of the transistor is driven harder into conduction, i.e. as the ratio of collector current I_C to base current I_B becomes significantly less than the transistor current gain h_{FE}.

For hard bottoming (low $V_{CE(\text{sat})}$), a rough guide is

$$\frac{I_C}{I_B} < \frac{h_{FE}}{5}.$$

And in the type of circuit shown in fig. 1.21, where the base current is derived via a simple resistor to supply, we select

$$\frac{R_B}{R_L} < \frac{h_{FE}}{5}.$$

Hence, in fig. 1.21, assuming a typical current gain h_{FE} of 150 for the 2N3053 transistor,

$$\frac{R_B}{R_L} < \frac{150}{5} = 30.$$

Hence, with

$$R_L = 50\Omega,$$

we choose

$$R_B < 30 \times 50\Omega = 1.5\text{k}\Omega.$$

In this circuit then, if the 50 Ω lamp load is to be efficiently turned on by the transistor, we should select a base resistor of less than 1.5 kΩ. Should this not be possible, e.g. if R_B is a light-dependent resistor with a minimum resistance of 10 kΩ, then a Darlington pair of transistors would be used to increase current gain.

If a transistor is operated near its maximum collector current rating then, in order to keep $V_{CE(\text{sat})}$ down to a fraction of a volt, the base current drive may need to be as high as $I_C/10$ to allow for the reduced h_{FE}.

It might be surprising to learn that $V_{CE(\text{sat})}$ can be much less than the base–emitter drop V_{BE}, which is approximately 0.6 V for a silicon transistor. This is because, under saturated conditions, the collector–base junction goes into forward conduction. We therefore have two forward-biased junctions in opposition, so that their p.d.s tend to cancel. This ability of the bipolar transistor to saturate with a very small voltage drop between collector and emitter makes it a very useful switching device. Many of the most important applications of electronics employ switching circuits, including the vast field of digital electronics.

In the switching mode, a transistor is operating either with virtually zero collector current (turned off) or with virtually zero collector voltage (turned on). In either of these conditions, the power dissipation in the device is very small. The only point where significant power dissipation occurs is at the time when switch-over occurs, where both collector–emitter voltage and collector current are finite.

A small transistor such as a 2N3053, with a maximum permissible power dissipation of under one watt, can switch a load of several watts. Care should be taken to work within the maximum collector voltage and current ratings; the switchover should also be as fast as possible to avoid excessive dissipation.

2

The field-effect transistor

2.1 Introduction

In the last chapter, it was emphasized that one of the main properties of the bipolar transistor is that it is a current-controlled amplifying device: the output current is controlled by a small input current. In the case of the field-effect transistor (FET) it is the input *voltage* which controls the output current. the current drawn by the input is usually negligible (it can be less than 1 pA). This is a great advantage where the signal comes from a device such as a capacitor microphone or piezo-electric transducer, which is unable to supply a significant current.

FETs are basically of two types: the junction field-effect transistor or JFET and the insulated gate field effect transistor or IGFET. The latter is more commonly known by a name describing its construction: metal-oxide semiconductor field-effect transistor (MOSFET) or MOS transistor.

2.2 The JFET

2.2.1 Construction

The n-channel JFET is shown in diagrammatic form in fig. 2.1(*a*) together with its circuit symbol in fig. 2.1(*b*). A tiny bar of n-type silicon has an ohmic (non-rectifying) contact on each end. At a point along the bar a region of p-type silicon forms a pn junction. In normal operation, the junction is reverse-biased. The lower contact on the bar is called the source and the upper contact the drain. As these names suggest, the electron current flows from source to drain and is controlled by the voltage applied to the p-region, called the gate.

An alternative type of construction is the p-channel device where the gate

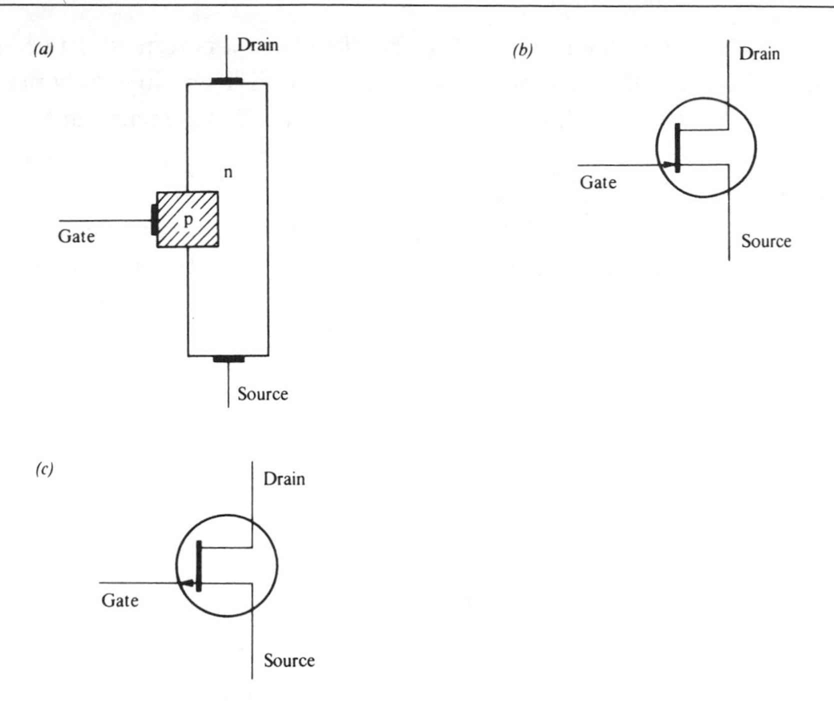

Fig. 2.1. The JFET: (*a*) n-channel type: outline construction; (*b*) n-channel type: circuit symbol; (*c*) p-channel type: circuit symbol.

is made of n-type material. Fig. 2.1(*c*) shows the circuit symbol for a p-channel JFET.

2.2.2 JFET operation

The operation of the JFET depends upon variations in the size of the depletion layer at the reverse-biased gate junction. Fig. 2.2 shows an n-channel FET in a test circuit and includes an outline of the depletion layer. The p-type gate is much more heavily doped than the n-type bar, so that the depletion region exists almost entirely in the bar. The gate carries a negative bias voltage (V_{GS}) relative to the source which gives rise to the particular shape of the depletion region shown: this is wider at the top than the bottom, because the drain is held more positive than the source. The wider the depletion layer, the narrower the channel there is available for the flow of electrons from source to drain, since the depletion region itself, being devoid of current carriers, behaves like an insulator. Hence, for a given drain–source voltage, the drain current is dependent upon the input voltage V_{GS}. In practice, there is normally a p-type gate on both sides of the n-type bar so that the channel is formed between two depletion layers. Unlike the bipolar transistor, the FET employs only majority carriers for its operation. It is therefore sometimes called the unipolar transistor and is less susceptible than the bipolar type to temperature changes and nuclear radiation, since these chiefly affect minority carriers.

In the p-channel JFET, which is less common than the n-channel type, the majority carriers in the channel are holes. In operation, the drain is therefore held negative with respect to the source; the gate, in order to be reverse-biased, is positive with respect to the source.

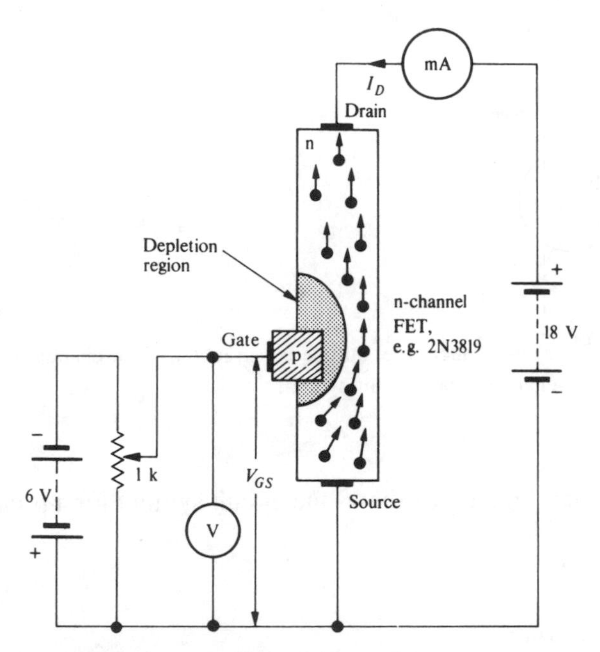

Fig. 2.2. n-channel JFET in a test circuit. The FET is drawn in diagrammatic form to show electron flow through channel.

2.3 The MOSFET

Fig. 2.3(*a*) shows the basic construction of an n-channel MOSFET. The drain and source are n-type regions formed in the p-type silicon bar, which is known as the *substrate*. The gate is a metal electrode insulated from the silicon bar by a layer of silicon oxide.

The MOSFET is drawn connected in a simple circuit with the drain positive with respect to the source. Under these conditions, no supply current flows, because the drain–substrate pn junction is reverse-biased. Even if the drain

(a)

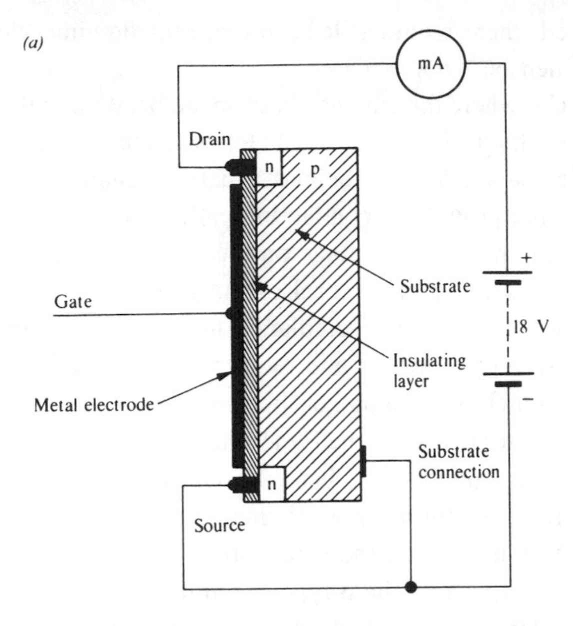

(b)

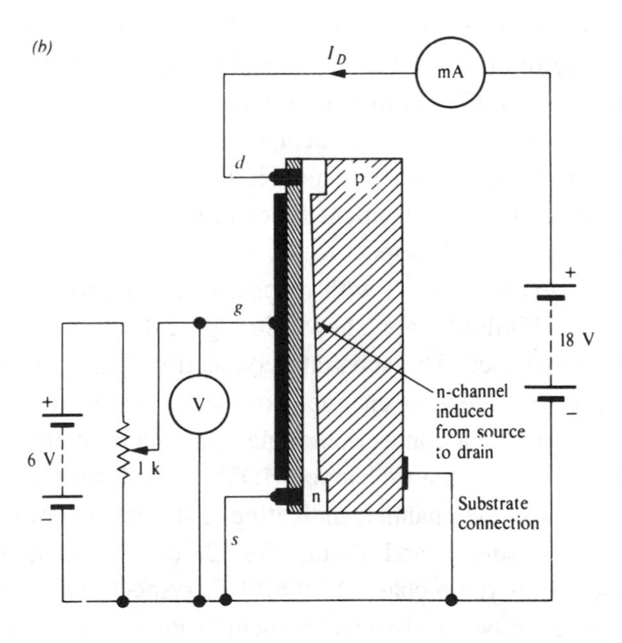

Fig. 2.3. The n-channel MOSFET: (*a*) zero gate bias, (*b*) positive gate induces n-type channel in substrate.

and source were reversed, there would still be no current flowing, since the source junction would then be reverse-biased.

Now consider fig. 2.3(*b*), where the gate has been made positive with respect to the source. The field of the positive gate repels holes in the p-type substrate away from the insulating layer, leaving behind a narrow channel of n-type silicon. This narrow channel provides a conducting path from source to drain.

In this way, given a certain positive voltage on the gate to make the device conduct, the drain current is under the control of the gate voltage.

Thus the characteristics bear some resemblance to the JFET, except that the latter conducts with zero gate voltage and requires a certain negative gate voltage to cut off the channel. To distinguish between these modes of operation, the MOSFET just described is termed an *enhancement* device and the JFET a *depletion* device. Depletion type n-channel MOSFETs can be manufactured by ensuring that there is normally a residual n-type channel between source and drain, even with no bias on the gate. This residual channel can be produced by diffusing impurities into the p-type substrate, but it is not usually necessary to go to this length: the silicon-oxide insulating layer contains trapped positive ions which induce an n-type channel in the substrate. Negative or positive gate voltages can be used on a depletion MOSFET since there is no gate junction to be kept reverse-biased. The effect of a positive gate voltage is to enhance further the drain current by broadening the induced n-channel. The n-channel depletion MOSFET is therefore a versatile device which may be operated in both depletion and enhancement modes.

MOSFETs of p-channel type are also available. These have an n-type substrate and are normally of the enhancement type. In such devices, the drain is normally operated negative with respect to the source and the drain current is zero until the gate is made negative with respect to the source.

Circuit symbols for MOSFETs are shown in fig. 2.4. Fig. 2.4(*a*) is a depletion-type n-channel device. The substrate connection (often marked 'b' for 'bulk') carries an arrow showing channel polarity. The substrate is normally connected to the source, a connection sometimes made internally. Fig. 2.4(*b*) shows an enhancement-mode n-channel MOSFET and differs from fig. 2.4(*a*) by showing breaks in the channel, indicating that there is normally no conducting path between source and drain. Fig. 2.4(*c*) and 2.4(*d*) show depletion- and enhancement-type p-channel MOSFETs respectively; notice the reversal of the substrate arrow to distinguish them from n-channel devices. In all the MOSFET symbols, the gate is clearly shown insulated from the channel.

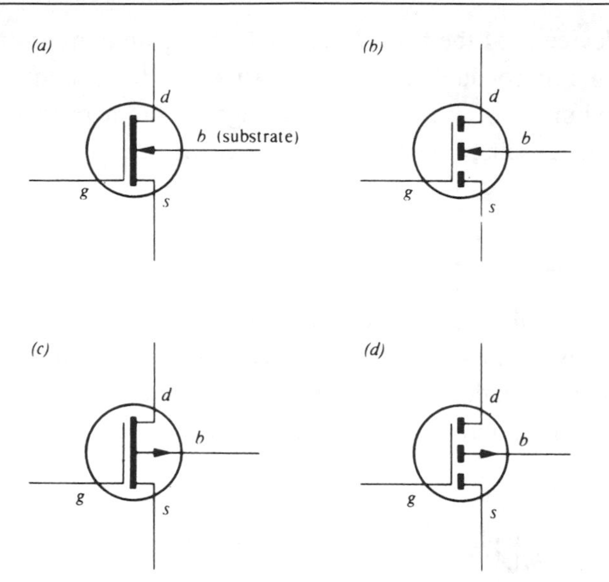

Fig. 2.4. MOSFET circuit symbols: (*a*) n-channel depletion type, (*b*) n-channel enhancement type, (*c*) p-channel depletion type, (*d*) p-channel enhancement type.

2.4 FET transfer characteristics

Just as we plotted the graph of collector current versus base current for the bipolar transistor in fig. 1.15, so we can plot graphs showing the transfer characteristics of the various types of FETs. In this case the graph is of drain current against gate-source voltage. Fig. 2.5 shows three such graphs for n-channel FETs; each graph shows drain current reducing as the gate potential goes more negative. Fig. 2.5(*a*) is the transfer characteristic of a JFET which, of course, must operate in the depletion mode. Fig. 2.5(*b*) is obtained with a depletion-type MOSFET showing that operation in the enhancement mode is practicable. Fig. 2.5(*c*) applies to an enhancement-type n-channel device. Similar characteristics are obtained from p-channel devices except that the gate–source polarity is reversed.

2.5 Transconductance

How do we express the performance of a FET? With the bipolar transistor, the current gain is easily understood and has the virtue of being a simple ratio without units. In the case of the FET we have to express the fact that the *drain current* I_D is controlled by gate–source *voltage* V_{GS}. Thus an expression for the 'gain' of the device is of the form I_D/V_{GS}, which, being current divided by voltage, has the units of conductance. The term used is *transconductance*, usually given the symbol g_m or y_{fs}, given by change in drain current (ΔI_D) divided by change in gate–source voltage (ΔV_{GS}) (see fig. 2.5(*a*)). Therefore we have

$$g_m = \frac{\Delta I_D}{\Delta V_{GS}}.$$

If I_D is in mA and V_{GS} is volts, g_m is in millimho or millisiemens (mS). The unit is often written, however, as mA/V; microsiemens (μS) are also used.

Thus if a FET has a transconductance of 3 mS, a change in the gate–source voltage of 1 V causes a drain current change of 3 mA.

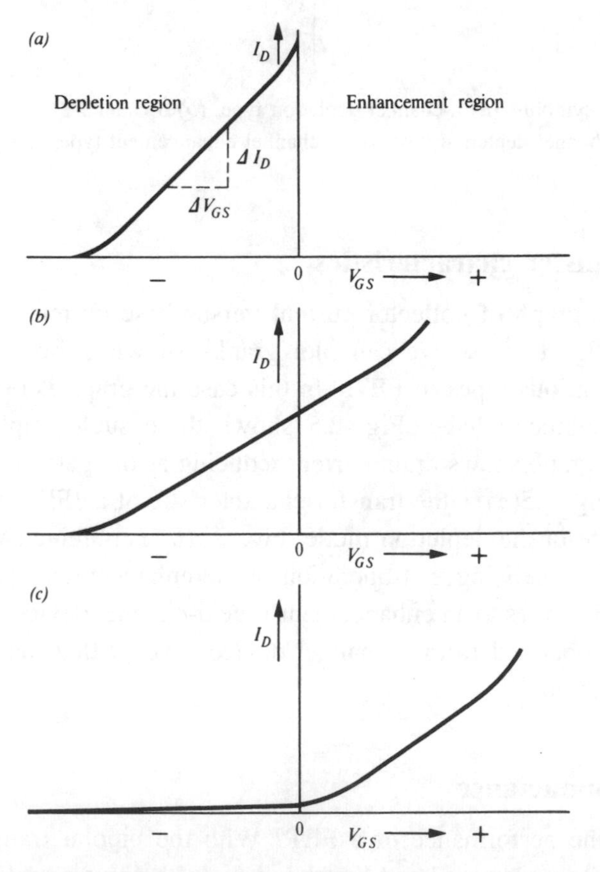

Fig. 2.5. Typical transfer characteristics of n-channel FETs. Drain current I_D is plotted against gate–source voltage V_{GS} for (*a*) JFET, (*b*) depletion-type MOSFET, (*c*) enhancement-type MOSFET.

2.6 A voltage amplifier using a FET

As with the bipolar transistor, in designing a voltage amplifier, we need to convert the output current of the FET into an output voltage. This requires a load resistor. Fig. 2.6 shows a rudimentary voltage amplifier using the inexpensive general purpose n-channel JFET type 2N3819. In order to obtain a reasonable voltage gain, a fairly high value of R_L is required (22 kΩ) and this in turn requires a high supply voltage V_{DD} (18 V). In this simple circuit the gate–source junction is kept reverse-biased by a small $1\frac{1}{2}$ V battery. This is inconvenient and is poor design, since it assumes that all specimens of the 2N3819 require the same bias voltage for a given drain current. This is far from true; in fact the circuit may not work with some examples of the device.

Fig. 2.7 is an improved circuit which incorporates *automatic* gate bias. The source is maintained positive with respect to earth by resistor R_S, whilst the gate is tied to earth via resistor R_G (despite the high value of R_G, the gate is at earth potential because it draws negligible current). In this way, the gate is actually negative with respect to the source. Furthermore, the bias is dependent on the source current. If the source current rises, the voltage drop across R_S rises, thus increasing the gate bias towards cut-off and tending to reduce

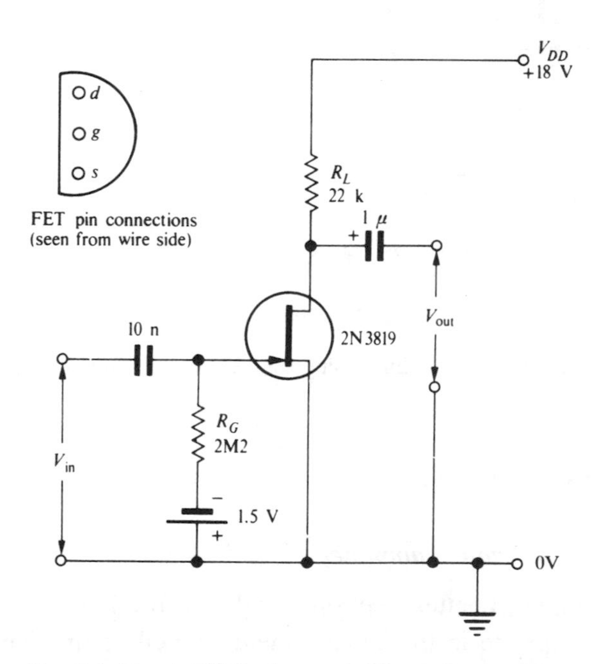

Fig. 2.6. Simple FET voltage amplifier using a gate-bias battery.

the source current. In this way the circuit can tolerate wide variations of bias requirements in the 2N3819 and has always been found to work satisfactorily.

This large capacitor C_S prevents any a.c. signal appearing on the source and reducing gain by negative feedback. The voltage gain of the circuit is usually between 20 and 30.

If the gate is disconnected from resistor R_G and the drain voltage monitored with a d.c. voltmeter directly connected between drain and earth rail, the circuit will be found to function as a crude but sensitive electrometer which detects electrically charged items placed near the gate. This is rendered possible by the fact that the gate draws a negligible current and so can change its potential under the influence of an adjacent charge.

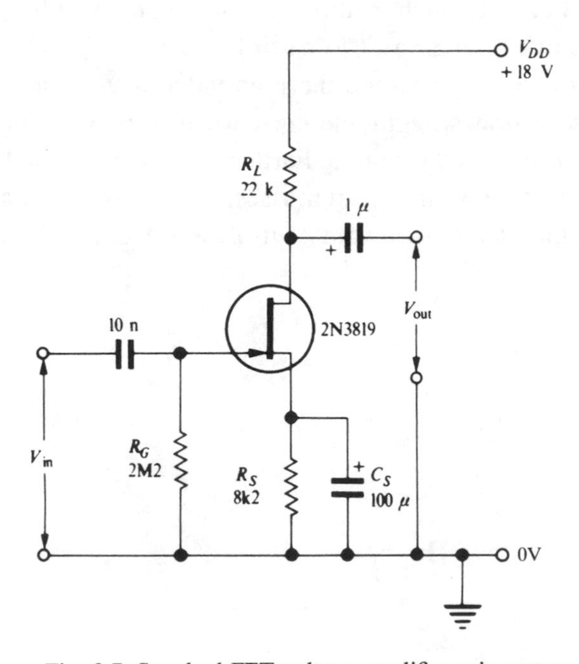

Fig. 2.7. Standard FET voltage amplifier using source resistor to provide gate bias.

2.7 MOSFETs in practice

2.7.1 High input resistance amplifier

A MOSFET will make an even better electrometer than a JFET. Furthermore, because of its ability to operate in the enhancement as well as the depletion mode, an n-channel MOSFET will work well in amplifier circuits of the simple type shown in fig. 2.6, but without the bias battery, R_G being connected directly to earth. Because the gate leakage current is very small (<1 pA), R_G can be as high as $10^{12}\Omega$ if required, whereas the maximum value with a JFET is about $10^8\Omega$ (100 MΩ).

2.7.2 Power applications

The MOSFET is by no means limited to small-signal applications. Power MOSFETs, some capable of handling hundreds of amps, are widely used in power amplifiers and power control.

It is worth mentioning in this connection the MOSFET/bipolar transistor integrated combination known as the *insulated gate bipolar transistor* (*IGBT*). The IGBT combines the voltage-operation of the MOSFET with the small potential drop of a saturated bipolar transistor switch. A 5 V digital logic pulse can thus, for example, control a motor drawing hundreds of amps.

2.7.3 Static precautions

The low-leakage gate insulation means that MOSFETs are very vulnerable to static charges, which can build up high gate voltages and break down the insulation. For this reason, MOSFETs are supplied with the leads short-circuited by a metal clip or piece of conducting foam plastic. This short circuit should preferably not be removed until the device is safely soldered into circuit. If there is a risk of excessive gate voltages arising in the circuit then a protection network (e.g. 'back-to-back' Zener diodes) should be incorporated between gate and earth (see section 10.13). This technique does unfortunately degrade the very high input resistance of the MOSFET. Some MOSFETs actually incorporate protection diodes and are not therefore as susceptible to breakdown as the unprotected devices.

2.7.4 Integrated circuits

The simplicity of a MOSFET means that it occupies very little area on the silicon substrate, so that many thousands of transistors can be included in a single *integrated circuit*. Complex digital chips such as microprocessors and memories therefore rely on MOS technology. *Complementary* MOS (*CMOS*) circuits exploit the extremely high input resistance of the devices to give the bonus of extremely low power consumption compared with the equivalent bipolar transistor circuits. These topics are discussed further in chapters 13 and 14 which cover the principles and applications of digital circuits and microcomputers.

3

Thermionic valves and the cathode-ray tube

3.1 Introduction

The thermionic valve was the first active (amplifying) element in electronics. Although obsolete for most small-scale amplification, the valve still finds a place where high voltages must be handled or high-power high-frequency signals are involved. In addition, the inherent characteristics of valve audio amplifiers are popular with many audiophiles. In particular, the graceful behaviour of the circuits near overload can give a subjectively clean sound at high levels. It is therefore useful for the electronic engineer to have at least a rudimentary knowledge of the valve and its circuitry. This chapter gives a brief account of valve (vacuum tube) circuits, including a description of the one thermionic device which is still extensively used: the cathode-ray tube.

3.2 Thermionic emission

In the early 1880s, Thomas Edison, having developed the carbon-filament lamp, turned his attention to the blackening of the glass bulb which occurred after some hours of use. In an attempt to intercept some of the blackening particles, he sealed a metal plate inside one of his lamps and was surprised to find that, if the plate was made positive with respect to the filament, it drew a current. For twenty years, no one was to know that this 'Edison effect' current was due to electrons emitted by the hot filament being captured by the positively charged plate. The term *thermionic* emission was coined to describe this thermal liberation of free particles, literally *thermal ions*. Although the word ion has come to mean an atom which has lost or gained an electron, its original significance was much broader, meaning simply a particle free to travel (literally 'going', from the Greek).

3.3 The thermionic diode

The blackening of lamps was investigated by Ambrose Fleming round about the same time as Edison was working on the subject, but Fleming carried the work a stage further in his search for an improved detector of Marconi's radio waves. In 1904 he patented his 'oscillation' valve, so called because it allowed current to pass in one direction only.

Fig. 3.1 shows a diagram of the diode (two-electrode) valve, as Fleming's invention came to be called. The incandescent filament is surrounded by a cylindrical *plate*, normally termed the *anode* because it is usually held positive with respect to the filament. Similarly, the filament is usually called the *cathode*. The circuit symbol for the diode is shown in fig. 3.2, the filament and plate being clearly represented.

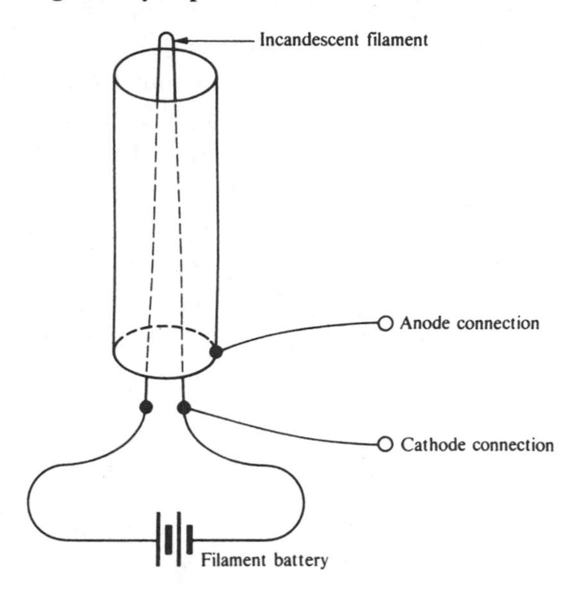

Fig. 3.1. Outline of Fleming's thermionic diode. Filament and plate are enclosed in an evacuated glass bulb.

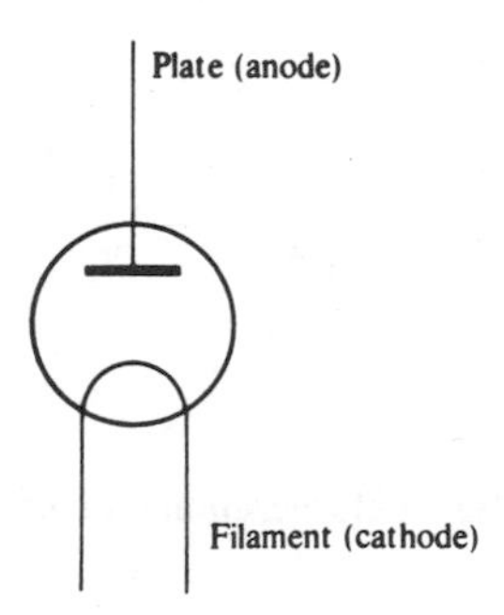

Fig. 3.2. Circuit symbol for thermionic diode.

The current–voltage characteristic of a thermionic diode is shown in fig. 3.3 where the resemblance to the pn junction characteristic of fig. 1.11 may be noted. In the thermionic diode, unlike the pn junction, a small forward current flows when there is zero voltage across the device. This is because the electrons are emitted from the filament with a definite velocity, having been shaken loose by the violent atomic vibrations at high temperature. A small proportion of these electrons reaches the anode even when there is no assisting field.

As the anode is made slightly positive, more electrons are drawn towards it. Not all the electrons emitted reach the anode, however, since the large cloud of electrons between cathode and anode appears as a negative space charge which has a repellent effect on the emitted electrons. This inhibiting effect of the space charge may be likened to that of the depletion layer in a semiconductor junction. As the anode is made increasingly positive, the effect of the space charge is overcome and more and more electrons reach the anode.

If the anode is made negative with respect to the filament, the emitted electrons are repelled back to the cathode, and no current flows at all when the anode is several volts negative. Notice that if the vacuum is good, there are no *minority carriers* around to cause a reverse leakage current. Any reverse current which exists is due to slight traces of gas in the evacuated envelope and to leakage across the outside of the glass. This is very different from the behaviour of the semiconductor pn junction, where the presence of minority carriers of thermal origin is inherent in the operation of the device.

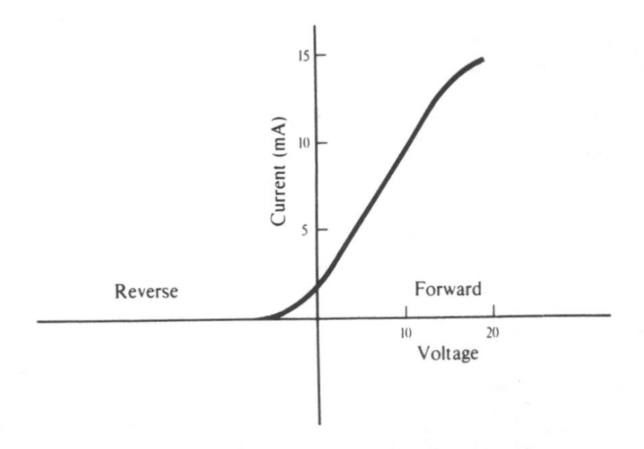

Fig. 3.3. Typical current–voltage characteristic of a small thermionic diode.

3.4 The triode valve

To produce a thermionic amplifying device, some sort of additional electrode must be added to the diode so that the anode current may be controlled. This was first done by Lee de Forest in 1907, who built a valve with a wire mesh, or *grid*, between the filament and anode. This three-electrode device, or *triode*, is shown in fig. 3.4 and its circuit symbol in fig. 3.5.

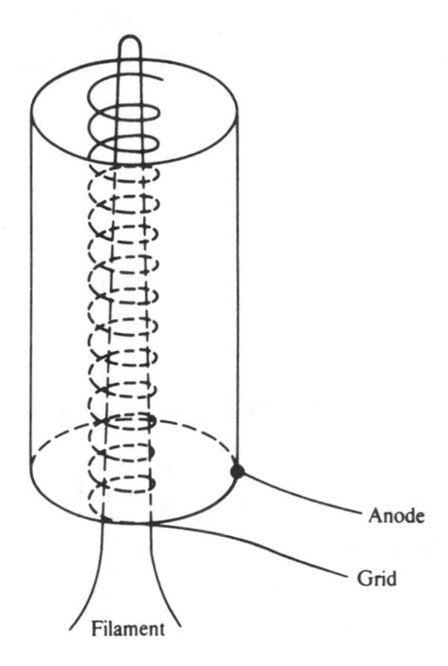

Fig. 3.4. Outline construction of the triode valve.

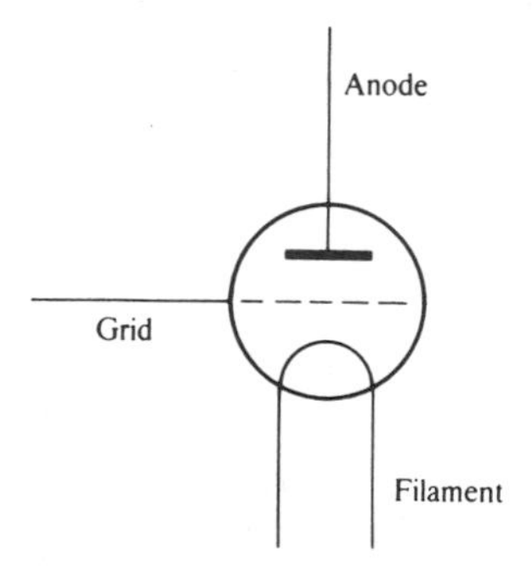

Fig. 3.5. Triode valve circuit symbol.

The grid is normally held negative with respect to the filament, under which condition it repels some of the emitted electrons back to the filament, allowing only a certain fraction to reach the anode through the spaces in the wire mesh. The more negative the grid is made, the more powerful its repellent field becomes and the lower the anode current falls. Eventually, the *cut-off* point is reached when none of the emitted electrons reaches the anode and the current falls to zero. This behaviour is similar to the FET: both the triode and FET produce an output *current* controlled by an input *voltage*. A typical triode valve transfer characteristic is shown in fig. 3.6; it is interesting to compare it with fig. 2.5(*a*) for the FET. As with the FET, the transfer property of a valve is specified by its transconductance, g_m, where

$$g_m = \frac{\text{change in anode current}}{\text{change in grid voltage}}$$

$$= \frac{\Delta I_a}{\Delta V_g} \text{ (usual units mA / V).}$$

Transconductance is sometimes inappropriately called *mutual* conductance; in fact it is from this term, used particularly of valves, that the symbol g_m is derived. A mutual relationship implies an interdependence which is not applicable to either the valve or FET: although grid or gate voltage controls anode or drain voltage, the reverse does not apply and the term transconductance (literally conductance across) is to be preferred.

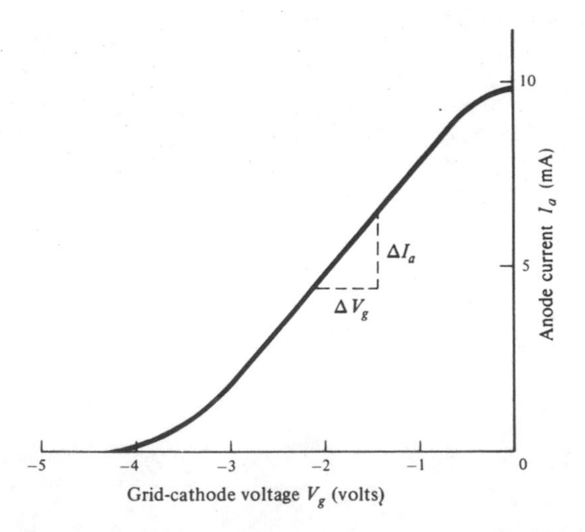

Fig. 3.6. Typical transfer characteristic of a small triode. The grid is not normally taken positive.

3.5 Cathode developments

As already mentioned, early valves used white-hot tungsten filaments. A filament temperature of 2300 K was necessary to obtain adequate electron emission; such *bright emitter* valves consumed considerable power for filament heating and had a relatively short life. It was soon discovered that a filament coated with oxides of barium and strontium emits copious supplies of electrons at a mere 1000 K (red heat). Thus, the 'dull emitter' valve with its oxide-coated cathode has become standard.

Early valve equipment used d.c., both for the *high tension* (HT) anode supply and for the filament supply (low tension or LT). It is very convenient, however, if the cathode can be heated by a.c. since this is readily available directly from a mains transformer. There are two problems associated with a.c. heated filaments. Firstly, the temperature of the filament may fluctuate in sympathy with the a.c. frequency, giving rise to a 100 Hz fluctuation in electron current with 50 Hz a.c. Secondly, because the input grid voltage is applied relative to the cathode, a proportion of the a.c. filament voltage will appear in the input signal, producing a 50 Hz 'hum' component.

These two problems are overcome in the *indirectly heated* cathode which is used in almost all small valves. As the name suggests, the cathode is electrically insulated from the heating filament, avoiding the direct injection of a.c.

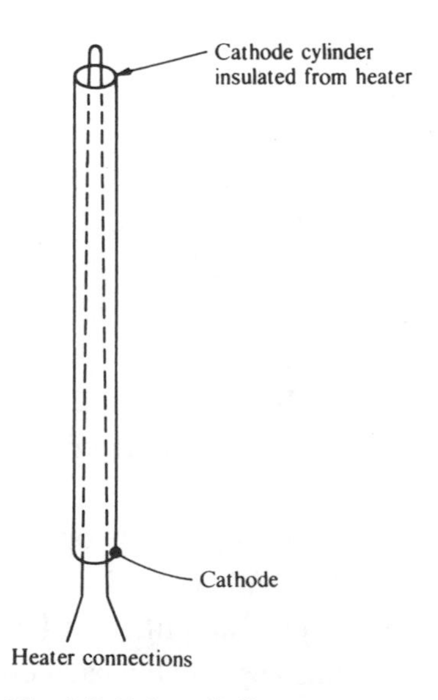

Fig. 3.7. Enlarged diagram of an indirectly heated cathode.

into the input circuit. The heater is normally of tungsten wire coated with a refractory insulator and enclosed in a thin hollow nickel cylinder which forms the cathode and is coated with the requisite barium and strontium oxides for efficient emission. This whole structure has a much higher thermal capacity than a simple filament and thus avoids thermal modulation of the emitted electrons by the a.c. supply. A disadvantage of this thermal capacity is that the cathode takes from 10 to 30 seconds to reach operating temperature, leading to the 'warm up' time associated with valve equipment. Fig. 3.7 shows an indirectly heated cathode and fig. 3.8 the circuit symbol of an indirectly heated triode. The heater symbol is frequently omitted in circuit diagrams.

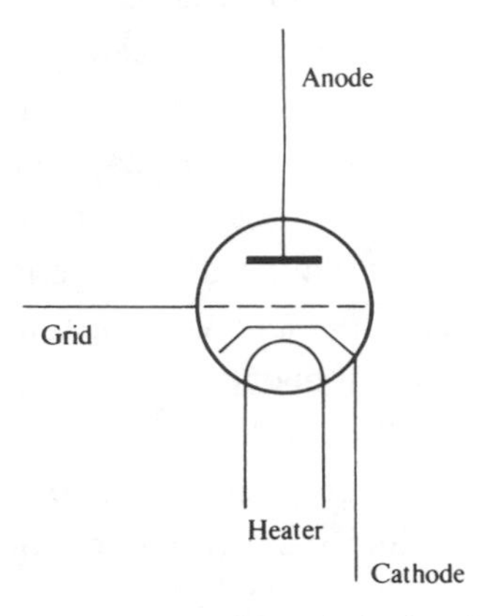

Fig. 3.8. Circuit symbol of an indirectly heated triode.

3.6 The triode voltage amplifier

The triode may be used in a voltage amplifier circuit similar in form to the FET circuit of fig. 2.7. Fig. 3.9 shows such a circuit. The valve used in this case is the ECC83 (12AX7) double triode, two triodes being included in the one envelope for economy. Either half of the double triode may be used by referring to the pin connections (B9A base) in fig. 3.10. The two heaters may be operated in series from 12.6 V by connecting between h_1 and h_2 or, in the more usual way, in parallel by joining h_1 and h_2 together and feeding the 6.3 V supply between that point and h_{tap}.

As with the other voltage amplifiers we have considered, the output voltage signal is developed across a load resistor R_L. The cathode resistor, R_K, serves a function similar to the source resistor in the FET amplifier of fig. 2.7; by dropping a steady voltage (typically 2 V) and thus keeping the cathode slightly positive with respect to earth, the grid is biased correctly for linear operation

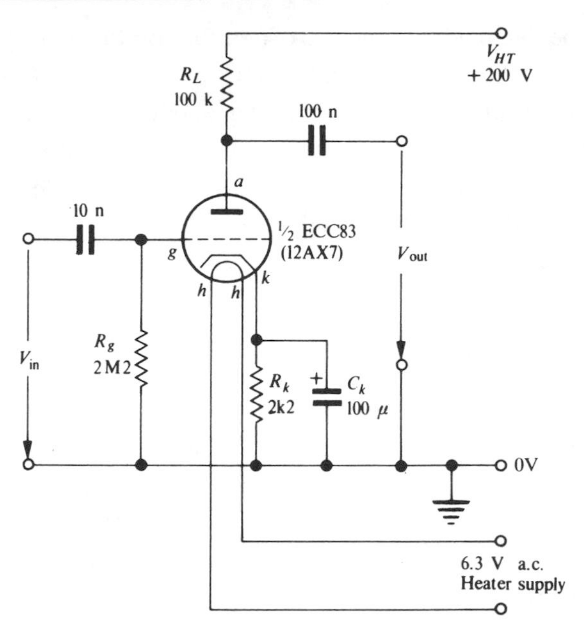

Fig. 3.9. Triode voltage amplifier.

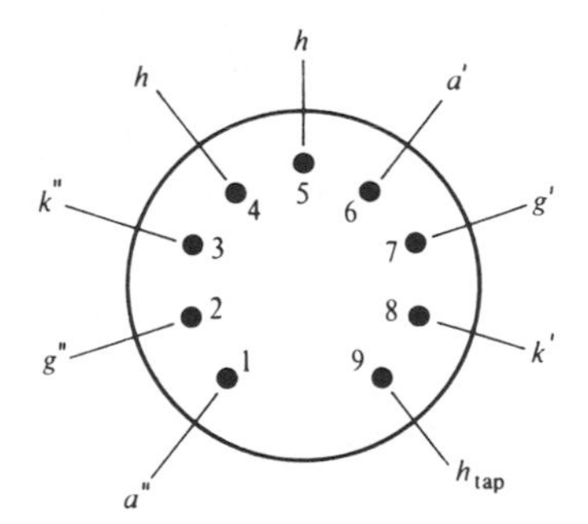

Fig. 3.10. ECC83 base connections (B9A base seen from below). Superscript ′ refers to first triode and superscript ″ to the second triode.

on the transfer characteristic (of the form of fig. 3.6). This circuit feature is known as automatic grid bias. The 100 µF bypass capacitor prevents a.c. signals appearing across the cathode resistor; these would reduce gain by opposing the input signal (negative feedback).

Like the FET, the valve is a voltage-operated device which draws negligible input current. It therefore has an inherently high input impedance and is suitable for amplifying the output of piezoelectric and capacitor microphones. The need for a 200 V HT supply is a disadvantage, but it does mean that large amplitude signals can be handled before cut-off or bottoming occurs. The characteristics of a triode are rather different from a FET or bipolar transistor in that, instead of cut-off and saturation occurring suddenly as the output swings up to HT rail or down to earth, a gradual distortion of the waveform occurs as the amplitude is increased. Nevertheless, a circuit such as fig. 3.9 can produce output signals of the order of 100 V peak-to-peak before serious distortion occurs. A typical value for the voltage gain is 30.

3.7 The tetrode and pentode

When the triode became extensively used for amplification in radio work in the 1920s, it was soon realized that its performance at high frequencies, above a few tens of kilohertz, left a good deal to be desired. At these frequencies the gain fell rapidly and some amplifiers would oscillate, generating spurious signals themselves. The problems of high-frequency amplification are discussed more fully in chapter 7, where it will be seen that such problems affect transistors as well as valves. The main cause of the high-frequency shortcomings of the triode is the capacitance between anode and grid. To overcome this problem, a second grid is introduced between the control grid and anode. This second grid or *screen grid* serves as an electrostatic screen between anode and grid. It is held at a positive d.c. potential similar to that of the anode in order to maintain the electron flow, but is connected to earth via a capacitor so that, as far the a.c. signal is concerned, it is an earthed screen. Thus we have the tetrode valve, its circuit symbol being shown in fig. 3.11.

When electrons strike a valve anode, they tend to dislodge other electrons

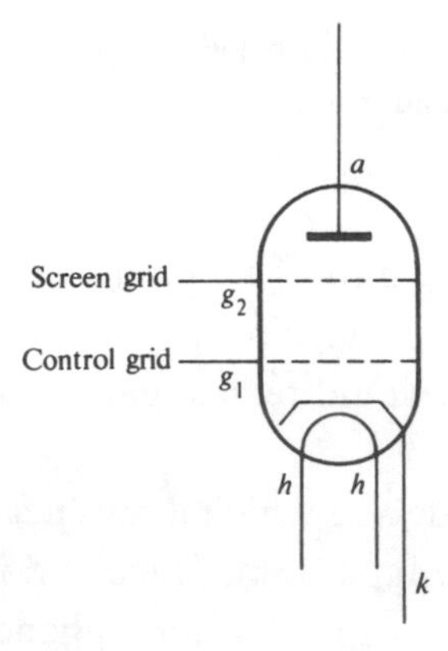

Fig. 3.11. The tetrode valve.

and cause what is known as *secondary emission*. One disadvantage of the tetrode is that these secondary electrons can be drawn to the screen grid, thus robbing the anode of current and giving rise to an undesirable 'kink' in the anode characteristic. One way of overcoming this snag is to form the electrons travelling towards the anode into a concentrated beam, using special beam-forming plates. The effect of this is to cause a powerful negative space charge which repels the secondaries back to the anode. Such valves are known as beam tetrodes and are used for audio power output stages (e.g. KT88, 6L6).

An alternative solution to secondary emission is the introduction of yet another grid – a *suppressor* grid between screen grid and anode. The suppressor grid is normally connected either to the cathode or to earth so that it repels secondary electrons whilst allowing the more energetic electron stream to pass through from screen grid to anode. This five-electrode valve is the pentode; its circuit symbol is shown in fig. 3.12.

Although the pentode was initially developed to fulfil the needs of high-frequency amplification, it turns out to have generally more useful characteristics than the triode, with the exception of a slightly higher noise level. The pentode has therefore been extensively used for amplification at high and low frequencies. It is useful to note that, although the triode and FET share the property of being three-electrode devices, the characteristics of the FET bear closer similarity to those of the pentode than they do to the triode.

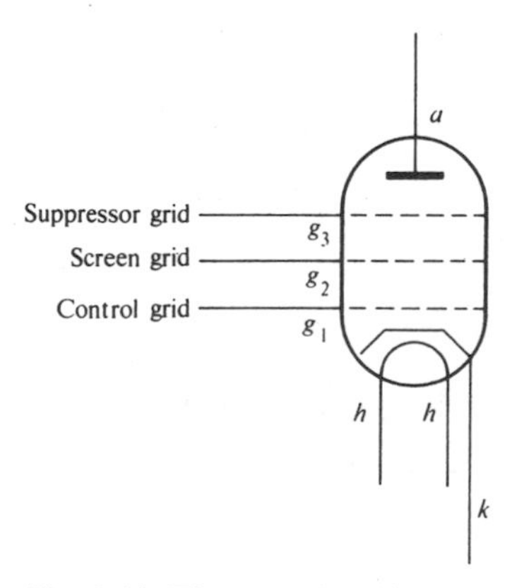

Fig. 3.12. The pentode valve.

3.8 The pentode voltage amplifier

A voltage amplifier circuit using an EF86 low-noise pentode is shown in fig. 3.13. This circuit gives a voltage gain of approximately 300 and is representative of many circuits found in valve audio equipment. Notice the connection between suppressor grid (g_3) and cathode, the screen grid (g_2) HT supply via a 1 MΩ resistor, and the 100 nF screen grid bypass (decoupling) capacitor short-circuiting a.c. signals to earth.

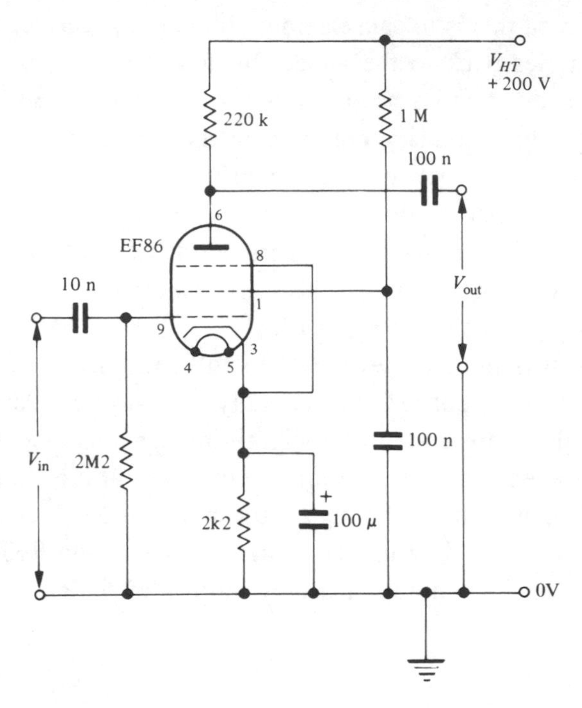

Fig. 3.13. Pentode voltage amplifier. Pin connection numbers refer to the B9A valve base as in fig. 3.10.

3.9 Valve switching circuits

If a valve is to be used for switching, the pentode is generally preferred to the triode because it has a much more definite bottoming point when it is switched on.

When designing a switching circuit to cope with more than 500 V or so, it is worth considering a pentode valve for the job rather than a transistor, particularly in an experimental rig when unexpected high voltage surges can occur. A surge which exceeds the maximum rated voltage of a transistor will usually destroy it at once, whereas a valve, lacking vulnerable junctions, is very tolerant of such mistreatment and can even survive an internal flash-over.

When using valves in switching circuits, a data book should be consulted and a valve selected which can handle the appropriate current. Anode and screen current should be limited to the specified maximum values by resistors. A valve is *on* when its grid-cathode voltage is zero. The negative grid-voltage at which the anode current cuts off depends on the valve, but is usually in the range −3 V to −40 V.

3.10 The cathode-ray tube

3.10.1 Construction and operation

One thermionic device which is unlikely to become obsolescent in the near future is the cathode-ray tube (CRT). The CRT is used in the oscilloscope for the study of electrical signals and of course, as the picture tube in a TV set, computer monitor or radar.

The CRT consists of three basic elements: an electron gun which provides a beam of electrons, a beam deflection system, which may be electrostatic or magnetic, and a fluorescent screen which emits visible light where the electron beam falls. The essential features of a CRT with electrostatic deflection are shown in fig. 3.14.

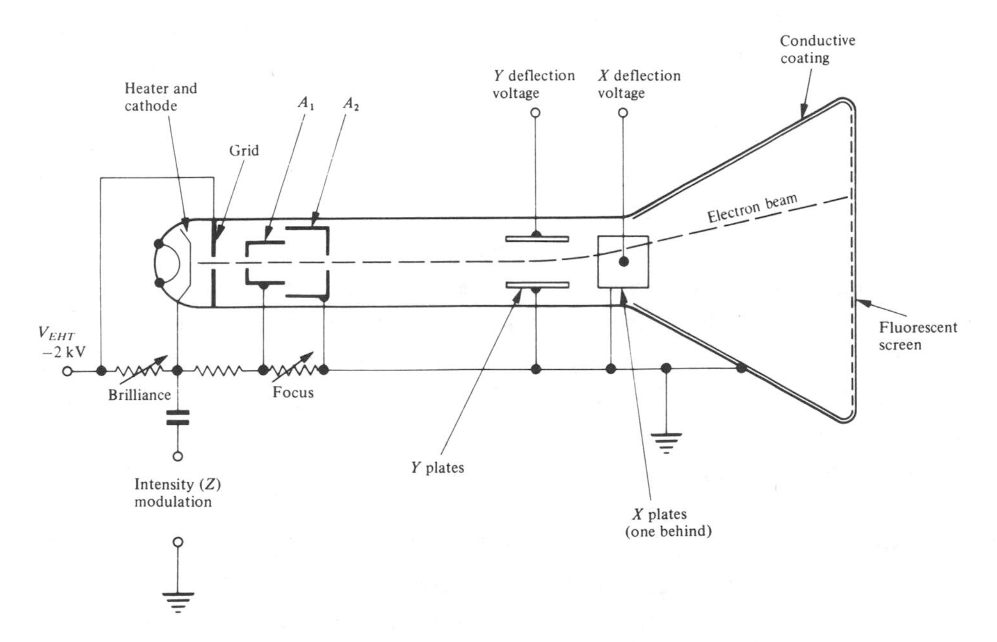

Fig. 3.14. Cathode-ray tube with electrostatic deflection. Associated simplified circuitry shows brilliance and focus controls.

The cathode emits electrons and these are drawn towards the first anode A_1 which is several thousand volts positive with respect to the cathode. The flow of electrons is controlled by the grid, on which the negative bias is determined by the brilliance control. The electron beam shoots through the hole in the centre of the first anode and also through the second anode which is slightly more positive than the first.

The purpose of the two anodes is to produce a local electric field with curved lines of force so that the electrons in the beam all converge to the same spot on the screen. The potential difference between A_1 and A_2 is adjusted by means of the focus control until a sharply-focused spot is obtained on the screen. This two-anode combination can be considered as an electron lens. A magnetic field can be applied in a similar manner to produce a magnetic lens, and is used with some CRTs for focusing. It is also exploited with great effect in the electron microscope, where a combination of electron lenses can be used to produce very high magnification with resolution a thousand times better than the optical microscope.

From the anodes of the CRT, the electron beam travels through the deflection plates, to which voltages may be applied to deflect the beam vertically in the case of the *Y* plates and horizontally with *X* plates. From the deflection plates the beam lands on the fluorescent screen or *phosphor*.

At first sight, there is nowhere for the electrons to go when they hit the screen and it might be thought that a negative charge would build up. In practice this is not the case because the energy in the electron beam is sufficient to cause secondary electrons to be '*splashed*' out of the screen; these are then collected by the conductive coating on the tube walls. There is, in fact, usually so much charge lost from the screen that it maintains itself several volts positive with respect to the second anode.

Although electrostatic deflection is standard in most oscilloscopes, it is not suitable for large CRTs used for television. These tubes, with their large screens (up to 900 mm diagonal) require a high electron beam energy to give adequate brightness (EHT is typically 25 kV). Their very large (110°) deflection angle would require enormously high deflection potentials if an electrostatic system were used. Magnetic deflection is standard for this application. Fig. 3.15 shows a typical magnetic deflection arrangement where pairs of coils are used to generate the deflecting field. Notice that the axis of the coils is *perpendicular* to the deflection direction, unlike electrostatic deflection plates, whose axis is *parallel* to the deflection direction. This difference emphasizes the contrasting behaviour of electrons in electrostatic and magnetic fields.

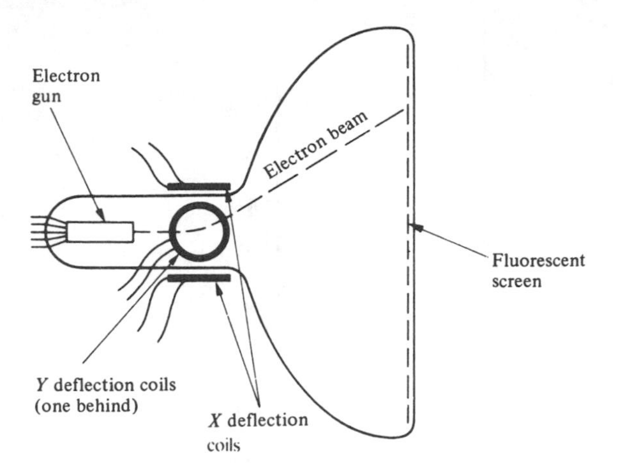

Fig. 3.15. Essentials of a magnetic deflection system as used with television tubes.

3.10.2 Screen phosphors

A variety of phosphors is available for CRTs. The most efficient in terms of optical brightness for a given beam current are usually green or blue-green and these are the colours usually seen on oscilloscope tubes. Some phosphors exhibit a very rapid decay of light output when the electron beam is switched off (short *persistence*) whilst others have a long persistence of many seconds. This slow decay can be valuable for the examination of low frequency and transient phenomena; older radar screens also employed long-persistence tubes. Most displays of transient phenomena today, however, take advantage of digital memory circuits for signal storage. By reading out the signal at high speed, short persistence phosphors can be used without flicker. As they are generally brighter than long persistence phosphors and less likely to discolour in use, a better picture is obtained.

For television picture tubes, phosphors of short persistence are used to avoid smearing on moving images.

To display colour pictures or graphics, red, green and blue phosphors which correspond as nearly as possible to the three additive primary colours are used. In the most common colour tube, the *shadowmask* tube, the screen consists of a regular pattern of groups (*triads*) of phosphor dots, each group containing a red, a green and a blue dot arranged in triangular formation. There are three electron guns in the tube, one covering the red phosphor dots, one the green and one the blue. As might be expected, a very refined arrangement is needed to ensure that each gun excites only its own phosphor dots. This is achieved

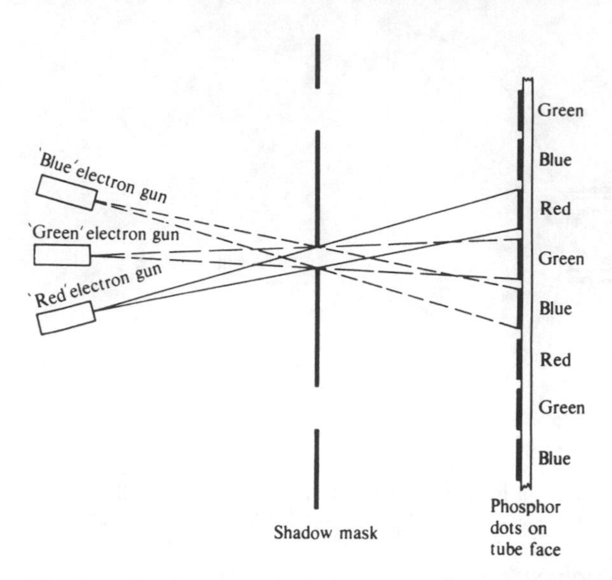

Fig. 3.16. Section through shadowmask colour picture tube to show principle of operation (not to scale).

by the shadowmask, a large metal plate mounted just behind the screen with a very precise pattern of holes or slots drilled in it. There is one hole for each group of three phosphor dots. Fig. 3.16 shows, on a deliberately distorted scale, one group of phosphor dots with its associated shadowmask hole and the three electron guns. As can be seen in the diagram, the shadowmask allows the red gun to fire only at the red dot, the green gun to fire only at the green dot and the blue gun to fire only at the blue dot. By controlling the relative intensities of the three electron beams, any colour can be produced anywhere on the tube face.

4

Negative feedback

4.1 Introduction to feedback principles

The concept of negative feedback is fundamental to life. A simple experiment will illustrate this point: close your eyes and then bring your index fingers together so that they touch at the tips. You will probably miss. By closing your eyes you have broken a feedback loop which is vital to most human actions; in order to perform an operation accurately we must be able to see what we are doing and thus apply any small corrections as and when necessary. In effect, we are taking the output (the action) and feeding it back to the input (the mental 'instruction' or intention) in such a way that the output is made equal to the input. In other words, the action is forced to correspond exactly with the intention.

Examples of negative feedback can also be found in the field of mechanical engineering. One of the clearest examples is the governor which is used to control the speed of rotating machinery. The most spectacular form of governor used to be found on the old steam engines which were the prime source of motive power in the 19th century. The governor is shown in basic form in fig. 4.1 and consists of a vertical shaft geared up to the main flywheel shaft of the engine, carrying weights on a flexible linkage.

As the speed of the engine, and hence of the governor shaft, increases, centrifugal force causes the weights to fly outwards on their linkage. The linkage is connected via a system of levers to the main steam valve so that, if the speed increases, the movement of the governor weights throttles back the main steam supply to the engine. Conversely, a tendency to slow down, perhaps due to increased load, will allow more steam in to boost the speed back to normal. The speed will thus settle down at a happy medium and be largely independent of variations in the load on the engine. Similar governors, though in a more elaborate form, control the speed of today's steam turbines which drive the alternators in power stations.

In this example, as in the physiological illustration discussed first, the system is kept under control by feeding a measure of its output back to the input. Mechanical control systems such as the governor are often known as *servo-systems* (literally, slave systems) and are fundamental to industrial automation. They form the basis of the science of cybernetics.

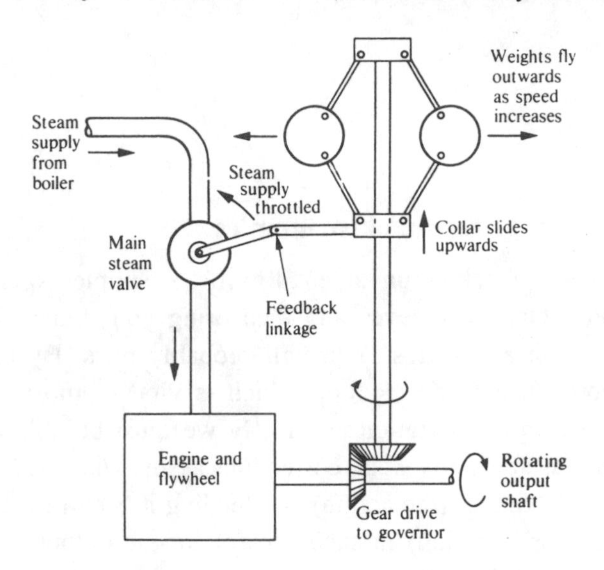

Fig. 4.1. Schematic outline of steam-engine governor showing the feedback principle.

4.2 Negative feedback in electronics

Just as the steam engine needs the controlling influence of the governor, so most electronic amplifiers require electrical negative feedback if their gain is to be accurately predictable and remain constant with changes to temperature, signal frequency and signal amplitude.

We have already seen in chapter 1 that the quiescent collector current of a transistor amplifier stage may vary with the transistor current gain. In chapter 6 we shall look at the parameters which control the voltage gain of an amplifier stage and see that the voltage gain is dependent upon variables like quiescent operating point and supply voltage. When many transistors and resistors are fabricated together in an integrated circuit amplifier, component variations can result in considerable uncertainty in the overall voltage gain. For example, a typical value for the voltage gain of the popular '741' type of integrated circuit

(IC) is 200 000 but the manufacturers' data sheets show that some specimens may have a gain as low as 20 000.

Negative feedback provides the answer to this problem of gain variation.

4.3 An amplifier with feedback

An examination of a basic feedback circuit together with one or two simple calculations will make clear the effect of negative feedback. Fig. 4.2 is a block diagram of an amplifier of voltage gain A_0 with a feedback loop incorporating an attenuator which feeds a fixed fraction, β, of the output back to the input. In this general case, we shall keep the polarity of amplifier gain and feedback positive, the feedback signal being *added* to the input signal. Having worked through the calculation we can then change the sign of either the feedback signal or the amplifier gain in order that the feedback be made negative.

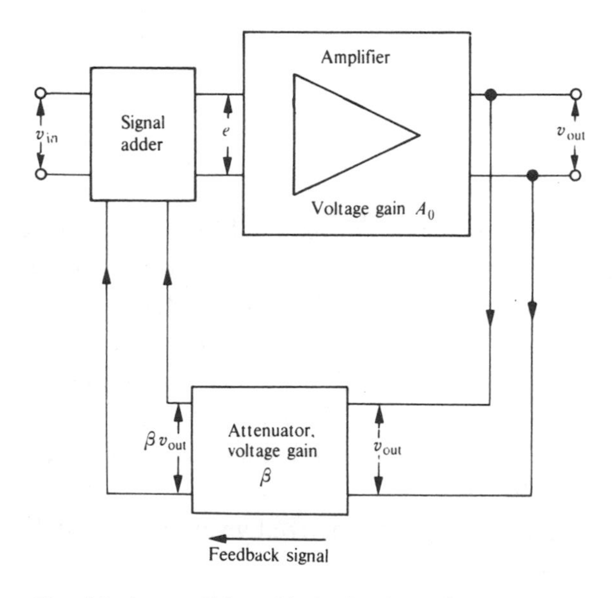

Fig. 4.2. An amplifier with feedback applied.

In fig. 4.2, we can specify the effective voltage gain, A, of the amplifier with feedback. This is given simply by the ratio of the output voltage to input voltage.

$$A = \frac{v_{out}}{v_{in}}.$$

Now we shall consider the signal e at the input of the basic amplifier of gain A_0:
input to basic amplifier,

$$e = v_{\text{in}} + \beta v_{\text{out}},$$

also, we know

$$v_{\text{out}} = A_0 e,$$

therefore

$$v_{\text{out}} = A_0 (v_{\text{in}} + \beta v_{\text{out}}).$$

Rearranging,

$$v_{\text{out}} (1 - \beta A_0) = A_0 v_{\text{in}},$$

$$\frac{v_{\text{out}}}{v_{\text{in}}} = \frac{A_0}{1 - \beta A_0}.$$

Hence, voltage gain

$$A = \frac{A_0}{1 - \beta A_0}. \tag{4.1}$$

Equation (4.1) is the general equation for an amplifier with feedback. The basic gain, A_0, is often referred to as the *open-loop* gain and the gain with feedback, A, as the *closed-loop* gain. As equation (4.1) stands, the feedback is positive and it may be seen immediately that when $\beta A_0 = 1$, an interesting situation develops in that A becomes infinite. Infinite gain implies that the amplifier will give an output signal with no input, and this is exactly what happens. Positive feedback is the basis of oscillators (signal generators) and is discussed more fully in chapter 12.

For negative feedback, we can make β negative by subtracting the feedback from the input instead of adding.

Then

$$A = \frac{A_0}{1 + \beta A_0}. \tag{4.2}$$

Now if, as is usually the case,

$$\beta A_0 \gg 1 \quad \left(A_0 \gg \frac{1}{\beta} \right),$$

then we can neglect the '1' in the denominator and

$$A \approx \frac{A_0}{\beta A_0},$$

i.e. voltage gain

$$A \approx \frac{1}{\beta}. \tag{4.3}$$

This is a most significant equation because, for the first time, we have 'designed' an amplifier with a precisely determined voltage gain. As long as the open-loop gain is much larger than the closed-loop gain (e.g. a hundred times greater) then the closed-loop gain is independent of the amplifier characteristics and dependent only on β. This feedback fraction, β, usually depends upon just two resistors in a potential divider. Resistors are the most stable components in electronics; their value can be precisely specified to any desired accuracy and is unlikely to change with time. Negative feedback extends these attributes of accuracy and stability to the gain of the entire amplifier.

4.4 Negative feedback and frequency response

No amplifier gives the same gain at all frequencies. As will be discussed further in chapter 7, the gain of any amplifier begins to fall at high frequencies, largely as a result of its internal stray capacitance. When an amplifier exhibits excessive variation of gain with signal frequency, it is said to have a poor frequency response. This deficiency is sometimes referred to as frequency distortion, but must not be confused with non-linear distortion, which will be discussed in section 4.5. Negative feedback can correct a poor frequency response as long as the open-loop gain remains greater than the closed-loop gain; equation (4.3) will then apply, and gain will be independent of frequency. Fig. 4.3 shows the graphs of the gain of a '741' IC amplifier plotted against frequency. The top line shows the open-loop gain; the enormous fall in gain at high frequencies is in fact deliberately introduced by an internal capacitor for stability reasons. The lower curves illustrate the way that negative feedback flattens the frequency response but at the expense of gain: frequency responses are plotted for closed-loop gains of 1000, 100 and 10 and are so level that they can be drawn with a ruler up to the region where the closed-loop gain approaches the open-loop figure.

Although the overall loss of gain with negative feedback may appear serious, it is in fact easy to connect two negative feedback amplifiers in cascade (i.e. one after the other) and thus return to a gain figure similar to the original open-loop gain but with a much improved frequency response.

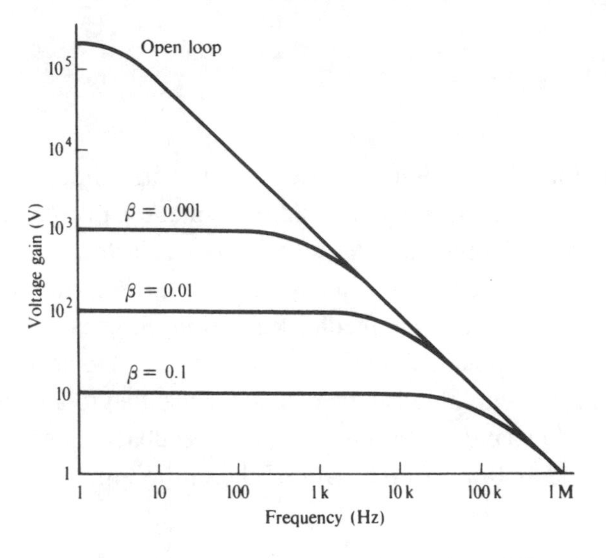

Fig. 4.3. Frequency response of the 741 IC amplifier with different feedback factors.

4.5 Non-linear distortion

4.5.1 Introduction

Anyone who has listened to a portable radio when the batteries need replacing has heard non-linear distortion. Distortion occurs when an amplifier does not present a perfect magnified replica of the input wave form but changes its shape in some way because of a non-linear transfer characteristic.

Fig. 4.4 shows a pure sine wave together with two versions of the same waveform after it has been subject to different forms of non-linear distortion. These varieties of distortion arise because the gain of the amplifier is in some way dependent upon the instantaneous signal amplitude. In fig. 4.4(*b*) the amplifier gain is dropping at large positive or negative signal excursions ('clipping distortion'), whilst in fig. 4.4(*c*) it is when the signal is very low in amplitude, near the zero crossing, that the gain falls. Such amplifier defects are grouped together under the general heading of non-linear distortion or amplitude distortion and can be considered as an error in the output of the amplifier. For obvious reasons, the example of fig. 4.4(*c*) is termed *crossover*

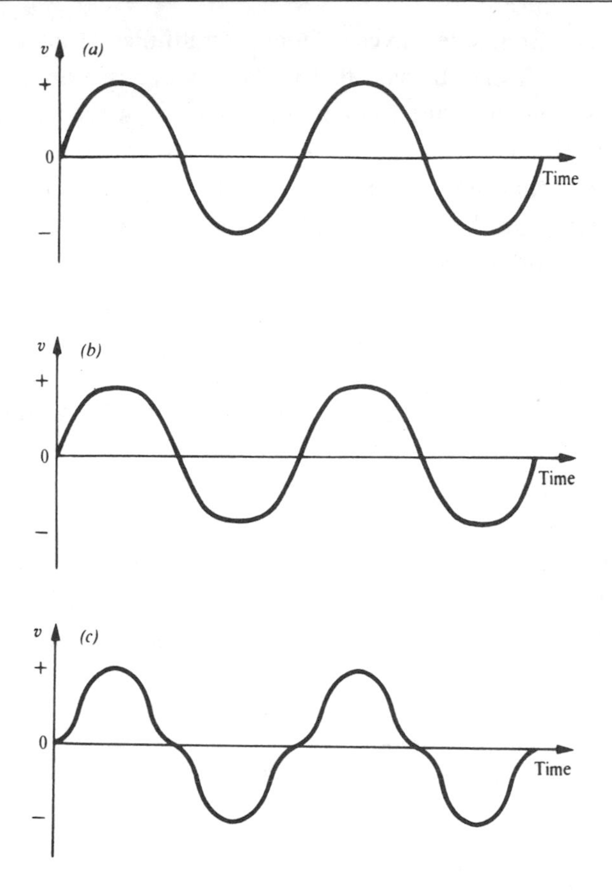

Fig. 4.4. (*a*) Pure sine wave. (*b*) Clipping distortion. (*c*) Crossover distortion.

distortion and is a common fault in badly designed or badly adjusted power amplifiers.

Clearly, if negative feedback can render the gain of an amplifier independent of variables such as transistor characteristics and signal frequency, then it ought to cure non-linear distortion by making gain independent of signal amplitude. This is indeed the case as long as the amplifier open-loop gain remains much higher than the closed-loop gain. It is appropriate to look at this important subject more clearly.

4.5.2 Harmonic distortion measurement

The use of a sine waveform to illustrate distortion in fig. 4.4 was not just an arbitrary choice. Fourier analysis shows that any repetitive waveform, of any shape, can be synthesized from sine waves at integral multiples of the original frequency. In other words, it may be said that the sine wave is a unique type of signal because it contains only one frequency. Any other wave shape contains a series of frequencies, known as *harmonics*. The first harmonic is called the *fundamental* and determines the repetition frequency of the waveform. The second, third, etc., multiples of the fundamental are simply called 'harmonics'; it is their amplitude and phase which determine the shape of the complex waveform.

When a sine wave suffers non-linear distortion in an amplifier, the amplifier is in effect adding harmonics to the original sine-wave signal. Distortion is therefore conveniently measured by feeding a pure sine wave into the amplifier and determining what proportion of the total output signal is represented by harmonics. This method of expressing non-linear distortion is called harmonic distortion measurement.

To measure the harmonic content, a frequency rejection circuit is connected to the amplifier output and tuned to reject the fundamental. Only the harmonics then remain and the total harmonic distortion (THD) is usually expressed as a percentage as follows:

$$\text{THD} = \frac{\text{r.m.s. harmonic voltage}}{\text{total r.m.s. output voltage}} \times 100\%.$$

4.5.3 *Intermodulation distortion measurement*

Another effect of non-linearity in an amplifier is observed when two signals are amplified together: in addition to the original signals, the output will contain a signal of frequency equal to the sum of the original frequencies and another signal at the difference frequency. In other words, if signals of 800 Hz and 900 Hz are fed to a non-linear amplifier, the output will contain, in addition to the amplified 800 Hz and 900 Hz signals and harmonics, new signals of frequencies 100 Hz and 1700 Hz. These sum and difference frequencies are called *intermodulation products* and are particularly objectionable in audio work because, unlike the lower-order harmonics, they are not musically related to the original frequencies. Intermodulation can be used as a measure of the non-linearity in an amplifier by expressing percentage *intermodulation distortion* (IMD) as follows:

$$\text{IMD} = \frac{\text{r.m.s. intermodulation products}}{\text{total r.m.s. output voltage}} \times 100\%.$$

IMD measurements require, of course, the rejection of two fundamental frequencies instead of the one frequency used in THD measurements and may use either a spectrum analyzer to display each modulation product separately or a demodulator like a radio detector to measure all the products together. It is worth noting that, in principle, no extra information is obtained in IMD measurement compared with THD measurement; they are both different ways of expressing non-linearity in an amplifier. In practice, however, it may be impossible to make accurate harmonic distortion measurements at the higher frequencies when the harmonics extend beyond the pass-band of the amplifier or the measuring instrument. In such a case, intermodulation measurements can give a more accurate picture of any non-linearity at high frequencies.

4.5.4 Calculation of r.m.s. distortion products

For a complete picture of distortion products, whether they are harmonics of a single frequency, or sum and difference frequencies from two signals, the individual frequencies can be picked out with a tunable narrow band filter, or spectrum analyzer. Such an instrument is expensive, but it enables the harmonics to be distinguished from noise, which may be significant when measurements are made at low output levels. The spectrum analyser also indicates whether the distortion is mainly the lower order harmonics (second and third) or the higher orders which are more objectionable subjectively in audio work. If harmonics are measured separately and the r.m.s. voltage levels measured for the second, third, fourth, etc. harmonics are V_2, V_3, V_4 etc., then

$$\mathrm{THD} = \sqrt{\left(\frac{V_2^2 + V_3^2 + V_4^2 + \ldots}{V_1^2 + V_2^2 + V_3^2 + V_4^2 + \ldots}\right)} \times 100\%,$$

where V_1 is the level of the fundamental and is likely to be the only significant term in the denominator. Intermodulation products may be combined in a similar way to give the r.m.s. value and hence a percentage for IMD.

4.5.5 Distortion and negative feedback

From the above description, non-linear distortion can be seen as an unwanted addition by the amplifier to the original signal. The calculation which follows shows that negative feedback reduces distortion by the same factor as it reduces gain.

Consider the amplifier of fig. 4.5 which has an open-loop voltage gain A_0

and a distortion content D_0 in the output before feedback is applied, i.e. without feedback,

$$v_{out} = A_0 v_{in} + D_0.$$

Now the negative feedback loop is connected feeding back fraction $-\beta$ of the output to the input.

If e is the signal voltage across the amplifier input terminals with feedback, then

$$v_{out} = A_0 e + D_0,$$

where

$$e = v_{in} - \beta v_{out}.$$

Therefore

$$v_{out} = A_0 (v_{in} - \beta v_{out}) + D_0.$$

Rearranging,

$$v_{out}(1+\beta A_0) = A_0 v_{in} + D_0.$$

Therefore

$$v_{out} = \frac{A_0}{1+\beta A_0} v_{in} + \frac{D_0}{1+\beta A_0}, \tag{4.4}$$

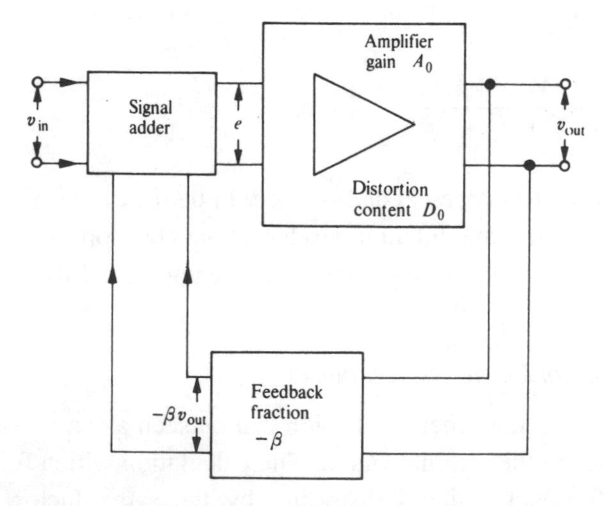

Fig. 4.5. Distorting amplifier with negative feedback.

or,

$$v_{out} = Av_{in} + D, \tag{4.5}$$

where $A = A_0/(1 + \beta A)$ is the closed-loop voltage gain and $D = D_0/(1 + \beta A_0)$ is the distortion in the output with negative feedback.

Thus we see that, when negative feedback is applied to an amplifier, distortion is reduced by the factor $(1 + \beta A_0)$, but, in return, the input signal must be increased by the factor $(1 + \beta A_0)$ in order to maintain the output signal. This may at first sight appear a pointless exercise until it is realized that virtually all the distortion is generated in the final stage of the amplifier, the one handling large signals. The use of a second amplifier to boost the input by the factor $(1 + \beta A_0)$ need not significantly contribute to the total distortion since only small signals are being handled at that stage of the proceedings. Voltage gain is cheap and well worth trading for low distortion. It is difficult to design a power amplifier with less than 1% total harmonic distortion under open-loop conditions, but, with negative feedback, figures lower than 0.01% are common in high-quality audio amplifiers.

Although negative feedback appears to be the panacea for all amplifier ailments it is important to note that it is only effective as long as the open-loop gain A_0 remains much greater than the closed-loop gain A. This may not be the case at high frequencies where capacitance, particularly significant in large power transistors, has a shunting effect. For this reason distortion often increases above 10 kHz in audio amplifiers. A second, perhaps rather obvious, case of open-loop gain suddenly dropping is when the amplifier overloads and the output stage is either cut off or saturated on part of the signal waveform. Under these conditions the output signal no longer responds to the input: the gain has fallen to zero and no amount of negative feedback can correct the resulting distortion. It is therefore a characteristic of feedback amplifiers that the overload point is very sharply defined and the resulting onset of distortion extremely sudden.

A less obvious example of falling open-loop gain is concerned with the crossover distortion illustrated in fig. 4.4(*c*). In a badly designed amplifier, crossover distortion can be so severe that the gain actually falls to zero at the cross-overs. Negative feedback is then no palliative and the resultant distortion is particularly disturbing in audio work because it generates high-order harmonics (up to the 7th, 8th, and 9th or higher) at low output levels. The 'natural order' in electronics is reversed and the signal actually becomes more distorted as its amplitude is reduced.

4.6 Instability and negative feedback

In a negative feedback amplifier it is assumed that βA_0 is negative at all frequencies. Should βA_0 become positive, we return to the general feedback equation for closed-loop gain

$$A = \frac{A_0}{1 - \beta A_0},$$

where, if $\beta A_0 = 1$, A becomes infinite and the amplifier oscillates.

It is possible for βA_0 to become positive in an amplifier, usually as a result of A_0 changing sign at high frequencies where the signal phase shifts produced by stray capacitances add up to a total 180° phase shift between input and output. The maximum phase shift which can be produced by one resistor and one capacitor is 90°, so a common method of overcoming the phase-shift problem is with one relatively large *compensation* capacitor which dominates the high-frequency characteristic and ensures a smooth first-order roll-off in gain (in a first-order roll-off, the gain falls by 6 dB for each doubling of frequency. Such a roll-off indicates a phase shift of 90°). At frequencies high enough for the stray capacitances to shift the phase round to 180°, roll-off due to the compensating capacitor ensures that the open-loop gain is so small that $\beta A_0 \ll 1$ and instability cannot arise.

It is this type of compensation that gives rise to the severe high-frequency roll-off produced in the open-loop frequency response of the 741 IC (fig. 4.3).

The above approach to high-frequency stability is an application of an important principle known as the *Nyquist stability criterion*, which is relevant to all feedback systems. More information on this subject is included in section 7.7 in the chapter on high-frequency amplification.

4.7 Current feedback

The discussion of negative feedback in this chapter has concentrated on voltage feedback since this is the most common form.

Current feedback is sometimes encountered and has the same basic characteristics as voltage feedback in that it stabilizes gain and reduces distortion. The most common form of current feedback can be applied to an amplifier stage which incorporates an emitter or source resistor (e.g. figs. 1.20 and 2.7). Normally the emitter or source resistor is bypassed by a large capacitor in order that no signal voltage appears across it, but if the bypass capacitor is removed then the a.c. voltage across the emitter or source resistor appears in

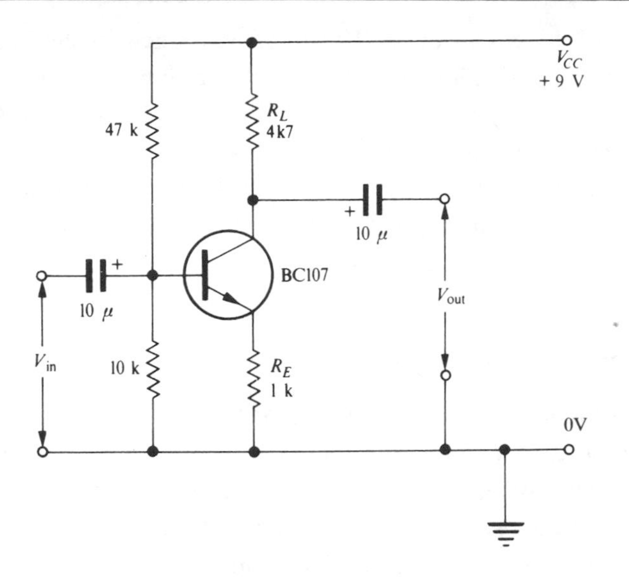

Fig. 4.6. Unbypassed emitter resistor gives current feedback.

series with the input signal. Fig. 4.6 shows a bipolar stage with an unbypassed emitter resistor; here it may be seen that the input, V_{in}, no longer has 'direct access' to the base–emitter junction, but has to go via R_E. Now across R_E, in addition to the steady p.d. due to the quiescent emitter current, is an a.c. voltage proportional to the a.c. signal component in the emitter current. In this way, a voltage signal proportional to the output *current* is fed back to the input. If we make the very reasonable approximation that the collector current and emitter current are equal, then the fraction of the output fed back is given by $\beta = R_E/R_L$. Hence, closed-loop voltage gain

$$A \approx \frac{1}{\beta} \approx \frac{R_L}{R_E}.$$

This type of current feedback is one of the simplest ways of exploiting the desirable properties of negative feedback in a single amplifier stage.

It is useful, in passing, to notice that the signals across R_E and R_L are 180° out of phase, that across R_E being in phase with the input. Two outputs of opposing phase may therefore be taken from R_L and R_E. If R_L and R_E are made equal then the two voltage signals are equal in magnitude and the circuit is often referred to as a *phase splitter*.

Sometimes R_E is partially bypassed by using two resistors in series, only one of which is shunted by a capacitor. The total resistance in the emitter circuit is decided by the d.c. conditions in the amplifier stage and the unbypassed portion chosen to give appropriate β for the required closed-loop gain.

Although voltage feedback and current feedback have in common the reduction of distortion and stabilization of gain, they differ in their properties when the output impedance of an amplifier is considered, but a discussion of this difference must wait until the next chapter.

4.8 Experiments with negative feedback

An IC amplifier forms the ideal basis for some negative feedback experiments. Fig. 4.7 shows the basic circuit of an amplifier with series voltage feedback. A fraction β of the output voltage signal is fed back in series with the input, but out of phase so that it opposes the incoming signal. In fig. 4.7, the triangle is the conventional symbol for an amplifier. The connections marked '+' and '−' are the input terminals and the third connection, at the triangle apex, is the output. For the sake of clarity, the power supply is not included in the diagram. The amplifier, like most IC amplifiers, has a 'single-ended' output and a 'differential' input. In other words, the output signal appears between the single output terminal and earth whilst the input responds to the potential *difference* between the two terminals. The terminal marked '+' is the non-inverting input: a positive signal applied relative to the other terminal gives a positive output. The terminal marked '−' is the inverting input: a positive signal relative to the other terminal gives a negative output. Incidentally, if

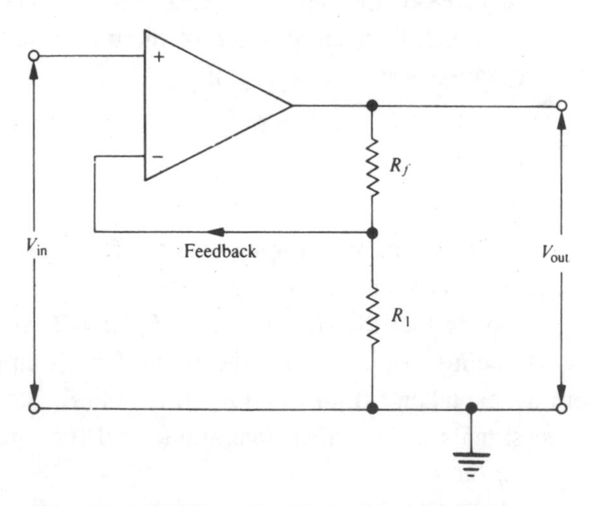

Fig. 4.7. Amplifier with series voltage feedback.

the same signal is applied, relative to earth, to both inputs at the same time, no output should be produced. Differential amplifiers are discussed in more detail in chapter 8.

The differential input makes the application of negative feedback very simple: a fraction of the output is fed back to the inverting input whilst the input signal V_{in} is applied between the non-inverting input and earth. The potential divider, R_f and R_1, determines β:

$$\beta = \frac{R_1}{R_1 + R_f}.$$

Thus, if the closed-loop gain A is much less than open-loop gain A_0, we have

$$A = \frac{1}{\beta},$$

$$A = \frac{R_1 + R_f}{R_1}.$$

Although series feedback is chosen for this example because of its simplicity, it is not the only way of feeding back a given fraction of the output voltage. Shunt feedback, where the feedback voltage is in parallel with the input signal is also commonly used and is discussed in chapter 11.

Fig. 4.8 shows the full working circuit, using a type 741 amplifier, for experiments with negative feedback. It is clearly derived from fig. 4.7 with the addition of a power supply rail and coupling capacitors for d.c. isolation. The 220 kΩ resistors, R_2 and R_3, look after the quiescent operating point by holding the potential of the non-inverting input midway between earth and the supply rail. The presence of capacitor C_3 in the feedback divider ensures that β is unity as far as the d.c. conditions are concerned, so that the output follows the d.c. level of the non-inverting input and sits at half the supply voltage; the circuit is thus able to handle both positive- and negative-going signals.

The voltage gain of the feedback amplifier may be measured by feeding the input from a signal generator and comparing input and output signals on an oscilloscope; the specified values of R_f and R_1 give a closed-loop gain of 100. To examine the amplifier without feedback, resistor R_1 in the feedback divider may be short-circuited; this disables the feedback at signal frequencies but does not upset the quiescent d.c. conditions. The voltage gain without feedback will be found to be very high (of the order of 10^5) at low frequencies in the region of 100 Hz. It will, however, fall at higher frequencies and the frequency response should be similar to the top curve in fig. 4.3. An unexpected fall in gain below 100 Hz will also be apparent. This is due to the finite reactance of capacitor C_3 at low frequencies which introduces a little negative feedback.

With the feedback in circuit, the value of R_f may be changed to give various closed-loop gains and the effect on the frequency response measured; fig. 4.3 is a guide to the results to be expected.

The supply voltage, V_{CC}, may be set as low as 6 V or as high as 36 V without upsetting the operation of the circuit. The open-loop gain of the 741 amplifier will, however, vary considerably with supply voltage; this fact allows the gain stabilizing properties of negative feedback to be demonstrated. With $\beta = 0.01$, the change in closed-loop gain with variation in supply voltage is negligible, but changes in open-loop gain will be observed with R_1 short-circuited.

The assessment of the reduction in non-linear distortion with feedback is difficult unless a distortion meter or spectrum analyser is available to eliminate the fundamental and measure the harmonics produced in the amplifier. Failing a suitable meter, a double-beam oscilloscope may be used to compare the

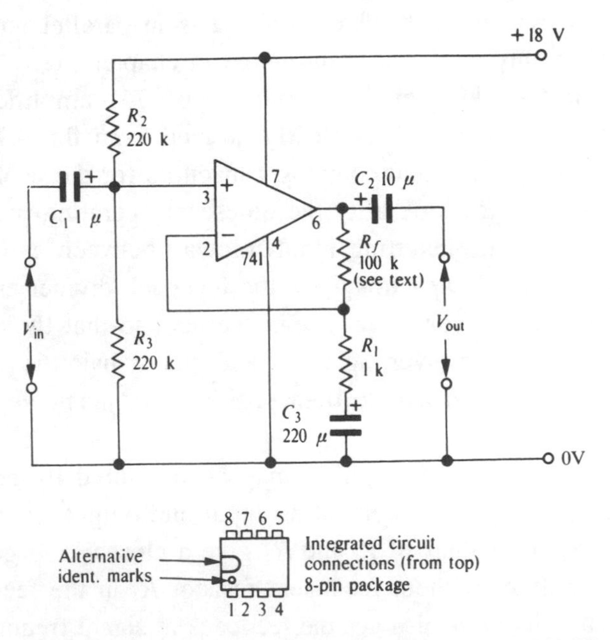

Fig. 4.8. Experimental feedback amplifier using a 741 IC. Short-circuit R_1 to measure open-loop gain.

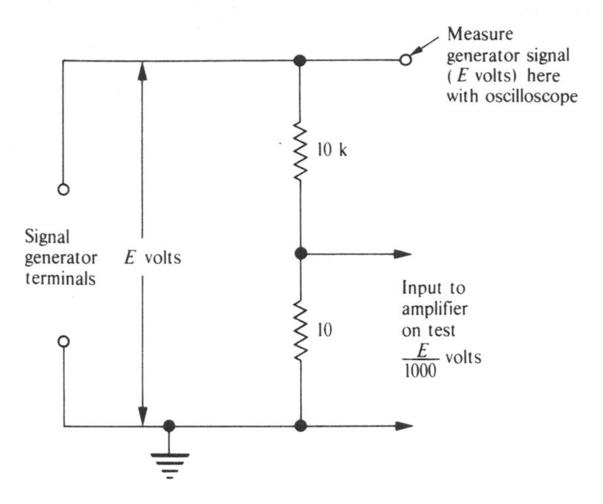

Fig. 4.9. Potential divider to facilitate input signal measurements on a high-gain circuit like fig. 4.8.

shape of input and output sine waves and a qualitative assessment of non-linearity made both with and without feedback.

When measuring the very high open-loop gain, difficulty will probably be experienced in measuring the tiny voltage signal (<1 mV) at the amplifier input. Many oscilloscopes are insufficiently sensitive to give a measurable trace. A potential divider connected to the signal generator output will solve the problem; fig. 4.9 shows the simple circuit. With the resistor values shown, the voltage fed to the amplifier is one thousandth of the output of the signal generator. The oscilloscope can be readily used to measure the latter voltage level, so that, if close-tolerance resistors are used in the divider, the exact input to the amplifier can be determined.

5

Impedance matching

5.1 Introduction

The subject of impedance matching is frequently surrounded by an undeserved aura of mystery. When a circuit fails to work as well as expected, impedance mismatch is often blamed in a rather vague sort of way, but correcting the problem may be regarded more as witchcraft than science. The purpose of this chapter is to outline the principles and practice of impedance matching and, in so doing, to attempt to dispel any associations of mystery and magic!

5.2 Input impedance

Any electrical device which requires a signal for its operation has an input impedance. Just like any other impedance (or resistance in d.c. circuits), the input impedance of a device is a measure of the current drawn by the input with a certain voltage across it.

For example, the input impedance of a 12 V light bulb rated at 0.5 A is 12/0.5 Ω, or 24 Ω. The bulb is a clear example of impedance because we know that there is nothing but a filament to consider. The input impedance of a circuit such as bipolar transistor amplifier might seem to be more complicated. At first sight, the presence of capacitors, resistors and semiconductor junctions in a circuit makes the input impedance difficult to assess. However, any input circuit, however complicated, may be resolved into the simple impedance shown in fig. 5.1. If V_{in} is the a.c. input signal voltage and I_{in} the a.c. current drawn by the input, then input impedance

$$Z_{in} = \frac{V_{in}}{I_{in}} \ \Omega. \tag{5.1}$$

In the majority of circuits, the input impedance is resistive over most of

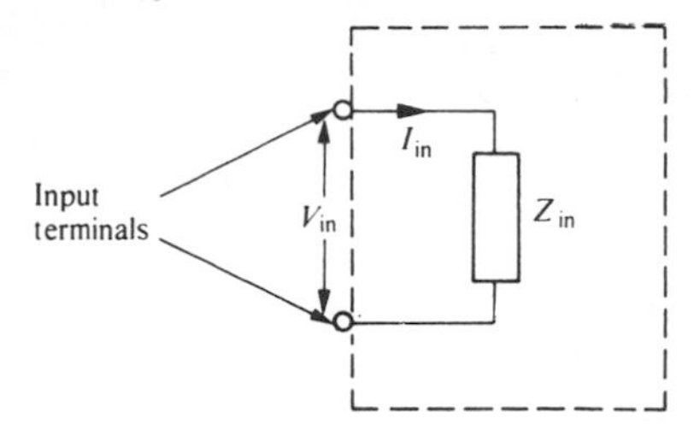

Fig. 5.1. A pair of input terminals illustrating the meaning of input impedance, Z_{in}.

the frequency range, there being negligible phase difference between input voltage and input current. In such cases, the input appears as the circuit of fig. 5.2 and Ohm's law applies, the complex algebra and vector diagrams of reactive circuits being unnecessary. It is important to note, however, that a resistive input impedance does not necessarily mean that a d.c. signal can be used to measure the input resistance; there may be reactive components in the way (e.g. a coupling capacitor) which, whilst insignificant at moderate a.c. frequencies will prevent d.c. measurements from being made on the input.

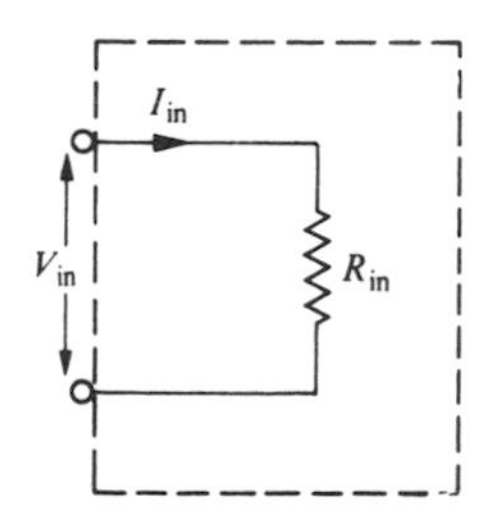

Fig. 5.2. A device with a resistive input impedance.

5.3 Measurement of input impedance

The measurement of input signal voltage is easy, using an oscilloscope or a.c. voltmeter. The a.c. input current, however, may not be easy to measure, particularly in the case of a high input impedance. The most convenient method for input impedance measurement is shown in the circuit of fig. 5.3. A resistor of known value R Ω is connected between the signal generator and the input circuit to be measured. The signal voltages V_1 and V_2 on either side of the resistor are then measured on an oscilloscope or high impedance a.c. voltmeter.

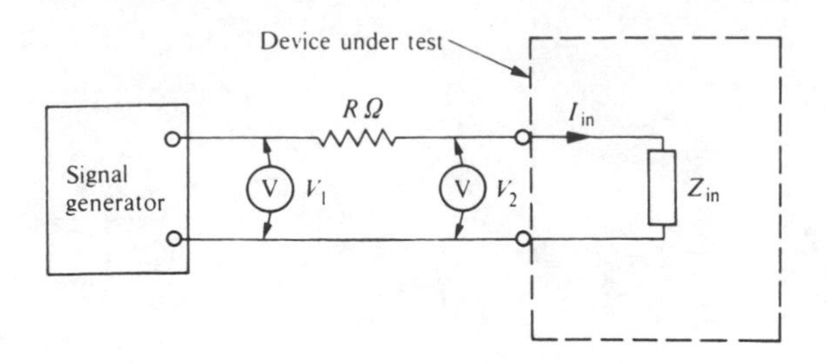

Fig. 5.3. Measurement of input impedance.

Then, by Ohm's law, if I_{in} is the a.c. input current (amps), the voltage drop across the resistor R is given by

$$V_1 - V_2 = RI_{in} \text{ volts.}$$

Therefore

$$I_{in} = \frac{V_1 - V_2}{R} \text{ amps.}$$

Now

$$Z_{in} = \frac{V_2}{I_{in}} \ \Omega$$

hence

$$Z_{in} = \frac{RV_2}{V_1 - V_2} = \frac{R}{V_1 / V_2 - 1} \ \Omega. \tag{5.2}$$

If the circuit under test is an amplifier, then the measurement of V_1 and V_2 is often most conveniently made at the amplifier output, measuring V_1 with the signal generator directly connected to the input and V_2 with the resistor R in series with the input. Since only the ratio of V_1/V_2 appears in the expression for Z_{in}, the amplifier gain does not upset the measurement. The output voltage of the generator is assumed to remain constant throughout the measurement. To take a very simple case, if the connection of 10 kΩ in series with the input causes the amplifier output voltage to drop by a half, then $V_1/V_2 = 2$ and $Z_{in} = 10$ kΩ.

5.4 Output impedance

A common example of the concept of output impedance is the way that the lights on a car go dim when the starter motor is operated. The heavy current drawn by the starter causes a voltage drop inside the battery, reducing its terminal voltage and dimming the lights. The voltage drop occurs across the *output resistance* of the battery, better known, perhaps, as its internal resistance or *source resistance.*

We can broaden this idea to include all output circuits, both a.c. and d.c., which invariably have a certain output impedance associated with a voltage generator. That this simple description applies to even the most complicated circuit is confirmed by Thévenin's theorem, which states: any network of impedances and generators having two output terminals may be replaced by the series combination of one impedance and one generator. Here, a 'generator' is assumed to be an ideal voltage-producing device which continues to produce a constant voltage even when current is drawn from it. Thévenin's description of an output circuit is shown in fig. 5.4, Z_{out} being the output impedance and V the open-circuit output voltage.

It is relevant to point out at this stage that, in discussing input and output impedance, a concept has been introduced for the first time: the *equivalent circuit.* Figs. 5.1, 5.2 and 5.4 are all equivalent circuits. They do not necessarily represent the actual components and connections in the circuits being examined, but are convenient representations which are helpful in understanding the way the circuits behave.

If we now return to fig. 5.4, we see that, if the output terminals are loaded by a resistor, or by the input terminals of another circuit, part of the voltage generator output voltage V will be dropped in the internal impedance Z_{out}.

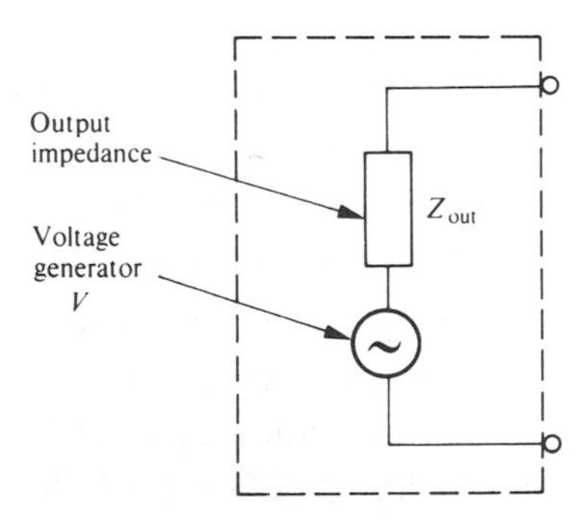

Fig. 5.4. Thévenin's equivalent circuit, applicable to any pair of output terminals.

Output circuits usually have a resistive output impedance over most of the frequency range and the equivalent circuit of fig. 5.5 will then apply, where R_{out} is the output resistance.

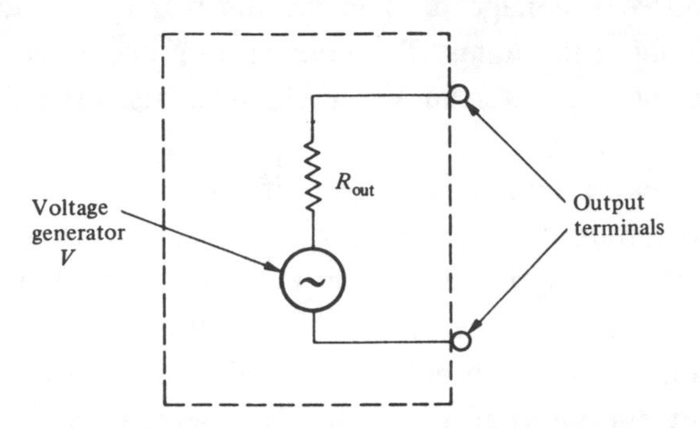

Fig. 5.5. An equivalent circuit applicable to most devices: resistive output impedance.

5.5 Measurement of output impedance

One simple method of measuring output impedance can be deduced from fig. 5.4. If the output terminals are short-circuited, the resultant short-circuit current flowing, I_{sc} (amps) can be measured and, since this is the current produced by voltage V in impedance Z_{out},

$$Z_{out} = \frac{V}{I_{sc}}\Omega. \quad (5.3)$$

The voltage V produced by the generator is measured at the output terminals under 'open circuit' conditions, i.e. negligible output current. Thus, output impedance may readily be expressed as the ratio of open-circuit voltage to short-circuit current.

Having discussed this basic method of output impedance measurement, it must be said that there are snags inherent in measuring the short-circuit output current of most circuits. Usually, the operating conditions will be so upset by the short circuit that a true reading will be unobtainable; in some cases components may be damaged by the abnormal loads they are asked to carry. A clear example would be the application of the short-circuit method to measure the output impedance of the a.c. mains! Despite these practical drawbacks, the method may justifiably be used in a theoretical derivation of the output impedance of a circuit and is used in this way later in the chapter.

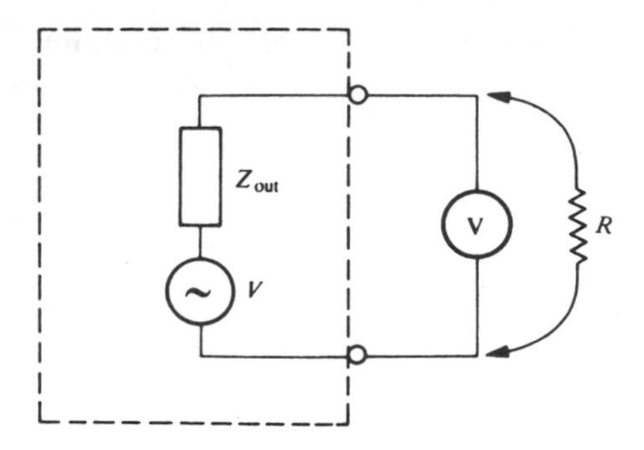

Fig. 5.6. Measurement of output impedance using a shunt resistor.

A practical method for the measurement of output impedance is shown in fig. 5.6. Here the open-circuit output voltage is measured with a high-impedance voltmeter or oscilloscope, and then a load of known resistance R Ω is shunted across the output terminals. The reduced output voltage with the load present is once again read from the meter. Z_{out} can be calculated as the ratio of potential drop to output current.

If open-circuit output voltage is V and output voltage with load R is V',

voltage drop across Z_{out} with load

$= V - V'$ volts,

output current with load $= \dfrac{V'}{R}$ amps,

therefore

$$Z_{out} = \frac{R(V - V')}{V'}\ \Omega. \tag{5.4}$$

5.6 Impedance matching for optimum voltage transfer

In the majority of electronic circuits, we are concerned with voltage signals. We have already discussed some *voltage* amplifiers; that universal measuring instrument, the oscilloscope, measures *voltage* signals; the sensitivity of microphones and phono cartridges is expressed in terms of *voltage* output. Even so-called 'power' amplifiers are best regarded as special voltage amplifiers with the ability to deliver high output currents. Hence, in the majority of cases when we are connecting one piece of circuitry to another, we require the maximum transfer of *voltage* signal. It is this requirement of *maximum voltage transfer* that is normally observed in impedance matching. Given this criterion, impedance matching is no problem.

Fig. 5.7 shows two circuit 'boxes' connected together: for optimum voltage transfer, V_{in} should be as nearly equal to V as possible. Writing down the equation for V_{in},

$$V_{in} = \frac{VZ_{in}}{Z_{out} + Z_{in}}$$

and

$$V_{in} \approx V \quad \text{if} \quad Z_{in} \gg Z_{out}.$$

In other words, for the best voltage transfer between two circuits, the output impedance of the first circuit should be much lower than the input impedance of the second; as a rule, $Z_{in} > 10\,Z_{out}$. It is for this reason that an item of test equipment, such as a signal generator, which is designed to feed a voltage signal to other circuits, is designed with a low output impedance (typically $<100\ \Omega$). On the other hand, an oscilloscope intended for studying voltage signals from a circuit under test is made with a high input impedance (typically $1\ M\Omega$).

If the conditions for optimum impedance matching are not observed, and a signal source feeds an input impedance comparable with the source output impedance, the most common result is simply a loss of signal. Such a situation arises when two bipolar transistor amplifier stages like the one in fig. 1.19 are connected one after the other (cascaded). Both the input and output impedance of this bipolar stage are of the same order (usually several thousand ohms – see section 6.3) and this means that some 50% of voltage signal is lost in the coupling between two stages. The FET amplifier of fig. 2.7, on the other hand, with its very high input impedance and moderate output

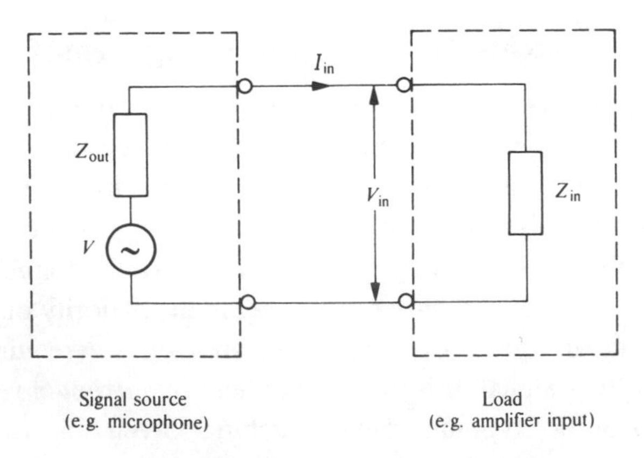

Fig. 5.7. Illustration of impedance matching between two devices.

impedance, is much better from the point of view of impedance matching; negligible signal is lost in coupling two such stages together.

There are one or two instances where special care needs to be taken with impedance matching, because too low a load impedance affects not only voltage gain but also frequency response. This occurs when the output impedance of the signal source is not purely resistive but is reactive and therefore varies with frequency. A common example is the capacitor microphone where the output impedance is expressed, not in ohms, but in picofarads, a typical value being in the region of 50 pF. In order to maintain good bass response, the input impedance in the amplifier needs to be high compared with the reactance of 50 pF at frequencies as low as 20 Hz. In practice this requires an input impedance of the region of 200 MΩ which is usually provided by a FET amplifier mounted within the microphone case.

5.7 Impedance matching for optimum power transfer

Although maximum voltage transfer is the usual criterion for impedance matching, there are occasional instances where we require maximum power transfer.

We can consider the circuit of fig. 5.7 from the viewpoint of maximum power transfer. For simplicity, Z_{out} and Z_{in} will be considered resistive and equal to R_{out} and R_{in} (ohms) respectively.

Then, if power in $R_{in} = W$ watts,

$$W = V_{in} I_{in} = \frac{V_{in}^2}{R_{in}} \text{ watts}$$

but

$$V_{in} = V \frac{R_{in}}{R_{out} + R_{in}} \quad \text{(potential divider)},$$

therefore

$$W = \frac{V^2}{R_{in}} \frac{R_{in}^2}{(R_{out} + R_{in})^2}$$

$$= \frac{V^2}{R_{in}\left(1 + \dfrac{R_{out}}{R_{in}}\right)^2}. \tag{5.5}$$

Power W is a maximum when denominator $R_{in}(1 + R_{out}/R_{in})^2$ is at its minimum value.

We can obtain the value of R_{in} for this condition by differentiating the denominator with respect to R_{in} and setting the result to zero, i.e.

$$\frac{d}{dR_{in}}\left[R_{in}\left(1+\frac{R_{out}}{R_{in}}\right)^2\right]=0.$$

Evaluating the expression,

$$\frac{d}{dR_{in}}\left[2R_{out}+R_{in}+\frac{R_{out}^{\;2}}{R_{in}}\right]=0$$

giving

$$1-\frac{R_{out}^{\;2}}{R_{in}^{\;2}}=0,\ \text{i.e.}\ R_{in}^{\;2}=R_{out}^{\;2}$$

Hence, for maximum power in R_{in}, $R_{in} = R_{out}$.

This result is known as the maximum power theorem: for maximum *power* transfer from source to load, the load resistance should be equal to the output resistance of the source. The theorem proof may be followed through with reactive as well as resistive components for Z_{out} and Z_{in}, i.e.

$$Z_{out} = R_{out} + jX_{out},$$

$$Z_{in} = R_{in} + jX_{in}.$$

It is then found that, for maximum power transfer, as well as $R_{in} = R_{out}$ we must have the condition $X_{in} = -X_{out}$, in other words, if one impedance is capacitative, the other should be inductive.

Having discussed the maximum power theorem, it is necessary to point out that its importance in electronics is minimal. The use of a load impedance equal to the generator output impedance automatically implies that as much power is dissipated in the generator as in the source. This is an inefficient situation, and one which could lead to overload of the components involved. Where an input signal is arriving via a transmission line, such as a long telephone cable, the load impedance is usually chosen to be equal to the internal *characteristic impedance* of the cable, but this is a very specialized example of power transfer and has as much to do with eliminating reflected signals within the line as with signal transfer to the amplifier.

Electrical circuits are frequently used as analogues of mechanical and acoustical systems, to simplify analysis of the latter. The maximum power theorem then proves to be a valuable concept for optimizing energy transfer in such systems.

5.8 Impedance matching for optimum current transfer

Occasionally we require an impedance match which provides maximum current into an input circuit. Referring once again to fig. 5.7, the maximum input current I_{in} will be achieved by making the total resistance in the circuit as small as possible. Assuming Z_{out} is fixed, we therefore aim for the lowest possible value of Z_{in}. This rather unusual condition is thus seen to be the direct opposite of the usual voltage transfer requirement.

5.9 Impedance matching for minimum amplifier noise

5.9.1 Signal-to-noise ratio

Noise is always present in electronic circuits. It is audible in a radio set between stations and when listening to a weak signal. Electrically speaking, noise is a random voltage fluctuation, which can be heard through a loudspeaker as hiss.

It is electrical noise which sets the ultimate lower limit on the measuring ranges of electronic instruments, limits the useful sensitivity of a radio receiver, and can intrude during quiet musical passages on an analogue audio cassette.

Being a random phenomenon, noise is not confined to a single frequency, but is present at all parts of the spectrum. The noise power produced by a circuit is in fact usually proportional to the frequency bandwidth.

Signal-to-noise ratio is a useful measure of the 'intelligibility' of the required signal in a system in relation to the noise. It is expressed as the ratio of signal power P_s to noise power P_n

$$\text{S/N ratio} = \frac{P_s}{P_n}.$$

It is commonly stated in decibels:

$$\text{S/N ratio} = 10 \log_{10} \frac{P_s}{P_n} \text{ dB}. \tag{5.6}$$

Noise and signal are present together at the output, and both 'see' the same impedance, so we can express the power ratio in terms of the more convenient ratio of root mean square (r.m.s.) signal voltage V_s to r.m.s. noise voltage V_n:

$$\text{S/N ratio} = 10 \log_{10} \left(\frac{V_s}{V_n}\right)^2 \text{ dB}$$

$$= 20 \log_{10} \frac{V_s}{V_n} \text{ dB.} \tag{5.7}$$

In the specification of equipment, we often require a measure of the maximum available signal-to-noise ratio. This is obtained by measuring the maximum available r.m.s. signal output voltage $V_{so(\max)}$ and comparing with the r.m.s. noise voltage present at the output, V_{no}:

$$\max \text{S/N ratio} = 20 \log_{10} \frac{V_{so(\max)}}{V_{no}} \text{ dB.} \tag{5.8}$$

As we shall see, it is usually necessary to specify the output impedance of the signal generator to obtain a meaningful value for signal-to-noise ratio.

5.9.2 Thermal noise

Any piece of wire produces a certain amount of electrical noise owing to thermal agitation of the atoms. This is known as *thermal* noise or *Johnson* noise.

Nyquist showed from thermodynamic principles that the r.m.s. noise e.m.f. V_n across a resistor R is given by

$$V_n = \sqrt{(4kTR\Delta f)}, \tag{5.9}$$

where k is Boltzmann's constant ($k = 1.380 \times 10^{-23}$ J K^{-1}), T is resistor temperature in K (= 273 + °C), Δf is the frequency bandwidth of the measurement circuit (Hz); R is the resistor value (Ω).

Substituting typical values for the full audio bandwidth (≈20 000 Hz) and room temperature (≈300 K) gives

$$\text{r.m.s. noise e.m.f. } V_n = 1.8 \times 10^{-8} \times \sqrt{R} \text{ volts,}$$

e.g.

$$\text{if } R = 10\,\text{k}\Omega, \sqrt{R} = 100\,\Omega^{1/2}, \text{ then}$$

$$V_n = 1.8 \times 10^{-8} \times 10^2 \text{ V}$$
$$= 1.8 \mu\text{V}.$$

5.9.3 Noise in transistors

Noise is produced when current flows in transistors. There are three main noise sources:

(*a*) *Thermal noise* arises from the finite resistance of the semiconductor material.

(*b*) *Shot noise* occurs whenever current carriers cross a barrier such as a junction. Each carrier causes a slight transient current surge as it travels across the junction, the combined effect being a random current fluctuation. Shot noise power is directly proportional to current; its effect is greatest when the junction has a high internal impedance, such as a reverse-biased collector–base junction.

(*c*) *Flicker noise* or $1/f$ *noise* is due to random variations in diffusion processes in the transistor. As the name suggests, the power spectrum of the $1/f$ noise is inversely proportional to frequency, and so consists mainly of low-frequency energy. It is the dominant noise source below about 1 kHz in bipolar transistors.

5.9.4 Noise figure

The existence of thermal noise means that it is impossible ever to have a perfectly 'clean' signal with an infinite signal-to-noise ratio.

The lowest noise level we can hope for is that given by the theoretical thermal noise voltage of the signal source resistance. In practice, the apparent input noise level is inevitably higher than this because of noise contributed by the amplifier. This degradation of signal-to-noise ratio is specified by the *noise figure* or *noise factor* (NF) of the amplifier, which is defined as the ratio of signal-to-noise power in the input signal (P_{si}/P_{ni}) to the signal-to-noise power at the amplifier output (P_{so}/P_{no}), i.e.

$$\text{NF} = \frac{P_{si}/P_{ni}}{P_{so}/P_{no}}.$$

Noise figure is conveniently expressed in terms of r.m.s. signal and noise voltages V_s and V_n respectively:

$$\text{NF} = \left(\frac{V_{si}/V_{ni}}{V_{so}/V_{no}}\right)^2.$$

The i and o subscripts indicate input and output as above. Now V_{so}/V_{si} is the amplifier voltage gain, A_V, therefore

$$NF = \left(\frac{V_{no}}{A_V V_{ni}}\right)^2. \tag{5.10}$$

V_{no}/A_V is the r.m.s. noise voltage which would have to be applied to the input of a noiseless amplifier of voltage gain A_V in order to give a noise output of V_{no}. The term V_{no}/A_V is called the ***total noise referred to input***, $V_{ni(\text{total})}$. This concept is valuable because it eliminates the amplifier gain from our definition of noise figure:

$$NF = \left(\frac{V_{ni(\text{total})}}{V_{ni}}\right)^2,$$

where V_{ni} is the r.m.s. noise voltage present in the source of the input signal. Noise figure is usually expressed in decibels:

$$NF = 20 \log_{10} \frac{V_{ni(\text{total})}}{V_{ni}} \text{ dB}. \tag{5.11}$$

A noiseless amplifier would have a noise figure of 0 dB. Good low-noise amplifiers usually achieve noise figures less than 3 dB.

If we consider an amplifier in isolation, the r.m.s. noise voltage V_{ni} actually arriving at the input arises solely from thermal noise in the internal resistance (output resistance) of the signal source. Total noise referred to input, $V_{ni(\text{total})}$, is made up of contributions from thermal noise, V_{ni} and from the noise produced by the amplifier transistors.

Now a transistor not only gives rise to an input noise voltage but also to an input noise current. Fig 5.8(a) shows the equivalent circuit of an amplifier input with noise generators included. The instantaneous noise voltage is e_n and instantaneous noise current i_n; we shall be using the mean square values of the signals $\overline{e_n^2}$, $\overline{i_n^2}$ and V_{ni}^2 (V_{ni} is already specified as an r.m.s. value).

It is necessary to add that $\overline{e_n^2}$, $\overline{i_n^2}$ and V_{ni}^2 are all proportional to the bandwidth of the amplifier. (It has been said in this context that the wider the window is opened, the more dirt flies in.) We shall adopt the common practice of using unity bandwidth ($\Delta f = 1$ Hz) at a specified point in the frequency range, say 1 kHz. With a fixed bandwidth, thermal noise V_{ni}^2, is constant at all parts of the spectrum, but $\overline{e_n^2}$ and $\overline{i_n^2}$ vary because of $1/f$ noise. The noise figure measured with a bandwidth of 1 Hz is called the *spot noise figure*; the frequency at which the measurement is made should also be stated.

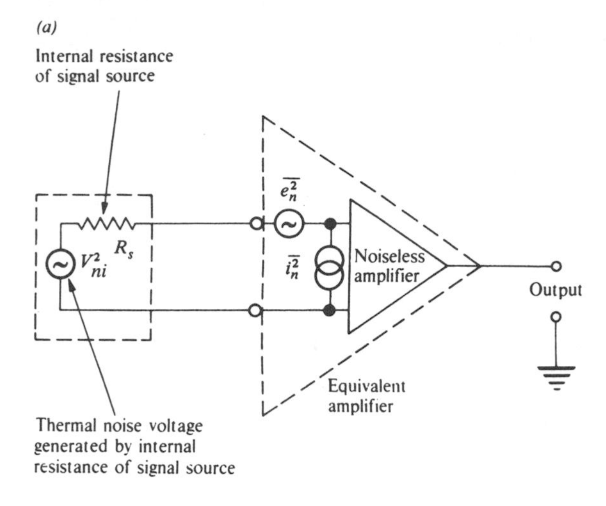

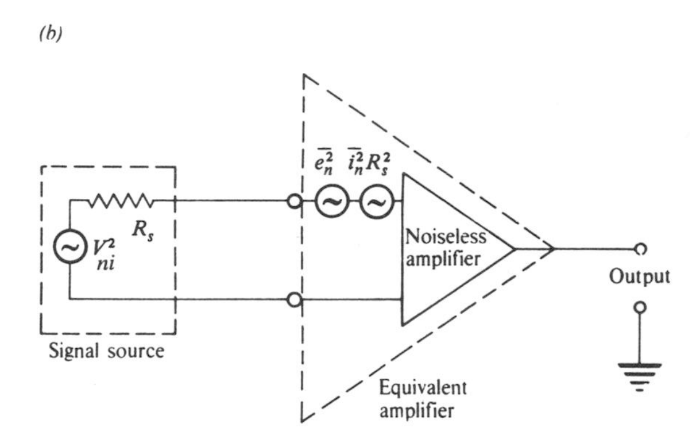

Fig. 5.8. Equivalent circuit of an amplifier showing equivalent noise generators at input: (*a*) voltage generators and current generator, (*b*) current generator converted into its equivalent voltage generator.

To calculate the noise figure, we require the total noise voltage at the input. Knowing the source resistance, R_s, we can replace the current generator with an equivalent voltage generator $\overline{i_n^2}R_s^2$ (fig. 5.8(*b*)) and proceed with our addition.

In adding together the various noise components to obtain $V_{ni(\text{total})}$, it must be remembered that the three generators are independent and have a random phase relationship. That is to say that the signals are *uncorrelated*. The only way to sum such signals is by adding the mean squares

$$V^2_{ni(\text{total})} = V^2_{ni} + \overline{e^2_n} + \overline{i^2_n}\, R^2_s.$$

$$\text{NF} = \frac{V^2_{ni(total)}}{V^2_{ni}},$$

therefore

$$\text{NF} = 1 + \frac{\overline{e^2_n} + \overline{i^2_n}\, R^2_s}{V^2_{ni}}. \tag{5.12}$$

As we would expect, NF is always greater than unity.

Now we know that V^2_{ni} is wholly due to thermal noise in R_s and is therefore equal to $4kTR_s$ ($\Delta f = 1$ Hz), therefore

$$\text{NF} = 1 + \frac{\overline{e^2_n} + \overline{i^2_n}\, R^2_s}{4kTR_s} \tag{5.13}$$

It is now possible to calculate a value for R_s to give minimum noise figure by differentiating equation (5.13):

$$\frac{\text{d(NF)}}{\text{d}R_s} = \frac{1}{4kT}\left(-\frac{\overline{e^2_n}}{R^2_s} + \overline{i^2_n}\right). \tag{5.14}$$

NF is a minimum when

$$\frac{\text{d(NF)}}{\text{d}R_s} = 0,$$

i.e.

$$\text{when } R^2_s = \frac{\overline{e^2_n}}{\overline{i^2_n}}.$$

Thus, for minimum NF, optimum source resistance is given by

$$R_{s(\text{opt})} = \left(\frac{\overline{e^2_n}}{\overline{i^2_n}}\right)^{1/2}. \tag{5.15}$$

If e_n is in volts and i_n in amps, R_s is in ohms.

5.9.5 Noise figure and the bipolar transistor

Noise voltage and current, e_n and i_n, are largely governed by conditions in the first stage of the amplifier. In a bipolar stage, they both increase as quiescent collector current increases, but i_n increases faster than e_n, so that $R_{s(\text{opt})}$ falls at higher collector currents. The lowest noise figures in a bipolar transistor (typically 1 dB) are obtained at very low collector currents, around 10 μA. The value of $R_{s(\mathrm{opt})}$ is then relatively high, in the tens of kΩ. The input transistor should be a 'low noise' type, which means that its $1/f$ noise is small. Needless to say, a useful low-noise transistor must also exhibit adequate current gain at very low collector currents.

The BC109 is sometimes used in low-noise circuits; it is very similar to the common BC107, but selected for high current gain.

Fig. 5.9 shows typical plots of noise figure against generator source resistance for a small silicon transistor such as the BC109. The noise bandwidth is 30 Hz–20 kHz, so that the figures indicate the audiofrequency performance of the device. The lowest noise figure is given by $I_c = 10$ μA, generator resistance between 10 kΩ and 100 kΩ being appropriate. Notice, however, that if generator resistance is only 1 kΩ, then a 100 μA collector current will be a better choice.

Clearly, for lowest possible noise figure, a low transistor collector current is desirable. But what if our generator source resistance is much lower than the few kilohms required for minimum NF with such a circuit? One answer is to step up the impedance with a transformer, as discussed in section 5.11. Another is to parallel up a number of input transistors, sometimes embodied in an IC.

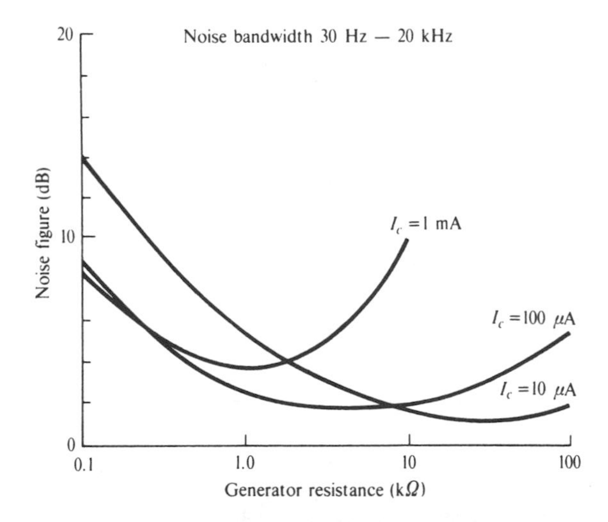

Fig. 5.9. Typical plot of noise figure against generator source resistance for a small silicon transistor such as BC109.

5.9.6 *Practical low-noise amplifier*

Fig. 5.10 shows a useful low-noise amplifier circuit which is typical of the input stage of a high-quality audio amplifier. Collector current in T_1 is of the order of 100 μA. Source resistances from 1 kΩ to 20 kΩ are satisfactory, and the noise figure can be better than 2 dB. Optimum source impedance is about 5 kΩ. Circuit gain is heavily stabilized with negative feedback via R_2. The values quoted give a voltage gain of 100. The circuit can be readily modified to incorporate frequency response equalization by replacing R_2 with a resistance–capacitance combination. Such an arrangement is used in audio amplifiers to equalize the disc characteristic. The d.c. feedback is run separately (R_4) thus stabilizing quiescent conditions.

The reader may be concerned that there now appear to be two criteria for impedance matching in a voltage amplifier. Maximum voltage transfer demands that the source resistance be much lower than amplifier input impedance, whilst minimum noise requires a specific source resistance which may not be very low. Are these criteria in conflict?

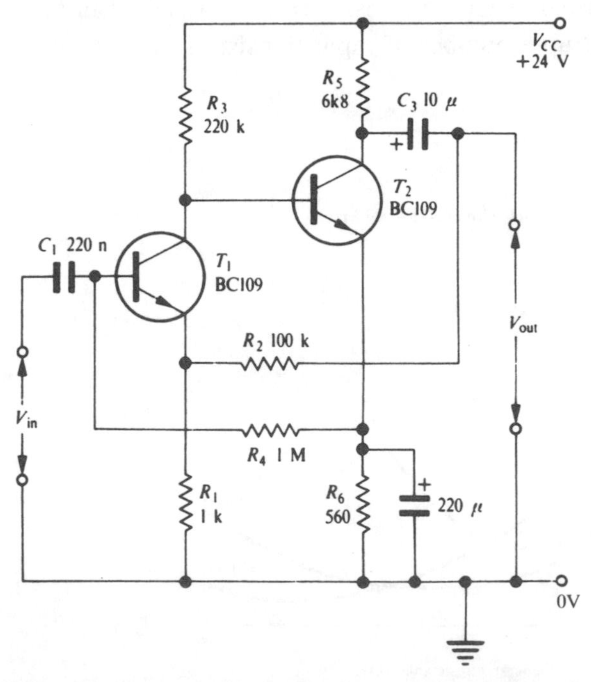

Fig. 5.10. Low-noise two-transistor voltage amplifier.
Voltage gain $\approx (R_2 + R_1)/R_1 \approx 100$. A 5 kΩ source resistance gives minimum noise figure.

Fortunately they are not, and the input impedance of an amplifier is normally much higher than its optimum source resistance.

5.9.7 Noise in FETs

The dominant noise sources in a FET are

(*a*) thermal noise in the channel and

(*b*) $1/f$ noise, which is particularly significant at frequencies below 1 kHz.

Shot noise is negligible, because the only junction is at the gate, and that carries only a tiny leakage current.

The thermal noise of the channel can be viewed as though there were an additional series resistance in the input circuit generating the usual $\sqrt{(4kTR\Delta f)}$ r.m.s. noise voltage. The value of this equivalent additional resistance can be shown to be of the order of $1/g_m$; so, with typical values of 1000 μS to 3000 μS for g_m, the additional resistance will be of the order of hundreds of ohms. The additional noise will therefore only be significant compared with thermal noise in the signal source if the impedance of the latter is very low indeed.

It is $1/f$ noise which dominates the scene at audiofrequencies in the FET. This can vary greatly from one device to another. In general, the input noise current is extremely low in the audiofrequency range, so the best noise figures can be obtained with high source resistances.

For a JFET, the optimum range of source resistance is usually between 1 MΩ and 10 MΩ, where noise figures are normally lower than 1 dB. Very good results can be obtained in the audio range with source resistance from 50 kΩ to 100 MΩ. Fig. 5.11 shows a low-noise audio voltage amplifier with a FET input. A good sample of 2N3819 will be satisfactory; any exhibiting excessive noise should be rejected. Input impedance is equal to R_1, which can be increased to several hundred megohms if required.

The FET is followed by a standard bipolar amplifier using a Darlington pair to avoid excessive loading on the 200 kΩ drain resistor R_4. Overall negative feedback via R_5 and R_3 stabilizes voltage gain. The adjustable bias resistor R_2 enables a wide range of FET pinch-off voltages to be accommodated.

The MOSFET, having no junctions at all, is free from shot noise. Its $1/f$ noise voltage may, however, be as much as 100 times higher than in the JFET. Very low input noise current and extremely low input leakage current imply that high values of signal source impedance can be used: values of 100 MΩ or more are necessary to give reasonable noise figures at audiofrequencies. At high frequencies, in the tens and hundreds of MHz, the noise characteristics of both the JFET and MOSFET change considerably. The $1/f$ noise voltage is no longer significant and the input noise current increases markedly owing to the capacitative coupling from channel to gate. Optimum signal source impedance may then be as little as one thousandth of the low-frequency value.

5.10 Introduction to impedance changing

Now that we have looked at the criteria for impedance matching, it is appropriate to consider the methods available for changing an output impedance if it is to feed an existing load and does not satisfy the matching criterion. Two examples will illustrate problems which may be encountered.

(*a*) We need to match a microphone of 30 Ω impedance to a low-noise bipolar transistor amplifier input similar to the circuit of fig. 5.10. The amplifier has an input impedance much higher than the microphone impedance, so voltage transfer is all right. However, for optimum noise figure, the amplifier needs to see a source impedance between 1 kΩ and 10 kΩ. We therefore need

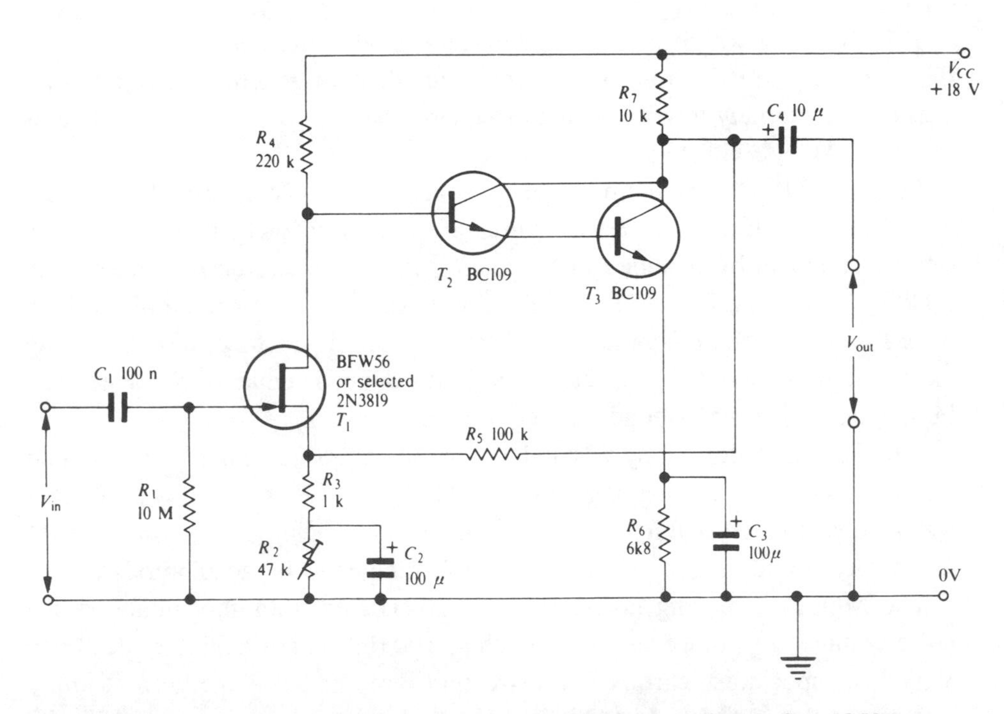

Fig. 5.11. Low-noise voltage amplifier with FET input. Adjust R_2 for +10 V on collector of T_3 with no signal. Voltage gain $\approx 1/\beta = (R_5 + R_3)/R_3 \approx 100$. Gives low-noise performance with source resistances of 50 kΩ – 10 MΩ.

to convert the 30 Ω microphone impedance to a value in the 5 kΩ region. A step-up transformer would be used to accomplish this.

(*b*) We have a signal generator with a high output impedance of 10 kΩ. This generator is to be used to test a number of transistor amplifiers, having a variety of input impedance ranging from 1 kΩ to 20 kΩ. Suppose we want each amplifier in turn to be fed with the same voltage signal without having to reset the generator output. We must convert the output impedance of the generator from 10 kΩ to something under 100 Ω. An emitter follower would be used here.

5.11 Impedance changing by transformer

The familar application of a transformer is that of changing a.c. voltages. Its application as an impedance converter is only a little less obvious. Consider the transformer in fig. 5.12. Here a voltage V_{in} on the primary winding is stepped up to a higher voltage V_{out}. A load R_L is connected to the secondary and draws a current I_{out}. The voltage *ratio* V_{out}/V_{in} is equal to the turns ratio:

$$\frac{V_{out}}{V_{in}} = n = \frac{N_2}{N_1} \tag{5.16}$$

where N_1 is the number of turns on the primary winding and N_2 the number on the secondary.

Now it is reasonable to assume that there is a negligible loss of power in the transformer, so that the power in the primary is equal to the power in the secondary circuit, i.e.

$$V_{in} I_{in} = V_{out} I_{out} \text{ (assuming } V \text{ and } I \text{ in phase).}$$

Now, from Ohm's law,

$$I_{in} = \frac{V_{out}^2}{V_{in} R_L}$$

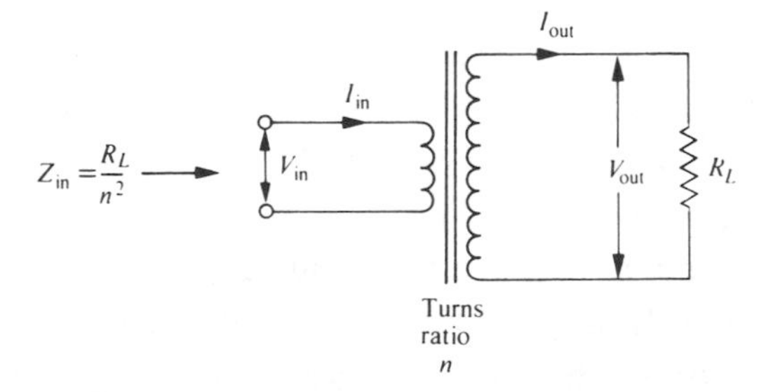

Fig. 5.12. The transformer as an impedance changer.

therefore

$$I_{out} = \frac{V_{out}}{R_L},$$

and

$$V_{in} I_{in} = \frac{V_{out}^2}{R_L}.$$

If we look at the primary of the transformer, it appears to have a certain input resistance R_{in} given by

$$R_{in} = \frac{V_{in}}{I_{in}}$$

and, substituting for I_{in},

$$R_{in} = \frac{V_{in}^2 R_L}{V_{out}^2}, \tag{5.17}$$

i.e.

$$R_{in} = \frac{R_L}{n^2}.$$

Thus we see that the primary of the transformer appears as a resistance equal to the load resistance on the secondary divided by the square of the turns ratio. R_{in} is the *reflected impedance* of R_L in the transformer primary winding.

It is perhaps surprising to see a transformer winding, which is after all a coil, appearing as a resistance rather than an inductance. This is because implicit in our calculation is the assumption that the inherent reactance of the primary winding, which would appear in parallel with R_{in}, is infinite. This is reasonable if the transformer has an adequate number of turns per volt on a suitable core, but it does set a low-frequency limit below which the transformer efficiency is lowered by the primary shunt reactance.

Our second major assumption is implicit in the statement that the voltage ratio V_{in}/V_{out} is identical to the turns ratio n. It is that the magnetic flux caused by the secondary current is equal and opposite to the flux caused by the primary current. In this way, the net flux in the primary coil is zero and no inductive effects are produced by the secondary current. In practice, some flux leakage occurs and gives rise to 'leakage inductances' which appear in series with the primary and secondary; these may be neglected except at high frequencies where, together with winding capacitance, they cause losses and phase shifts which determine the upper limit of the transformer frequency response.

To summarize, the transformer can convert a high voltage signal at low current into a low voltage signal at high current or vice versa. It is therefore performing an impedance-changing function and is particularly useful where a weak signal, such as that from a microphone, is to be matched into an amplifier for the best possible signal-to-noise ratio. If a signal source of resistance R_s is to be matched to an amplifier requiring an optimum source resistance R_{in}, the transformer turns ratio required is given by:

$$n = \left(\frac{R_s}{R_{in}}\right)^{1/2}.$$

Thus in example (*a*) in section 5.10, if we are matching a 30 Ω microphone to present a source resistance of 5000 Ω to the amplifier, the required step-up transformer turns ratio is

$$n = \left(\frac{5000}{30}\right)^{1/2}$$

$$= (167)^{1/2}$$

$$= 12.9.$$

A ratio between 1 : 10 and 1 : 15 would be chosen.

As well as having the correct ratio, the matching transformer should be specified to handle the required frequency range and signal levels, thus ensuring adequate primary inductance and negligible leakage reactance.

5.12 The emitter follower

5.12.1 Emitter follower circuit design

If an output impedance is to be reduced for the purpose of optimum voltage transfer to the next circuit, then a transformer will be of little advantage, for, in reducing the output impedance, it will also step down the voltage. A much more satisfactory solution to the problem is the use of a transistor in the *emitter follower* (*common collector*) circuit. In this circuit, of which fig. 5.13 is a typical example, the voltage gain is just a little less than unity. However, because the transistor gives a current gain, the emitter follower lowers the output impedance of any signal source connected to the input.

As the name 'common collector' suggests, the transistor collector is connected straight to the supply line which, as far as the signals are concerned, is the same thing as the earth (common) rail, because power supply outputs are always designed to present a very low impedance to signals. The output load resistor, R_L, is in the emitter circuit, whilst the input signal is fed in between base and earth in the usual way.

Before we consider the behaviour of a.c. signals in the emitter follower, it is appropriate to consider the d.c. quiescent (no signal) conditions. As with the common-emitter amplifier of chapter 1 we must ensure that the output signal is able to swing both positively (towards supply rail) and negatively (towards earth). For maximum output swing capability the quiescent emitter voltage should be midway between earth and supply, i.e. at about 4.5 V in this example.

With a value of 4.7 kΩ selected for R_L, the circuit of fig. 5.13 is designed for a 1 mA quiescent emitter current. This figure of 1 mA is here chosen fairly arbitrarily, as it was for the common-emitter amplifier. The choice of quiescent current is dependent upon the current signal which is demanded from the output – the greater the current swing, the greater the quiescent current required to preserve linear operation. This topic is further examined in section 5.17.

The base bias resistor, R_B, feeds sufficient base current into the base–emitter junction to maintain the required emitter current. In this example, we have assumed a value of 200 for the d.c. current gain, h_{FE}, of the BC107. Thus, a 1 mA emitter current requires 1/200 mA base current (5 μA) to maintain it.

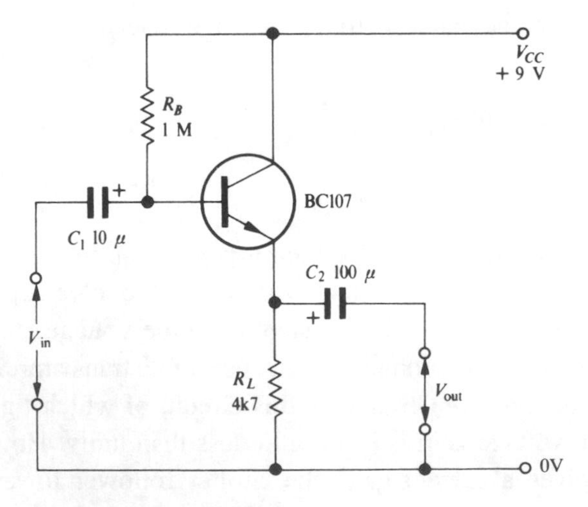

Fig. 5.13. Emitter follower.

This current is provided by R_B whose value is calculated from Ohm's law, assuming the base voltage to be similar to the emitter voltage (4.5 V). Voltage across $R_B \approx 9 - 4.5$ V; current in $R_B = 5 \times 10^{-6}$ A; then,

$$R_B = \frac{9-4.5}{5\times10^{-6}}\ \Omega.$$

$$R_B = 900\ \text{k}\Omega.$$

A value of 1 MΩ is selected as a preferred value reasonably near to the calculated value.

Although this method of biasing the transistor for the current-operating point could hardly be simpler, it does exhibit a degree of self-regulation to compensate for any variations in transistor current gain, h_{FE}. We first encountered the need for this compensating action in the voltage amplifier in chapter 1, and it applies equally to the emitter follower.

Consider, in fig. 5.13, what happens if the transistor has a current gain much higher than the assumed value of 200. The emitter current will be higher than 1 mA and hence more than 4.5 V will be dropped in R_L. As a result, the voltage across R_B will be less than 4.5 V and the base current will be less than the intended 5 μA.

In this way, although variations in h_{FE} will result in some variations in the quiescent emitter voltage, the transistor can never cut off or saturate and the circuit will always work, though very high- or low-gain specimens of transistor will result in restricted positive or negative output voltage swing capability.

5.12.2 a.c. signals in the emitter follower

To look at the behaviour of a.c. signals, we shall redraw the circuit showing only those elements which are significant as far as the a.c. signals are concerned. This a.c. view of the circuit is in fig. 5.14. Bias resistor R_B is shown in dotted form because we shall consider its effect only after we have followed the signal through the transistor. The connection between V_{CC} and earth indicates that, as far as the a.c. signal is concerned, the supply appears as a short circuit.

The emitter follower input is fed by a signal generator of open-circuit e.m.f. e and source resistance (output resistance) R_s, the actual voltage signal appearing at the input being v_{in}. This signal gives rise to an a.c. base current i_b, emitter current i_e and collector current i_c. Across the load, R_L, is developed the output voltage signal v_{out}. These a.c. signal voltages and currents, are normally denoted by the lower case v and i as distinct from the upper case V and I which refer to d.c. voltage and current.

It is a valuable property of linear circuits that we can consider a.c. signals independently of the steady d.c. voltages and currents which exist in the circuit; this is an example of the *superposition principle*, which only applies to linear systems.

There is a vital difference between the emitter follower and the common-emitter amplifier. Whereas the input and output of the common-emitter amplifier are separated by the reverse-biased collector–base junction, the emitter follower input and output are linked by the forward-biased base–emitter junction.

As will be shown in the next chapter (section 6.2), this junction has a low resistance, r_e, the actual value of r_e being given by

$$r_e = \frac{25}{I_E}\,\Omega, \qquad [(6.4)]$$

where I_E is d.c. emitter current in mA. With the 1 mA emitter current of our present example, r_e works out to 25 Ω, so that the input is effectively connected to the output via a resistance which is small compared with the emitter load R_L.

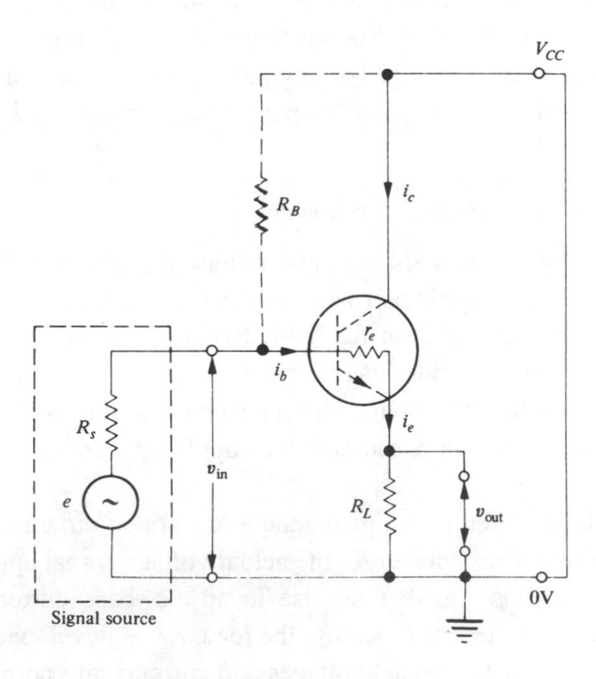

Fig. 5.14. Emitter follower as seen by an a.c. signal.

We can see in fig. 5.14 that r_e and R_L constitute a potential divider for the signals, so that voltage gain A_V always turns out slightly less than unity:

$$A_V = \frac{R_L}{R_L + r_e}. \tag{5.18}$$

In this example,

$$R_L \approx 5\ \text{k}\Omega$$

and

$$r_e \approx 25\ \Omega,$$

then

$$A_V = \frac{5000}{5000 + 25} = 0.995.$$

With a voltage gain so close to unity, the emitter voltage *follows* the base voltage very closely indeed, this action giving its name to the circuit.

5.12.3 Emitter follower input impedance

Consideration of the a.c. currents and voltages shown in fig. 5.14 enables us to calculate the input impedance of the emitter follower. If we neglect stray capacitance (a reasonable assumption except at very high frequencies), input and output impedances are purely resistive.

$$\text{Input resistance } R_{\text{in}} = \frac{v_{\text{in}}}{i_b}.$$

Now

$$i_c = h_{fe}\, i_b,$$

where h_{fe} is the a.c. (small signal) transistor current gain.

Emitter current,

$$i_e = i_c + i_b$$

$$= i_b\,(h_{fe} + 1),$$

hence

$$i_b = \frac{i_e}{h_{fe} + 1},$$

therefore

$$R_{\text{in}} = \frac{v_{\text{in}}}{i_e}(h_{fe} + 1), \tag{5.19}$$

but voltage gain is approximately unity, i.e.,

$$v_{\text{in}} \approx v_{\text{out}},$$

therefore

$$R_{\text{in}} = \frac{v_{\text{out}}}{i_e}(h_{fe} + 1). \tag{5.20}$$

Now

$$\frac{v_{\text{out}}}{i_e} = R_L,$$

therefore

$$R_{\text{in}} = R_L(h_{fe} + 1). \tag{5.21}$$

Since $h_{fe} \gg 1$,

$$R_{\text{in}} \approx R_L h_{fe}. \tag{5.22}$$

The input impedance of an emitter follower is therefore equal to the emitter load resistance multiplied by the transistor current gain.

It is important to note that, if the emitter follower output is loaded by another circuit, then the effective emitter load is given by the parallel combination of R_L and the external impedance. A change in the load on an emitter follower therefore causes a corresponding change in the input impedance.

We have so far neglected the effect of the bias resistor R_B on the input impedance. Since R_B is connected between the emitter follower input and the supply rail V_{CC}, it is effectively in parallel with the input, V_{CC} being equivalent to earth for a.c. signals. Hence the true input impedance of the emitter follower is given by the parallel combination of $R_L h_{fe}$ and R_B, therefore

$$R_{\text{in}} \approx \frac{R_B R_L h_{fe}}{R_B + R_L h_{fe}}. \tag{5.23}$$

5.12.4 Emitter follower output impedance

To calculate the output impedance Z_{out} of the emitter follower, the most basic measurement method discussed earlier can be used:

$$Z_{\text{out}} = \frac{\text{open-circuit e.m.f.}}{\text{short-circuit current}}. \qquad [(5.3)]$$

Open-circuit e.m.f. is given by v_{out}, which we have shown is approximately equal to v_{in}. Also, referring again to fig. 5.14, $v_{in} \approx e$, where e is the open-circuit voltage of the signal generator. Hence $v_{out} \approx e$. The short-circuit emitter current will be determined by the current in the base circuit, transistor junction resistance being neglected. With R_L short-circuited to a.c. signals, the base current is given by $i_{b(sc)} = e/R_s$. Hence we have the short-circuit emitter current:

$$i_{e(sc)} = \frac{e}{R_s}(h_{fe} + 1) \approx \frac{e}{R_s} h_{fe}.$$

Now

$$Z_{out} \approx \frac{e}{i_{e(sc)}}.$$

Substituting for $i_{e(sc)}$,

$$Z_{out} \approx \frac{R_s}{h_{fe}}. \tag{5.24}$$

The emitter follower, therefore, reduces the output impedance of a generator by a factor equal to the transistor current gain. If the generator output impedance is very high (comparable with $R_L h_{fe}$), then the value of R_L must be included in parallel with the calculated emitter follower output impedance. If the generator feeding the emitter follower has a negligible source resistance ($R_s = 0$), then the output impedance is given by the effective resistance r_e of the base–emitter junction.

5.12.5 Darlington pair

If the required impedance transition is greater than can be achieved with one transistor, two transistors may be used in the Darlington connection, whereby the emitter current of the first transistor forms the base current of the second. The current gain of the pair is thus the product of the two individual gains.

Fig. 5.15 shows a Darlington pair. Depending on the output load, the input impedance may be as high as 10 MΩ. More than two transistors can be used in the Darlington connection, but unfortunately a law of 'diminishing returns' applies to the total current gain: the first transistor is usually working at such a low collector current that it is inefficient, and current gain falls drastically. If, however, the final transistor is a power device, being called upon to handle currents of several amps, then up to four transistors may be profitably used in the Darlington connection.

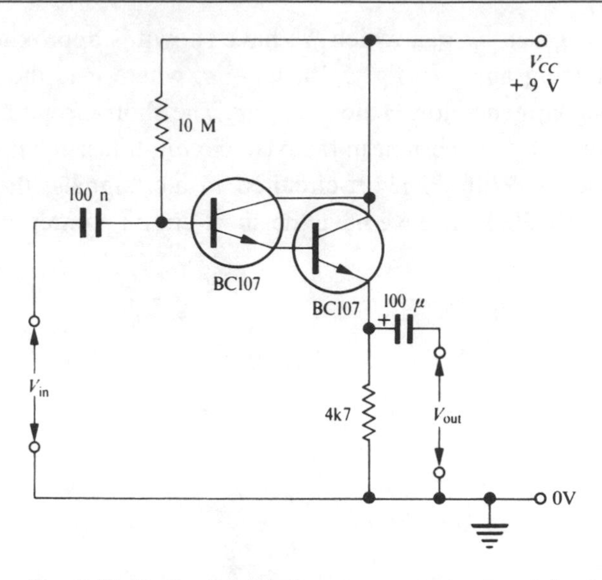

Fig. 5.15. Darlington pair increases current gain, giving high input impedance.

5.12.6 Improved d.c. stability

Whilst the single base bias resistor provides adequate control of the d.c. conditions for many applications, improved stability of operating point can be obtained by using a potential divider to set the base potential (fig. 5.16). The 10 kΩ and 12 kΩ resistors hold the base potential slightly higher than $V_{CC}/2$ above earth; the emitter thus settles at $V_{CC}/2$, allowing for the 0.6 V drop in the base–emitter junction.

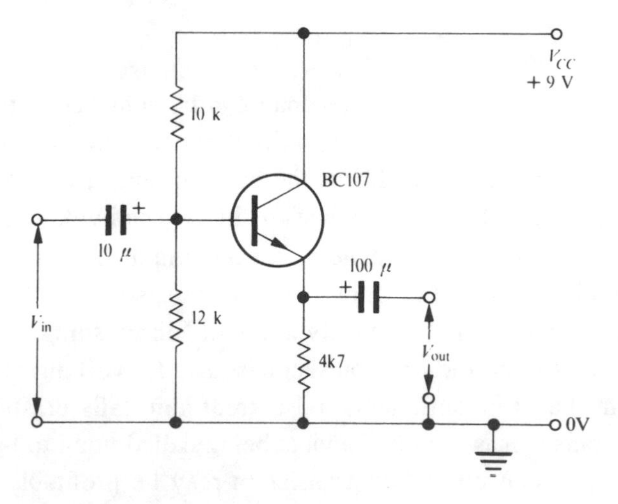

Fig. 5.16. Emitter follower with d.c. stabilization by base potential divider.

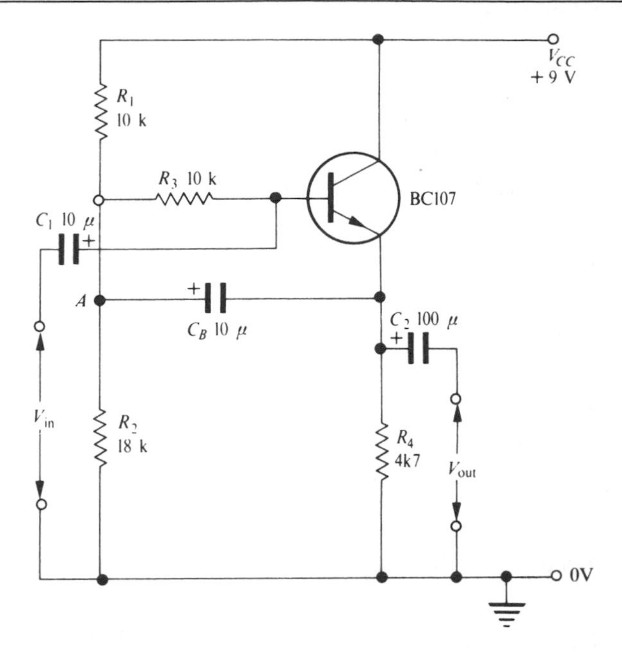

Fig. 5.17. Bootstrap capacitor C_B increases effective value of R_3 and gives high input impedance.

The main disadvantage of this stable arrangement is that the base potential divider shunts the input and reduces the input impedance to the region of 5 kΩ. This snag can be overcome by the expedient of *bootstrapping* (fig. 5.17). Bootstrapping is an application of positive feedback whereby the effective value of a resistor is greatly increased as far as a.c. signals are concerned.

In fig. 5.17, the d.c. base potential is determined by R_1 and R_2 in a similar way to the circuit of fig. 5.16. The bootstrap capacitor C_B, however, feeds back the output signal to the junction of the resistors at point A, thus making that point swing up and down in phase with the input signal. Thus R_3 draws very little current from the input and its effective resistance appears correspondingly greater. The circuit is pulling up and down its input network 'by its own bootstraps'. The factor by which the effective value of R_3 is increased depends on the voltage gain A_V of the emitter follower, since the a.c. potential difference across R_3 is now $(v_{in} - A_V v_{in})$, and the current is therefore reduced by the factor $v_{in}/(v_{in} - A_V v_{in})$. Thus bootstrapping with a voltage gain of 0.99 will produce a hundredfold increase in the effective value of R_3. Very high input impedances may be obtained by using a Darlington pair with bootstrapping.

5.12.7 Feeding long cables

The low output impedance of the emitter follower is useful, not only for optimizing signal voltage transfer between two circuits, but also for reducing problems associated with long connecting cables between units.

There are two chief snags associated with long cables: induced extraneous signals, and shunt capacitance. The most common extraneous signal is mains hum: the mains wiring in a building radiates an a.c. electrostatic field which is often picked up on connecting cables. It is common practice to use shielded (coaxial) cable where low-level signals are to be carried, the earthed braid which surrounds the 'live' conductor providing an electrostatic screen. Unfortunately, shielded cable increases the problem of shunt capacitance. With a typical self-capacitance as high as 200 pF per metre, such a cable behaves as a fairly large capacitor across the output circuit. High frequencies are thus in danger of attenuation because of the shunting effect of the low reactance of the cable capacitance.

A low output impedance feeding the cable minimizes both hum pickup and attenuation due to cable capacitance. The reason is apparent from the output equivalent circuit of fig. 5.5. A low impedance feed presents a low value of shunt impedance from the signal wire to earth. The internal impedance of an emitter follower output is typically less than 50 Ω, swamping the shunting effect of cable capacitance. Such a low feeding impedance also makes a conductor a very inefficient aerial and unlikely to pick up mains hum.

Further reduction of extraneous pickup can be obtained using a balanced line technique and a differential input, discussed further in chapter 8.

5.13 The source follower

5.13.1 Circuit design

By connecting the load in the source circuit of a FET, we have the common drain or source follower circuit shown in fig. 5.18. The source follower, like the emitter follower, is an impedance-matching circuit with a voltage gain close to unity but with the added advantage of the inherently high input impedance of the FET.

Fig. 5.19 shows the source follower with a.c. voltages and currents shown in the signal paths, just as fig. 5.14 showed them with the emitter follower. The voltage gain of the source follower is just less than unity. A glance at fig. 5.19 will help to demonstrate this; there is only the small signal, v_{gs}, between gate and source of the FET, so that v_{in} and v_{out} will only differ by

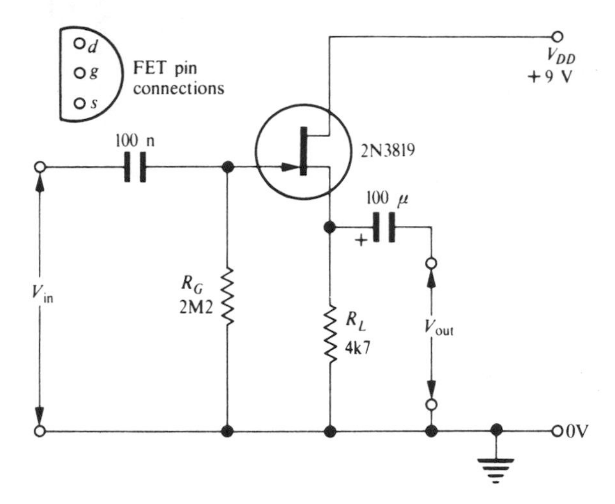

Fig. 5.18. Basic source follower circuit.

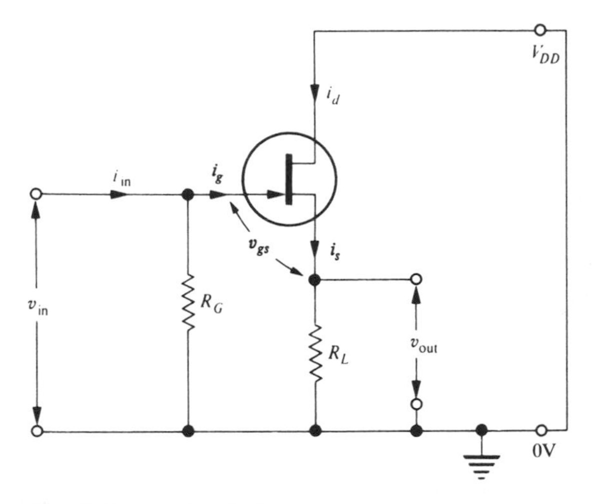

Fig. 5.19. a.c. signal view of source follower.

this small amount and give a gain of about 1. A more detailed examination will confirm this and give us a relationship for voltage gain.

Since the gate current i_g is negligible, we know that the a.c. drain current, i_d, is equal to the source current i_s. Now the drain current i_d is related to the gate–source voltage v_{gs} by the transconductance g_m:

$$i_d = g_m v_{gs},$$

therefore since

$$i_d = i_s,$$

$$i_s = g_m v_{gs}.$$

Now from Ohm's law,

$$v_{out} = R_L i_s$$
$$= R_L g_m v_{gs}.$$

But $v_{gs} = v_{in} - v_{out}$, as pointed out above, therefore

$$v_{out} = R_L g_m (v_{in} - v_{out}).$$

Therefore rearranging,

$$v_{out} = R_L g_m v_{in} - R_L g_m v_{out},$$
$$v_{out}(1 + R_L g_m) = R_L g_m v_{in},$$

therefore voltage gain

$$A_V = \frac{v_{out}}{v_{in}} = \frac{R_L g_m}{1 + R_L g_m}. \tag{5.25}$$

Since $R_L g_m \gg 1$,

$$\frac{v_{out}}{v_{in}} \approx 1.$$

In the circuit of fig. 5.18, taking $g_m = 2$ mA/V for the 2N3819,

$$A_V = \frac{4700 \times 2 \times 10^{-3}}{1 + (4700 \times 2 \times 10^{-3})}$$
$$= \frac{9.4}{10.4},$$

therefore $A_V \approx 0.90$.

Thus, we see that in practice the voltage gain of the source follower is not as close to unity as that of the emitter follower, but the loss is not usually significant.

5.13.2 *Input and output impedance of the source follower*

The gate of the FET draws a negligible current, so the whole of the input current i_{in} flows in the gate resistor R_G. The input impedance of the source follower is therefore equal to R_G. With most junction FETs, R_G may be as high as 200 MΩ before the gate current itself becomes significant. In the case of a MOSFET, with its insulated gate, input impedances up to 10^{15} Ω can be achieved.

To calculate the output impedance of the source follower, we again adopt the expedient of measuring open-circuit voltage and dividing by short-circuit current. Looking again at fig. 5.19, and assuming approximately unity gain, open-circuit voltage is given by $v_{out} \approx v_{in}$.

Now if the output is short-circuited, the whole v_{in} appears between gate and source. Hence, the short-circuit source current is given by:

$$i_{s(sc)} = g_m \, v_{in}, \tag{5.26}$$

$$\text{output impedance } Z_{out} = \frac{v_{out}}{i_{s(sc)}} = \frac{v_{in}}{g_m \, v_{in}}$$

$$= \frac{1}{g_m}. \tag{5.27}$$

If g_m is in mA/V, then Z_{out} is in kΩ.

This is an interesting result because it shows that, unlike the case of the emitter follower, the source follower output impedance is independent of the signal source connected to the input. For a typical transconductance of 2 mA/V, Z_{out} is therefore 500 Ω.

Comparison of source follower and emitter follower shows that the latter is capable of producing a lower output impedance than the former. The source follower is, however, unrivalled when it comes to matching a signal source of very high internal impedance such as a capacitor microphone.

5.13.3 Improved operating point for the source follower

The simple source follower circuit of fig. 5.18 has a very limited output voltage swing capability. This is because of an inappropriate quiescent d.c. condition. Ideally, the source would sit midway between earth and supply, at about +4 V, but, with the gate tied to earth, it cannot rise higher than about +2 V or the device will be cut off.

The d.c. conditions can be improved by tapping in the gate resistor only part of the way down the source resistor, as shown in fig. 5.20. In this circuit, the FET develops its 1–2 V gate bias across the 1 kΩ resistor, leaving the remainder of the source resistor (3.3 kΩ) to drop approximately 3 V so that the source sits at the ideal voltage midway between earth and supply.

An added bonus with this simple modification is that the 2.2 MΩ gate resistor is partially bootstrapped, its lower end swinging with about three-quarters of the amplitude of the output signal. Its effective value is thus increased about four times, giving an input impedance in the region of 10 MΩ. By simply increasing the gate resistor R_G, input impedances of hundreds of megohms are readily obtained without any change in the d.c. conditions.

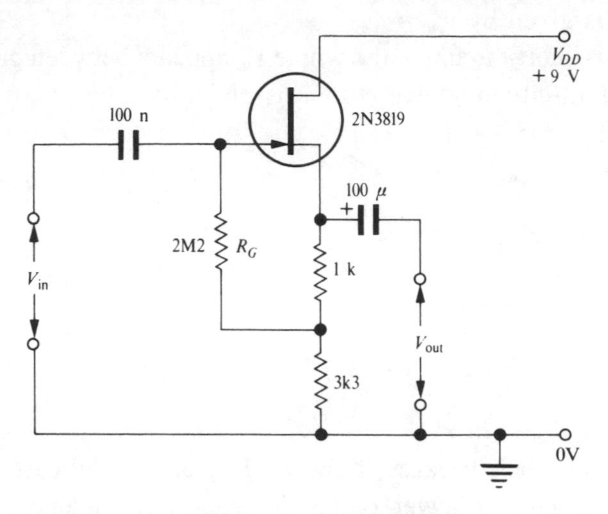

Fig. 5.20. Source follower with improved d.c. conditions.

An alternative method of arranging the quiescent d.c. conditions in the source follower is to hold the gate positive with respect to earth by means of a potential divider. Fig. 5.21 shows such a circuit with the potential divider bootstrapped to give a high input impedance in the region of 100 MΩ. The

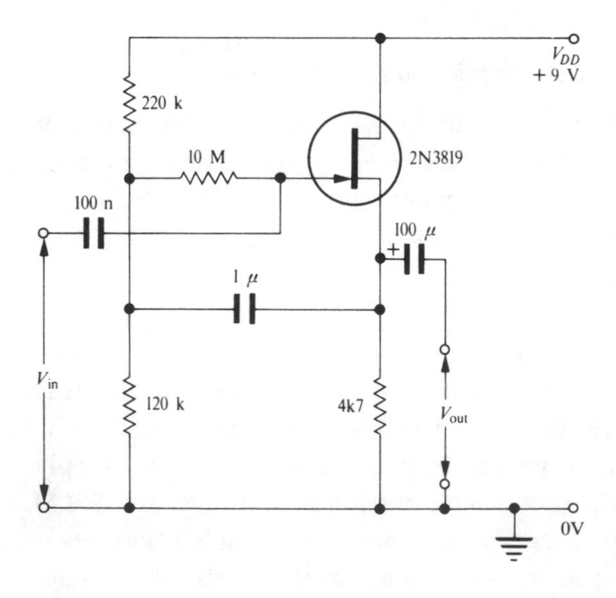

Fig. 5.21. Bootstrapped source follower. $Z_{in} \approx 100$ MΩ.

gate is held 3 V positive above earth and the gate–source bias will adjust itself so that the source is 1–2 V more positive than the gate, i.e. 4–5 V above earth, which is an ideal operating point for a 9 V supply.

5.14 Voltage and power gain

Although the emitter follower and source follower give voltage gains of less than unity, they produce a considerable power gain because the available output current is much greater than the input current. Thus, these active circuits may be contrasted with the passive transformer which, when used to lower an output impedance also involves a voltage step-down and, of course, no power gain.

5.15 Negative feedback and output impedance

We saw in the last chapter that, if negative voltage feedback is applied to an amplifier of open-loop voltage gain A_0, with fraction β fed back to the input (fig. 5.22), the closed-loop voltage gain A, is given by

$$A = \frac{A_0}{1+\beta A_0}. \qquad [(4.2)]$$

Hence, if a voltage signal v_{in} is fed to the amplifier input,

$$v_{out} = Av_{in}$$

$$= \frac{A_0 v_{in}}{1+\beta A_0}.$$

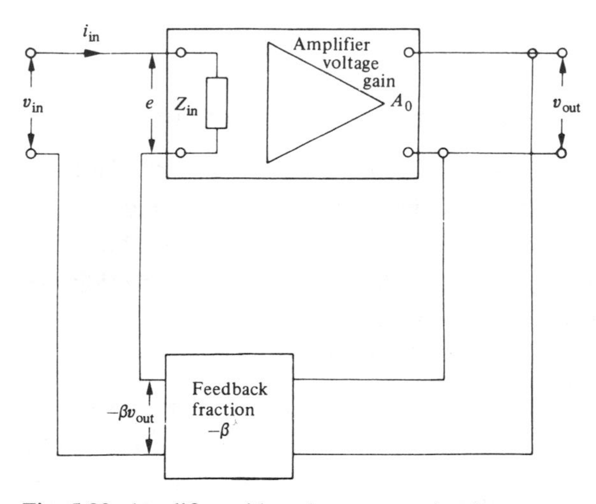

Fig. 5.22. Amplifier with series-connected voltage negative feedback.

Now, we obtain an interesting result if we calculate the output impedance of the feedback amplifier by expressing the ratio of open-circuit voltage to short-circuit current. When the output is short-circuited, the output voltage falls to zero, so that the feedback factor also falls to zero and the amplifier reverts to open-loop operation. Thus, if Z_0 is the output impedance of the basic open-loop amplifier, the short-circuit output current is given by

$$i_{sc} = \frac{A_0 v_{\text{in}}}{Z_0}.$$

Now, the output impedance Z, with feedback, is given by

$$Z = \frac{v_{\text{out}}}{i_{sc}}$$

$$= \frac{A_0 v_{\text{in}}}{(1+\beta A_0)} \times \frac{Z_0}{A_0 v_{\text{in}}}$$

therefore

$$Z = \frac{Z_0}{1+\beta A_0}. \tag{5.28}$$

Thus we see that the output impedance of an amplifier with voltage feedback is reduced by the factor $1/(1+\beta A_0)$, the same factor by which the gain is reduced. By stabilizing the voltage gain, the negative feedback holds the output voltage relatively constant despite variations in the load current.

It should be mentioned that the reverse effect occurs with negative *current* feedback. The feedback holds the output current constant, resulting in an *increased* output impedance.

5.16 Negative feedback and input impedance

In discussing the effect of negative feedback on input impedance, we shall concentrate on one specific case – series voltage feedback. Fig. 5.22 is a schematic diagram of such an arrangement. Here the basic amplifier has an input impedance Z_{in} and a voltage gain A_0. The usual feedback fraction, $-\beta$, of the output, v_{out}, is fed back in series with the input.

Thus voltage signal across Z_{in} is given by

$$e = v_{\text{in}} - \beta v_{\text{out}}.$$

Therefore input current

$$i_{in} = \frac{v_{in} - \beta v_{out}}{Z_{in}}$$

$$= v_{in}\left(\frac{1-\beta A}{Z_{in}}\right).$$

Now, if Z'_{in} is the input impedance of the amplifier with feedback,

$$Z'_{in} = \frac{v_{in}}{i_{in}}.$$

Substituting for i_{in},

$$Z'_{in} = \frac{v_{in} Z_{in}}{v_{in}(1-\beta A)}$$

$$= \frac{Z_{in}}{1 - \dfrac{\beta A_0}{(1+\beta A_0)}},$$

therefore

$$Z'_{in} = Z_{in}(1+\beta A_0). \tag{5.29}$$

Hence, in this case, with series voltage feedback applied, the input impedance of the amplifier is increased by the factor $(1 + \beta A_0)$. At this point it is appropriate to mention that both the emitter follower and source follower can be considered as amplifiers with the whole of the output fed back in series with the input ($\beta = 1$). This represents an alternative approach to the calculation of their near-unity voltage gain and high input impedance.

Integrated circuit (IC) amplifiers are often used in the *voltage follower* mode. As the name implies, the voltage follower is simply a sophisticated version of the emitter or source follower. With $\beta = 1$, the high open-loop gain of the IC amplifier ensures a voltage gain which is exactly unity for all practical purposes. The very high value of A_0 produces a high input impedance and low output impedance. A practical voltage follower circuit is included in section 11.4.

The effect of shunt voltage negative feedback is different from series feedback: input impedance is actually reduced. This leads to the valuable concept of the *virtual earth*, which is discussed in section 11.5.

5.17 Power output of emitter followers

5.17.1 Load current and quiescent current

It is very important to realize that a low output impedance does not necessarily imply any ability on the part of a circuit to supply high currents to a low-impedance load.

For example, although the simple emitter follower circuit of fig. 5.13 can exhibit an output impedance of less than 50 Ω, it cannot be expected to feed a load of this impedance with any significant voltage swing. Even one volt peak-to-peak swing in 50 Ω involves a current swing of 1/50 A (20 mA). Since we only have 1 mA quiescent current, the absolute maximum current swing in the load is limited to ±1 mA, or 2 mA peak-to-peak.

Amplifiers which are designed to be able to supply large voltage swings to low-impedance loads such as loudspeakers are usually called power amplifiers. However, this name must not be taken as implying that such circuits are fundamentally different from those discussed already: power amplifiers are nothing more than 'tough' voltage amplifiers, able to supply high output currents.

It is perfectly possible to scale up the quiescent current in an emitter follower circuit in order to have available an increased output current swing. The resistors can all be scaled down in value by the same factor and, if necessary, a transistor of higher power rating substituted for the BC107. The use of a high quiescent current is wasteful of power, however, and a single emitter follower stage is not the most efficient way of feeding power to a load.

5.17.2 Class B and class AB push–pull operation

As we have seen, the reason for maintaining a quiescent current in the emitter follower is in order that the load current can swing both up and down to cope with the positive and negative half-cycles of the a.c. signal. Now consider the circuit of fig. 5.23 where we have two complementary transistors: the npn handles the positive half-cycles and pnp the negative half-cycles. The ZTX450–ZTX550 transistor pair specified is able to handle collector currents of up to 1 A safely, so that loads down to about 6 Ω impedance can be safely driven. This type of circuit is said to operate in *class B*; each transistor handles only one signal polarity and the quiescent current can therefore be zero. It is sometimes known as *push–pull output* because T_2 'pushes' the negative half-cycles from below and then hands over to T_1 to 'pull' the positive half-cycles from above.

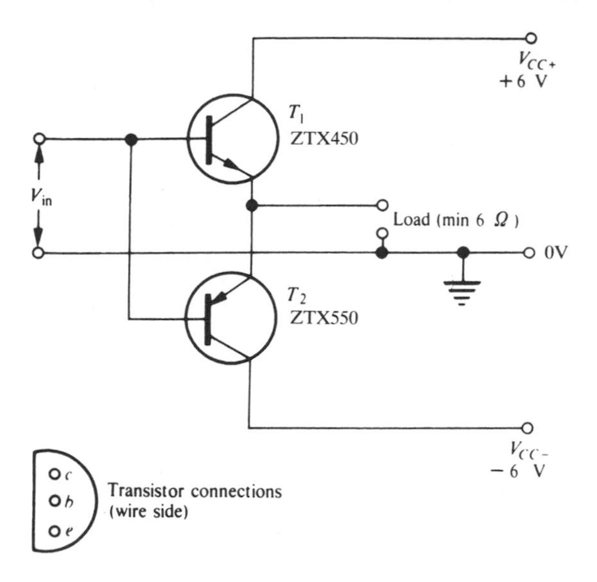

Fig. 5.23. Rudimentary class B complementary emitter follower. Alternative transistors are 2N3053 (npn) and 2N4037 (pnp). Connections for these are same as BC107.

A class B circuit is defined as one where the transistor operating point is chosen so that conduction only occurs for half the input cycle. This is contrasted with the class A mode, which applies to most voltage amplifier circuits, where the transistor is conducting at all points in the input cycle. A third mode, class C, is defined as one where the transistor conducts for *less* than half the input cycle, but this is only used in radio transmitter output stages, in conjunction with a sharply-tuned resonant circuit.

We can define the power conversion efficiency of an amplifier stage as the percentage of the total power drawn from the d.c. supply rails which is actually delivered as a.c. power to the load. A class A stage cannot exceed a theoretical maximum efficiency of 50%, whilst a class B stage can be shown to reach 78% and a class C can approach 100% efficiency.

As might be anticipated from its stark simplicity, the circuit of fig. 5.23 is inadequate for any application where low distortion is required. Each transistor requires about 0.6 V across its base–emitter junction before it begins to conduct significantly, so that small signals (<1 V p–p) are not transmitted at all. The circuit will work with larger signals, but produces the *crossover distortion* shown in fig. 5.24. Near the zero crossing of the wave, the gain drops to zero, resulting in the severe 'kink' shown. In audio work, this type of distortion

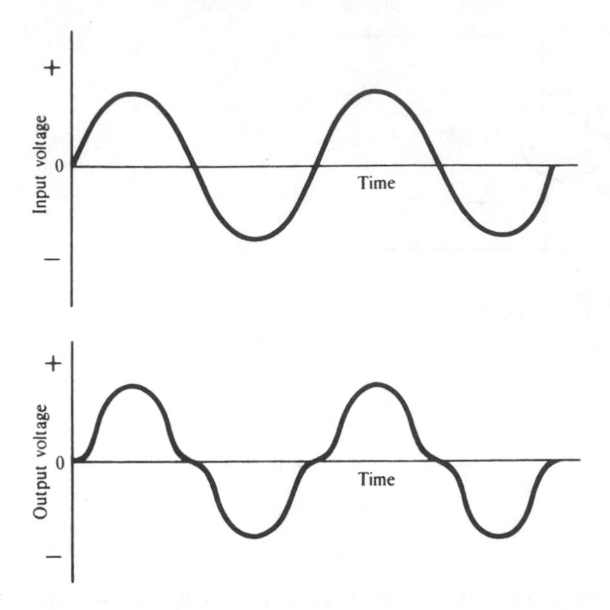

Fig. 5.24. typical input and output waveforms from fig. 5.23, showing severe crossover distortion.

is particularly objectionable because its effect increases as the signal level decreases.

Although the simple circuit of fig. 5.23 is suitable for certain servocontrol applications, it needs improvement for acceptable audio results. Some bias must be supplied to the transistors in order that the base–emitter junctions are ready to conduct on the smallest input signals. To obtain linear performance, it is usual to bias the transistors so that, even with no input signal, they are conducting a small quiescent collector current. For small signals, the stage is then operating in class A, with both transistors conducting, whilst, for larger signals, only one transistor will be operating on each half of the waveform. This mode of operation is termed class AB and is the usual arrangement in audio power amplifiers.

Fig. 5.25 shows our simple class B circuit modified for class AB operation. Instead of being connected together directly, the two bases are separated by a pair of forward-biased diodes, D_1 and D_2, which provide just sufficient bias to make the transistors conduct under quiescent conditions. Emitter resistors R_3 and R_4 provide a degree of current feedback to improve d.c. stability.

As with all class AB amplifiers, the quiescent conditions are closely dependent on the forward p.d. of the transistor base–emitter junction (V_{BE}). A small

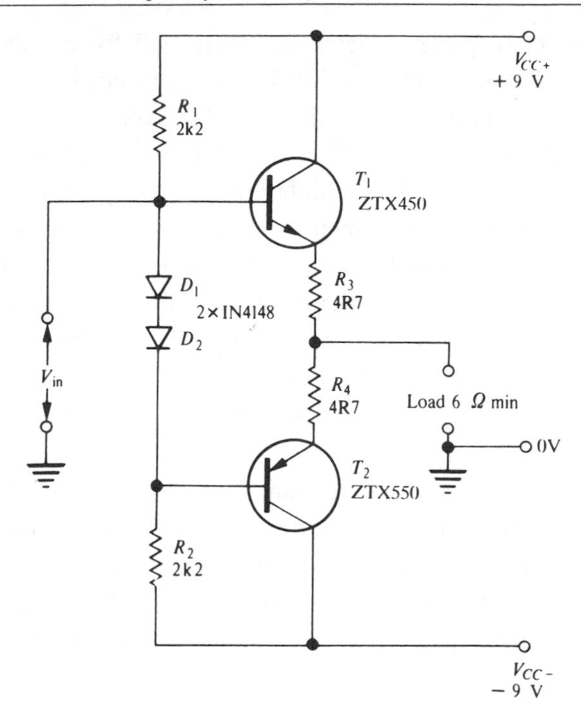

Fig. 5.25. Class AB complementary emitter follower.

change in V_{BE} due to a change in temperature can produce a large change in quiescent collector current. The use of diodes to supply the bias provides temperature compensation: if the ambient temperature rises, producing a fall in V_{BE}, then the diode p.d. will also fall, keeping the base current, and hence the collector current, relatively constant. Ideally the diodes should be sited close to the transistors in order to be subject to the same temperature fluctuations. Quiescent collector current should be in the region of 5 mA–10 mA.

In order to accommodate variations in transistor V_{BE} it is desirable to be able to adjust quiescent current manually to the required value. The modification of fig. 5.26(*a*) provides this facility. The bias voltage is derived from a transistor (T_3) in a 'V_{BE} multiplier' circuit shown in isolation in fig. 5.26(*b*). This latter part of the circuit is really a small voltage amplifier, with its own V_{BE} as input, and having negative feedback applied via the potential divider R_x and R_y. Here, feedback fraction $\beta = R_y/(R_x + R_y)$. the output V_{BIAS}, is therefore given by

$$V_{\text{BIAS}} \approx \frac{V_{\text{BE}}}{\beta} = \frac{V_{\text{BE}}(R_x + R_y)}{R_y}.$$

In the circuit of fig. 5.26(*a*), potentiometer R_5 is adjusted for a quiescent current of 5–10 mA. Temperature compensation is again present because the V_{BE} of T_3 will change by the same factor as the V_{BE} of T_1 and T_2.

The circuits of figs. 5.25 and 5.26 are able to deliver about half a watt of audio power into an 8 Ω loudspeaker. Where higher powers are required power transistors are used, usually in the complementary Darlington arrangement of fig. 5.27. This circuit will safely give at least 3 A to the load. Quiescent current should be set to about 40 mA. If the complementary Darlington pair (e.g. T_1 and T_4) is compared with the npn Darlington pair of fig. 5.15 it will be realized that the advantage of the complementary circuit is that the bias p.d. needs only to be $2V_{BE}$ instead of the $4V_{BE}$ that would otherwise be required; this leads to improved thermal stability.

Resistors R_7 and R_8 across the base–emitter junctions of the power transistors improve high-frequency response by providing a path for the current carriers that would otherwise linger in the base region on turn-off.

By conducting away the collector–base leakage current (I_{CBO}) they also increase the collector–emitter breakdown voltage of the power transistors.

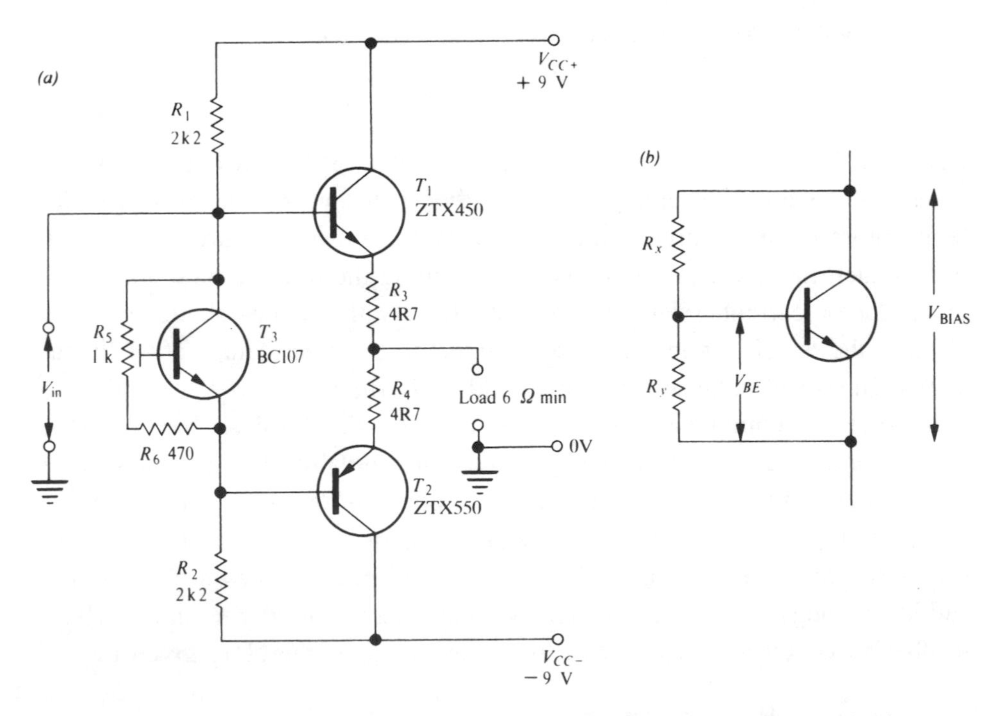

Fig. 5.26. (*a*) class AB complementary emitter follower with bias provided by a V_{BE} multiplier. The latter is shown separately in (*b*).

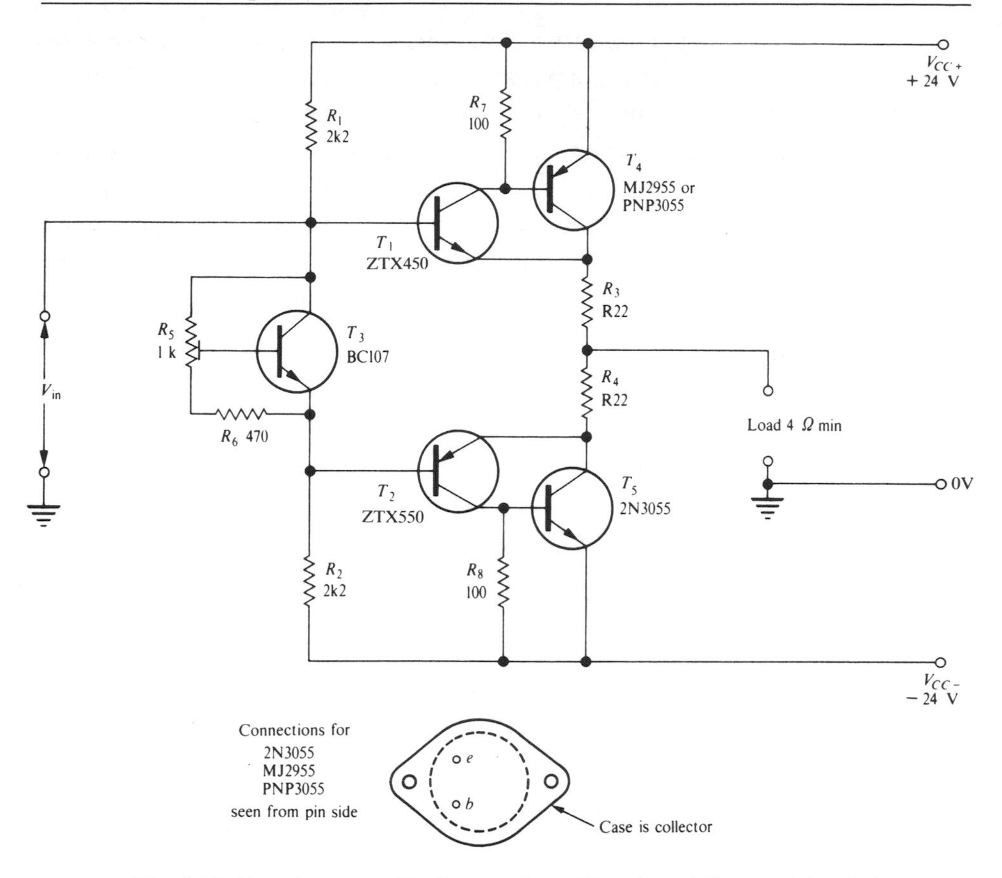

Fig. 5.27. Complementary Darlington class AB emitter follower, giving 3 A output capability.

5.17.3 A typical audio power amplifier

To illustrate an application of the high-power emitter followers discussed above, fig. 5.28 is included. This is the circuit of a typical audio amplifier capable of delivering 30 W into a 4 Ω load at low distortion.

The basis of the output circuit is the complementary Darlington emitter follower arrangement of fig. 5.27. This is driven by voltage amplifier T_3 with a split collector load, R_9, and R_{10}. Resistor R_9 is bootstrapped to the output by C_5, which greatly boosts its effective resistance and hence increases open-loop voltage gain and reduces distortion by providing a nearly constant current load. R_6 is adjusted to give a quiescent current of about 40 mA.

The input stage of the amplifier is a differential pair (T_1 and T_2). This type of circuit is discussed in detail in chapter 8; it provides excellent d.c. stability (notice that no coupling capacitors are used between stages) and incorporates the inverting and non-inverting input arrangement similar to the 741 IC mentioned in section 4.8. The input signal is fed to the non-inverting input, whilst negative feedback is taken from the output via potential divider R_5 and R_4 back to the inverting input. The presence of C_3 means that feedback fraction β is unity at d.c., minimizing drift and ensuring that the output terminals have a negligible d.c. potential difference. An output coupling capacitor is not, therefore, required. At signal frequencies, β is given in the usual way:

$$\beta = \frac{R_4}{R_5 + R_4},$$

giving an overall voltage gain

$$\frac{V_{\text{out}}}{V_{\text{in}}} \approx \frac{1}{\beta} = \frac{R_5 + R_4}{R_4} = \frac{10\,000 + 470}{470}$$

$$\approx 22.$$

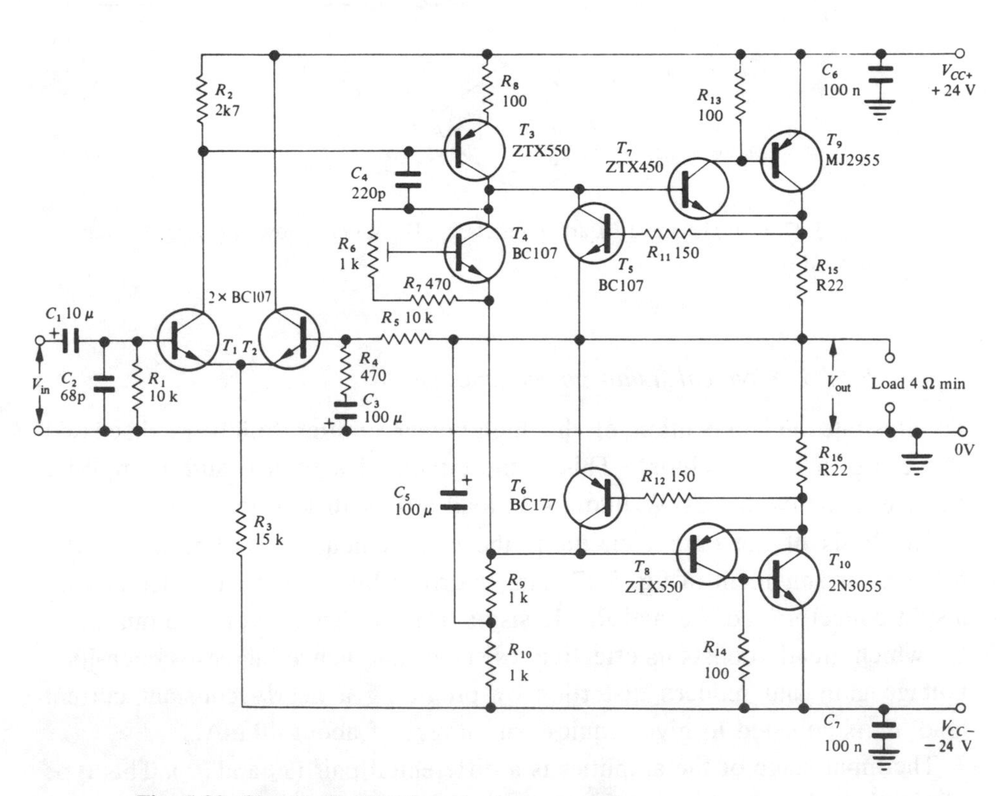

Fig. 5.28. Circuit of a 30 W audio power amplifier.

In other words, for 11 V r.m.s. output, which corresponds to a mean power of 30 W in a 4 Ω load, the required input signal is 0.5 V.

Transistors T_5 and T_6 are a useful precaution against a short-circuited load. In such a condition, the output stage sees virtually zero load impedance which would lead to a dangerously high current in the output transistors. T_5 and T_6, however, monitor the p.d. across the 0.22Ω emitter resistors R_{15} and R_{16}. When this p.d. exceeds 0.6 V, T_5 and T_6 turn on and partially short-circuit the signal voltage at the input of the complementary Darlingtons. With the values quoted for R_{15} and R_{16}, this automatic limiting begins to occur at peak emitter currents of about 3 A.

The output transistors are called upon to dissipate up to 10 W each at full output and should be mounted on a proprietary finned heat sink (see section 9.11).

Badly designed or wrongly constructed power amplifiers are prone to oscillate at ultrasonic frequencies. At the very least, this phenomenon causes mysterious distortion and excessive power dissipation; at worst, the spurious high-frequency output can burn out loudspeakers and transistors. The circuit of fig. 5.28 incorporates a compensation capacitor C_4, mentioned in section 4.6 as a prerequisite for stability in a feedback amplifier. The shunt capacitor C_2 at the input also assists in reducing stray coupling at radio frequencies. C_6 and C_7 decouple the power supply lines, presenting a low impedance at radio frequencies and thus minimizing unwanted coupling between amplifier stages.

Despite these precautions, instability may still occur unless the layout is compact: a Veroboard construction is recommended. Inevitably, the output transistors must be mounted separately from the circuit board, on their own heat sink, but the wires to them should be short and preferably of a 10 A rating to minimize potential drops.

5.17.4 High-power FETs

The high input impedance of FETs makes them potentially attractive as output stages in a power amplifier because they can be driven directly from a high impedance voltage source. They are also inherently rather more linear devices than the bipolar transistor (see section 6.10) and so are capable of still lower distortion. Most important of all, their greater thermal robustness requires only minimal overload protection for high reliability. High-power FETs are available in a wide range of types with power dissipation ratings from 1 W to over 100 W. They are often known as V-MOS and HEXFET devices, these terms referring to their internal construction.

5.17.5 IC power amplifiers

For practical power amplification up to 50 W or so, the most cost-effective solution is usually an IC. The close electrical and thermal coupling of components normally gives very consistent results. Component catalogues and data books can be consulted for the most appropriate type, which ranges from the 8-pin DIL LM386 giving 600 mW to the TDA2050 in a TO220 package capable of 50 W output when mounted on a heat sink.

6

Semiconductor device characteristics

6.1 Introduction

We have now considered some of the main principles of electronic circuits and have studied them in sufficient detail to understand the significance of many of the component values specified in practical circuits. Although we can now guarantee that our simple bipolar transistor and FET amplifiers will work in practice (i.e. we know how to set up the correct d.c. conditions), there are gaps in the treatment so far. We have not, for instance, considered how to calculate the voltage gain of the common-emitter or common-source amplifiers. Although negative feedback is normally employed to set the voltage gain precisely, we need to have at least an approximate idea of the basic open-loop gain of an amplifier to predict the closed-loop performance exactly. This chapter, therefore, deals in detail with device characteristics and their relation to circuit performance. The pn junction diode is considered first, then the bipolar transistor, followed by the FET.

6.2 pn junction characteristics

The basic characteristics of the pn junction were considered qualitatively in chapter 1; we shall now look at the equation of the current–voltage relationship sketched in fig. 6.1. Consideration of the number of current carriers able to climb the potential barrier represented by the depletion layer gives the following current–voltage equation

$$I = I_0\left(\exp\frac{eV}{kT} - 1\right), \tag{6.1}$$

where

I is the current in the diode (amps)

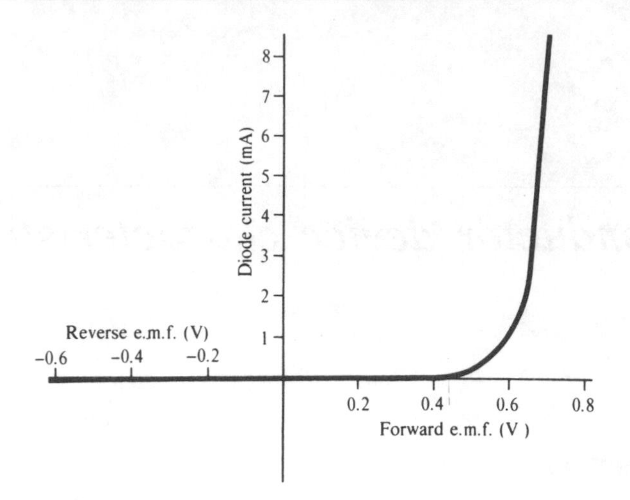

Fig. 6.1. Current–voltage characteristic of pn junction.

V is the applied e.m.f. (volts)
e is the charge on the electron (coulombs)
k is Boltzmann's constant (joules/kelvin)
T is the junction temperature (kelvin)

Setting the applied e.m.f., V, to a large negative value reveals the electrical significance of the constant term I_0,

$$\exp \frac{eV}{kT} \to 0,$$

leaving $I = -I_0$.

I_0 is therefore the thermal leakage current which, at constant temperature and under conditions of reverse bias, is constant for reverse voltages of greater than 0.1 V magnitude. I_0 is commonly called the *reverse saturation current* (fig. 6.2).

In the forward direction, if $V > 0.1$ V, $\exp(eV/kT) \gg 1$, so that

$$I \approx I_0 \exp\left(\frac{eV}{kT}\right). \tag{6.2}$$

We can substitute values for e and k and assume that T is room temperature (20 °C or 293 K). The constant e/kT then turns out to be approximately equal to 40 V^{-1} so that the current–voltage relationship becomes

$$I \approx I_0 \exp(40V). \tag{6.3}$$

It would be useful to be able to express a single value for the forward

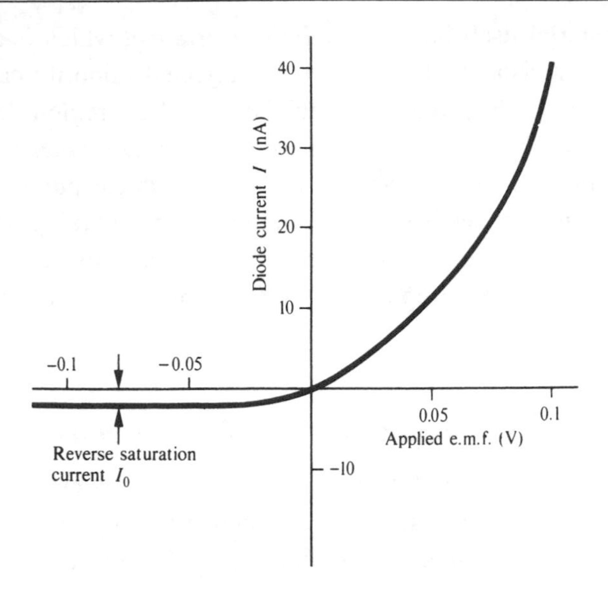

Fig. 6.2. Enlarged view around origin of fig. 6.1 showing reverse saturation current I_0.

resistance of a diode, but, since the characteristic is non-linear, an expression of the form V/I, as for a resistor, is not relevant. The term used is the *dynamic*, or *incremental*, resistance defined as

$$r = \frac{dV}{dI}.$$

Differentiating equation (6.3) gives

$$\frac{dI}{dV} = 40 \times I_0 \exp(40V)$$

$$= 40I,$$

therefore

$$\text{dynamic resistance } r = \frac{dV}{dI} = \frac{1}{40I}\ \Omega,$$

or, if I is in mA,

$$r = \frac{25}{I}\ \Omega. \tag{6.4}$$

Hence, the dynamic resistance of a diode is inversely proportional to the forward current. A forward current of 1 mA gives a dynamic resistance of about 25 Ω.

In practice, for real diodes, the above theory is something of a simplification.

The semiconductor material itself has a finite *bulk resistance*, which is added to the junction dynamic resistance. In addition, the injected minority carriers (that is to say the electrons in the p-region and the holes in the n-region) tend to increase the effective resistance of the diode. Despite these additional factors, equations (6.1)–(6.4) are sufficiently accurate for most practical purposes. In particular, if we are considering the base–emitter junction of a working transistor, there is less deviation from the simple theory than with a discrete diode: injected minority carriers are swept up and away from the base region by the action of the collector.

6.3 Bipolar transistor input and transfer characteristics

As we have already seen in section 1.4, the bipolar transistor is essentially a current-operated device. It is the base *current* which primarily controls the collector current, so that the first step in examining the behaviour of a transistor voltage amplifier is to find out what base current is produced by our input voltage signal. In other words, we need to know the dynamic resistance of the base–emitter junction *as viewed from the base*. This parameter is usually given the symbol h_{ie} and, if v_{be} represents a small a.c. input voltage signal and i_b the resulting a.c. base current,

$$h_{ie} = \frac{v_{be}}{i_b}.$$

We can derive the value of h_{ie} from the pn junction equation (6.2) which can be applied to the base–emitter junction of a transistor as follows:

$$I_E \approx I_0' \exp \frac{eV_{BE}}{kT}. \tag{6.2a}$$

I_E is the d.c. emitter current (note that, because $I_E \approx I_C$, equation (6.2a) may also be used as an expression for I_C). I_0' is the emitter current which flows when $V_{BE} = 0$, and is the result of leakage current in the collector–base junction (see section 1.4.4). V_{BE} is the d.c. base–emitter p.d.

We can determine the dynamic resistance r_e of the emitter–base junction by differentiating equation (6.2a), as we did in the case of the simple diode to get equation (6.4). This gives:

$$\frac{dI_E}{dV_{BE}} = r_e \approx \frac{25}{I_E}, \text{ at room temperature,} \tag{6.5}$$

where I_E is in mA.

This is the junction resistance as 'seen' from the emitter. We know from the consideration of the emitter follower input impedance in section 5.12.3 that, because the base current is smaller than the emitter current by a factor $1/(h_{fe}+1)$, then any impedance in the emitter circuit is effectively multiplied by $(h_{fe}+1)$, when 'seen' from the base, i.e.

$$h_{ie} \approx (h_{fe}+1)\,r_e$$

and, for

$$h_{fe} \gg 1,$$

$$h_{ie} \approx h_{fe} r_e.$$

Therefore, substituting for r_e,

$$h_{ie} \approx \frac{25h_{fe}}{I_E}\ \Omega, \tag{6.6}$$

where I_E is in mA.

In practice, an accurate estimate of h_{ie} must also include the effective resistance r_b between the external base lead and the functional base inside the transistor. The value of r_b is made up of the sum of two resistances: one is the ohmic resistance of the base region ($r_{bb'}$) and the other is an equivalent resistance $r_{b'}$ which allows for variations in effective base width with collector–base p.d., sometimes called the base-spreading resistance. In a small silicon transistor, r_b is typically in the region 500 Ω to 1000 Ω. The complete equation for h_{ie} is then as follows:

$$h_{ie} = \frac{25h_{fe}}{I_E} + r_b, \tag{6.7}$$

where I_E is in mA.

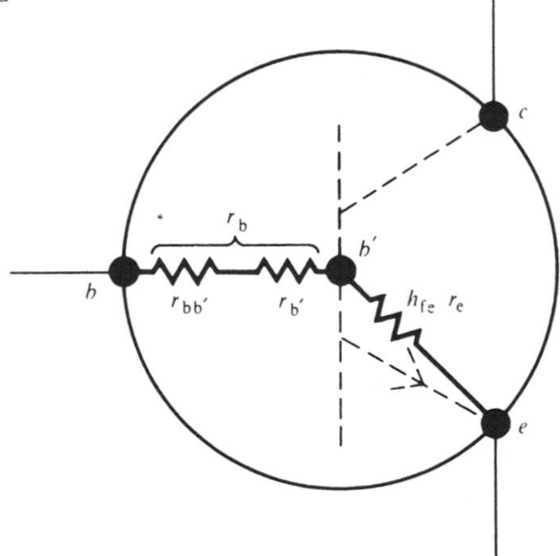

Fig. 6.3. Equivalent circuit for a transistor base–emitter junction, as 'viewed' from the base.

The equivalent circuit represented by this equation is shown in fig. 6.3. With emitter currents of 1 mA or less, r_b can usually be neglected in calculating the input resistance of a common-emitter amplifier.

Having found input resistance, h_{ie}, we can calculate a.c. base current i_b resulting from a small a.c. voltage input signal v_{be}:

$$i_b = \frac{v_{be}}{h_{ie}}.$$

This a.c. base current gives rise to an a.c. collector current

$$i_c = h_{fe} i_b,$$

therefore

$$i_c = h_{fe} \frac{v_{be}}{h_{ie}}. \tag{6.8}$$

Equation (6.8) shows that we can predict the collector *current* signal i_c which results from a base–emitter *voltage* input signal v_{be}. This can be simplified still further by defining a *transconductance*, g_m, as we did for the FET in chapter 2 so that $i_c = g_m v_{be}$. Transconductance is then given by

$$g_m = \frac{i_c}{v_{be}}. \tag{6.9a}$$

Substituting for i_c (equation (6.8)),

$$g_m = \frac{h_{fe}}{h_{ie}}, \tag{6.9b}$$

but we already have $h_{ie} \approx h_{fe} r_e$ (neglecting r_b). Hence,

$$g_m \approx \frac{1}{r_e} \tag{6.10}$$

$$= \frac{I_E}{25} \text{ siemens } (I_E \text{ in mA}) \text{ at room temperature}$$

$$= 40 I_E \text{ mA/V}$$

and since $I_E \approx I_C$,

$$g_m \approx 40 I_C \text{ mA/V}, \tag{6.11}$$

where I_C is d.c. collector current in mA. Thus, we have a transistor parameter which does not vary from one transistor to another, but depends only upon the collector current.

Having established a value of transconductance for the bipolar transistor, we are well on the way to finding what output voltage signal to expect for a given voltage input applied to an amplifier. We must first, however, establish what happens in the transistor as far as the output signal is concerned.

6.4 Introduction to output characteristics

We have so far described the behaviour of the bipolar transistor in terms of current gain h_{fe} and transconductance g_m. We have not yet, however, considered the effect of variations in collector voltage and have in fact assumed that collector current is substantially independent of such changes. To what extent is this true? In a voltage amplifier, the collector voltage will be swinging up and down with the signal: if this fact in itself causes variations in collector current, then it must be incorporated in any calculation. The FET can similarly be examined to see whether its drain current varies significantly with drain voltage, all other conditions being held constant. The device output characteristics (collector characteristics of the bipolar transistor and drain characteristics of the FET) give the required information about device behaviour and at the same time provide a background to the complete picture of signals in a voltage amplifier.

6.5 Collector characteristics

6.5.1 Measurement method and results

The measurement circuit of fig. 6.4 provides a means of plotting transistor common-emitter output characteristics. The base current I_B is set at a constant value with RV_1 and measured on micro-ammeter M_1. Potentiometer RV_2 is then used to set a range of values of collector–emitter voltage V_{CE} measured on voltmeter M_3, the resultant collector current I_C being measured on milli-ammeter M_2.

The form of the resultant graph of collector current against collector voltage is shown in fig. 6.5(*a*). Here it can be clearly seen that, as the collector–emitter voltage is increased from zero, there is initially a sharp rise in collector current as the collector increases in efficiency until V_{CE} reaches about 0.6 V when there is a levelling of the curve; further increase in V_{CE} has a negligible effect on collector current. It is interesting to note that the sharp *knee* of the curve

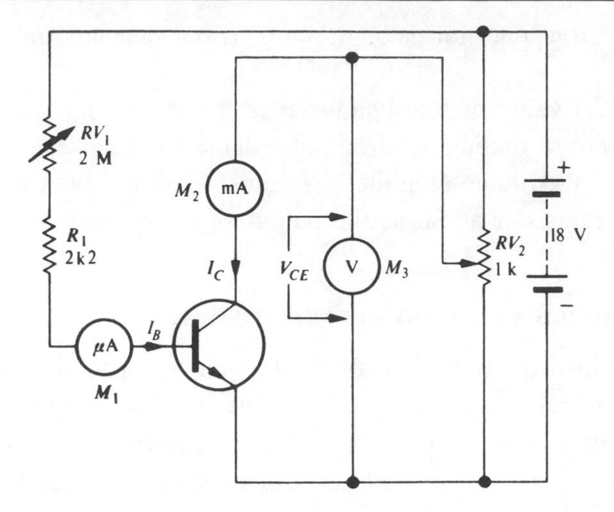

Fig. 6.4. Measurement circuit for plotting collector characteristics of bipolar transistor.

occurs when the collector–base voltage is around zero, indicating that the collector is near maximum efficiency as soon as the collector–base junction attains the slightest reverse bias.

By plotting curves for different values of base current, a family of output characteristics can be drawn for a transistor. Fig. 6.5(*b*) shows such a set of curves, where the essentially *constant collector current* over a range of collector voltages is clearly seen. The slight upward slope of the characteristics is due to a small improvement in current gain at high collector voltages. This is due to the wider collector–base depletion layer making the base region effectively narrower and allowing less recombination. The upward slope to the curves is more marked at high collector currents because, although the *fractional* improvement in h_{FE} with increased collector voltage is similar at all collector currents, this results in a greater *absolute* increase in collector current where the collector current was high to begin with.

Although for most purposes the transistor can be considered a near-perfect constant-current generator, the slope of the characteristic curves is sometimes included in precise calculation and is normally termed h_{oe}, the common-emitter output conductance:

$$h_{oe} = \frac{\mathrm{d}I_C}{\mathrm{d}V_{CE}}. \tag{6.12}$$

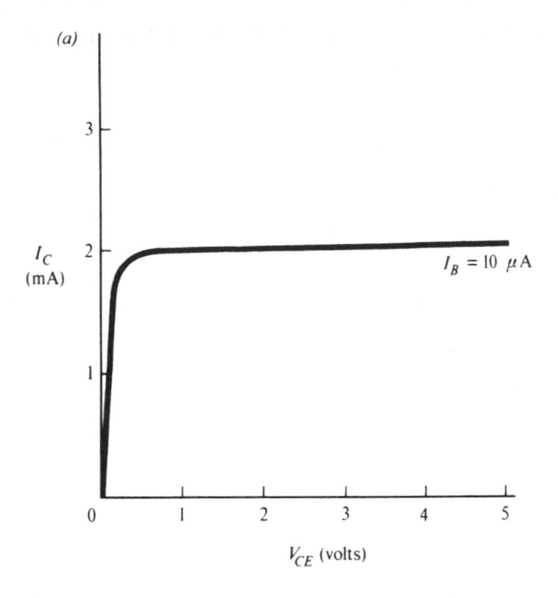

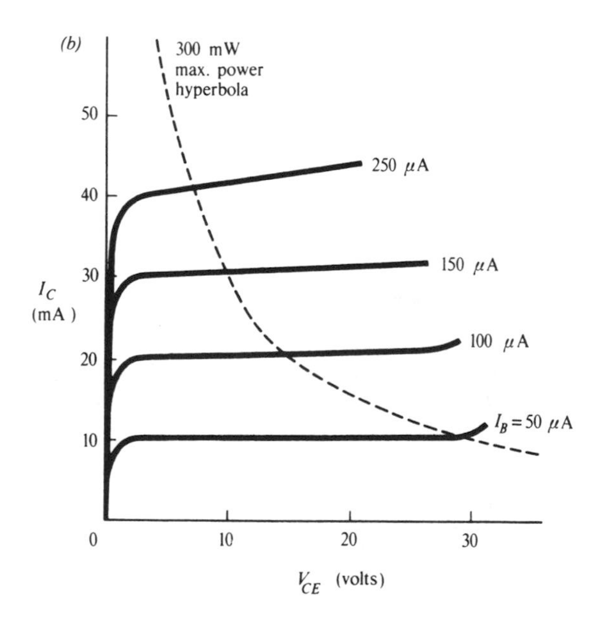

Fig. 6.5. Typical collector characteristics for a small silicon transistor. (*a*) A single characteristic for 10 μA base current. Collector current of 2 mA indicates h_{FE} of 200. (*b*) A family of characteristics with maximum power hyperbola superimposed.

Fig. 6.6 exploits the constant-current characteristic of the transistor in a simple battery charger circuit for nickel–cadmium cells. It is based on the bias circuit of the common-emitter amplifier discussed in section 1.6.6. The current in the load, which would normally be the cells to be charged, can be adjusted between zero and 1 A by setting the base voltage on the potentiometer. As the cells charge, their terminal voltage rises, but the transistor maintains the current constant. With the addition of a timer to disconnect the battery, after the calculated charging period, fast charging can be completed in under an hour without damage to the cells.

To return to the transistor characteristics in fig. 6.5(*b*), the dotted curve is known as the *maximum-power hyperbola* and is important because it is the locus of the maximum power rating of the transistor, given by the simple hyperbolic equation $W = V_{CE}I_C$. The hyperbola in fig. 6.5 is drawn for 300 mW, the maximum power rating P_{max} of the BC107. At no time must the product of V_{CE} and I_C exceed P_{max}; in other words, the characteristic curves to the right of the hyperbola are in a sense hypothetical, since the transistor must not be operated in that region. Such plots are normally obtained by pulsing the transistor so that P_{max} is not exceeded for long enough to cause the transistor to burn out. The maximum power hyperbola is of considerable importance in power amplifier design.

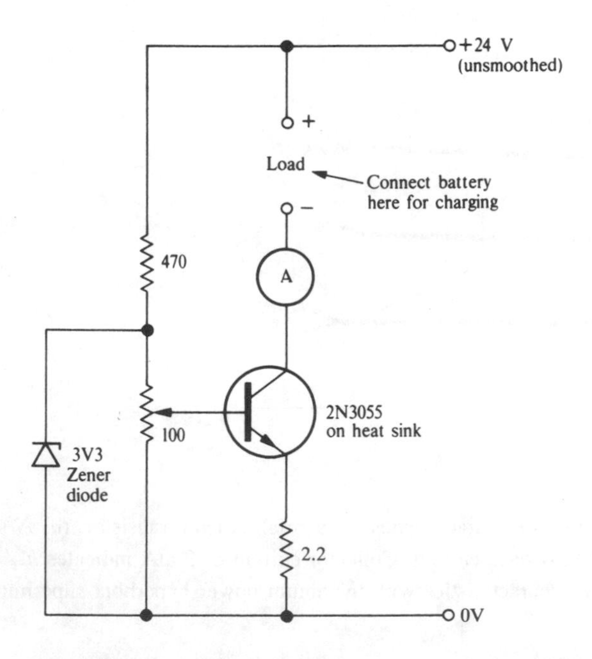

Fig. 6.6. Circuit which exploits constant-current collector characteristics of transistor: a charger for nickel–cadmium batteries. Maximum current is 1.0 A.

The slight upward curvature of the characteristics towards $V_{CE} = 30$ V is important because it shows the trend towards avalanche breakdown in the reverse-biased collector–base junction, which normally sets the limit on the maximum permissible collector–emitter voltage.

6.5.2 Load line and saturation

In fig. 6.7(*a*) a set of collector characteristics is shown for the range of voltages and currents applicable to the common-emitter amplifier stages considered in chapter 1. The output side of such a circuit is drawn in fig. 6.7(*b*). On the characteristics is plotted the straight line *XY*. This is the *load line* for a 9 V supply (V_{CC}) and 4.5 kΩ collector load (R_L) and represents the path which the collector voltage and current must follow with this value of load resistor and supply voltage V_{CC}.

The equation of the load line is simply an expression of the relation of the p.d. across R_L to the collector current:

$$V_{CE} = V_{CC} - R_L I_C;$$

rearranging,

$$I_C = \frac{-1}{R_L} \cdot V_{CE} + \frac{V_{CC}}{R_L}. \tag{6.13}$$

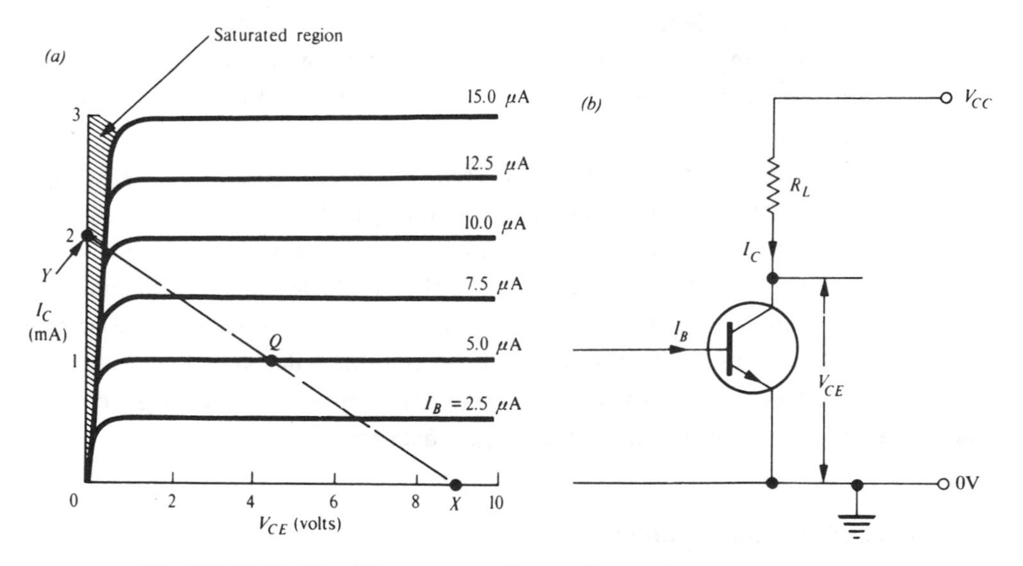

Fig. 6.7 (*a*) Family of low-power collector characteristics with a load line (*XY*) drawn in for $V_{CC} = 9$ V; $R_L = 4.5$ kΩ. (*b*) Output section of relevant common-emitter amplifier.

Equation (6.13) gives the straight line with a negative gradient of magnitude $1/R_L$ and a positive intercept on the current axis at V_{CC}/R_L. The load line is very easily plotted from the two points X and Y which correspond to maximum collector voltage and maximum collector current respectively. Point X represents the cut-off condition when the collector current is zero; there is therefore no voltage drop in the collector load resistor and the collector sits at full supply voltage V_{CC} (9 V). The other load line point, Y, represents the hypothetical condition of zero collector voltage, when the full supply is dropped across the collector load. Under this condition,

$$I_C = \frac{V_{CC}}{R_L}$$

$$= \frac{9\text{V}}{4500\,\Omega}$$

$$= 2\,\text{mA}.$$

Having drawn the load line, we can select the optimum quiescent operating point, Q, of $V_{CE} = 4.5$ V and $I_C = 1$ mA to allow the maximum positive and negative swing as we did in chapter 1. The advantage of examining the swing of collector voltage and current superimposed on the device characteristics is that the exact practical limits of that swing can be seen. Clearly, the output can go right up to 9 V (cut-off), though, as will be shown shortly, some distortion may be apparent. The lower limit of the linear output voltage swing is where the load line no longer intersects one of the characteristics. Below this point is the shaded region which represents a bottomed or saturated transistor, the collector current being no longer under the control of the transistor but limited only by the collector load resistor and supply voltage.

In the saturated region, the collector current is given by $I_{C(\text{sat})} = V_{CC}/R_L$, the transistor being driven into saturation by pushing in a base current greater than $I_{C(\text{sat})}/h_{FE}$. Transistors used for switching operate alternately in the saturated region and at cut-off. In saturation, the collector–base junction is actually forward-biased and, since the potential differences of the collector–base and base–emitter junctions are then roughly equal and opposite, very low collector–emitter voltages ($V_{CE(\text{sat})}$) may be obtained (typically <0.2 V). In general, the higher the base current, the lower $V_{CE(\text{sat})}$ becomes.

6.5.3 Maximum ratings

It is in the design of power output stages (discussed in section 5.17) that special attention to the output characteristics can be most helpful.

The reader may at first wonder why collector characteristics should be relevant to power output stages, since virtually all such circuits are emitter followers. Fortunately we do not require a separate set of 'emitter characteristics': collector current and emitter current are so very nearly the same in modern high-gain transistors that the one set of characteristics applies equally well to common-emitter and common-collector operation.

It is important that the load line should never intersect the maximum power hyperbola; this would indicate that excessive power dissipation was occurring at some point in the signal cycle. The load line intercept on the V_{CE} axis (point X in fig. 6.7(*a*)) will normally be at the supply rail voltage (V_{CC}).

It is clear that V_{CC} should always be less than the specified maximum permissible collector–emitter voltage ($V_{CE(\max)}$) for the transistor. Finally, the load line intercept on the I_C axis (point Y in fig. 6.7(*a*)) should be lower than the maximum collector current rating $I_{C(\max)}$ of the transistor.

The observance of these conditions sets an upper limit to the supply voltage, a lower limit to the load resistance R_L, and hence an upper limit on available power output. It does, however, ensure a long life for the output transistors.

Although maximum power dissipation and maximum I_C and V_{CE} for a particular circuit are most conveniently determined graphically from the device output characteristics and load line, these values can still be obtained without the curves. $V_{CE(\max)}$ and $I_{C(\max)}$ present no problem: these are the load line intercepts, given by V_{CC} and V_{CC}/R_L respectively. Maximum power dissipation, $P_{\max}$, involves a little more calculation. Consider the circuit in fig. 6.7(*b*). Instantaneous power dissipated in the transistor is:

$$P = V_{CE} I_C$$

$$= V_{CE}\left(\frac{V_{CC} - V_{CE}}{R_L}\right)$$

$$= \frac{1}{R_L}(V_{CC}V_{CE} - V_{CE}^2). \tag{6.14}$$

To obtain $P_{\max}$, we differentiate to obtain $\mathrm{d}P/\mathrm{d}V_{CE}$ and equate $\mathrm{d}P/\mathrm{d}V_{CE}$ to zero.

$$\frac{\mathrm{d}P}{\mathrm{d}V_{CE}} = \frac{1}{R_L}(V_{CC} - 2V_{CE}) = 0$$

Thus maximum power dissipation occurs when

$$V_{CE} = \frac{V_{CC}}{2}.$$

This gives the important result that maximum instantaneous power dissipation in the transistor occurs when it is dropping exactly half the supply voltage.

Substituting $V_{CE} = V_{CC}/2$ in equation (6.14) gives

$$P_{\max} = \frac{1}{R_L}\left(\frac{V_{CC}^2}{2} - \frac{V_{CC}^2}{4}\right),$$

$$= \frac{V_{CC}^2}{4R_L}. \tag{6.15}$$

In a class AB push–pull amplifier like fig. 5.28 with twin complementary power supplies, if V_{CC} is the magnitude of each supply, then $P_{\max}$ is the maximum instantaneous power dissipation of *each* transistor.

6.6 FET drain characteristics

A family of output characteristics can be drawn for a common source FET by plotting drain current against drain–source voltage at different values of gate–source voltage. The test circuit of fig. 6.8 will enable the curves to be plotted. The required gate–source voltage V_{GS} is set by RV_1 and measured on voltmeter M_1. Potentiometer RV_2 is then used to set a range of values of drain–source voltage V_{DS}, measured on voltmeter M_2, the resultant drain current I_D

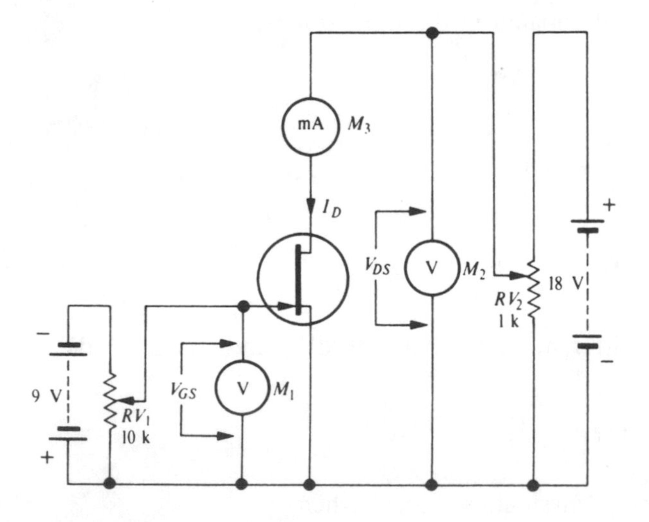

Fig. 6.8. Measurement circuit for plotting drain characteristics of a FET.

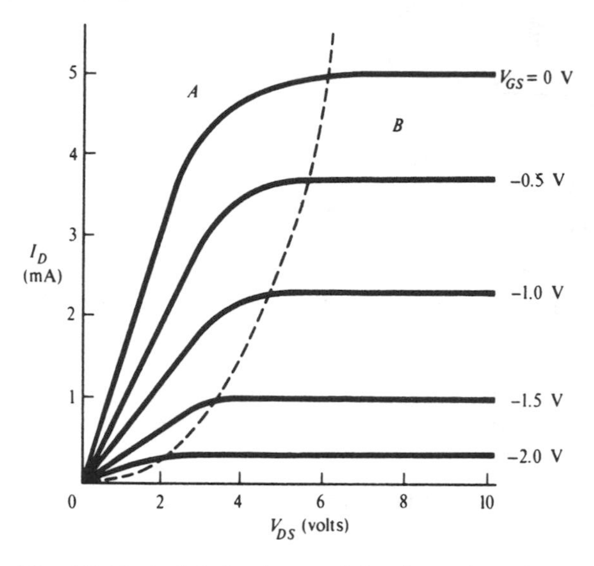

Fig. 6.9. Typical drain characteristics for n-channel FET, showing variable-resistance region (*A*) and pinch-off region (*B*).

being measured on milliammeter M_3. A typical set of FET characteristic curves is shown in fig. 6.9.

It is clear from fig. 6.9 that, very broadly, the FET drain characteristics are similar to the bipolar transistor collector characteristics in that they represent a constant-current source over most of the voltage range. In other words, if the gate–source bias is fixed at −1 V, increasing V_{DS} from 5 V to 15 V has a negligible effect on the drain current. This, perhaps rather surprising, fact applies to the area of the curves to the right of the dotted line, this area being known as the pinch-off region.

To understand the operation of the FET in this region, we need to consider fig. 6.10, where a diagrammatic FET is connected to drain and gate supplies. Initially we shall assume that $V_{DS} = 0$ and that the gate voltage V_{GS} is sufficient to establish a depletion layer extending across part of the width of the silicon bar, leaving the remainder as a conduction channel between source and drain. Now, as we increase V_{DS} from zero, at first the channel behaves simply as a resistor of value determined by the width of the channel left by the depletion layer. As V_{DS} reaches a few hundred millivolts, it begins to contribute to the reverse bias on the gate and the depletion layer spreads out, particularly at the more positive (drain) end of the channel, until only a very narrow conduc-

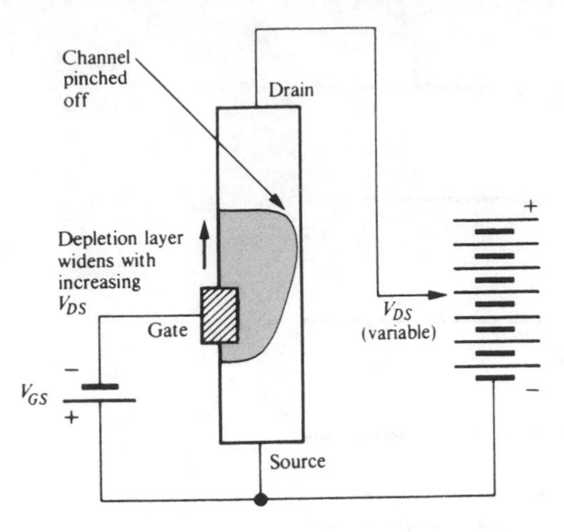

Fig. 6.10. The origin of pinch-off in the FET.

tion channel remains. Further increase in V_{DS} then serves to constrict the channel still further, almost exactly balancing the expected increase in I_D due to the increased voltage. This is the pinch-off region; the transition from the *resistive* region is clearly seen in fig. 6.9 as the curves turn over to become nearly horizontal (constant current). As with the bi-polar transistor, there is a small positive slope to the 'constant' current region, this slope being expressed as the *drain conductance*, g_d, or y_{os} so that

$$g_d = \frac{\mathrm{d}I_D}{\mathrm{d}V_{DS}}.$$

Also used is r_d the dynamic *drain resistance*, where

$$r_d = \frac{1}{g_d}.$$

6.7 The FET as a voltage-controlled resistor

At low values of V_{DS}, we notice an interesting result with the FET which is not obtained with the bipolar transistor. Here the gate voltage has the effect of varying the resistance of the drain–source channel. This is clear in area A in fig. 6.9 as a different slope for each characteristic near the origin. Reference to figs. 6.5 and 6.7 indicates that the effect does not apply to the bipolar transistor.

These characteristics mean that the FET can be used as a voltage-controlled

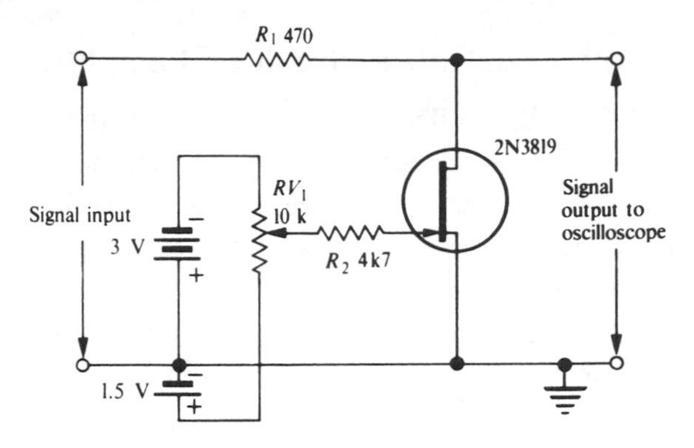

Fig. 6.11. Demonstration of the variable resistance properties of a FET.

variable resistor for controlling the amplitude of a.c. or d.c. signals, a facility which is exploited in automatic gain control systems.

Fig. 6.11 shows an experimental voltage–controlled attenuator circuit which is designed to be fed from the low impedance output of a signal generator.

The signal input should be kept below 500 mV if waveform distortion is to be reasonably low. The 3 V and 1.5 V batteries and RV_1 provide a variable voltage which controls the attenuation of the circuit. The gate may be taken a fraction of a volt positive in order to obtain maximum attenuation; any accidental forward current flow is limited to a safe value by R_2. It is a convincing test to substitute a bipolar transistor for the FET and check that it will not work as a voltage-controlled attenuator without gross distortion.

In many attenuator applications, the low input impedance of fig. 6.11 is a disadvantage, but R_1 can in fact be increased in 10 kΩ or even 100 kΩ if a smaller peak-to-peak output is acceptable. Various negative feedback circuits which feed part of the output signal back to the gate, have been devised to overcome the distortion due to the curvature of the FET characteristics.

There are many applications for the voltage-controlled attenuator. It can be a useful remote gain control; only d.c. need be handled by the control cables, avoiding hum and loss of signal. Automatic gain control (AGC) can be arranged by using a control voltage derived by rectifying the output signal of an amplifier. Such an arrangement will keep the output signal constant with widely varying input levels. Voltage-controlled attenuators are the basis of the limiters and volume compressors which are extensively used in broadcasting and recording practice. The well-known 'Dolby' noise reduction technique uses FETs as voltage-controlled attenuators.

6.8 Common-emitter equivalent circuit and amplifier gain

We have now established from the collector characteristics that the bipolar transistor is a near-perfect constant-current source. This fact facilitates the calculation of the output voltage signal developed across the collector load resistor in a voltage amplifier. Fig. 6.12(*a*) shows the output section of an amplifier stage with load R_L in the collector circuit of a transistor; i_c is the a.c. component of the collector current. The equivalent circuit, fig. 6.12(*b*), shows the transistor as a constant current generator producing current i_c in load R_L. By Ohm's law,

$$v_{out} = -i_c R_L. \tag{6.16}$$

Fig. 6.13 shows the equivalent circuit of both input and output sections of a common-emitter amplifier. This form of equivalent circuit is known as the *hybrid* π type, which leads to the *h-parameters* such as h_{ie} etc.

We know that collector current i_c is related to the base–emitter voltage v_{in} by the transconductance g_m (equation (6.9a)), i.e.

$$i_c = v_{in} g_m \tag{6.17}$$

and we already have from equation (6.16)

$$v_{out} = -i_c R_L.$$

Now we can calculate the voltage gain A of the stage:

$$A = \frac{v_{out}}{v_{in}}.$$

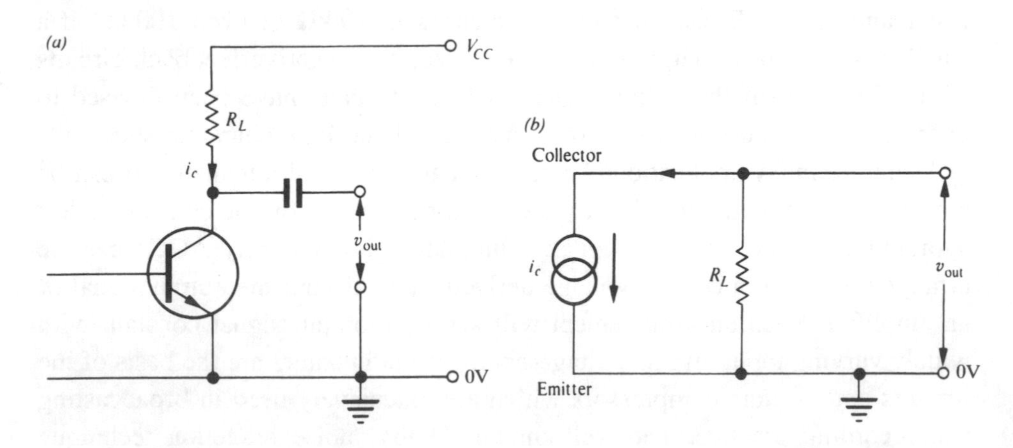

Fig. 6.12. (*a*) Output side of common-emitter amplifier. (*b*) Its equivalent circuit.

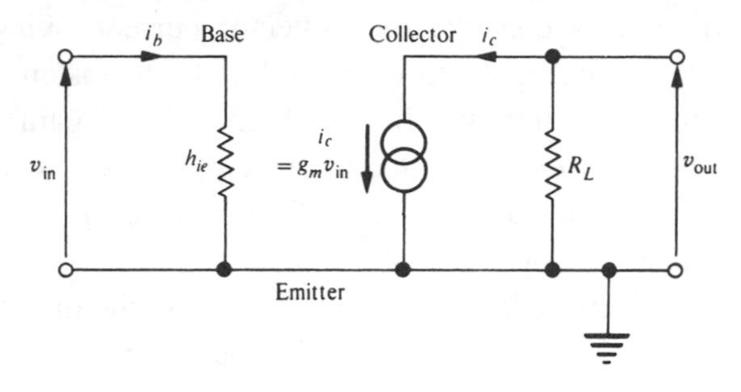

Fig. 6.13. Equivalent circuit of common-emitter amplifier (simplified hybrid π).

Substituting for v_{out} from (6.16) and (6.17),

$$A = \frac{-v_{in} g_m R_L}{v_{in}}, \tag{6.18}$$

therefore

$$A = -g_m R_L, \tag{6.19}$$

where g_m is in mA/V and R_L in kΩ.

The negative sign indicates a phase reversal; that is, a positive-going input produces a negative-going output. In this simple calculation, which is sufficiently accurate for most practical purposes, we have neglected the slight slope of the collector characteristics, h_{oe} (output conductance). The presence of this slope means that the constant-current generator is not perfect; however, we can easily correct the equivalent circuit by drawing in a resistor of value $1/h_{oe}$ in parallel with the load resistor R_L. It is as though the collector circuit of the transistor had a finite internal resistance of value $1/h_{oe}$, whereas a true constant-current generator has an infinite internal resistance. Typical values of $1/h_{oe}$ are 50 kΩ to 100 kΩ, so that, for values of R_L of 5 kΩ or below, collector slope resistance can be ignored.

If we express g_m in terms of mean collector current I_C (from equation (6.11)), and evaluate voltage gain A, we have at room temperature

$$A = 40 I_C R_L, \tag{6.20}$$

where I_C is in mA and R_L in kΩ.

This makes the calculation of voltage gain simple, as long as we know the mean collector current. For example, if $I_C = 1$ mA and $R_L = 5$ kΩ, voltage gain $A = 200$. Now this calculation assumes that the a.c. component of the collector current, i_c is very tiny compared with the d.c. (mean) collector current, I_C. If this is not the case and there is considerable collector current swing then transconductance g_m varies, as we might expect, according to the instantaneous collector current. Thus, the amplifier will have a high voltage gain as the collector voltage swings near zero (high collector current) and a low voltage gain at the crest of the wave near cut-off (low collector current). The result is amplitude distortion, illustrated in fig. 6.14.

The same effect may be viewed in a different way as variation of input resistance h_{ie} with base current, h_{ie} being low for high base currents and high for low base currents. One cure of the distortion therefore is to swamp the h_{ie} changes with series resistance in the input circuit. An unbypassed emitter resistor will have a similar effect, introducing current feedback and increasing the input impedance. Voltage feedback can, of course, also be used to reduce the distortion. All these palliatives for the inherent non-linearity of the bipolar transistor unfortunately reduce circuit gain. One of the neatest solutions to the non-linearity problem is the differential amplifier, discussed in chapter 8, which forms the basis of most linear ICs.

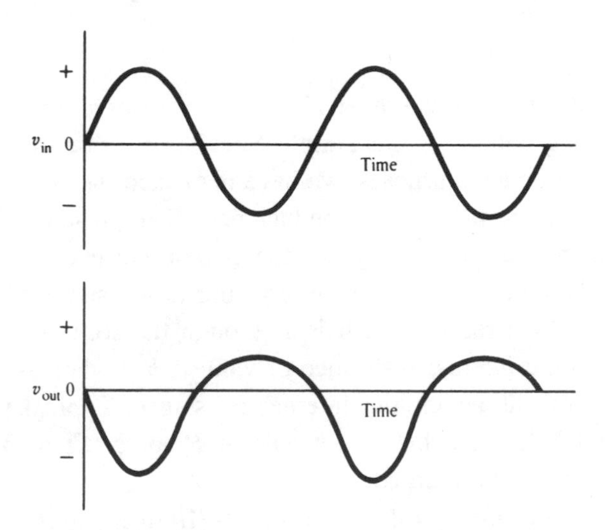

Fig. 6.14. Input and output waveforms showing distortion in common-emitter amplifier at large voltage swings. Input waveform is here magnified to assist waveshape comparison.

6.9 FET common source equivalent circuit and amplifier gain

Fig. 6.15 shows the essentials of a JFET amplifier and fig. 6.16 its equivalent circuit. Here, the input resistance is normally equal to the resistor value used

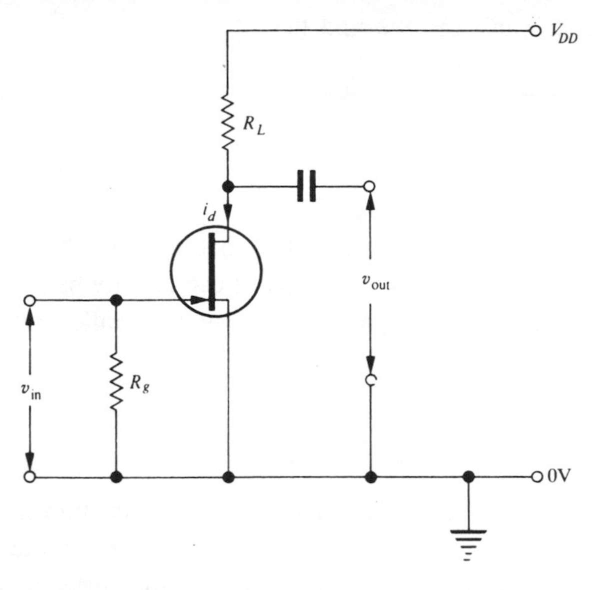

Fig. 6.15. Basic common-source amplifier (gate bias components omitted for clarity).

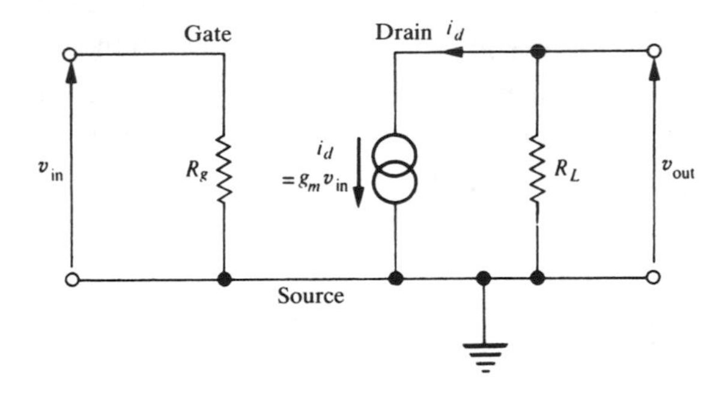

Fig. 6.16. Equivalent circuit of common-source amplifier.

in the gate circuit, the input resistance of the JFET itself being of the order of hundreds of megohms. The output equivalent circuit is similar to that of the bipolar transistor, the drain current being given by

$$i_d = g_m v_{gs} = g_m v_{in}$$

and the output voltage by

$$v_{out} = -i_d R_L = -g_m v_{in} R_L .$$

Thus, as with the bipolar transistor, we find that,

$$\text{voltage gain } A = \frac{v_{\text{out}}}{v_{\text{in}}}$$

$$\approx -g_m R_L .$$

The drain slope resistance of the FET (r_d, or $1/g_d$) is typically 100 kΩ, much higher than the usual range of values of R_L, so that it can usually be neglected. It can, however, be included in the equivalent circuit in parallel with R_L if required.

6.10 Variation in FET transconductance

In section 6.3 we saw that the transconductance of the bipolar transistor is a remarkably consistent parameter, being equal to $40I_C$ mA/V at room temperature if collector current I_C is in mA. Ignoring second-order effects such as base resistance, this simple relationship is true for all types of bipolar transistor and, unlike current gain h_{fe}, does not vary from one specimen to another. Now, with the FET, transconductance varies from one device type to another, usually lying between 0.5 mA/V and 5 mA/V. Furthermore, transconductance varies with drain current, as will be shown in the following calculation.

A theoretical treatment of the operation of the FET in the pinch-off region shows that drain current I_D bears a square-law relation to gate–source voltage V_{GS} according to the following equation:

$$I_D = I_{DSS}\left(1 - \frac{V_{GS}}{V_P}\right)^2 . \tag{6.21}$$

I_{DSS} is the drain current which flows when $V_{GS} = 0$. V_P is the pinch-off voltage, which is defined either as the gate–source voltage required to pinch-off the channel entirely so that I_D falls to zero (also known as the cut-off voltage) or as the drain–source voltage required to bring the device into the pinch-off region with $V_{GS} = 0$. It can be shown that these two definitions of V_P are equivalent and lead to the same result.

Now

$$g_m = \frac{dI_D}{dV_{GS}} .$$

Differentiating the I_D equation with respect to V_{GS} gives

$$g_m = \frac{-2I_{DSS}}{V_P}\left(1 - \frac{V_{GS}}{V_P}\right)$$

$$= \frac{-2\sqrt{I_{DSS}}}{V_P} \cdot \sqrt{I_D}. \qquad (6.22)$$

Thus, for a given FET, g_m is proportional to the square root of I_D: if $g_m = 1$ mA/V at $I_D = 1$ mA, we can expect $g_m = 3$ mA/V at $I_D = 9$ mA.

In voltage amplifiers using a FET, the drain current is swinging up and down with the a.c. signal, and g_m will be varying according to the square root of the instantaneous drain current. Thus, the position is similar to the bipolar transistor case, where g_m is directly proportional to instantaneous collector current. The result in both cases is a distorted output waveform if the signal is large, though the square-law response of the FET introduces only the second harmonic of the waveform, whilst the exponential response of the bipolar transistor produces the full gamut of harmonics. The non-linearity of the FET is thus inherently easier to 'tame' than that of the bipolar transistor.

7

Amplification at high frequencies

7.1 High-frequency considerations

Amplifiers designed to operate at frequencies above about 10 kilohertz (kHz) come under the general classification of high-frequency circuits, where the designer must allow for certain capacitance effects which are neglected at low frequencies. High frequencies, often called radio frequencies (RF), may be grouped into the following rough classifications:

10 kHz–100 kHz	Upper audio frequencies and very low frequency (VLF) broadcasting
100 kHz–2 megahertz (MHz)	Long and medium wave (MF) AM radio
2 MHz–30 MHz	Short-wave (HF) radio
30 MHz–300 MHz	VHF TV and FM radio
300 MHz–2 gigahertz (GHz)	UHF TV and mobile (cellular) telephones

Frequencies higher than 2 GHz (2000 MHz) are normally termed microwaves, and because of the extremely short wavelength, travel in straight lines like light. This property gives rise to their application in radar, where objects are detected by the reflection of microwaves.

Although their directional properties make them unsuitable for normal terrestrial broadcasting, microwaves are extensively used in the transmission of signals for telecommunications and satellite broadcasting. In these applications, the high frequency of microwaves makes available plenty of bandwidth and hence many signal channels from one transmitter, whilst the directional properties of the antennas can virtually eliminate external interference. Microwave circuits comprise an increasingly important branch of electronics, being applied widely in 'wireless' digital interconnections. They represent extreme cases of high frequency behaviour where precise interconnection dimensions are as critical as the components themselves. Whilst the

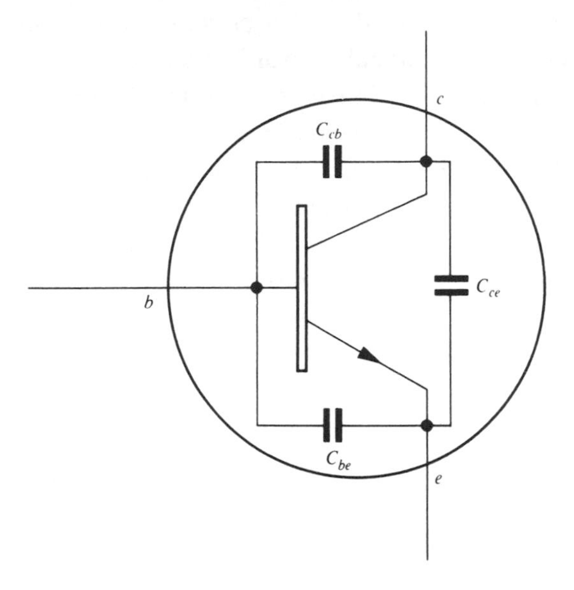

Fig. 7.1. Bipolar transistor, showing junction capacitances.

general RF principles outlined in this chapter apply equally to microwave techniques, practical microwave circuits lie outside the scope of this book.

Between 10 kHz and 2 GHz, conventional circuit techniques can be applied as long as the various capacitances inherent in transistors and the associated circuits are included in the circuit calculations. For instance, the properties of the bipolar transistor at high frequencies are largely determined by the three internal capacitances shown in fig. 7.1. The least serious of these is the collector–emitter capacitance, C_{ce}, being small in value, and readily included in the tuned collector load often employed in high-frequency amplifiers. As will be shown in the next section, C_{be} reduces high-frequency current gain, whilst C_{cb} is prominent in restricting the high-frequency performance of a voltage amplifier.

7.2 High frequencies and the bipolar transistor

7.2.1 Base–emitter capacitance

We saw in section 1.3.10 that a reverse-biased pn junction, such as a collector–base junction, behaves like a capacitor, where the capacitance is dependent on the junction area and the width of the depletion layer. A forward-biased pn junction such as a base–emitter junction also possesses capacitance, C_{be}, which appears in parallel with its normal forward resistance. The effective capacitance of a forward-biased pn junction arises from two basic causes. The first is simply the capacitance of the depletion layer, which is narrower in a forward-biased junction than in a reverse-biased one and therefore has a higher capacitance; the second component of capacitance arises from the finite speed of the minority carriers as they diffuse across the junction. These carriers, because the diffusion is relatively slow, appear to be temporarily stored in the semiconductor material when the external applied signal changes quickly; the effect as far as the external signal is concerned is akin to the storage of charge by a conventional capacitor. A sudden decrease in applied signal voltage means that the carriers in the base of the transistor must be removed before the collector current changes. The effective capacitance of the base–emitter junction of a small silicon transistor is typically of the order of 100 pF to 1000 pF; this capacitance has a considerable shunting effect on the base current at high frequencies and results in a fall in a.c. current gain (h_{fe}) at high frequencies.

7.2.2 *Transition frequency*

As signal frequency is increased, there comes a point where the reactance of the base–emitter capacitance is comparable with the base–emitter resistance h_{ie}, and much of the base current which should be performing the normal task of controlling the collector current is instead flowing in the base–emitter capacitance C_{be}. The result is a fall in current gain. The *cut-off* frequency, $f_{h_{fe}}$, is reached when the reactance of C_{be} is equal to the input resistance, h_{ie}, and the current gain thus falls by the factor $1/\sqrt{2}$ (3 dB). Above $f_{h_{fe}}$ most of the 'base' current is actually flowing in C_{be} and every time the frequency is doubled the current gain falls by half. Expressed on a logarithmic scale, h_{fe} falls by 6 dB for each octave (doubling) of frequency. If the graph is extrapolated, we can deduce the frequency at which h_{fe} falls to unity; this is termed the *transition frequency*, f_T, above which the transistor is of little use as an amplifier. Transition frequency f_T is also called the gain–band-width product, since, in the region between $f_{h_{fe}}$ and f_T, the product of current gain and signal frequency is roughly constant and equal to f_T, i.e.

$$f_T \approx h_{fe0} \times f_{h_{fe}}. \qquad (7.1)$$

A plot of h_{fe} against frequency for a typical small silicon transistor is shown on log scales in fig. 7.2. Transition frequency f_T is here 200 MHz and cut-off frequency $f_{h_{fe}}$ is 1 MHz.

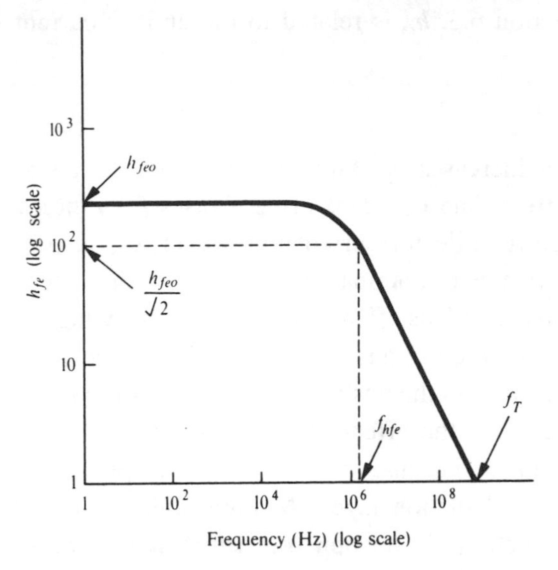

Fig. 7.2. Graph of small signal current gain h_{fe} against frequency f. Cut-off frequency (f_{hfe}) and transition frequency (f_T) are shown.

7.2.3 Variation of transition frequency with collector current

A simple equivalent circuit for the input of a transistor at high frequencies is shown in fig. 7.3. Clearly, the greater the proportion of base current i_b which flows in C_{be} and the less the proportion which actually flows in the junction resistance h_{ie}, the lower will appear the current gain of the transistor. In other words, for good high-frequency performance, the reactance of C_{be} should be greater than h_{ie}: or,

$$\frac{1}{2\pi f C_{be}} \gg h_{ie}.$$

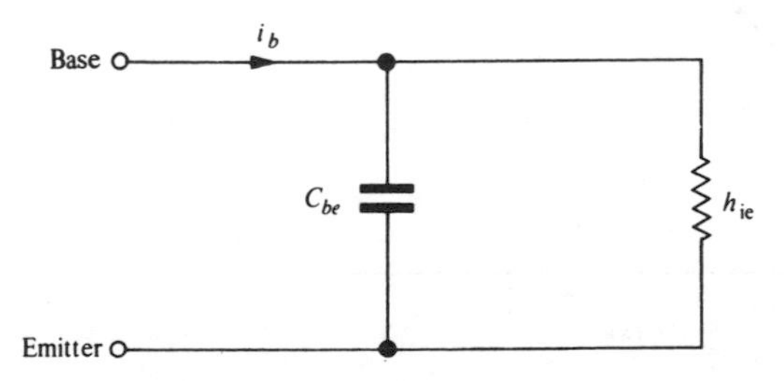

Fig. 7.3. Equivalent circuit of input of a common-emitter amplifier showing base-emitter capacitance.

As we have seen in section 6.3, h_{ie} is related to the emitter current I_E:

$$h_{ie} \approx \frac{25h_{fe}}{I_E}\ \Omega. \qquad [(6.6)]$$

Thus, as emitter current increases, h_{ie} falls and therefore carries a greater proportion of the base current than C_{be}. This effect causes f_T to increase with emitter current (and therefore collector current). However, C_{be} also increases with collector current and begins to dominate the picture when a phenomenon known as *base stretching* occurs. This effect occurs at relatively high collector currents (≈50 mA in a small silicon transistor) and is due to the influence of the base region minority carriers on the position of the collector–base depletion layer. At low collector currents, the effective thickness of the base is much less than its physical thickness because of the space occupied in the base region by the collector–base depletion layer. At high collector currents, the flow of mobile charges (electrons in an npn transistor) across the base and through the depletion layer is sufficient to upset the charge distribution equilibrium and cause the whole depletion layer to move deeper into the collector and away from the emitter. Thus, the base becomes effectively thicker and carrier transit time is increased. This appears as an increase in C_{be}, which then begins to dominate the high-frequency picture and causes f_T, after its initial

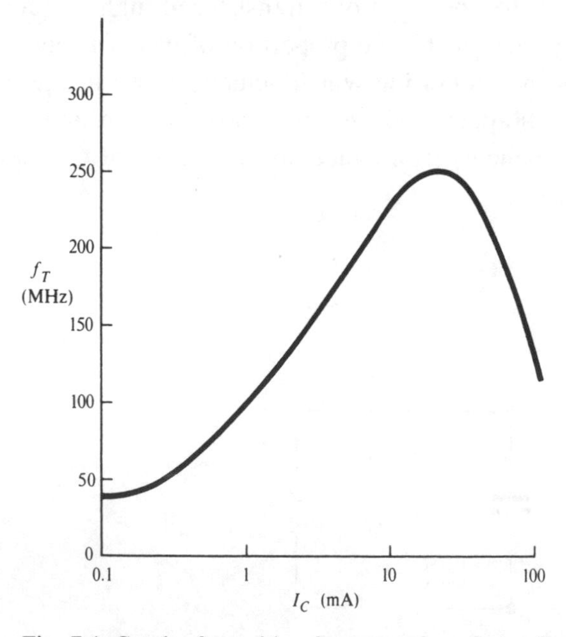

Fig. 7.4. Graph of transition frequency f_T against collector current I_C.

rise with increasing collector current, to fall off again at high collector currents. The resultant plot of f_T against collector current I_C for a small silicon transistor is shown in fig. 7.4, where it can be seen that optimum high-frequency performance is obtained at collector currents between 5 mA and 50 mA.

7.3. Transistor circuit performance at high frequencies

7.3.1 Voltage amplifiers and Miller effect

So far we have seen that the current gain of the bipolar transistor falls at high frequencies because of the high base–emitter capacitance. The capacitance of the collector–base junction, C_{cb}, must now be considered. Because C_{cb} is very much smaller than C_{be} (typically 5 pF) it plays a negligible part in the high-frequency behaviour of current gain, where the collector voltage is assumed constant. In a voltage amplifier, however, the collector swings with the output signal and C_{cb} becomes much more significant than might at first be imagined. The problem with C_{cb} is that it appears between the input and output terminals of the amplifier and, as is shown below, when seen from the input terminals, C_{cb} appears to be very much larger than its actual value.

Fig. 7.5(a) shows an amplifier of voltage gain $-A$, the negative sign

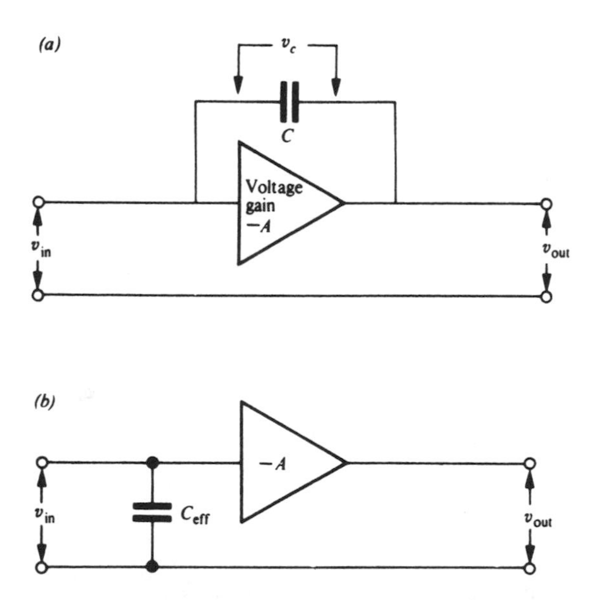

Fig. 7.5. The Miller effect. (*a*) Voltage amplifier with capacitance C between input and output. (*b*) Equivalent circuit where C appears as $C_{eff}(\approx CA)$ across input.

indicating the usual phase reversal between output and input. Capacitor C is connected between output and input and represents the feedback capacitance which occurs between collector and base of a common-emitter amplifier or between drain and gate of a common-source amplifier. We are interested in the effective value of C as it appears looking into the amplifier input (C_{eff} in fig. 7.5(*b*)). We can deduce the value of C_{eff} by applying an e.m.f., v_{in}, to the input terminals and measuring the total charge, q, which flows into the circuit as a result.

$$\text{Since capacitance} = \frac{\text{charge on capacitor}}{\text{p.d. at capacitor terminals}}$$

$$C_{eff} = \frac{q}{v_{in}}. \tag{7.2}$$

Now in fig. 7.5(*a*) the charge q stored in the circuit is given by the product of the voltage across C, v_c, and the value of C, i.e.

$$q = v_c C.$$

Now

$$v_c = v_{in} - v_{out}.$$

But

$$v_{out} = -Av_{in}$$

therefore

$$v_c = (A+1)\, v_{in}$$

and

$$q = C(A+1)\, v_{in}. \tag{7.3}$$

Combining (7.2) and (7.3),

effective capacitance

$$C_{eff} = \frac{C(A+1)\, v_{in}}{v_{in}}$$

therefore

$$C_{eff} = C(A+1) \tag{7.4}$$

and if $A \gg 1$,

$$C_{eff} \approx CA. \tag{7.5}$$

Thus we have the *Miller effect*, where a capacitor connected between the input and output of an inverting amplifier appears, as far as the input signal is concerned, as though its value were increased by a factor equal to the amplifier gain.

7.3.2 Common-emitter amplifier at high frequencies

In the common-emitter amplifier, Miller effect adds to the effective value of the base–emitter capacitance by AC_{cb}, where A is the amplifier voltage gain ($A \approx g_m R_L$), and C_{cb} is the collector–base capacitance shown in the simplified equivalent circuit of fig. 7.6. The result is the equivalent circuit of fig. 7.7, where the total effective capacitance, C_T, in shunt with the input is given by

$$\begin{aligned} C_T &= C_{be} + AC_{cb} \\ &= C_{be} + g_m R_L C_{cb}. \end{aligned}$$

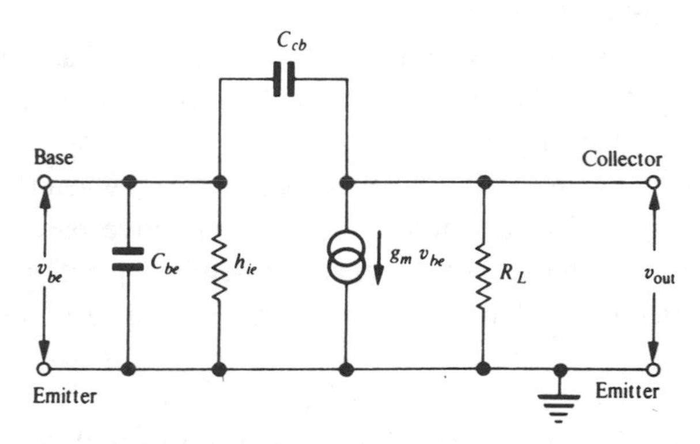

Fig. 7.6. Equivalent circuit of common-emitter voltage amplifier showing C_{be} and C_{cb}.

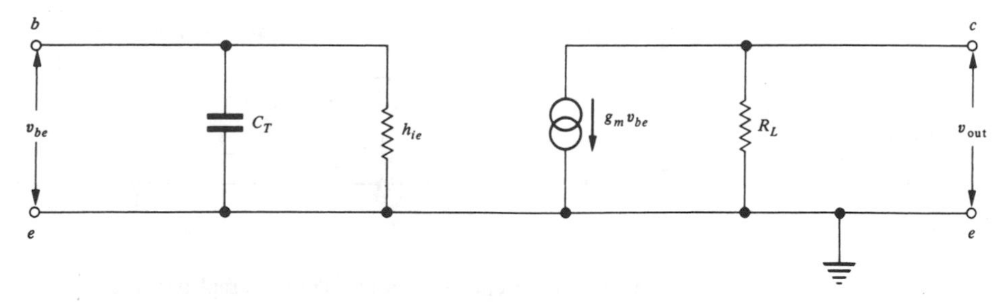

Fig. 7.7. Circuit of fig. 7.6 with total effective capacitance shown as C_T.

Thus the overall effect of internal capacitances in the common-emitter amplifier is to reduce the input impedance Z_{in} at high frequencies, given by the parallel impedance calculation

$$\frac{1}{Z_{\text{in}}} = \frac{1}{h_{ie}} + j\omega C_T$$

therefore

$$|Z_{\text{in}}| = \frac{h_{ie}}{\sqrt{(1 + \omega^2 C_T^2 h_{ie}^2)}}.$$

It might therefore seem that, as long as the output impedance of the signal source is very low then the actual measured voltage gain of the common-emitter amplifier will not deteriorate greatly at high frequencies. This is true to a limited extent, but, unfortunately, the falling input impedance means that the amplifier is giving progressively less *power* gain as the frequency increases. It is, after all, its ability to provide power gain which distinguishes an amplifier from a step-up transformer.

Even given a signal source of low output impedance, the voltage gain of a common-emitter amplifier falls with increasing frequency because of the inherent resistance of the base layer, $r_{bb'}$ (fig. 6.3). We have so far neglected $r_{bb'}$ in our high-frequency treatment, but it becomes significant when the input signal comes from a low resistance source. The total effective base–emitter capacitance, C_T, appears, of course, at the actual base–emitter junction, whilst the input voltage v_{be} can only reach the junction via the internal base resistance $r_{bb'}$.

A more precise equivalent circuit which divides h_{ie} into $h_{fe}r_e$ and $r_{bb'}$ is shown in fig. 7.8, where it is clear that $r_{bb'}$ appears in series with the output

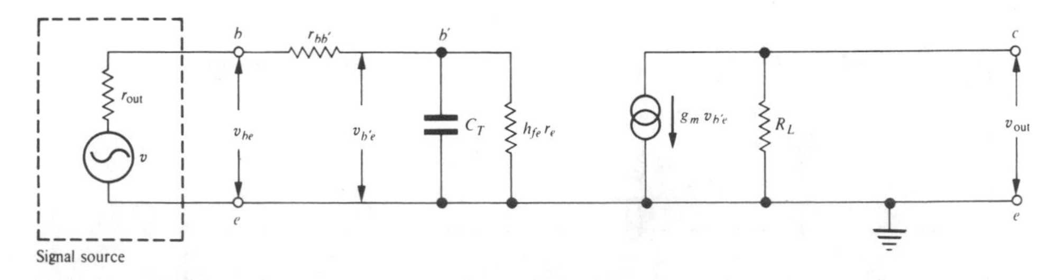

Fig. 7.8. More precise equivalent circuit of common-emitter amplifier. Input resistance h_{ie} is split into its component parts $r_{bb'}$ and $h_{fe}r_e$. Signal source with its internal resistance r_{out} is also shown.

Table 7.1 Mobility values

	Electrons	Holes
Si	0.14	0.05 ($m^2\ s^{-1}\ V^{-1}$)
Ge	0.39	0.19
GaAs	0.85	0.05

impedance, r_{out}, of the generator. At high frequencies, the total series resistance $r_{out} + r_{bb'}$ appears with effective base–emitter capacitance C_T as a low-pass filter, which produces a loss of signal amplitude and a phase shift. This type of RC circuit is discussed in detail later in this chapter (section 7.7).

To summarise, for a good high-frequency performance a transistor should be designed with a low base–emitter capacitance, C_{be}, a low base–collector capacitance, C_{cb} and a low internal base resistance $r_{bb'}$. Small physical size, using surface-mount technology (SMT), and especially within integrated circuit structures, helps keep capacitance values small. We have already noted that C_{be} arises chiefly because of the relatively slow movement of carriers in the base region. Carrier velocity depends on the electric field present and the *mobility* of the carrier in the semiconductor material. Mobility is defined as the carrier velocity (m/s) imparted by unit field strength (1 V/m). Mobility values for electrons and holes in silicon (Si), germanium (Ge) and gallium arsenide (GaAs) at room temperature are as shown in Table 7.1; the units being m/s per V/m.

Thus it is clear that an npn transistor, where the current crossing the base region consists of electrons, may be expected to have a better high-frequency response than a pnp device, where holes, which have a lower mobility, are the main carriers. Whilst silicon devices are generally preferred because of their low leakage current and ease of fabrication, it is interesting to note that germanium is potentially a faster performer, whilst the remarkable speed of gallium arsenide (GaAs) devices is due to the exceptionally high electron mobility.

7.3.3 High frequencies and the emitter follower

An emitter follower output stage is frequently used to preserve good high-frequency response where signals are fed along a cable. As already noted in section 5.12.7, high frequencies can suffer attenuation because of the shunting effect of the cable capacitance. Feeding the cable via an emitter follower minimizes this high-frequency attenuation: the low output impedance can drive high-capacity loads without significant potential drop. It is thus of interest to look at the high-frequency performance of the emitter follower itself.

We know already from chapter 5 that the most important characteristic of the emitter follower is not its voltage gain, which is nearly unity, but the ratio of its input impedance to its output impedance, given by the transistor current gain, h_{fe}.

It is therefore current gain which is the chief factor in determining emitter follower performance and it is only when this parameter falls in value at high frequencies that the input–output impedance transformation will be reduced. The performance of an emitter follower begins to deteriorate significantly above transistor 'cut-off' frequency, $f_{h_{fe}}$, where the current gain has fallen by 3 dB, or a factor of $1/\sqrt{2}$.

7.4 FETs at high frequencies

FETs are potentially capable of a very good high-frequency performance. We have seen that one of the limiting factors on the high-frequency response of the bipolar transistor is the slowness of dissipation of minority carriers in the base region, giving rise to the relatively high diffusion capacitance between base and emitter. There is no analogous effect in the FET and we need only consider the various 'natural' capacitances of the device. Fig. 7.9 shows a

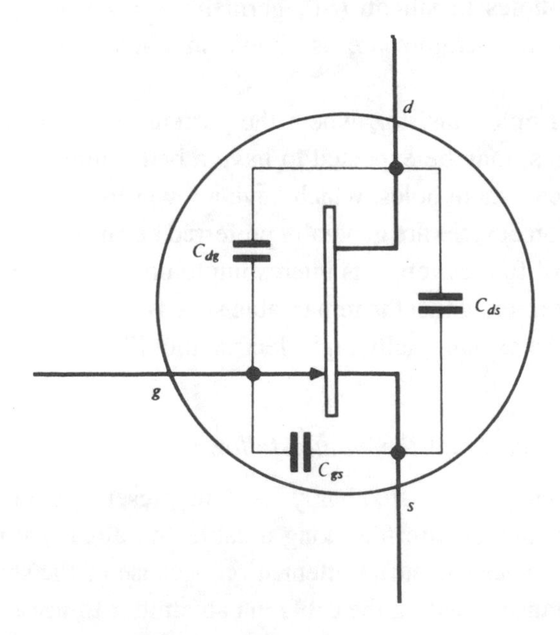

Fig. 7.9. JFET showing internal capacitances.

JFET with these capacitances, C_{gs}, C_{ds}, C_{dg} drawn in schematically; they are sometimes known as C_{iss}, C_{oss}, C_{rss} respectively, indicating short circuit input, output and reverse (feedback) capacitance. Typical values are as follows:

$$C_{gs} \approx 5\ \text{pF},$$

$$C_{ds} \approx 2\ \text{pF},$$

$$C_{dg} \approx 1\ \text{pF}.$$

Because of the Miller effect, C_{dg}, although the smallest in value, is the most important influence on high-frequency response. Fig. 7.10 shows a basic common-source amplifier equivalent circuit with the three capacitances included, and fig. 7.11 shows the effective value of C_{dg} lumped together with C_{gs} in the input circuit as C_T:

$$C_T = C_{gs} + (A+1)\,C_{dg},$$

where A is amplifier voltage gain.

Thus the dominant effect at high frequencies is the falling input impedance, given by $1/j\omega C_T$, which may amount to only a few tens of ohms at 100 MHz.

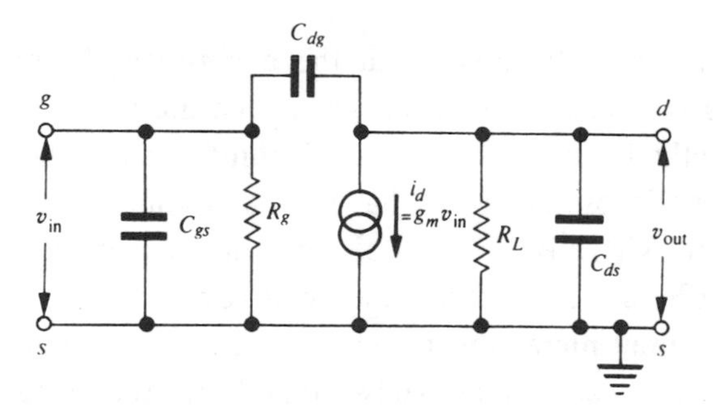

Fig. 7.10. High-frequency equivalent circuit of common-source amplifier.

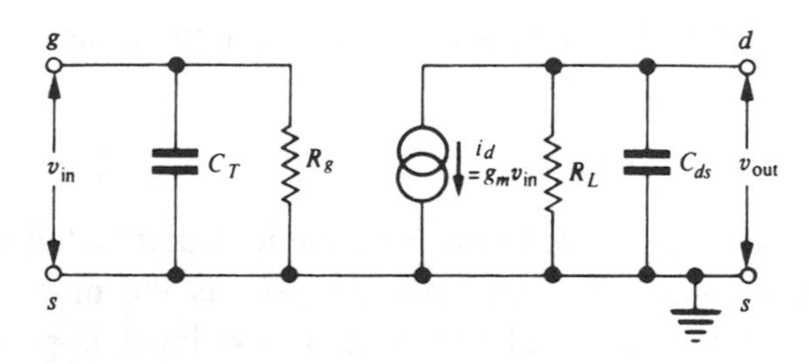

Fig. 7.11. Common-source amplifier with total effective input capacitance shown as C_T.

The effect of C_{ds} may usually be compensated by incorporating it as part of a tuned circuit in the drain; this avoids the shunting effect that it would have on a resistive load.

Internal capacitance is not the only factor which affects high-frequency gain. Precise calculations must take account of the fact that transconductance g_m, drain slope resistance r_d, and gate–source resistance r_{gs} all drop in value at high frequencies. At 100 MHz, transconductance and r_d may fall to half their low-frequency value and gate–source resistance may be only 10 kΩ instead of the hundreds of megohms expected at low frequencies. Capacitance, however is the chief reason for reduced high-frequency gain.

7.5 Special circuits for high frequencies

7.5.1 General

The single most significant barrier to useful high-frequency amplification is the Miller effect, whereby any capacitance existing between output and input appears effectively amplified by the voltage gain of the circuit. The circuits to be discussed all attempt to minimize Miller effect by reducing the effective feedback capacitance.

As a practical point, it should be noted that the experimental testing of high-frequency amplifiers is a specialized task, where the normal generator–amplifier–oscilloscope method can give very misleading results, not least because only the more exotic oscilloscopes have a flat frequency response above 50 MHz. In addition, virtually all oscilloscopes have an input capacitance in the region of 30 pF, and connecting leads will add to this; even with special RF scope probes some interaction between testgear and circuit is to be expected. The circuits in this section are shown chiefly for interest and not in the expectation that the reader can make accurate tests on them at high frequencies. Those willing to experiment, however, should be able to get useful amplification of radio or TV signals by using the appropriate circuit.

7.5.2 Common-base amplifier

The use of a bipolar transistor in the common-base mode, whereby the signal is fed into the emitter, provides the same voltage gain as the more usual common-emitter circuit. It has the disadvantage of a low input impedance, this being a factor of h_{fe} lower than in the common-emitter mode; the power gain is therefore lower by the same factor.

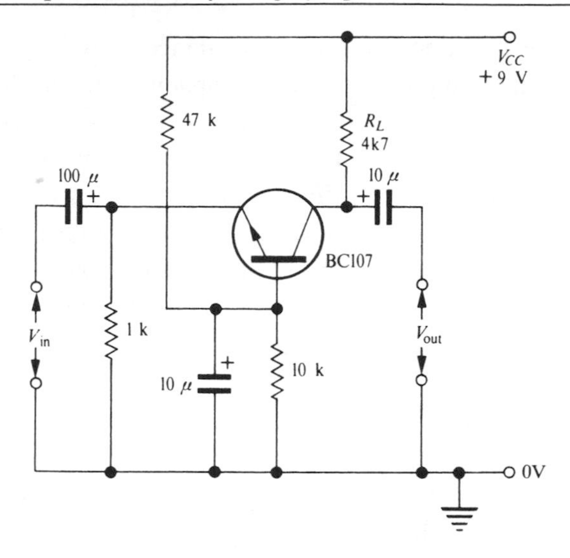

Fig. 7.12. Voltage amplifier with transistor in common-base mode.

The advantage of the common-base circuit at high frequencies is that the base acts as a screen between input and output and thus reduces the Miller effect to negligible proportions. The high-frequency fall-off is also minimized by the inherently low input impedance, the reactance of the base–emitter capacitance being less significant than in the case of the common-emitter amplifier.

Fig. 7.12 shows a practical common-base circuit. The calculation of component values for correct d.c. conditions is exactly the same as in the fully stabilized common-emitter amplifier of fig. 1.20. The stage operates with 1 mA quiescent collector current, the base being held at the appropriate potential to ensure that 1 V is dropped across the 1 kΩ resistor in the emitter. The base is decoupled to earth; i.e. all a.c. signals are bypassed via the 10 μF capacitor straight to earth. The output signal is taken from the 4.7 kΩ collector load in the usual way.

Fig. 7.13 shows a simplified equivalent circuit of the common-base stage, including internal capacitances. The first point to note is that input signal v_{in} feeds straight into the low emitter impedance, r_e, where

$$r_e \approx \frac{25}{I_E} \Omega. \qquad [(6.5)]$$

(I_E is the emitter current in mA.)

Secondly, the capacitance shunting the input impedance is only C_{be} and is not augmented by any Miller-amplified feedback capacitance. Thirdly the direction of the output current generator is the opposite to the common-emitter case. The result is that input and output voltage signals are in phase, unlike the common-emitter amplifier where they differ by 180°.

The practical effect of the low input impedance and negligible feedback capacitance in the common-base amplifier is that it gives useful voltage amplification up to frequencies as high as the transition frequency, f_T, of the transistor. The common-emitter amplifier on the other hand, is only useful up to $f_{h_{fe}}$ ($\approx f_T/h_{fe}$).

Although the low input impedance of the common-base amplifier can be a disadvantage in some circumstances, it is of the right order to terminate correctly a VHF or UHF aerial and feeder cable (approximately 70 Ω). The common-base stage may therefore be usefully employed as a signal pre-amplifier for weak television signals. Fig. 7.14 shows a circuit which can be used as a pre-amplifier for UHF television signals, giving at least 10 dB improvement in signal strength.

Here the BF180 silicon npn transistor is used; it is specifically intended for this sort of job with f_T approaching 700 MHz. The circuit is basically similar to that of fig. 7.13, except that a resonant circuit is used as a collector load. This is common practice in high-frequency amplification because the high impedance at resonance leads to a high effective R_L and thus a relatively high gain. A further advantage is that the collector–base capacitance and other stray capacitances on the output side do not act as attenuation but simply serve as a contribution to the tuning capacitance.

Because of the high frequencies involved (in the region 400–800 MHz) the

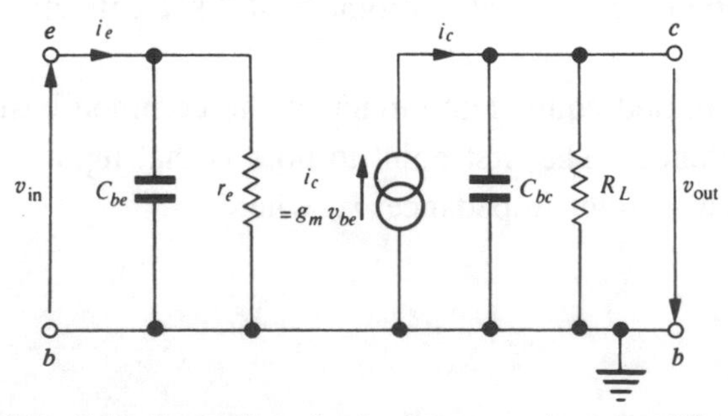

Fig. 7.13. Equivalent circuit of common-base amplifier.

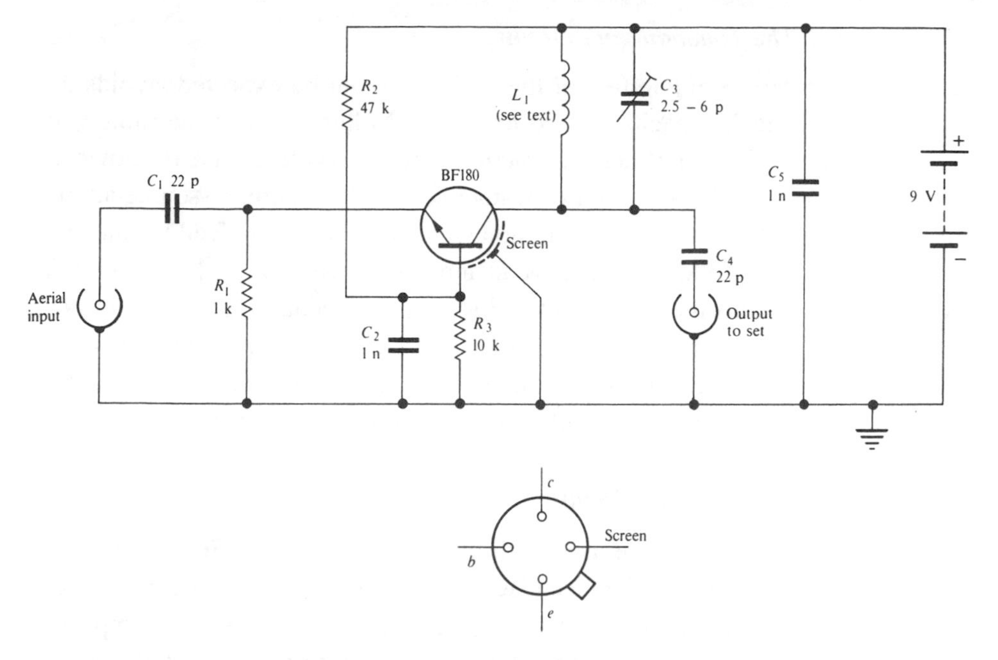

Fig. 7.14. Aerial pre-amplifier for UHF television signals.

inductance L_1 is only about 3 cm of 18–20 SWG wire bent into a semicircle. The output circuit has a low Q and therefore broad tuning; a setting for trimmer capacitor C_3 should be found whereby all local television channels receive similar amplification.

To avoid unwanted feedback from output to input, which can result in instability (unwanted oscillation), it is advisable to fit a metal screen between the input and output sections and enclose the whole circuit in a metal box. The screen should ideally run close to the transistor can and screen the emitter lead from the collector lead. The 1 nF decoupling capacitors may with advantage be of the 'feed-through' type which decouples a connection as it passes through a metal screen, the capacitor being mounted in the screen. This ensures that the decoupling occurs at exactly the right place and that no opportunity exists for the induction of unwanted signals in the wire.

The description of the inductor for this circuit emphasizes that, at frequencies of a few hundred MHz, a short piece of wire may appear as a significant impedance. Needless to say, all connections should be kept as short as possible.

7.5.3 The common-gate circuit

The common-gate configuration of the FET, as might be expected, avoids the drain–gate feedback capacitance and hence the Miller effect in the same way as the common-base amplifier. A practical common-gate circuit is shown in fig. 7.15, which will be seen as a rearrangement of the common-source amplifier of fig. 2.7. As far as the input signal is concerned, it is feeding into the output of a source follower and sees an input impedance $Z_{in} = 1/g_m$, which is typically 500 Ω; thus the advantages of high input impedance normally associated with the FET is thrown away in the common-gate mode. This fact, combined with a poor signal-to-noise ratio at high frequencies results in the common-gate amplifier finding little application in its own right.

7.5.4 The cascode circuit

This technique, which can be used with bipolar transistors or FETs is another way of effectively eliminating feedback capacitance, and thus Miller effect. In the FET case, the device is used in the usual common-source amplifier circuit, but, instead of the load and output terminal being connected straight to the drain, a second FET is interposed, this second device operating in common-gate mode. The technique is illustrated in fig. 7.16, which is a cascode amplifier designed as a useful pre-amplifier for Band II VHF (FM) radio (88–108 MHz) giving approximately 20 db gain. The lower FET, T_1, is acting as a common-source amplifier, but, instead of seeing a high impedance in its drain circuit, it sees the source of T_2, which appears as an impedance of $1/g_m$, T_2 being in common-gate mode.

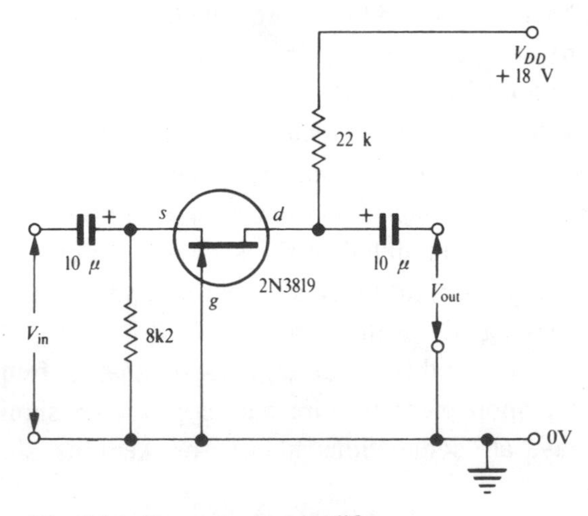

Fig. 7.15. Common-gate amplifier.

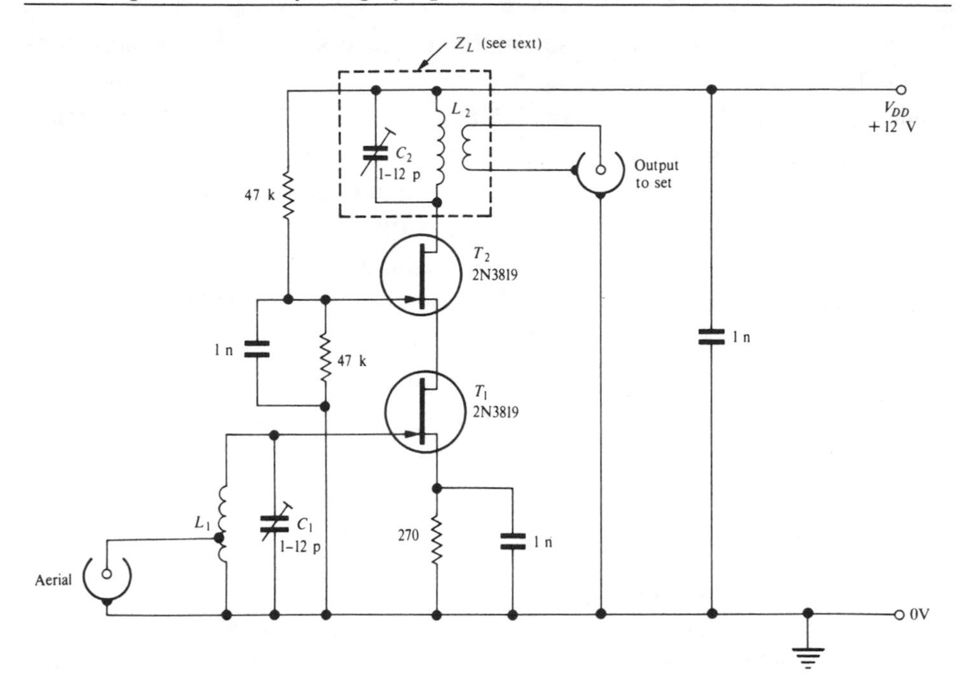

Fig. 7.16. Pre-amplifier for VHF radio using JFETs in cascode connection.

Thus, the voltage gain of T_1

$$\approx -g_m R_L$$

$$\approx -g_m \frac{1}{g_m} \quad \text{(assuming both FETs have same } g_m\text{)}$$

$$\approx -1.$$

Because of this unity voltage gain, there is a negligible signal voltage across the drain–gate capacitance of T_1 and no significant Miller effect.

The output load Z_L is in the drain circuit of T_2 and is, in this particular case, a tuned circuit to give a high impedance at the required frequency. Now, since the drain current of T_2 must be equal to the drain current of T_1, the output signal voltage is obtained at the drain of T_2, giving a voltage gain for the whole stage equal to $g_m Z_L$ in the usual way. Thus, it is seen that in a cascode stage the upper device, T_2, serves to hold the drain voltage of the lower device, T_1, constant, whilst at the same time faithfully conducting the drain current swing to the load Z_L. The normal common-source voltage gain and high input impedance is therefore obtained without the attendant high-frequency loss from the feedback capacitance of T_1, which has negligible signal across it.

Just as with the common-base amplifier, the cascode stage is best constructed with a screen between the two transistors to avoid coupling between input and output. To tune the amplifier to Band II (FM) radio frequencies, coil L_1 should have six turns of 18 SWG enamelled copper wire 1 cm in diameter. The aerial tap, which avoids excessive damping on the coil from the low aerial impedance, is one turn up from the earthy end. L_2 in the drain circuit has six turns of 1 cm diameter and has a coupling coil of one turn interwound at the supply rail end. Tuning to select the station required is achieved by adjusting the trimming capacitors C_1 and C_2. C_1 has the sharper tuning because the high input impedance of T_1 allows a high Q in the input tuned circuit. L_2 and C_2 are damped by the drain resistance, r_d, of T_2 and exhibit much broader tuning.

7.5.5 The dual-gate MOSFET

From the cascode circuit of fig. 7.16 it is a relatively small step to view the two FETs as one device with a single channel and two gates. This is the dual-gate MOSFET sometimes used in high-frequency amplification. Such a device may be readily substituted into fig. 7.16 as is shown in fig. 7.17. The

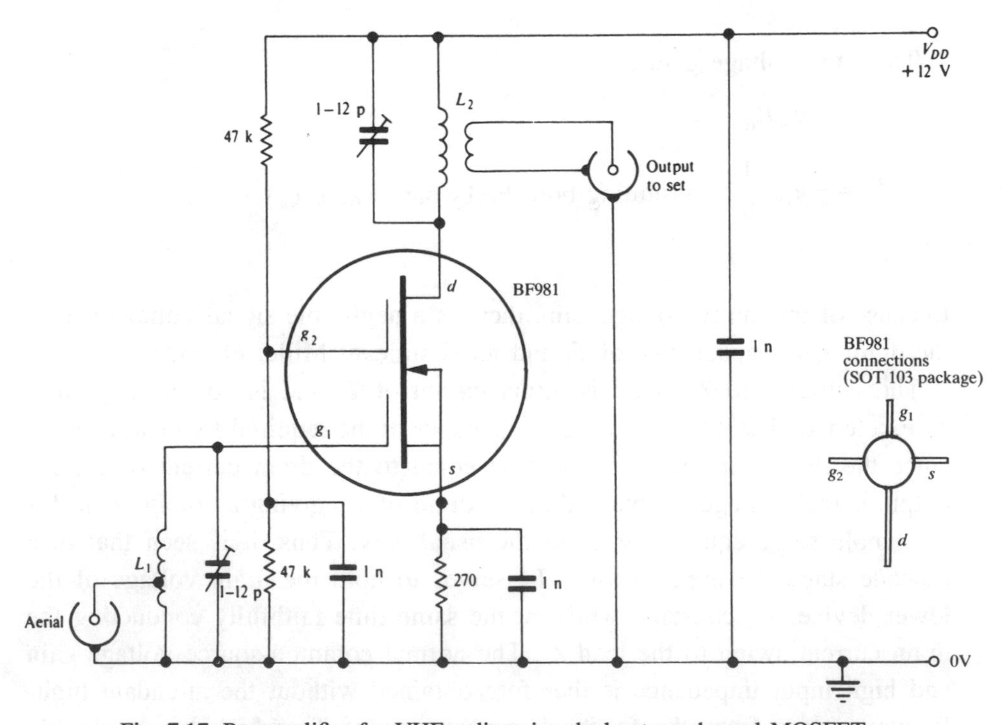

Fig. 7.17. Pre-amplifier for VHF radio using dual-gate n-channel MOSFET

dual-gate MOSFET is the solid-state equivalent of the tetrode (screen grid) valve; gate g_2 serves as an electrostatic screen between drain and gate g_1, thus drastically reducing feedback capacitance to a value as low as 0.02 pF, compared with 1 pF for the 'triode' FET. Coil details in fig. 7.17 are identical to those in fig. 7.16; the remarks about screening apply to both circuits and, in fact, to all R.F. work. Printed-circuit boards (PCBs) for R.F. applications are normally designed with an earthed copper *ground plane* on the opposite side of the board from the signal tracks, which assists in screening.

7.6 Wideband high-frequency amplifiers

There are some applications where the technique of using tuned circuits to boost gain at the required frequency, and at the same time 'mop up' stray capacitance as part of the resonant circuit, cannot be used. A common example is the Y-amplifier in an oscilloscope, which is commonly required to have a flat frequency response from d.c. to 100 MHz. Another example is the video amplifier handling TV picture signals which must have a bandwidth of 6 MHz. In such a case, a resistive collector or drain load must be used in each stage and this will cause problems where stray capacitance is concerned (fig. 7.18), even though the aforementioned techniques for avoiding Miller effect are fully exploited. The output stray capacitance, C_S, will have a severe attenuating effect in conjunction with R_L by reducing the effective load impedance (it is relevant to note here that a capacitance of 10 pF has a reactance of only 1600 Ω at 10 MHz).

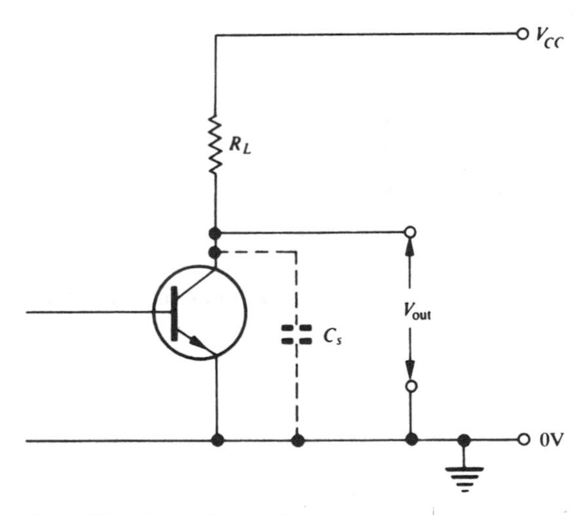

Fig. 7.18. The problem of stray capacitance at the output of a wideband amplifier.

One way to obtain a flat frequency response at high frequencies is simply to use a low value of load resistance, of the order of 100 Ω instead of 10 kΩ, but this involves cutting low-frequency gain in exchange for bandwidth.

An alternative, which allows a high load resistance without excessive high-frequency loss is to connect a small inductor in series with R_L (fig. 7.19). As the frequency is increased, the reactance of C_S falls, but it is balanced by the rise in reactance of the inductor L, until the two reactances are equal and resonance occurs. The resonant peak, although heavily damped by R_L, serves, in a well-designed circuit, to maintain the gain near the high frequency limit of the circuit. Since the inductor is effectively in parallel with the stray capacitance, this technique is known as *shunt peaking*.

Typical frequency response curves of a circuit with and without shunt peaking are shown in fig. 7.20. The appropriate value of L normally lies in the range between a microhenry and a millihenry, depending upon the value of R_L, the bandwidth required, and the magnitude of the stray capacitance. Final choice of L is often made by experiment.

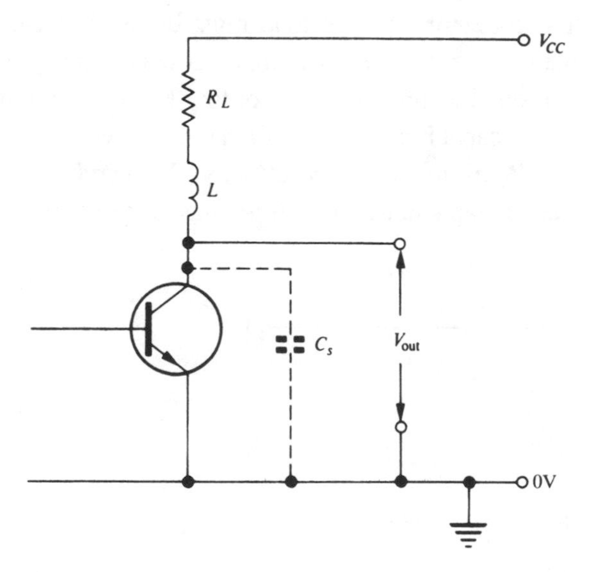

Fig. 7.19. Amplifier with shunt-peaking inductor, L, to extend high-frequency response.

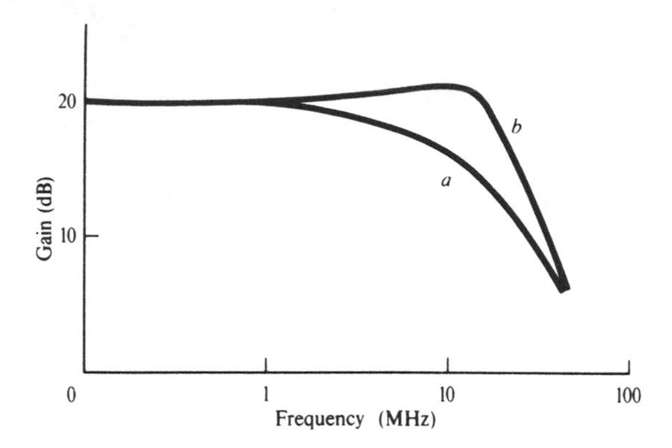

Fig. 7.20. Typical video amplifier frequency response: (*a*) without shunt peaking, (*b*) with shunt peaking.

7.7 Amplitude and phase response of low-pass filter

Fig. 7.21 shows the basic configuration of resistance and capacitance which we have seen is responsible for the decline in gain of amplifiers at high frequencies. Because low frequencies are transmitted and high frequencies are attenuated, such an arrangement is called a *low-pass* filter. Because it contains only one RC section it is known as a *first-order low pass* filter. The voltage 'gain' of the circuit is, of course, less than unity, and is given by

$$A = \frac{V_{out}}{V_{in}} = \frac{1/j\omega C}{(1/j\omega C)+R} = \frac{1}{1+j\omega CR}, \tag{7.6}$$

where $1/j\omega C$ is the reactance of C and $\omega = 2\pi f$, where f is the frequency.

Now, comparing the magnitudes of input and output and neglecting phase, the magnitude of the gain is

$$|A| = \left|\frac{V_{out}}{V_{in}}\right| = \frac{1}{\sqrt{(1+\omega^2C^2R^2)}}, \tag{7.7}$$

Fig. 7.21. Single RC low-pass filter.

and when $\omega^2 C^2 R^2 = 1$,

$$|V_{out}| = \frac{|V_{in}|}{\sqrt{2}};$$

this is the half-power or −3 dB point. Let the half-power frequency be f_1, then

$$4\pi^2 f_1^2 C^2 R^2 = 1$$

and

$$CR = \frac{1}{2\pi f_1}. \tag{7.8}$$

Now, substituting for CR in the original relation,

$$A = \frac{V_{out}}{V_{in}} = \frac{1}{1 + j(2\pi f / 2\pi f_1)}$$

$$= \frac{1}{1 + j(f/f_1)}. \tag{7.9}$$

This gives the amplitude response

$$|A| = \left|\frac{V_{out}}{V_{in}}\right| = \frac{1}{\sqrt{[1 + (f^2/f_1^2)]}}. \tag{7.10}$$

It is this last relation which gives an amplifier response curve its familiar shape of a flat portion followed by a steady decline. If $f \ll f_1$, then $A \approx 1$; if $f \gg f_1$, then $A \approx f_1/f$ and the gain is inversely proportional to frequency. Therefore every time the frequency doubles (increases by one octave) the voltage gain is halved (falls by 6 dB). When the frequency increases by a factor of ten, the gain drops by 20 dB. Thus, we find that, for this first-order RC circuit, the response curve at high frequencies is asymptotic to a slope of −6 dB/octave or −20 dB/decade. This is known as a first-order roll off, and is shown in fig. 7.22. The half-power frequency f_1 comes at the intersection of the −6 dB/octave asymptote and the 0 dB level; the further knowledge that the response is 3 dB down at f_1 enables the curve of fig. 7.22 to be drawn.

In section 4.6 in the chapter on negative feedback, it was pointed out that, for feedback to remain negative over the whole frequency range, phase shift must not approach 180°. It is therefore useful to look at the way the phase of the output signal from an amplifier changes with frequency and relate this to the variation in amplitude.

We can now return to equation (7.9),

$$A = \frac{V_{out}}{V_{in}} = \frac{1}{1 + j(f/f_1)},$$

and calculate the phase relationship between input and output at different frequencies, given that the phase angle ϕ can be obtained from the equation

$$\tan\phi = \frac{\text{imaginary part of } A}{\text{real part of } A}.$$

Equation (7.9) can be rewritten:

$$A = \frac{1 - j(f/f_1)}{1 + (f^2/f_1^2)} \qquad (7.11)$$

At low frequencies, $f \ll f_1$ and the imaginary term disappears, giving zero phase shift. When $f = f_1$, the real and imaginary parts are equal, giving a phase lag, ϕ, of 45° ($\tan\phi = 1$).

The phase response at higher frequencies is most readily calculated by multiplying top and bottom of equation (7.9) by $-jf_1/f$, giving

$$A = -j\frac{f_1}{f}\left(\frac{1}{1 - j(f_1/f)}\right). \qquad (7.12)$$

Then, when $f \gg f_1$, $[1 - j(f_1/f)] \approx 1$, therefore

$$A \approx -j\frac{f_1}{f}. \qquad (7.13)$$

The real term has disappeared, giving $\tan\phi \to -\infty$ and $\phi \approx -90°$.

With these results, we can draw the phase characteristics of the first-order low-pass filter (fig. 7.23). The two curves of fig. 7.22 and 7.23 together constitute a very useful description of the amplitude and phase performance of a circuit at various frequencies and hence are often drawn as a pair, when they are known as a *Bode plot*.

The two sets of information are sometimes combined into a polar graph called the *Nyquist plot*, much used in the design of servo-systems and other feedback applications. The Nyquist plot corresponding to the Bode plot of figs. 7.22 and 7.23 is shown in fig. 7.24. Polar coordinates are used, so that the gain can be explicitly drawn as a vector from the origin, the length of the vector representing the gain magnitude, and the angle between the vector and the horizontal axis the actual phase lag of the system. The Nyquist plot is the

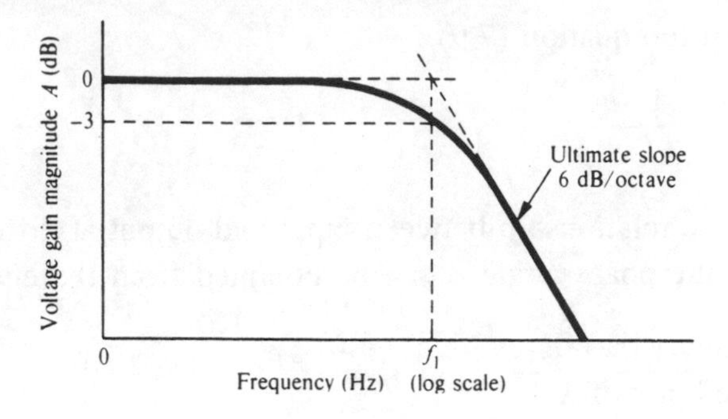

Fig. 7.22. Frequency response of a first-order low-pass circuit.

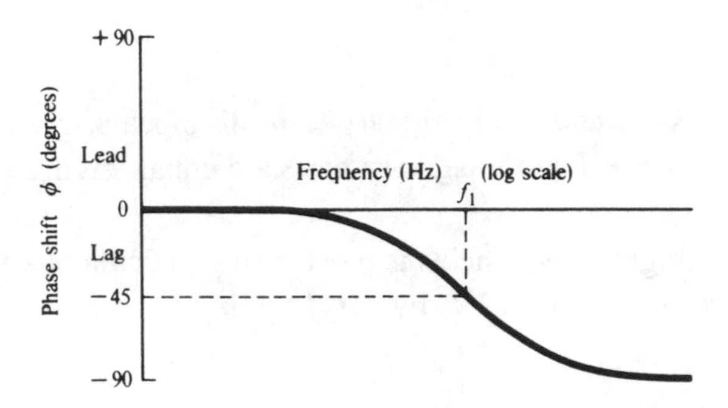

Fig. 7.23. Phase response of a first-order low-pass circuit.

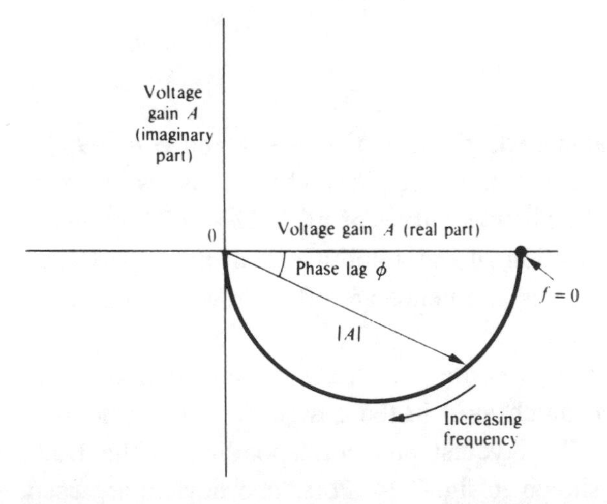

Fig. 7.24. Nyquist diagram corresponding to Bode plot of figs. 7.22 and 7.23.

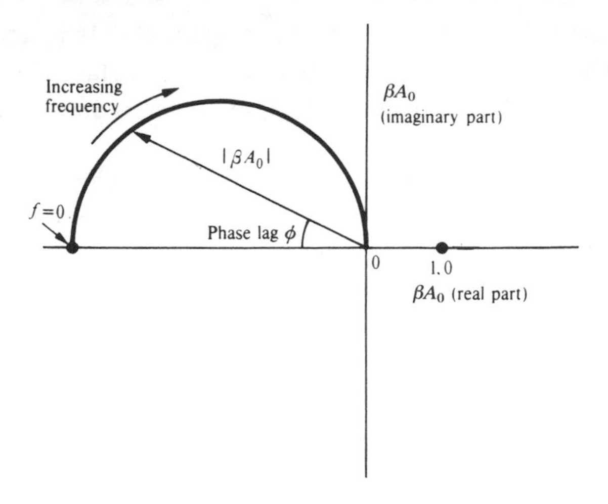

Fig. 7.25. Nyquist plot used to analyse a negative feedback system. Loop goes unstable if plot encloses point (1, 0).

path traced out by the tip of the vector as frequency is increased from zero to infinity. In the analysis of a feedback system, the length of the vector represents $|\beta A_0|$, as shown in fig. 7.25. The diagram appears back-to-front compared with fig. 7.24 because negative feedback requires a phase reversal in either β or A_0.

If the plot of fig. 7.25 ventures into the first or fourth quadrants, it is an indication that the feedback becomes positive at some frequencies and, if the plot encloses the point (1, 0) shown, oscillation will occur because positive feedback is occurring with a loop gain greater than unity. As is clear from fig. 7.25, the first-order system is ideal from the stability point of view since it cannot result in more than 90° phase shift.

Although the Nyquist and Bode plots can be extremely valuable in trouble-shooting a feedback system, in electronic circuits, the simple frequency response of fig. 7.22 is far and away the most commonly specified characteristic. As well as being the easiest diagram to plot, requiring only a signal generator and millivoltmeter for the measurements, it does in fact contain implicit phase information for these simple *minimum phase* circuits.

As we have seen, a first-order system produces a 6 dB/octave slope at high frequencies, and a maximum phase shift of 90°. We may deduce from this that in most circuits, whenever we see a response fall-off no steeper than 6 dB/octave, the phase shift will be no more than 90°. If a circuit involves two first-order systems, it produces the second-order slope of 12 dB/octave and the maximum phase shift will be 180°, though this angle will only be reached near zero gain. Similarly a third-order system gives an 18 dB/octave slope and a maximum phase shift of 270°. This latter system, of course is almost certain to be unstable when negative feedback is applied.

We can thus assess the maximum phase shift of a system, and therefore its stability with feedback, by studying the rate of fall of the frequency response graph at high frequencies. Fig. 7.26 shows the high-frequency response of a typical multi-stage integrated circuit amplifier, where three distinct slopes of

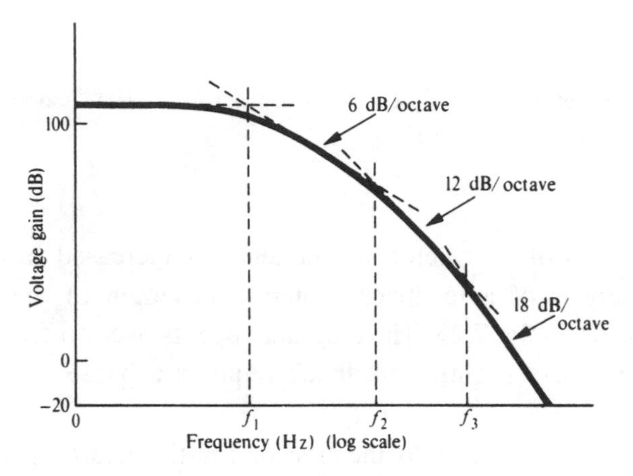

Fig. 7.26. Typical frequency response of a multistage amplifier showing break-points f_1, f_2, f_3 in high-frequency roll-off.

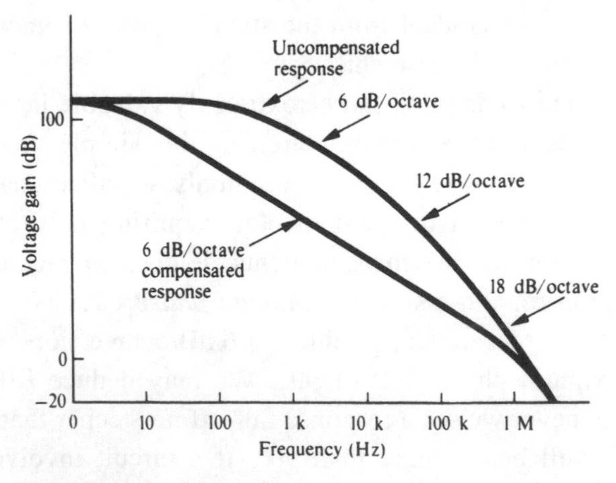

Fig. 7.27. Typical result of applying lag compensation to an amplifier.

HF fall-off may be distinguished. The region where the slope changes is known as a *break point* and it is clear that there are three such points at f_1, f_2, f_3. Note that the gain is still greater than unity (0 dB) over the 12 dB/octave and part of the 18 dB/octave region, indicating that negative feedback would provoke high-frequency oscillation. The cure for this problem is to apply external compensation networks to reduce the slope. This can be a complicated business if maximum bandwidth is to be preserved, each slope being tackled separately by producing the appropriate phase lead to compensate for the excessive lag. The most common method, however, is to use one capacitor appropriately placed in the circuit to swamp the high frequency scene with a dominant first-order phase lag, giving a healthy 6 dB/octave slope. Fig. 7.27 shows the result of such treatment, the first-order slope extending right down to the unity gain point. Practical examples of compensation will be met in section 11.11.

8

Low-frequency signals, d.c. and the differential amplifier

8.1 Introduction

So far in our discussion of amplifiers, a vital component in the design has been the coupling capacitor which transmits the a.c. signals but removes the steady d.c. voltage present at the input and output of each stage. This is necessary in order to avoid one stage upsetting the operation of adjacent ones.

A two-stage capacitor-coupled amplifier is shown in fig. 8.1 together with quiescent d.c. voltages. It is clear that C_2 is isolating the collector of T_1 (which needs to sit at 4.5 V for correct operation) from the base of T_2, which is only 0.6 V above the grounded emitter, being a forward-biased junction. Making a direct connection between stages, omitting C_2, would have the unfortunate result of clamping the collector of T_1 only about 0.6 V above 0 V and passing a 2 mA base current into T_2 through T_1 collector load, permanently bottoming T_2. The design would not be a success!

Coupling capacitors can, however, be eliminated by special d.c. amplifier design which is employed in virtually all present-day circuitry. There are two main reasons for this. The first, very practical, reason is that large capacitors cannot be fabricated on ICs, the maximum being a few tens of picofarads. The second reason is that the coupling capacitor inevitably leads to attenuation and phase shift at low frequencies: after all, there is no clear distinction between low-frequency a.c. and slowly changing d.c. and it is impossible to provide isolation from the latter without affecting the former.

Often the actual signals which are to be amplified are d.c. voltages: two examples are thermocouple amplifiers and servo controls, where the signal voltage may change so slowly that it is classified as d.c.

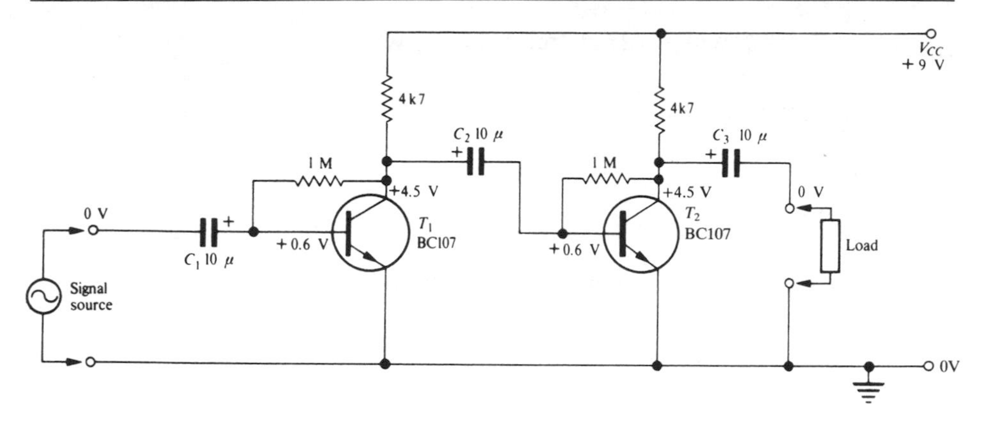

Fig. 8.1. Two-stage capacitor-coupled amplifier, showing typical quiescent d.c. voltages (measured relative to earth potential).

8.2 Low-frequency attenuation

Fig. 8.2 shows a coupling capacitor feeding a resistive load. At high frequencies, the reactance of the capacitor is negligible – it might as well be a piece of wire. At low frequencies, however, the picture is different and both attenuation and phase shift result; the circuit is therefore known as a *high-pass* filter. The following calculation shows the effect of the series capacitor. From fig. 8.2,

$$\text{gain } A = \frac{V_{\text{out}}}{V_{\text{in}}} = \frac{R}{R+(1/j\omega C)}$$

$$= \frac{1}{1+(1/j\omega CR)}, \tag{8.1}$$

therefore magnitude of gain

$$|A| = \frac{1}{\sqrt{[1+(1/\omega^2C^2R^2)]}} \tag{8.2}$$

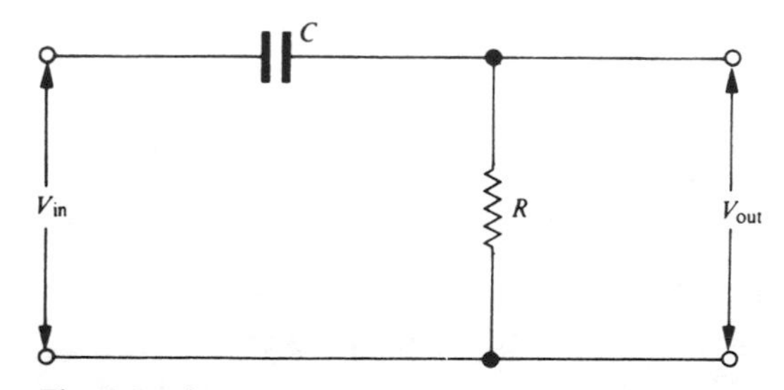

Fig. 8.2. First-order high-pass filter.

and $|A|$ is 3 dB down ($\div\sqrt{2}$) at frequency f_1, where

$$\frac{1}{4\pi^2 f_1^2 C^2 R^2} = 1,$$

i.e.

$$CR = 1/2\pi f_1.$$

Substituting for CR in equation (8.1), we have

$$A = \frac{1}{1 - j(f_1/f)}, \qquad (8.3)$$

therefore

$$|A| = \frac{1}{\sqrt{[1 + (f_1^2/f^2)]}}. \qquad (8.4)$$

Now if

$$f \gg f_1, \qquad |A| \approx 1$$

and if

$$f \ll f_1, \qquad |A| \approx \frac{f}{f_1};$$

i.e. $|A|$ falls by half when f is reduced by half. Thus we have a 6 dB/octave fall-off as we had in the high-frequency case (section 7.7).

Also, if, $f \ll f_1$ the complex gain is given by

$$A \approx j\frac{f}{f_1}.$$

If ϕ is the phase angle between V_{out} and V_{in},

$$\tan\phi = \frac{\text{imaginary part of } A}{\text{real part of } A} \approx \frac{f/f_1}{0}$$

$$\approx \infty \, (f \ll f_1)$$

Thus $\phi = 90°$, i.e. V_{out} leads V_{in} by 90°.

At the –3 dB point, $f = f_1$, $\tan\phi = 1$, and V_{out} leads V_{in} by 45°.

We can now sketch the frequency and phase response of this first-order high-pass filter in the form of a Bode plot (fig. 8.3).

If a negative feedback loop includes a high-pass filter, low-frequency phase shift is just as likely to provoke instability as high-frequency phase shift, but this is an avoidable problem. Whilst all amplifiers inevitably exhibit a declining gain, with an associated phase shift, at high frequencies, it is possible to design an amplifier with flat response down to zero frequency (d.c.); this characteristic is common to virtually all IC amplifiers. Such circuit designs completely avoid CR couplings like fig. 8.2 and are known as d.c. amplifiers, though they will normally handle a.c. perfectly well, subject only to the usual high-frequency roll-off. The abbreviation 'd.c.' may in fact be interpreted as 'direct-coupled' which refers to the absence of coupling capacitors.

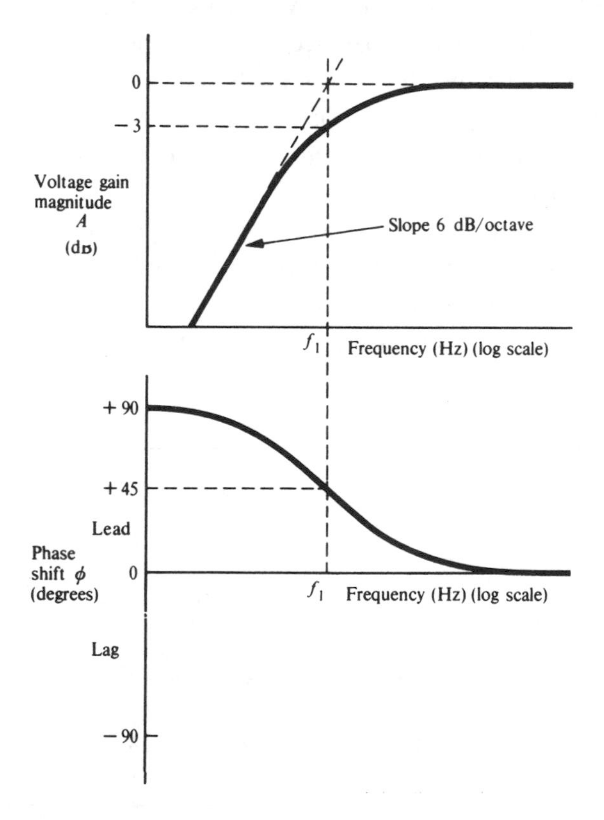

Fig. 8.3. Bode plot (frequency and phase response) for first-order high-pass filter.

8.3 Features of d.c. amplifiers

8.3.1 Design

The design of an amplifier without coupling capacitors restricts the range of permissible circuit voltages. In particular, it is very desirable that, when there is no signal, both the input and the output should be at earth potential. This, of course, means that the quiescent d.c. output voltage can no longer be set midway between earth (0 V) and supply rail (V_{CC}). This limitation may seem to present a problem: we have so far assumed that the output should sit at $V_{CC}/2$ in order to allow both positive and negative swings in the signal. The solution in the case of the d.c. amplifier is to use two balanced power supplies, one positive and one negative (known as *split rail* operation).

A simple two-transistor d.c. amplifier is shown in fig. 8.4. It is based on the a.c. amplifier of fig. 1.20, but uses balanced power supplies and complementary (npn and pnp) transistors. By means of the potential divider, consisting of R_4 and R_5 with R_3, the emitter of T_1 is held at a potential just slightly negative of earth (–0.6 V). Thus T_1 is correctly biased if its base is tied to earth via the input resistor R_1.

On the output side, we know that the collector of an npn transistor must be positive with respect to the base and yet, in this d.c. amplifier, we require the collector to be at earth potential if zero input is to give zero output. This paradox is resolved by T_2, a pnp transistor, which is interposed to shift the

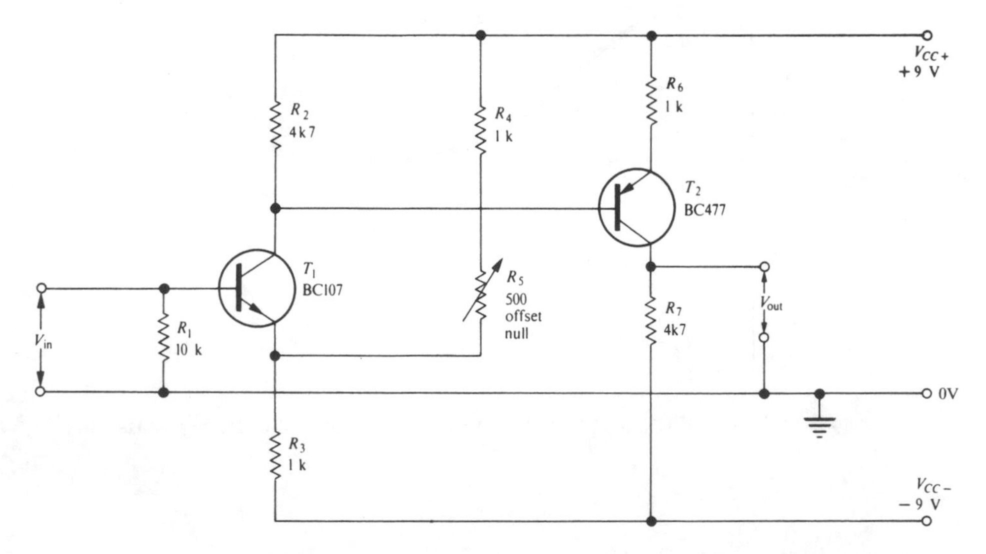

Fig. 8.4. Simple d.c. amplifier illustrating use of balanced power supplies.

quiescent output voltage back to zero and at the same time to perform further amplification.

T_2 is simply a d.c. stabilized amplifier stage working between supplies of +9 V and −9 V, with its base potential held, not by a potential divider, but by T_1 collector. The optimum working condition is obtained when T_2 quiescent collector potential is at earth (0 V), giving zero output voltage for zero input to T_1. If zero input does not result in zero d.c. output, the amplifier is said to exhibit an *offset voltage*; the purpose of variable resistor R_5, the *offset null control* is to adjust for zero output with zero input, rather as an analogue voltmeter zero-set screw is adjusted to bring the pointer to scale zero with no applied signal.

Working backwards through the amplifier, we require T_2 collector to be at 0 V, which implies a voltage drop of exactly 9 V in R_7. Hence the collector current of T_2 is given by (9/4700) A or 1.9 mA. A 1.9 mA emitter current gives a 1.9 V drop in the 1 kΩ emitter resistor, R_6, so that T_2 emitter will be sitting at (9–1.9) V (+7.1 V). In order to maintain these conditions, T_2, being a pnp transistor, must have its base held 0.6 V more negative than its emitter, that is at (7.1–0.6) V, or +6.5 V. This is a suitable quiescent operating voltage for T_1 collector, so that the two amplifier stages can be directly coupled. We know then that, if R_5 is adjusted for zero offset, the quiescent conditions in T_1 are automatically correct. This is just one of many circuit applications where complementary transistors are valuable.

The voltage gain of fig. 8.4 is easy to calculate, since both stages are common-emitter amplifiers without emitter decoupling capacitors, there being, of course, no purpose in using the latter in a d.c. amplifier. The approximate voltage gain of each stage is given by the ratio of the collector load to the emitter load (see section 4.7). The T_1 emitter load is effectively R_3 in parallel with the series combination of R_4 and R_5, the net value being approximately 550 Ω.

Thus,

$$\text{1st stage voltage gain} \approx \frac{4.7}{0.550} \approx 8.5,$$

$$\text{2nd stage voltage gain} \approx \frac{R_7}{R_6} \approx \frac{4.7}{1} \approx 4.7.$$

Thus, total voltage gain of the two stages

$$\approx 8.5 \times 4.7$$

$$\approx 40.$$

8.3.2 Input bias current

It will be found in operation that the setting of R_5 for zero offset is dependent upon the source resistance connected to the input. If R_5 is initially set with the input open-circuited (i.e. with just the 10 kΩ resistor R_1 as source resistance) and then a thermocouple, with negligible resistance, is connected to the input, R_5 will require resetting. The reason for this is that T_1 base is drawing its normal bias current (typically 5 μA) through the input circuitry. A current of 5 μA in 10 kΩ gives a voltage drop of 50 mV, and this appears as an input offset voltage when there is only the 10 kΩ in circuit. The offset disappears when the input is short-circuited or connected to a low-resistance source. Input bias current and offset voltage will be further discussed when we meet IC amplifiers in chapter 11.

8.3.3 Drift

One snag in the operation of a directly-coupled amplifier is that changes in circuit d.c. operating conditions are indistinguishable from the signal being amplified. Such changes usually occur as a result of temperature variations. For instance, in fig. 8.4, if the temperature increases, the potential drop in the base–emitter junctions will decrease and result in a slight increase in collector current in both transistors. The resulting change in offset voltage is known as drift; it is usually the input stage which is most susceptible, because it is subject to the most amplification.

Drift may be examined in the circuit of fig. 8.4 by connecting the output to a sensitive d.c. voltmeter or an oscilloscope with the Y-amplifier switched to d.c. With no input connected R_5 is adjusted for zero output and then transistor T_1 is warmed up between the fingers. A gradual drift in the output voltage will be seen; the direction of the change may be noted and checked to see whether it agrees with the above explanation.

Since electronic circuitry is not normally held at constant temperature, d.c. amplifiers invariably exhibit some drift. With appropriate circuit design, however, drift may be held down to a negligible level. Output drift depends on amplifier gain as well as internal drift. In the specification of an amplifier, in order to express drift independently of gain, drift is usually 'referred back to input' in microvolts per degree; this is the change in input voltage signal that would be required to give the same output change as that produced by drift. For example, if an amplifier has a gain of 100 and the output voltage drifts by 0.2 V for a temperature rise of 25 deg C, then

$$\text{drift referred to input} = \frac{0.2}{100 \times 25}\ \text{V/deg C}$$
$$= 80\ \mu\text{V/deg C.}$$

8.4 The differential amplifier

8.4.1 Basic circuit

The most straightforward way of reducing drift is to use some sort of a balanced amplifier, where voltage changes in one part of the circuit are balanced by equal and opposite changes in another part. This condition is most easily fulfilled in the differential (difference) amplifier. The name describes the way that the amplifier output is dependent on voltage *difference* between two parts of the circuit, so that if both parts drift to the same extent, there is no net drift in the output. A differential amplifier is shown in fig. 8.5; it is also known

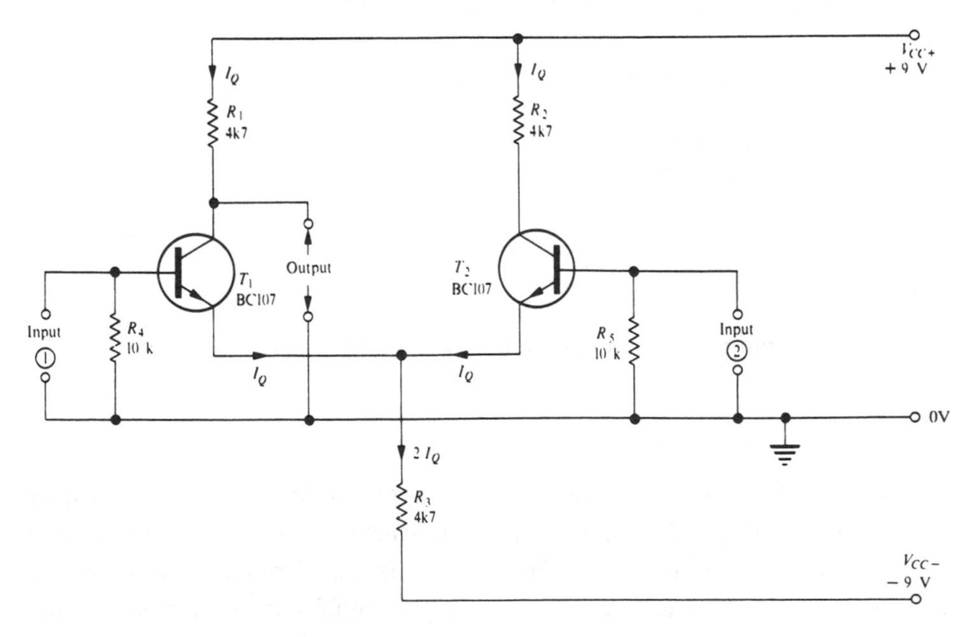

Fig. 8.5. Basic differential amplifier.

as the *long-tailed pair*, named after the large-value common 'tail' resistor, R_3.

The balanced configuration of the differential amplifier provides two sets of input terminals, input (1) and input (2). Usually just one output is used, and this would normally feed further d.c. amplifier stages. Some level-shifting arrangement, such as the pnp stage used in fig. 8.4, would normally be necessary, since the quiescent voltage at T_1 collector is approximately 4.5 V above earth. To see why this is so, we shall consider both transistors to be identical and the quiescent collector current in each one to be I_Q.

Then emitter current in each transistor $\approx I_Q$ and current in tail resistor $(R_3) \approx 2I_Q$.

Now since the base of each transistor is held at earth potential (via R_4 and R_5), emitter follower action will ensure that the emitters are also approximately at earth potential; (actually of course 0.6 V below earth).

Hence voltage drop in $R_3 \approx V_{CC}$ therefore

$$\text{current in } R_3\ (\approx 2I_Q) \approx \frac{V_{CC}}{R_3}$$

hence

$$I_Q \approx \frac{V_{CC}}{2\times 4.7}\ \text{mA.}\quad (R_3 = 4.7\ \text{k}\Omega)$$

Now quiescent collector voltage of each transistor

$$= V_{CC} - 4.7 \times I_Q \quad (R_1 = R_2 = 4.7\ \text{k}\Omega)$$

$$= V_{CC} - \frac{V_{CC} \times 4.7}{2\times 4.7}$$

$$= \frac{V_{CC}}{2}.$$

Thus the output is able to swing in both the positive and negative directions.

8.4.2 Voltage gain

In assessing the voltage gain of the differential amplifier, we shall consider two input conditions. The first condition is shown in fig. 8.6; input (1) and input (2) are both connected to the same signal v_{in}. Emitter follower action will ensure that an exact replica of the input signal appears across the tail

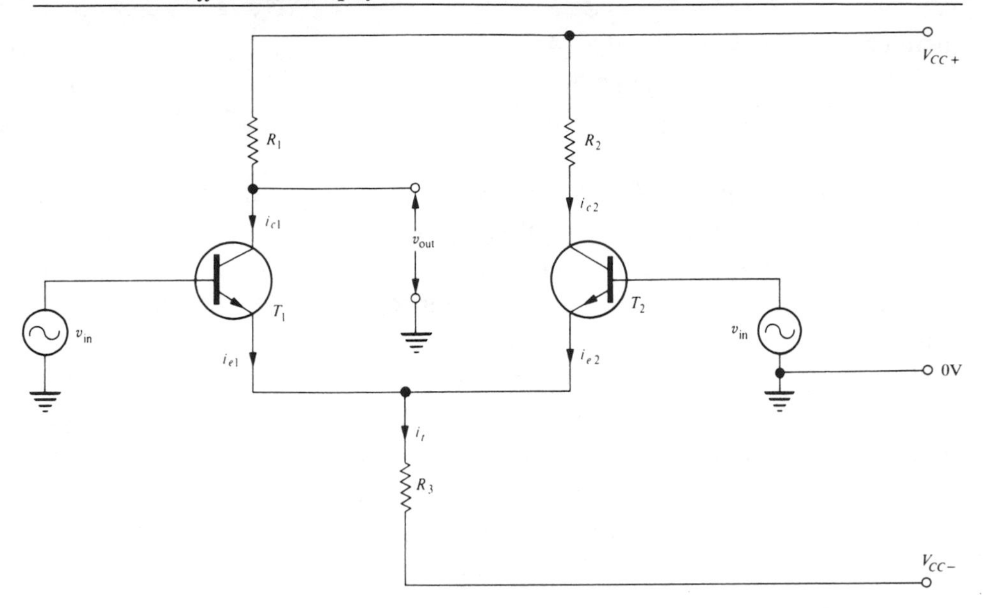

Fig. 8.6. Differential amplifier with two identical input signals (common mode).

resistor R_3. The total signal current in R_3 (i_t) will be given by

$$i_t \approx \frac{v_{in}}{R_3}. \tag{8.5}$$

Again, assuming the transistors to be identical, this current will be split equally between both devices so that signal collector current of T_1 = signal collector current of T_2, i.e.

$$i_t = i_{c1} + i_{c2} \quad \text{(assuming } i_c \approx i_e\text{)}$$

and since

$$i_{c1} = i_{c2},$$

$$i_t = 2i_{c1}.$$

Now,

$$v_{out} = -i_{c1}R_1 \quad \text{(neglecting } 1/h_{oe}\text{)}$$

$$= \frac{-i_t R_1}{2}$$

$$\approx -\frac{v_{in}R_1}{2R_3} \quad \text{(from 8.5)}$$

therefore

$$\text{voltage gain} = \frac{v_{\text{out}}}{v_{\text{in}}} \approx \frac{-R_1}{2R_3}. \quad (8.6)$$

In the case of fig. 8.5, $R_1 = R_3$ and, when both inputs are driven together with the same signal, the overall voltage gain is one half. This type of input is called a *common-mode* input and the corresponding gain the *common-mode gain*. The higher the value of the tail resistor R_3 compared with the collector load R_1, the lower the common-mode gain.

Now we shall consider the amplifier with a differential input, i.e. with a different signal on input (1) from that on input (2). Here we can no longer assume that the emitters are connected together with negligible resistance as we did in fig. 8.6, as this would imply that the inputs could not be different. In fact, of course, each transistor exhibits a dynamic emitter resistance r_e, as we discussed first in section 5.12.2 in connection with the emitter follower.

As was shown in section 6.3, r_e is directly related to the transconductance g_m of the transistor:

$$r_e = \frac{1}{g_m}, \text{ where } r_e \text{ is in k}\Omega \text{ if } g_m \text{ is in mA/V.} \quad [(6.10)]$$

Furthermore g_m is related to the mean emitter current, I_E, of the transistor thus

$$g_m \approx 40I_E, \text{ where } g_m \text{ is in mA/V if } I_E \text{ is in mA.} \quad [(6.11)]$$

Now, referring to the equivalent circuit of fig. 8.7, where the two inputs of the differential amplifier have different input signals, for the purpose of calculating the differential voltage gain we shall consider the signal current flowing in r_e for each transistor. It is reasonable to assume that R_3 is much larger in value than r_e so that any common-mode signal current (i_t) flowing in R_3 can be neglected in comparison with i_{e1} and i_{e2}.

Then,

$$i_{e1} \approx -i_{e2} \approx \frac{v_{\text{in(1)}} - v_{\text{in(2)}}}{2r_e} \text{ (by Ohm's law)}$$

and assuming $i_c \approx i_e$,

$$i_{c1} \approx -i_{c2} \approx \frac{v_{\text{in(1)}} - v_{\text{in(2)}}}{2r_e}.$$

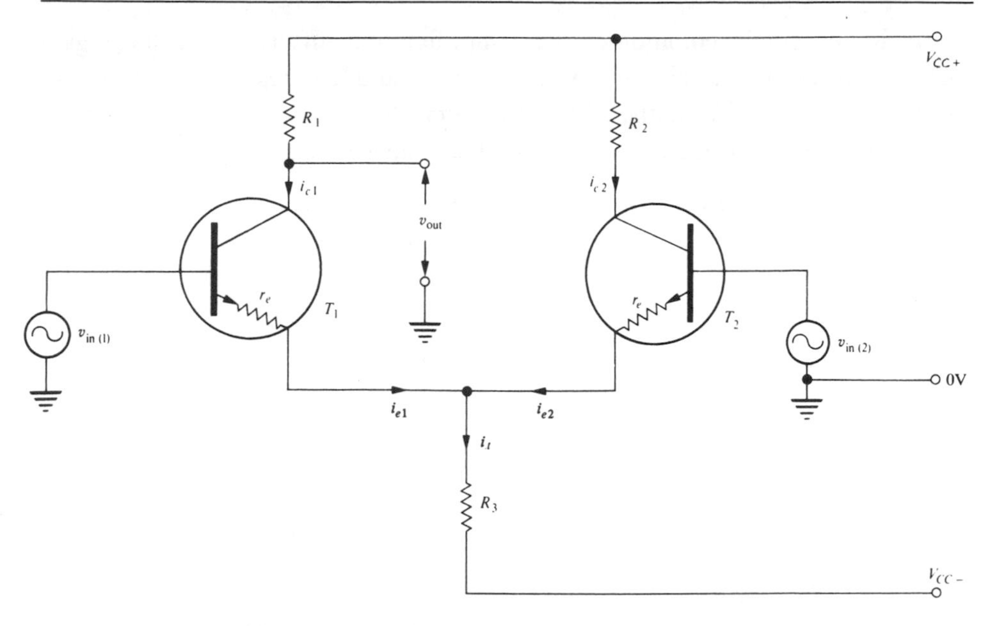

Fig. 8.7. Differential amplifier with two different input signals. Transistor symbols include equivalent emitter junction resistance r_e.

Now,

$$v_{out} = -i_{c1}R_1$$
$$\approx \frac{-R_1(v_{in(1)} - v_{in(2)})}{2r_e}.$$

Now, substituting $g_m = 1/r_e$,

$$v_{out} = \frac{-g_m R_1(v_{in(1)} - v_{in(2)})}{2}. \tag{8.7}$$

Thus we see that the differential amplifier responds to the *difference* in potential between its inputs. Note that if $v_{in(1)}$ is more positive than $v_{in(2)}$ the output is negative and if $v_{in(2)}$ is more positive than $v_{in(1)}$ the output is positive. Hence input (1) is called the inverting input and input (2) the non-inverting input. The differential voltage gain A_{VD} is given by

$$A_{VD} = \frac{v_{out}}{v_{in(1)} - v_{in(2)}}$$
$$\approx \frac{-g_m R_1}{2}, \quad g_m \text{ in mA/V}, R_1 \text{ in k}\Omega. \tag{8.8}$$

As in the single common-emitter amplifier, the differential voltage gain takes the form of a product of transconductance and load resistance. The effective transconductance of the differential amplifier, $g_{m(\text{eff})}$ appears as half the transconductance of either of the individual transistors, i.e.

$$g_{m(\text{eff})} \approx \frac{40I_E}{2}$$

therefore

$$g_{m(\text{eff})} \approx \frac{40I_T}{4} \text{ (tail current } I_T \text{ is the sum of two equal emitter currents)}$$

$$\approx 10I_T,$$

i.e.

$$A_{VD} \approx -10I_T R_1, \tag{8.9}$$

where I_E is the mean d.c. emitter current of each transistor (mA) and where I_T is the mean d.c. current in the tail resistor (mA).

We saw in section 6.8 that the transfer characteristic of a single common-emitter stage is markedly non-linear because of the exponential characteristic of the base–emitter diode. The transconductance is continuously changing with instantaneous signal level as the emitter current varies, thus giving rise to distortion in the signal. It is interesting to note that the differential amplifier, on the other hand, is inherently linear for small input signals. This is because, as the emitter current of T_1 increases, so the emitter current of T_2 must decrease, because the tail resistor R_3 is much greater than r_e and therefore behaves as a constant-current source. Any increases in g_m due to T_1 emitter current increasing is balanced by a corresponding decrease in g_m in T_2. This only applies for small signals, but the improved linearity of the differential amplifier over the common emitter amplifier is considerable and can be seen in fig. 8.8 where the two transfer characteristics are compared. The linear region for the differential amplifier extends over a differential input range of approximately ±25 mV.

8.4.3 Common-mode rejection and drift reduction

In the above discussion, we have seen that the differential amplifier exhibits a very low gain when both inputs carry the same signal (common-mode) but responds with a high gain to a potential *difference* between inputs (differential

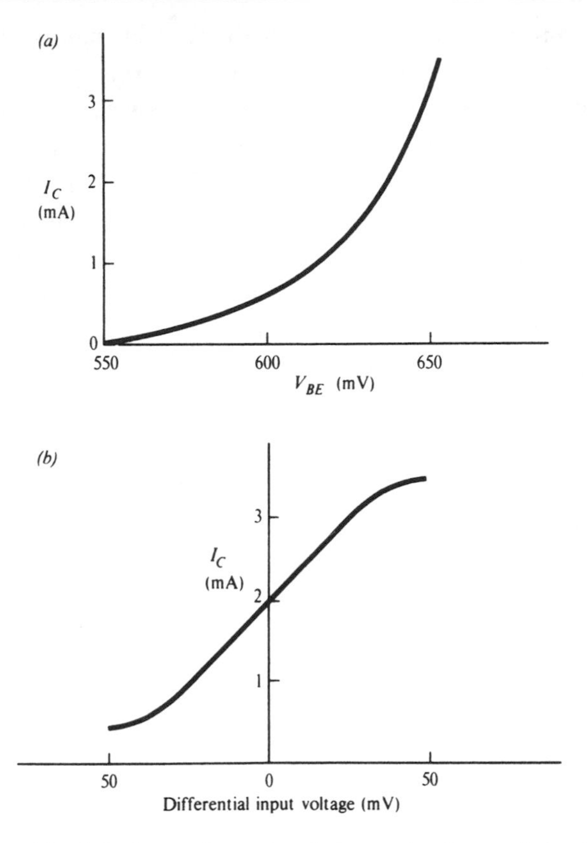

Fig. 8.8. Comparison of typical transfer characteristics for (*a*) single common-emitter amplifier, (*b*) differential amplifier. Collector current is plotted against input voltage in each case, and shows improved linearity for the differential amplifier.

mode). To summarize the calculations above, we have, for the basic amplifier of fig. 8.5,

$$\text{common-mode voltage gain } A_{VCM} \approx \frac{-R_1}{2R_3}$$

$$\approx -\frac{1}{2}$$

and differential voltage gain,

$$A_{VD} \approx -10 I_T R_1 \qquad \approx -10 \times 2 \times 4.7$$

$$\approx -100.$$

One effect of temperature variations on a transistor is to change the base–emitter voltage. In a simple d.c. amplifier such as fig. 8.4, this gives rise to a drift in the output voltage. In the differential amplifier, however, assuming both transistors are identical and subject to the same temperature variations, the same base–emitter voltage changes will appear at both inputs; thus any drift will be subject only to the low common-mode gain. The wanted signal is applied in the differential mode and is subject to the high differential gain.

Most signal sources have one terminal earthed; the easiest way to connect such a source in the differential mode is to earth one input of the differential amplifier and connect the other to the 'live' terminal of the signal source. The valuable drift rejection properties of the amplifier will still be present.

The standard method of assessing the 'quality' of a differential amplifier is the measurement of *common-mode rejection ratio* (CMRR), which is the ratio of differential voltage gain to common-mode voltage gain.

$$\text{CMRR} = \frac{A_{VD}}{A_{VCM}} \tag{8.10}$$

and therefore, in the case of the amplifier under discussion:

$$\begin{aligned} \text{CMRR} &\approx \frac{10 I_T R_1 \times 2R_3}{R_1} \\ &= 20 I_T R_3 \\ &\approx 200 \quad (I_T \text{ in mA, } R_3 \text{ in k}\Omega) \end{aligned}$$

with the values given. CMRR is usually expressed in decibels:

$$\text{CMRR} = 20 \log_{10}\left(\frac{A_{VD}}{A_{VCM}}\right) \text{dB.} \tag{8.11}$$

Note that the multiplying factor here is 20 because we are using voltage gains and not power gains.

Hence CMRR for our amplifier here is 20×2.3 dB = 46 dB. This is not a very impressive figure by modern standards, where ratios of 80–100 dB are common. The easiest way to improve CMRR is to increase the tail resistor R_3, but this involves using a high voltage on $-V_{CC}$ if the tail current is to be maintained. Fortunately, there is an ideal solution to the problem, in the shape of a transistor used as a constant-current source. A perfect constant-current source would appear as an infinite resistance; a transistor approximates the perfect constant-current source by appearing as $1/h_{oe}$, typically 100 kΩ (see section 6.8). The great advantage of the transistor over the equivalent resistor is that it can maintain a constant tail current without a high-voltage supply.

The constant-current source is essentially a d.c. stabilized common-emitter stage and is shown in the tail of a differential amplifier in fig. 8.9.. The base is held at approximately 1 V above the negative rail by the potential divider consisting of R_Y, R_Z and the diode D_1. After allowing for the 0.6 V base–emitter drop, this leaves 0.4 V to be dropped in the 200 Ω emitter resistor R_X. Therefore an emitter current of 0.4/220 A (approximately 2 mA) will result; the constant-current source thus draws 2 mA tail current, I_T, through the differential pair. The use of the diode D_1 in the bottom limb of the potential divider provides temperature compensation. The diode p.d. falls with increasing temperature, just like the base–emitter p.d., so that the applied base voltage conforms over a wide range of temperatures to the transistor's requirement for a 2 mA emitter current. In an IC, the diode can be made identical to the transmitter B–E junction, giving such ideal tracking that the circuit is known as a *current mirror*.

Given that the differential amplifier has a constant-current source in its tail, we can now produce a very simple qualitative explanation of its operation.

The two inputs cannot change the tail current I_T, they can only divert it

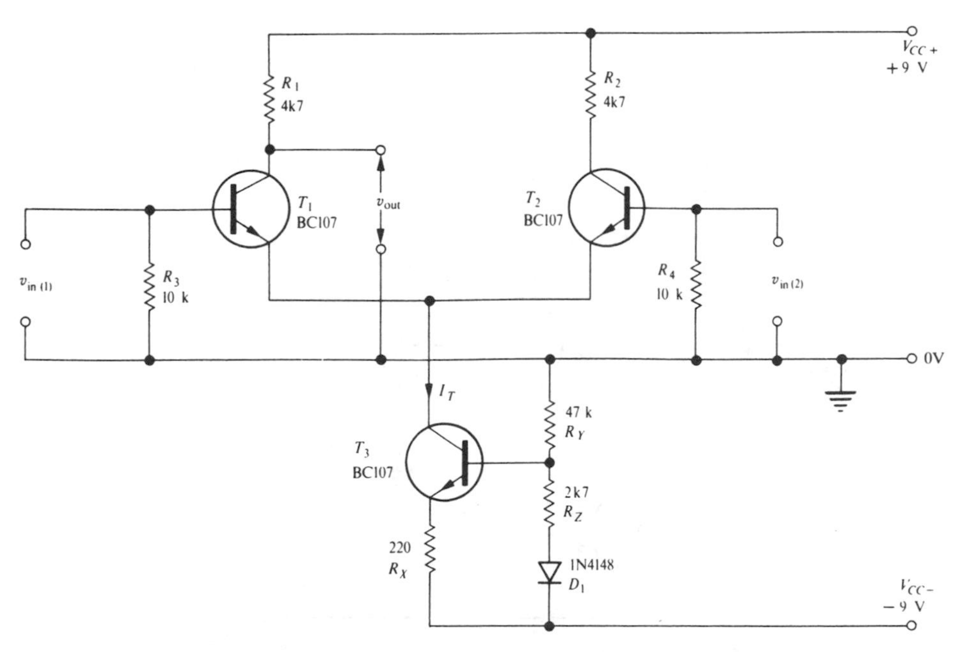

Fig. 8.9. Differential amplifier with transistor constant-current source in the tail.

from one transistor to another. Hence, if $v_{in(1)}$ and $v_{in(2)}$ are identical (common mode), neither collector current will change and no output will be obtained. The only way to obtain an output is if $v_{in(1)}$ and $v_{in(2)}$ differ, when a greater fraction of the tail current will be diverted to one transistor than the other. For example, if $v_{in(1)}$ is more positive than $v_{in(2)}$, T_1 collector current will increase at the expense of T_2 and a negative-going v_{out} will be obtained.

8.4.4 Balanced output

CMRR can be improved still further by taking the output from between the two collectors (fig. 8.10) instead of between one collector and earth. In this case, the common-mode rejection is limited only by the degree of balance between the transistors, the theoretical common-mode gain being zero if the two transistors and collector loads are identical. The double-ended output is particularly useful if the output load is floating free of earth; an example is a voltmeter, which can be connected between the collectors to make a simple electronic millivoltmeter with very low drift.

A double-ended output differential stage may be used to feed a second differential stage which itself has a single-ended output; this approaches the

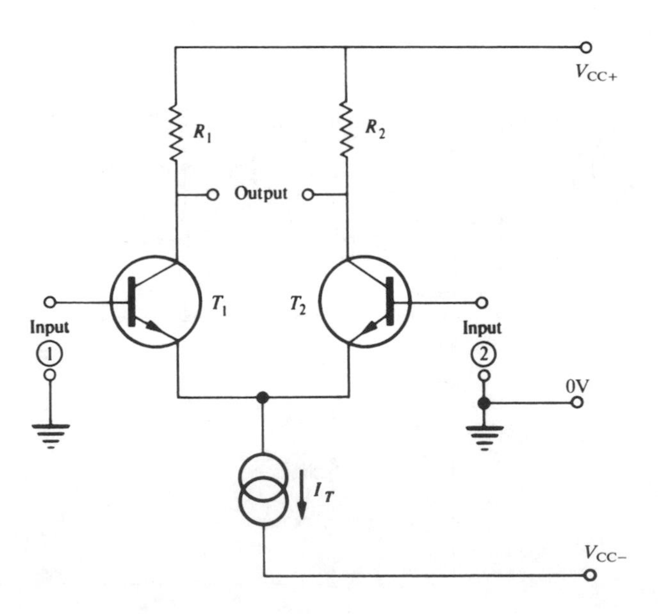

Fig. 8.10. Balanced output from a differential amplifier.

very high common-mode rejection of the double-ended output whilst providing the convenience of the single-ended output. Negative feedback techniques are commonly used to further enhance the common-mode rejection.

Because the outputs of the two transistors add together in the double-ended configuration, the gain is twice that of the single-ended output case, i.e.

$$A_{VD} \approx \frac{-g_m R_1}{2} - \frac{g_m R_2}{2}$$

$$\approx -g_m R_1 \quad (\text{since } R_1 = R_2),$$

where

$$g_m \approx \frac{40 I_T}{2}$$

$$\approx 20 I_T, \text{ where } I_T \text{ is the tail current,}$$

therefore

$$A_{VD} \approx -20 I_T R_1. \tag{8.12}$$

8.4.5 Voltage-controlled amplifier

It is interesting to note that the gain of a differential amplifier is directly proportional to tail current I_T. What is more, in the double-ended output differential amplifier, variations in tail current do not produce an offset in the output as they do in the single-ended case; the resulting change in quiescent collector voltage is the same for both transistors. Hence, the differential voltage gain may be independently controlled by varying the tail current of the differential pair. If v_{in} and v_{out} are the input and output voltage signals in a double-ended differential amplifier, then we have:

$$v_{out} \approx -20 R_1 (I_T \times v_{in}).$$

If now a second voltage input signal v_x is used to control I_T, by changing the potential of the base of the 'constant-current' tail transistor, then the amplifier output will be proportional to the product of v_{in} and v_x. This is the basis of the variable-transconductance analogue multiplier, an invaluable circuit building block (Gilbert cell) which finds application in voltage-controlled amplifier (VCA) and modulator chips. Circuits using a multiplier are described in section 11.19.

8.5 IC amplifiers

The ideal differential amplifier has a pair of identical transistors which are in good thermal contact so that they remain identical over the whole working temperature range. Furthermore the tail current should be independent of temperature, which implies that the constant-current transistor should have a temperature-compensated base voltage. These requirements are readily met in an IC where all components are in close thermal contact, being on the one chip of silicon. The manufacture of ICs is basically a photographic operation. Various masks are used to control the diffusion processes employed to make the p and n regions which form transistors, resistors and diodes. For this reason, although the diffusion processes themselves are subject to some random variation, giving slightly different component parameters from one batch of circuits to the next, it is easy to manufacture closely matched transistors and resistors on any one chip, just by ensuring that identical masks are used for them.

It is impossible to incorporate coupling capacitors of any value greater than a few tens of picofarads on an IC; therefore the circuits are normally designed as d.c. amplifiers. The differential amplifier is the standard IC input stage because it ensures low drift and makes available inverting and non-inverting inputs. Most IC output stages employ some form of emitter follower to give a low output impedance.

Fig. 8.11 is representative of the internal circuit of an IC amplifier. Beginning with the ubiquitous differential stage (T_1 and T_2), it incorporates a further voltage amplifier (T_3), and has a class AB complementary emitter follower output stage (T_4 and T_5). Diodes D_1 and D_2 provide sufficient bias for the output transistors to minimize crossover distortion (see section 5.17).

When studying manufacturers' circuit diagrams of ICs, it can sometimes be difficult to identify even a basic similarity with fig. 8.11. This is not because there is a fundamental difference, but because transistors are the cheapest components of all to fabricate on an IC: if it is possible to use a transistor instead of a resistor in any part of the circuit, a transistor will be used. Furthermore, because the various transistors are diffused into the one silicon substrate, the juxtaposition of diffused layers is sometimes such as to create unwanted 'parasitic' transistors and these must either be deliberately incorporated into the circuit or biased off so that their effect is minimal. The result of these design constraints is sometimes that the circuit diagram is difficult to interpret in terms of its discrete-component equivalent. The reader can be assured, however, that most IC amplifiers *behave* as though they looked like fig. 8.11.

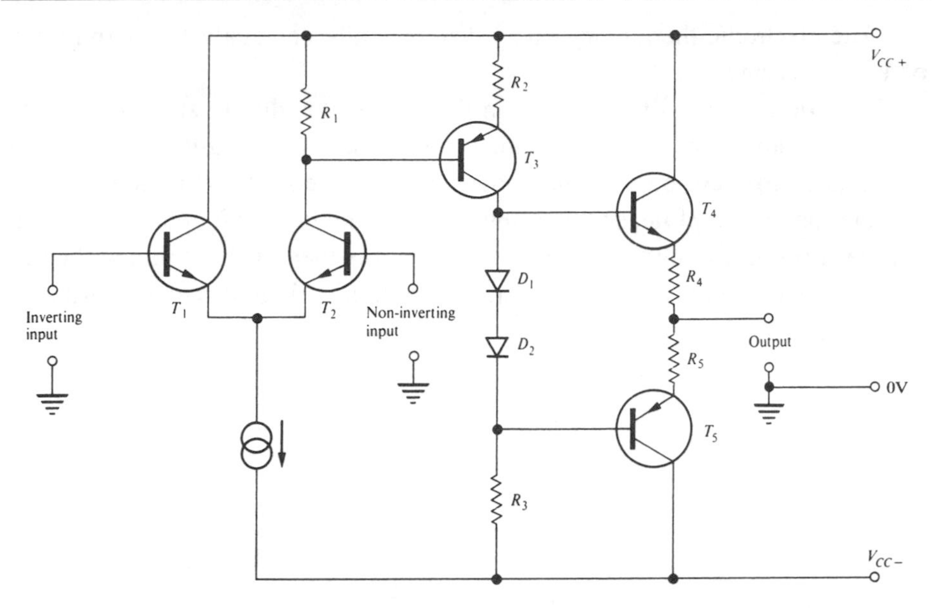

Fig. 8.11. Circuit diagram representative of a typical IC amplifier.

8.6 Electronic thermometer

A very useful electronic thermometer can be constructed from a thermocouple and an IC amplifier; the circuit is shown in fig. 8.12. A 741 IC operational amplifier is used; this has differential inputs, the negative feedback being returned to the inverting input with the input signal connected to the non-inverting input. Amplifier voltage gain is given by $-1/\beta$, which in this case is $-(R_f + R_1)/R_1$. Thus, β with the specified resistor values is approximately $-1/100$ and the voltage gain approximately 100; this may easily be increased to any value up to 10 000 or so by increasing the value of R_f.

The thermocouple itself may consist of wires of virtually any two dissimilar metals twisted together. Copper and constantan give a high output; copper and iron are also satisfactory. It is important to note that a thermocouple inevitably has *two* junctions, marked in fig. 8.12 as the one in the circle and the point X, where the constantan joins up again with the circuit. The electrical output of the thermocouple is dependent on the temperature *difference* between these two junctions. Thus, if one junction is warmed between the fingers, the instrument is measuring the difference between body temperature and room temperature.

For an absolute measurement of temperature, the second junction X should be held at a constant temperature, e.g. 0°C, obtained by immersion in melting ice. The electronic thermometer may then be calibrated against a thermometer of known accuracy.

Correction of any offset voltage in the 741 is straightforward because provision is made in the internal circuit for an offset null control, RV_1; this may be adjusted for zero output when both junctions are at the same temperature.

Note that the amplifier will not function with the input left open, i.e. without the thermocouple connected, because the input is then unable to draw the bias current for one half of the differential amplifier. This is discussed further in section 11.2.2.

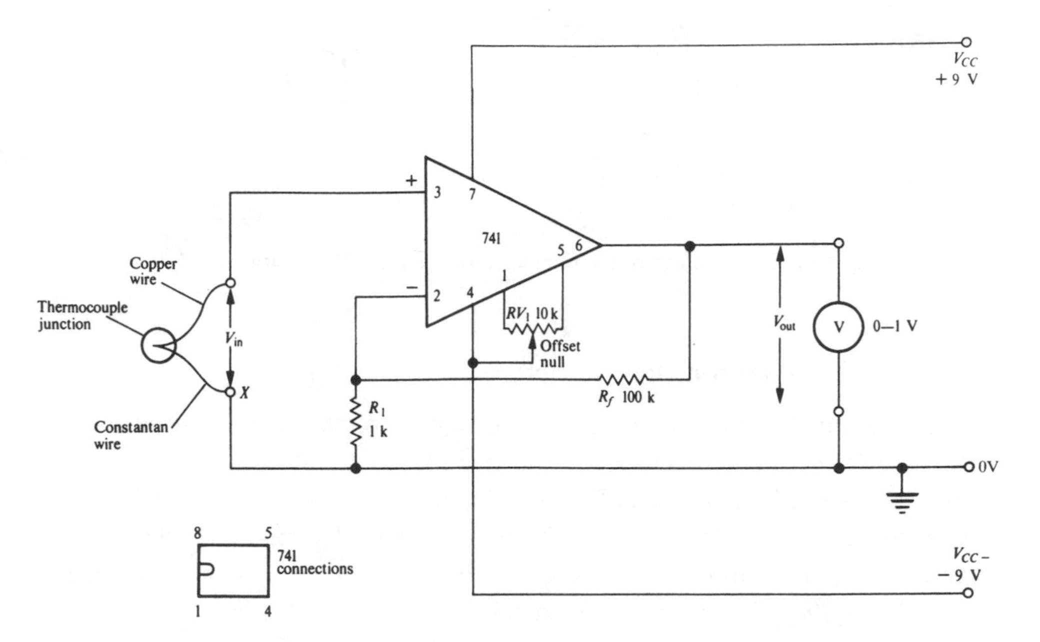

Fig. 8.12. Simple electronic thermometer.

8.7 Noise rejection with the differential amplifier

So far we have concentrated on the ability of the differential amplifier to minimize d.c. drift. We have seen that this property arises from the fact that it is only the potential *difference* between the two inputs which is amplified, any signal common to both inputs being rejected. This common-mode rejection can be very usefully applied in the reduction of hum and noise picked up in the connections between a signal source and amplifier.

One example is the electrocardiograph (e.c.g.) used in medical work to examine the electrical impulses produced by the heart. This requires that electrodes be placed on the patient's body to pick up the impulses.

Now anyone who has ever touched the input of an audio amplifier or oscilloscope with a finger will know that the human body picks up large 50 Hz a.c. signals from the mains, along with various noise signals from electrical appliances. It would therefore be expected that the tiny cardiac signals of only a few millivolts would be swamped by this hum and noise pick-up. The differential amplifier provides a solution.

Fig. 8.13 shows in outline form how a differential amplifier can be used in this sort of situation. The p.d. between the two electrodes corresponds to the wanted signal, but the large hum signal appears on both electrodes and is therefore in the common mode. The differential amplifier amplifies the wanted signal and rejects the hum picked up on the body and also in the connecting leads.

Differential amplifiers specially designed for medical purposes are called physiological amplifiers and have a high input impedance on both inputs and a good common-mode rejection.

There is a host of other applications for the differential amplifier, mostly where unwanted noise is picked up on connecting leads. This problem is often acute in an industrial environment, when sensitive electronic measuring equipment must be used adjacent to heavy electrical machinery such as an overhead crane which may be generating enormous noise signals. Many oscilloscopes are available with differential inputs and can be invaluable in such an environment. It is a common practice to feed the required signal between one input

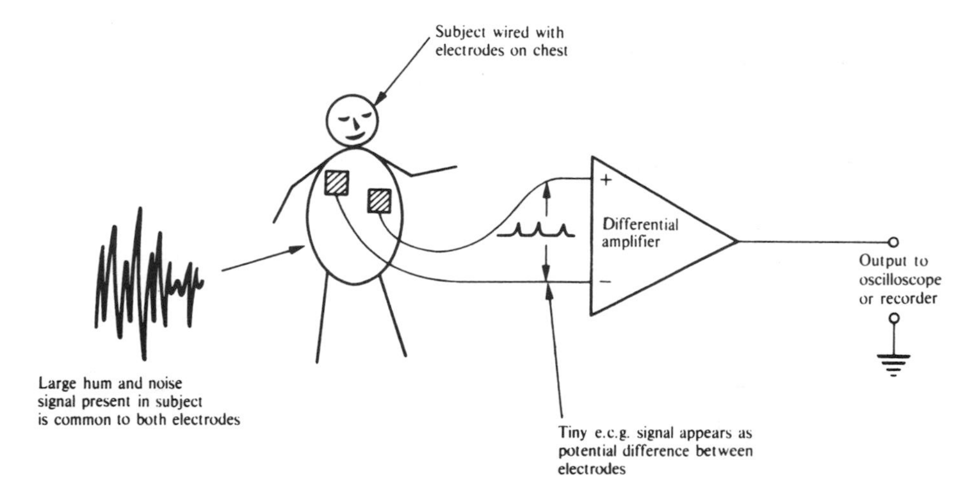

Fig. 8.13. Physiological application of the differential amplifier.

and earth and connect a 'dummy' connecting lead to the second input. This ensures that the induced noise is common to both inputs, and rejected, but the desired signal is amplified because it is in the differential mode.

8.8 Simple physiological amplifier

Fig. 8.14 shows a 741 IC connected as a differential amplifier suitable for the display of e.c.g. pulses on an oscilloscope. Whilst the circuit does not possess the high input impedance normally required in this application, it effectively demonstrates the rejection of induced hum and noise by a differential amplifier. As is common practice in e.c.g. amplifiers, frequency response is deliberately curtailed above a few tens of Hz to minimize high-frequency noise which may not be identical in the two inputs. The 1 nF capacitor across the 10 MΩ feedback resistor performs this response tailoring by increasing negative feedback at high frequencies.

The basic properties of the differential amplifier may be shown by touching one of the differential inputs – a large 50 Hz hum signal will be seen on the oscilloscope. Now touch the other input as well with the other hand, and notice the way the hum is cancelled by the differential amplifier. With the oscillo-

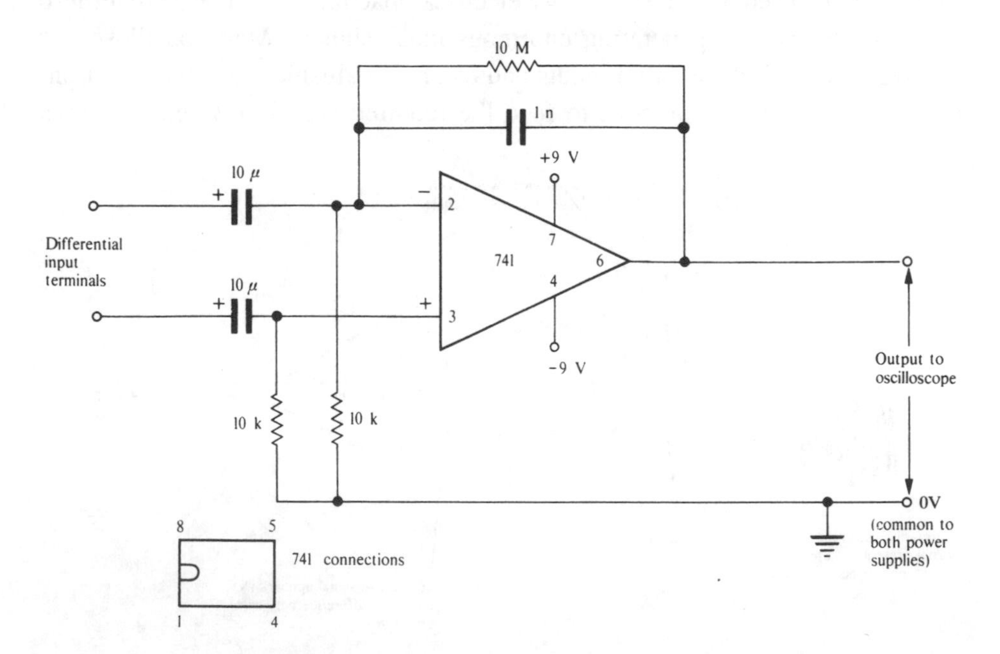

Fig. 8.14. Experimental electrocardiograph exploiting the differential inputs of the 741 IC amplifier.

scope Y-amplifier switched to a sensitivity of 0.1 V/cm and the timebase on a slow scan of about 0.1 s/cm you should now see the e.c.g. pulses from your heart as 'blips' on the screen. Moistening the hands with saline solution and grasping as large a contact area as possible will improve results.

8.9 Chopper d.c. amplifiers

In some critical instrumentation applications, even the differential amplifier has too much drift. In such cases, a radically different approach is used. This is to convert the d.c. to a.c. by chopping it up, amplify in an a.c. amplifier with inherent zero drift and then rectify the signal to recover the d.c. Such a system is shown in fig. 8.15.

Choppers were originally electromechanical, with chopping frequency limited to a few hundred Hz, but this had a serious effect on high-frequency response: as a general rule the maximum input frequency which can be handled is about a quarter of the chopping frequency. Today's electronic switches can of course chop at many MHz.

The use of the simple diode rectifier shown in fig. 8.15 has the property of giving a positive output for both positive and negative input signals. For some applications, such as the amplifier of a chart recorder, this can be an advantage, the modulus of the input signal being displayed. In most cases, however, signal polarity must be preserved at the amplifier output. Here a synchronous chopper is used so that input and output are switched through simultaneously and the instantaneous polarity of the input signal is transmitted through to the output; this arrangement is shown in fig. 8.16.

The high-frequency limitations of the chopper amplifier may be overcome, whilst at the same time exploiting its negligible drift characteristics, by using it in conjunction with a normal d.c. amplifier as a *chopper-stabilized* amplifier. This is also referred to as a *Goldberg amplifier*, after its inventor. There are various types of chopper-stabilized circuit, but one typical circuit arrangement

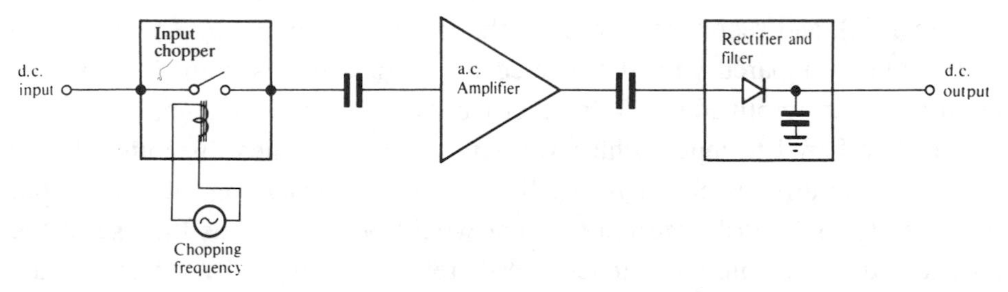

Fig. 8.15. Block diagram of chopper-type d.c. amplifier. Chopper shown as electromechanical for clarity.

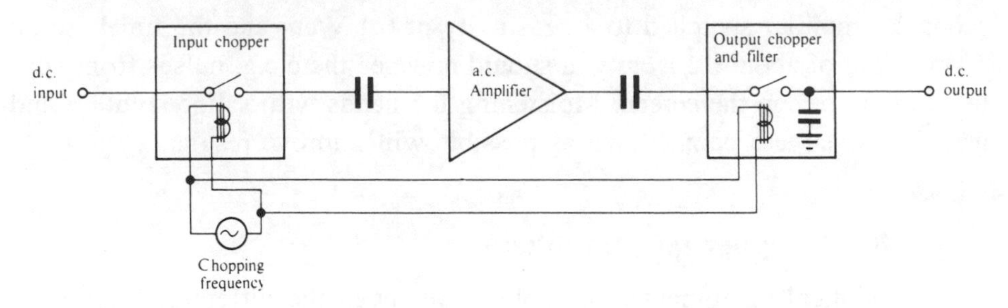

Fig. 8.16. Synchronous-chopper amplifier.

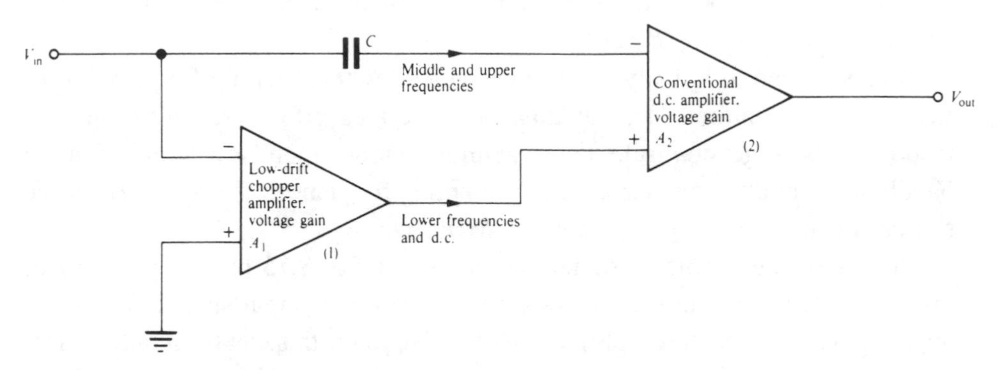

Fig. 8.17. Chopper-stabilized amplifier.

is outlined in fig. 8.17. Middle and higher frequencies are transmitted directly to the conventional amplifier via coupling capacitor C, but low-frequency and d.c. signals are blocked by C and forced to go via the chopper amplifier where they are subject to extra voltage gain A_1 before they arrive with the middle and higher frequencies at A_2.

Now if we make the reasonable assumption that the chopper amplifier (1) is drift-free, then the observed output drift, say e_o, must all have arisen in amplifier (2), and represents a drift at the input of A_2 of e_o/A_2. This is bound to be small compared with the wanted d.c. signal at this point because the latter has been amplified by A_1. In other words, looking at the complete amplifier, drift referred to input, which we have seen is the most meaningful way of expressing drift, is only e_o/A_1A_2. If A_1 were not present and the d.c. input fed directly to A_2, drift referred to input would be e_o/A_2. Thus the use of the auxiliary drift-free amplifier reduces drift referred to input by a factor equal to the amplifier gain A_1. It is, of course, clear that our open-loop middle- and high-frequency gain is only A_2 whilst d.c. and low-frequency gain is A_1A_2. However, when overall negative feedback is applied in the usual way, uniform gain is obtained over the full range so that the effect of the auxiliary amplifier is undetectable except in the very valuable reduction of drift.

Chopper-stabilized amplifiers have an important role in any instrumentation system where very low long-term drift is required, for example in strain gauge measurement. They are sometimes combined in one IC together with an analogue to digital converter.

9

Power supplies and power control

9.1 Power sources

The necessary d.c. supplies for electronic circuits may be drawn from batteries or obtained by rectification of the a.c. mains. Batteries have the advantage of portability and complete absence of a.c. components in their output. There is, however, a danger of leakage if exhausted batteries are accidentally allowed to stay too long in equipment; this may endanger many hundreds of pounds worth of circuitry through corrosion damage.

The e.m.f. of a battery is not usually constant throughout its life, that of the common low cost zinc chloride and alkaline batteries, falling from 1.6 V to 1.3 V over the useful life of the cell. Mercury oxide cells have a much better e.m.f. characteristic, remaining at 1.3 V over virtually the whole of their life and then falling off rapidly so that there is no doubt when the end of their life is reached; they are, however, expensive. Silver oxide cells have a similarly constant e.m.f. of 1.55 V.

Rechargeable nickel–cadmium (NiCd) cells are available in an enormous range of sizes, ranging from tiny 'button' cells to the large batteries used for electric traction. The smaller sizes are usually hermetically sealed so that there is no risk of leakage and no need for topping up. NiCd batteries and the newer nickel-metal-hydride (NiMH) type make an ideal power source for portable electronics, since the need for battery replacement is avoided; the charger may be incorporated into the instrument giving facilities for mains or battery operation. The e.m.f. of a nickel–cadmium cell falls from 1.3 V to 1.1 V over the useful discharge range.

Expensive but having a high capacity to weight ratio are lithium batteries, commonly used in cameras. Depending on type, the lithium cell has the unusually high and constant e.m.f. of 3.0–3.7 V. The newer lithium ion batteries combine these advantages in a rechargeable form.

The capacity of a battery is usually expressed in ampere-hours (AH), which is the product of the discharge current and the time for which the battery will give that current. Because the capacity of most batteries is greater at low load currents than high load currents, the normal rate of discharge is usually quoted along with the capacity. The discharge rate is often stated in terms of time required to discharge the battery completely; thus a car battery may have a capacity of 40 AH at the 10-hour rate, indicating that it can deliver a current of 4 A for 10 hours. Although in principle the same battery might be expected to give 40 A for 1 hour, in practice its capacity would be very much reduced at this 1-hour rate. The capacity of the sort of dry cells used to power small electronic circuits is often in the region of 3 AH at the 100-hour rate, indicating that a current of 30 mA is available for 100 hours.

Very few of the circuits in this book make unusual demands upon power supplies and most can therefore be powered by batteries. Some circuits, such as power amplifiers, require a greater output current capability than ordinary zinc chloride cells can provide, whilst occasionally a circuit may require a very stable voltage. Commercial circuits are usually powered from the a.c. mains in order to avoid battery replacement and ensure constant performance at all times. Batteries are used, of course, for portable equipment and also for those laboratory instruments where complete isolation from the mains is desirable.

9.2 Rectification of a.c.

In the majority of power supply units, a transformer is used to step down the a.c. mains to the required voltage, then the a.c. supply is rectified using diodes to give d.c. A simple rectifier circuit is shown in fig. 9.1.

The diode only allows current to flow when terminal A is positive, cutting off when A is negative. The waveform of the resulting d.c. voltage across the load is shown in fig. 9.2(b), where it may be compared with the original sinusoidal a.c. waveform in fig. 9.2(a). Since only the positive half-cycles are available, the d.c. consists of a series of unidirectional pulses. This is known as *half-wave* (HW) rectification.

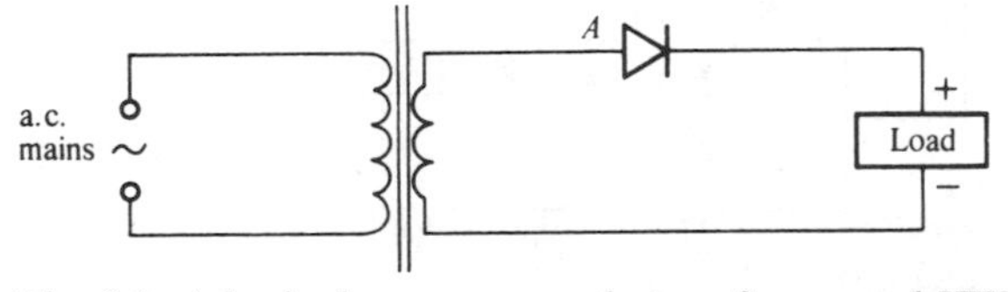

Fig. 9.1. A basic d.c. power-supply transformer and HW rectifier.

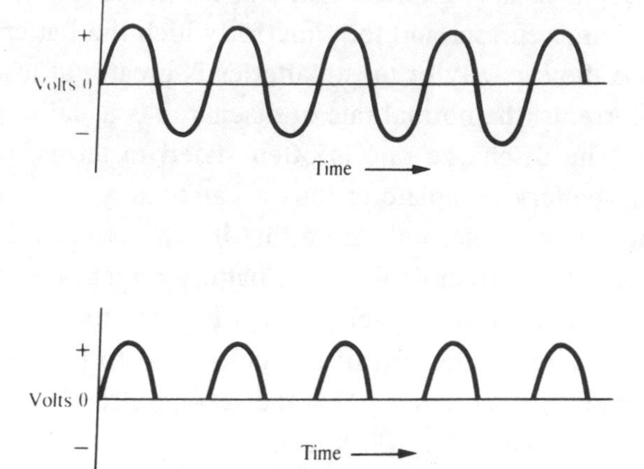

Fig. 9.2. Waveforms in HW rectifier circuit: (*a*) a.c. input waveform, (*b*) rectified unidirectional waveform across load.

Fig. 9.3 shows an improved rectifier circuit which makes use of the whole a.c. waveform and is therefore known as a *full-wave* (FW) rectifier. Because of the similarity of the diamond-shaped diode configuration to the Wheatstone bridge, this particular circuit is called a *bridge* rectifier. We can easily sort out its operation by considering what happens on successive half-cycles of the transformer output. When A is positive, D_1 conducts to make the top end of the load positive; at the same time B is negative and D_3 conducts to the bottom end of the load. On the next half-cycle, A is negative and B is positive so that D_2 conducts from B to the top end of the load and D_4 conducts from A to the bottom end of the load.

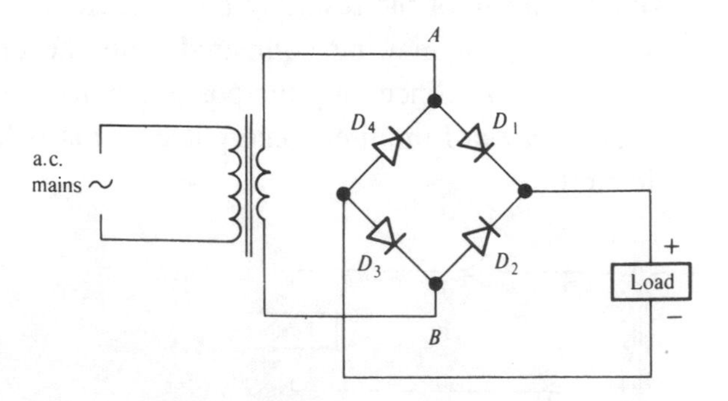

Fig. 9.3. FW bridge rectifier.

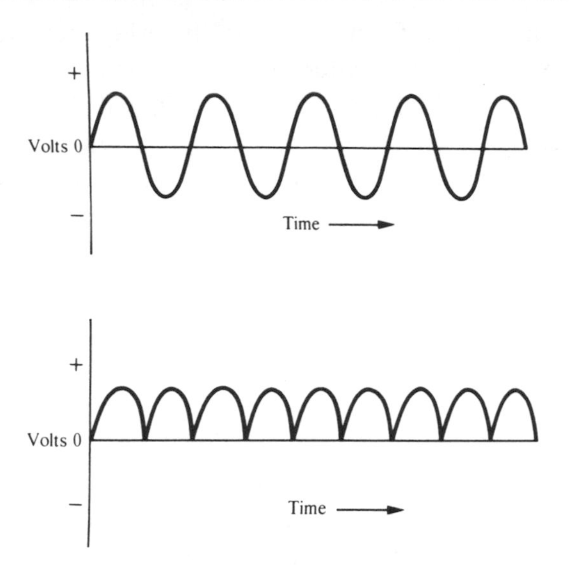

Fig. 9.4. Waveforms in FW rectifier circuit: (*a*) a.c. input waveform, (*b*) rectified unidirectional waveform across load.

The resultant waveform is shown in fig. 9.4, where it is clear that the FW d.c. waveform is of a more continuous nature than in the HW case. Notice that the frequency of the FW rectified waveform is double that of the original a.c., the negative half-cycle being inverted and located between adjacent positive half-cycles.

Fig. 9.5 shows a different type of FW rectifier. This uses a transformer secondary of double the normal voltage, but with a centre tap, O. As a result, the a.c. waveforms across AO and BO are always 180° out of phase; for this reason the circuit is called the *bi-phase* FW rectifier. On the first half-cycle, A is positive and D_1 conducts to the top end of the load. On the next half-cycle, A is negative and B is positive; hence D_2 conducts to the top end of the load. The circuit is really two HW rectifiers working together into one load.

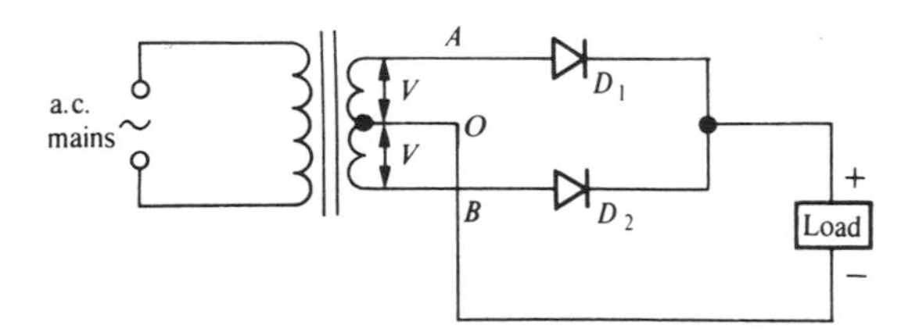

Fig. 9.5. Bi-phase FW rectifier.

Waveforms are as shown in fig. 9.4. Compared with the bridge rectifier, the bi-phase circuit saves two diodes but requires an extra transformer winding. It used to be popular in thermionic valve circuits where D_1 and D_2 were conveniently two diode valves in one envelope with a common cathode, but it is not frequently used now that solid-state diodes are available very cheaply: it is worth saving a transformer winding by using the bridge rectifier.

9.3 Smoothing

9.3.1 General

The waveforms of figs. 9.2 and 9.4, whilst unidirectional, cannot be classed as continuous d.c. because they still contain a large alternating component. They are suitable for battery charging, where the FW waveform of fig. 9.4 is preferred because of its higher average current, but they cannot be used in this raw state to supply electronic equipment.

To smooth the d.c., a reservoir capacitor, C, is connected across the rectifier output, as in fig. 9.6. The value of C for use with a 50 Hz supply typically ranges from 100 μF to 30 000 μF, depending on load current and degree of smoothing required.

The smoothing effect of the reservoir capacitor may be calculated by examining the charge and discharge periods in the waveform of fig. 9.7. The actual output voltage waveform is shown as a solid line, with its mean d.c. value as a dotted line. The corresponding half-cycles of a.c. on which the rectifier conducts are drawn in dashed form. The capacitor charges up near the peak of the a.c. half-cycle and then discharges into the load for the remainder of the cycle. Thus, for the majority of the time, the load current is being supplied by the reservoir capacitor, the rectifier simply 'topping up' the capacitor charge

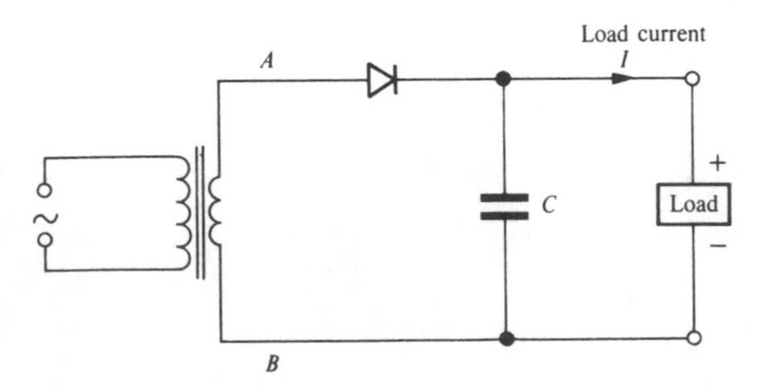

Fig. 9.6. HW rectifier with reservoir capacitor.

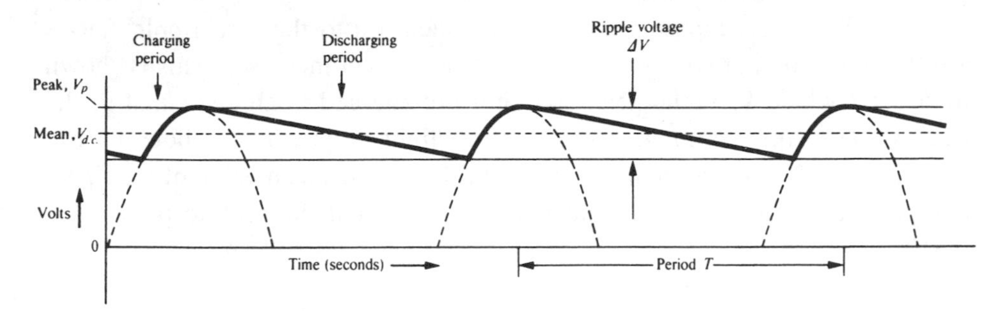

Fig. 9.7. Output waveform of HW rectifier with reservoir capacitor.

at the peak of each half-cycle. If the load current were zero, the capacitor would simply remain charged up to V_p, the peak value of the a.c. input (= $\sqrt{2} \times$ r.m.s. value).

9.3.2 *Ripple*

A finite load current, I, causes the capacitor voltage to fall by a certain amount, say ΔV, during the a.c. cycle. The resulting unevenness in d.c. output voltage is called ripple, and appears as an a.c. waveform of approximately triangular shape superimposed on the steady d.c. Peak-to-peak ripple is equal to the fall in capacitor voltage ΔV during the cycle, which can be calculated by considering the rate of discharge.

For this purpose we shall neglect the time taken for the capacitor to charge and assume that discharge occupies time T, i.e. the whole of each a.c. period. This is reasonable if the ripple is small compared with the d.c. output voltage.

The charge flowing out of reservoir capacitor C during each period of the a.c. is given by

$$\Delta Q = IT \text{ coulombs,}$$

where I is mean load current.

Fall in capacitor p.d. per period, and therefore peak-to-peak ripple is given by

$$\Delta V = \frac{\Delta Q}{C} = \frac{IT}{C} \text{ volts} \tag{9.1}$$

(I in amps, T in seconds, C in farads), or

$$\Delta V = \frac{I}{Cf} \text{ volts,} \tag{9.2}$$

where f Hz is frequency of a.c.

As well as superimposing an a.c. component on to the d.c., ripple causes a fall in mean output voltage, $V_{d.c.}$, as the load current increases. This is shown in fig. 9.7 where V_p is the peak capacitor voltage and is almost equal to the peak value of the a.c. input, there being a slight drop, usually about 1 V, in the rectifier. The difference between V_p and $V_{d.c.}$ is the mean value of the ripple component which, with a triangular wave, is one-half the peak-to-peak value ($\frac{1}{2}\Delta V$).

Hence mean d.c. output

$$V_{d.c.} \approx V_p - \frac{I}{2Cf} \text{ volts.} \tag{9.3}$$

The calculations above have assumed an HW rectifier which gives ripple of the same frequency as the a.c. input. When FW rectification is used, reference back to fig. 9.4(*b*) and to fig. 9.8 indicates that ripple frequency is then twice the a.c. input frequency.

Thus, with the FW rectifier, peak-to-peak ripple

$$\Delta V = \frac{I}{2Cf}, \tag{9.4}$$

where f is a.c. input frequency, and mean d.c. output

$$V_{d.c.} \approx V_p - \frac{I}{4Cf} \text{ volts.} \tag{9.5}$$

Hence, with the FW rectifier, ripple is half that occurring with the HW rectifier; this leads to the on-load voltage drop being half that experienced with the HW rectifier. For this reason, the FW rectifier is standard for most applications.

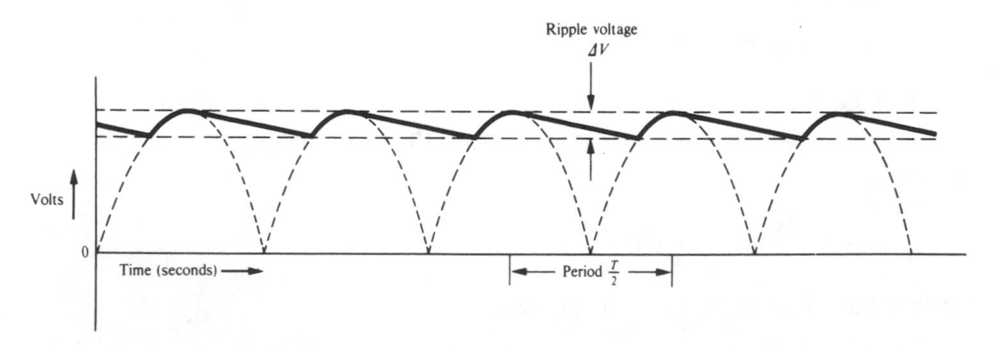

Fig. 9.8. Output waveform of FW rectifier with reservoir capacitor, showing reduced ripple voltage compared with fig. 9.7.

9.4 Load regulation

Because the output voltage of any practical power supply falls a certain amount when a load current is drawn, it is useful to specify this fall. *Load regulation* is the usual term. The load regulation of a power supply (sometimes just called regulation) is the percentage change in output voltage when the load current is increased from zero to full rated load. An alternative way of expressing the same information is simply to measure the output resistance of the supply: $R_{out} = \Delta V_{out}/\Delta I_{out}$, where ΔV_{out} represents the fall in output voltage resulting from the increase in load current ΔI_{out}.

9.5 Rectifier and capacitor ratings

The selection of diodes and capacitors for power supplies needs some care. A further look at the HW rectifier of fig. 9.6 will illustrate this. Consider the p.d. across the rectifier diode when end A of the transformer secondary is negative. The diode is not conducting, but the reservoir capacitor C at the diode cathode is charged up nearly to the positive peak value of the a.c., V_p. Since the diode anode is experiencing the negative value of V_p from the transformer, the total p.d. across the diode is $2V_p$ and it must therefore be able to withstand this reverse voltage without avalanche breakdown. The maximum reverse-voltage rating of a diode is called the *peak inverse voltage* (PIV or V_{RRM}).

Although theoretically a PIV rating of $2\sqrt{2} \times$ r.m.s. transformer voltage is adequate, it is wise to choose a PIV of at least four times the r.m.s. voltage; this allows for the presence of transient voltage pulses which may arrive at the transformer secondary as a result of mains-borne interference and which can be sufficient to break down the rectifier junction.

Rectifier current rating is usually specified as the maximum mean output current from the power supply, although in practice the current is supplied to the reservoir in short spurts of much greater magnitude. Most rectifiers are tolerant of such high peak currents provided they are of short duration, but some specify a maximum permissible value of reservoir capacitor. The reservoir capacitor should of course have a voltage rating equal to V_p ($=\sqrt{2} \times$ r.m.s. a.c. input). A requirement which may perhaps be less obvious is the *current* rating of the capacitor. Over most of the a.c. cycle the load current is in fact supplied by the capacitor, which must fulfil this function without overheating. The 'ripple current rating' of a capacitor is usually quoted in the manufacturer's specification and, in the case of electrolytic capacitors, may range from

200 mA for small general-purpose components to tens of amps for large purpose-built reservoir capacitors.

9.6 Voltage-multiplying circuits

Fig. 9.9 shows a circuit which gives a d.c. output equal to twice the peak transformer secondary voltage. It is really two separate HW rectifiers and reservoir capacitors working from the one transformer secondary. D_1 and C_1 give positive V_p relative to end B of the winding, and D_2 and C_2 give negative V_p. C_1 and C_2 thus behave like two batteries in series, giving a total of $2V_p$.

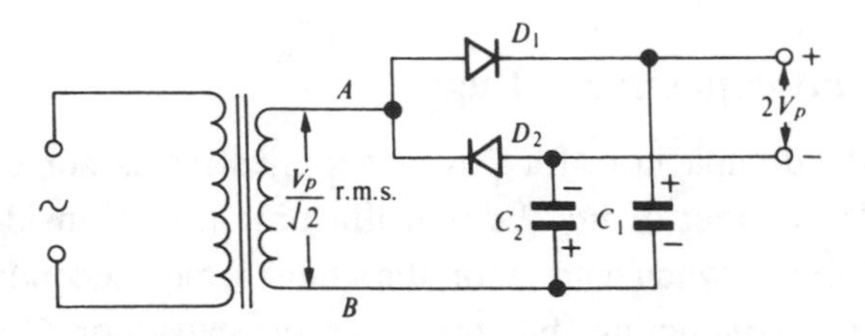

Fig. 9.9. Voltage doubler.

Fig. 9.10 is a slightly different voltage doubler circuit, which has the advantage for some applications that one output terminal is common with end B of the transformer secondary. To examine its operation, we shall consider successive half-cycles of the transformer output. When A is negative and B positive, D_1 conducts and charges C_1 to V_p; then, when A is positive and B negative, D_1 cuts off and D_2 conducts, sharing the charge on C_1 with C_2 and at the same time conducting to C_2 the positive half-cycle from the transformer, which is in series with the p.d. on C_1. The total charge on C_2 thus gives it a p.d. considerably higher than V_p. This process continues on successive cycles until, after a few cycles, C_2 is charged to $2V_p$.

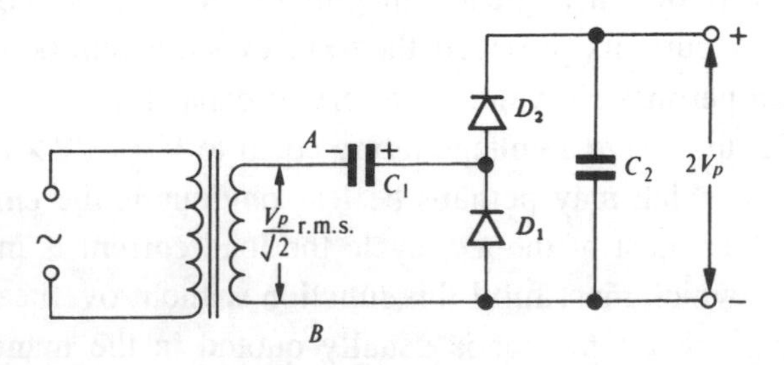

Fig. 9.10. Cascade voltage doubler.

The Cockroft–Walton voltage multiplier of fig. 9.11 is shown as a 'sextupler', but is capable of indefinite extension to give any even multiple of the peak input voltage. Comparison with fig. 9.10 shows that it is a number of cascade voltage doublers in series. When the a.c. input is switched on, C_2, C_4, C_6 gradually charge up until each has a p.d. of $2V_p$ and, since these are in series, the output is $6V_p$. Capacitors C_1, C_3, C_5 act as a.c. coupling capacitors, transmitting the a.c. wave to rectifiers D_2, D_4, D_6 so that C_2, C_4 and C_6 receive their charge.

Voltage doubler and multiplier circuits are at their best when supplying relatively small load currents such as in EHT supplies for devices like cathode-ray tubes, photomultipliers and Geiger–Muller tubes. At high loads, their ripple and load regulation are poor for two reasons. Firstly the reservoir capacitor is made up of a number of capacitors in series, so that its effective

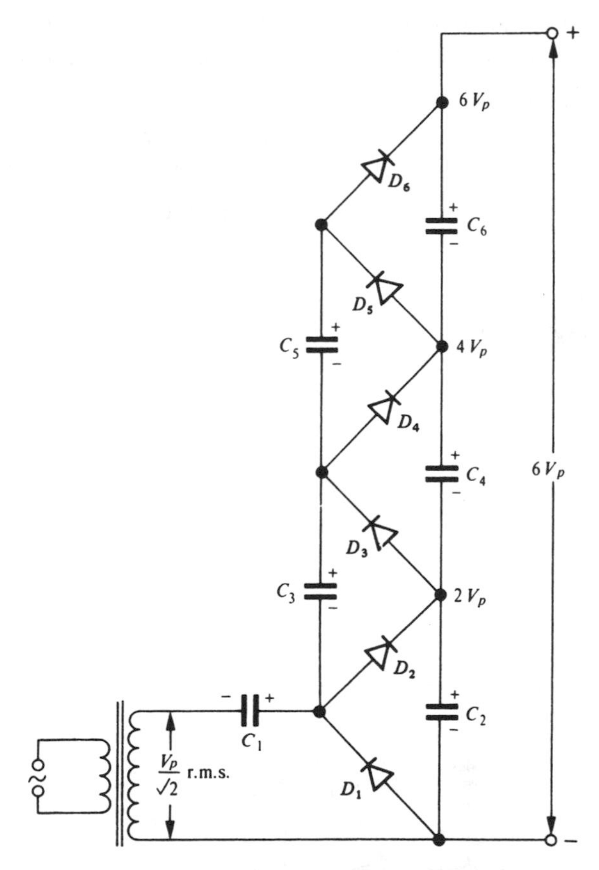

Fig. 9.11. Cockroft–Walton voltage multiplier.

value is reduced. Secondly, it takes several cycles of a.c. to recharge all the capacitors in a voltage multiplier, which contrasts with the standard FW rectifier where the reservoir is recharged twice per a.c. cycle.

9.7 Filter circuits

We have already seen that the d.c. on a reservoir capacitor has an a.c. ripple superimposed on it, except in the unusual circumstance of the load current being zero. For most purposes, this ripple must be reduced if the supply is to be acceptable for the supply of electronic circuits. This may be achieved by a simple low-pass filter such as the circuit of fig. 9.12.

The ripple waveform is approximately triangular in shape, but may be treated by ordinary a.c. theory if it is considered as a fundamental frequency with a series of harmonics. The exact proportions of fundamental and harmonics may be found by Fourier analysis, but this refinement is not necessary in the design of power supply filters. It is best to err on the safe side: if the fundamental is adequately attenuated, the harmonics, being of much higher frequency, will be reduced so must that they can be neglected. The ripple may thus be treated as though it were a sine wave of 100 Hz with FW circuits or 50 Hz with HW circuits. Then, in the case of the RC filter of fig. 9.12, if $V_{rip(res)}$ is the r.m.s. ripple voltage on the reservoir capacitor and $V_{rip(out)}$ the output ripple (frequency f).

$$V_{rip(out)} = \frac{1/j\omega C}{R+(1/j\omega C)} V_{rip(res)} \quad (\omega = 2\pi f).$$

In most cases $R \gg 1/\omega C$ so that

$$V_{rip(out)} \approx \frac{1/j\omega C}{R} V_{rip(res)},$$

i.e.

$$|V_{rip(out)}| \approx \frac{V_{rip(res)}}{R\omega C}. \tag{9.6}$$

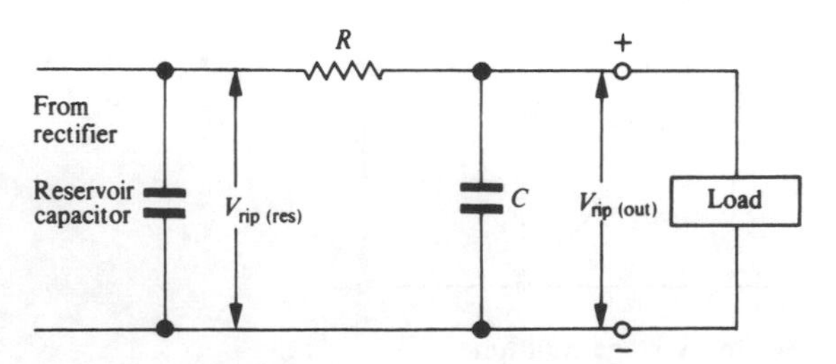

Fig. 9.12. Low-pass filter for ripple reduction.

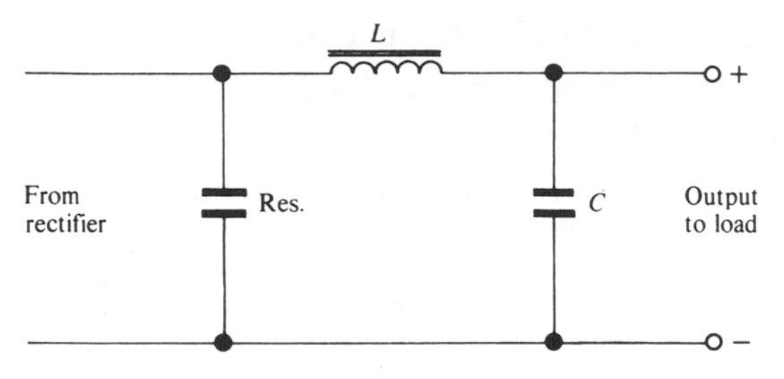

Fig. 9.13. Low-pass filter using a choke (inductor).

For HW rectification, f is supply frequency; in the FW case, f is twice supply frequency.

One snag of the RC filter is that the series resistor increases the output resistance of the supply, thus degrading the load regulation. The choke-capacitor (LC) filter of fig. 9.13 avoids this problem because the choke has a high impedance at ripple frequency, but a low d.c. resistance, thus reducing ripple but not greatly increasing output resistance.

For LC filter

$$V_{\text{rip(out)}} = \frac{1/j\omega C}{j\omega L + (1/j\omega C)} V_{\text{rip(res)}}.$$

Now in most cases $\omega L \gg 1/\omega C$,

$$V_{\text{rip(out)}} \approx \frac{1/j\omega C}{j\omega L} V_{\text{rip(res)}},$$

i.e.

$$|V_{\text{rip(out)}}| \approx \frac{1}{\omega^2 LC} V_{\text{rip(res)}}. \tag{9.7}$$

9.8 Decoupling

When several circuit stages are fed from one power supply, it is possible that unwanted feedback may occur between the stages via the power supply output impedance. It may be difficult to lower the supply output impedance sufficiently to prevent this, and it is a common practice to incorporate RC filters in the power supply to low-level amplifier stages.

This procedure is called *decoupling* and has the additional benefit of reducing ripple to the decoupled stages. A typical example of decoupling is shown with typical values in fig. 9.14; the value of R is limited by the permissible voltage drop and C is then chosen to give the necessary filtering action. The term decoupling is also used to refer to any use of a capacitor to shunt out a.c. signal components; for example the emitter resistor bypass capacitor, shown as C_E in the first stage of fig. 9.14, is called an emitter decoupling capacitor.

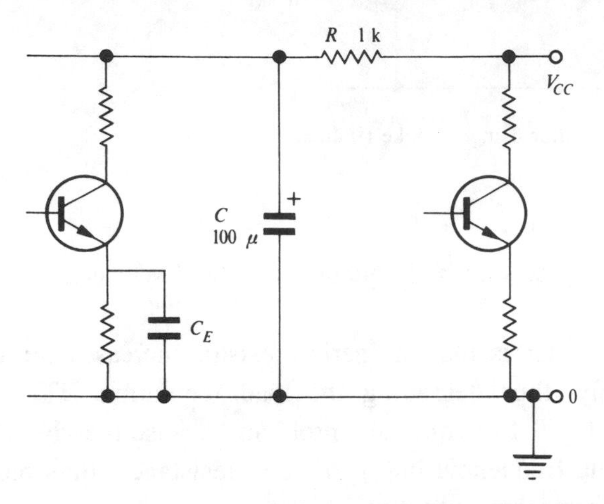

Fig. 9.14. Typical power supply decoupling using R and C. Emitter resistor bypass capacitor C_E is another example of decoupling.

9.9 Variable voltage supplies

9.9.1 Potentiometer

In experimental electronics, a power supply with an adjustable output voltage is a very useful piece of equipment. The simplest way to obtain a variable output voltage is to use a potentiometer, an example of which is shown in fig. 9.15. Here, the d.c. source is an 18 V battery, but it could equally well be a transformer, rectifier and reservoir capacitor. The potentiometer can be adjusted to give any required output voltage, but there is a serious snag: the presence of the potentiometer increases the output resistance of the supply so that the load regulation is very poor.

This may be illustrated by building the circuit of fig. 9.15, connecting a voltmeter to the output terminals and adjusting the potentiometer to give 10 V output. If a 1 kΩ load resistor is now connected across the output terminals; a dramatic fall in output voltage will occur.

This poor load regulation results from the high output resistance of the potentiometer, which is at a maximum when the slider is around the middle of the track. (The effective output resistance may be deduced from the fall in

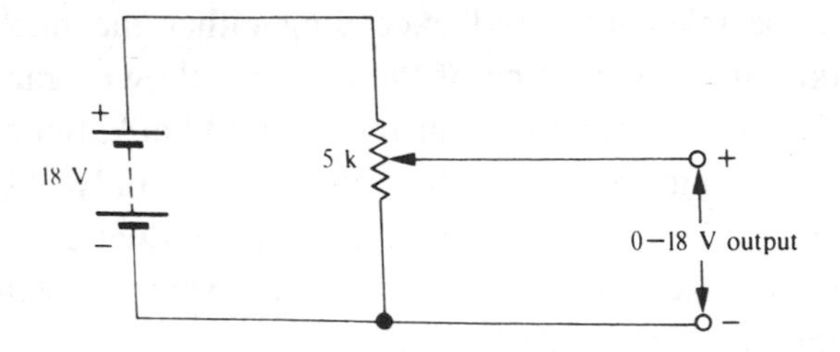

Fig. 9.15. Basic potentiometer circuit provides variable voltage supply.

output voltage with the 1 kΩ load. See section 5.5.) The output resistance can, of course, be reduced by using a lower value potentiometer, but, in order to give satisfactory regulation, the value would have to be so low that more power would be wasted in the potentiometer than is used in the load. A much more satisfactory solution is to use a transistor in the emitter follower configuration to reduce output resistance as described in the next section.

9.9.2 The emitter follower in power supply circuits

Consider the circuit of fig. 9.16. The potentiometer provides a variable voltage as before, but, this time, the output resistance is lowered by the emitter follower transistor.

The effect of the emitter follower may be assessed experimentally by once again loading the output with a 1 kΩ resistor and noting the fall in voltage compared with the previous case. The output resistance is reduced by a factor equal to the current gain, h_{FE}, of the transistor. The transistor emitter load is provided by the actual load on the power supply itself; the output voltage across the load is equal to the voltage applied to the base minus the base–emitter potential drop of approximately 0.6 V. Because of its position in the circuit, the emitter follower in power supplies is sometimes known as the *series pass* transistor.

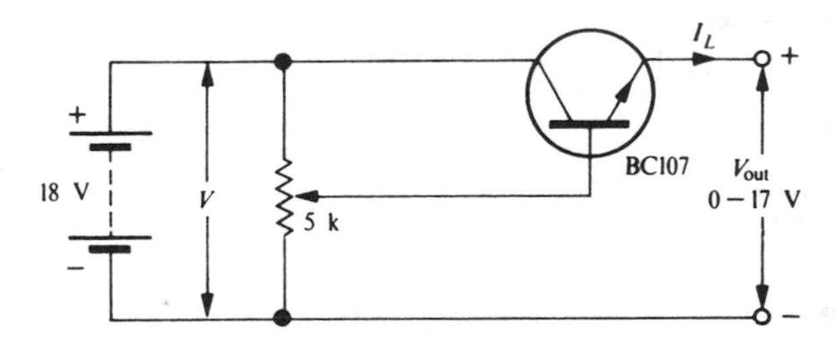

Fig. 9.16. Emitter follower lowers output resistance of potentiometer.

Care must, of course, be taken to avoid exceeding either the maximum current rating or the maximum power rating of the emitter follower transistor. In the case of the BC107, the maximum current rating is 300 mA, but a more serious limitation is the maximum power dissipation, which is only 360 mW. Now the power dissipation in the transistor is given by the product of the load current and the voltage drop across the transistor. For example, using the symbols in fig. 9.16, transistor power dissipation

$$W = I_L(V - V_{out})\text{ watts.}$$

Thus, if $I_L = 20\text{mA}$, $V = 18\text{V}$ and $V_{out} = 3\text{V}$,

$$\begin{aligned} W &= 20 \times (18 - 3)\text{ mW} \\ &= 300\text{ mW}. \end{aligned}$$

For a given load current, the lower the output voltage, the greater is the dissipation in the transistor. The circuit is more generally useful if a power transistor is added to form a Darlington pair, as in fig. 9.17. Here, the output resistance is further reduced by the current gain of the additional transistor. For instance, if the output resistance of the 5 kΩ potentiometer is of the order of 1 kΩ, this will be divided by the product of the transistor current gains, typically 200 for the BC107 and 30 for the 2N3055. According to simple theory, the output resistance will then be given by

$$\begin{aligned} R_{out} &= \frac{1000}{200 \times 30}\,\Omega \\ &\approx \frac{1}{6}\,\Omega. \end{aligned}$$

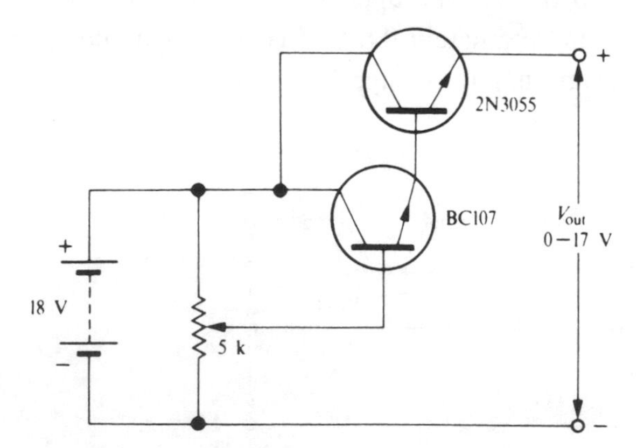

Fig. 9.17. Use of Darlington pair with power transistor to provide higher output current.

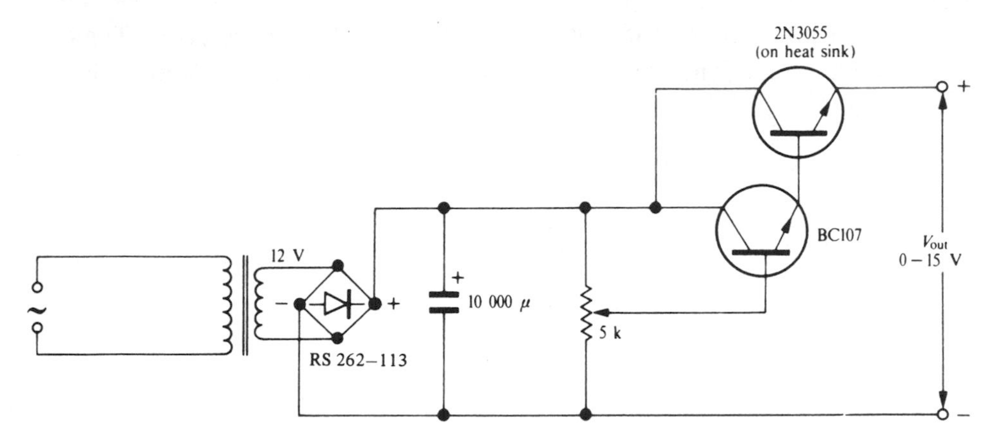

Fig. 9.18. Mains-operated variable-voltage power supply (maximum load 2 A).

Practical limiting factors on output resistance are the emitter resistance of the output transistor and the internal resistance of the battery.

The 2N3055 power transistor will dissipate about 3 W in free air, but its power handling is greatly increased by mounting it on a heat sink (see section 9.11). The circuit in fig. 9.17 may be fed from a transformer, bridge rectifier and reservoir capacitor to provide a general purpose variable voltage power supply for experimental applications (fig. 9.18).

9.10 Voltage stabilizers

9.10.1 General

The variable voltage circuits discussed so far have one disadvantage in common. No matter how high a current gain is provided in the emitter follower, there is a limit to the attainable load regulation, this being governed by the internal impedance of transformer and rectifier and size of reservoir capacitor.

Furthermore, any fluctuation in mains input voltage will be faithfully transmitted to the output so that, even if the load is constant, there will be variations in output voltage. These limitations are overcome in that class of circuits known as voltage stabilizers.

9.10.2 Basic Zener diode stabilizer

The reverse breakdown properties of the pn junction were discussed in chapter 1, where it was mentioned that Zener or avalanche diodes make use of high doping levels to obtain artificially low reverse breakdown voltages. Typical characteristics of a Zener diode are shown in fig. 9.19, where a 5 V breakdown is illustrated.

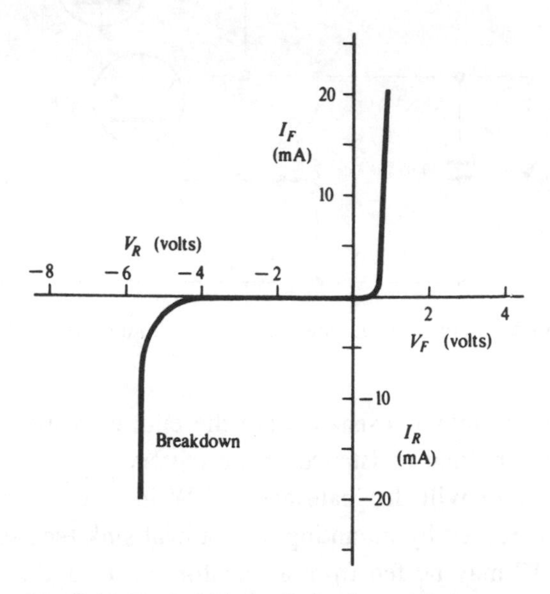

Fig. 9.19. Typical Zener diode characteristics.

The p.d. across the diode in the breakdown condition is almost constant over a wide range of currents; this property is exploited in the simple voltage circuit of fig. 9.20. Here the output voltage is the equal to the diode p.d. and is therefore constant over a wide range of input voltages.

Note that the input voltage to a stabilizer circuit must be at least two or

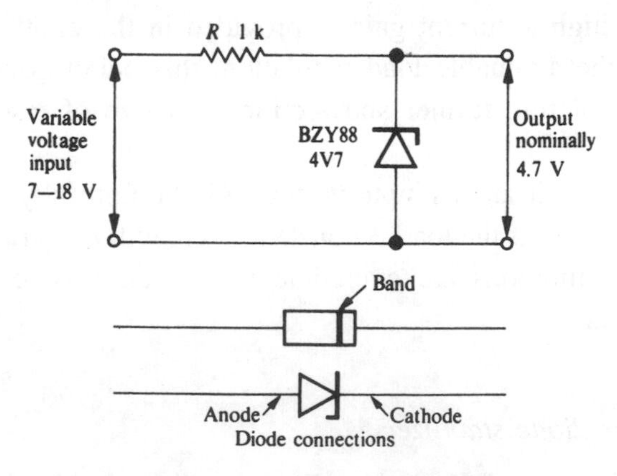

Fig. 9.20. Basic Zener diode voltage stabilizer.

three volts higher than the required output voltage in order to push adequate current through the Zener diode to ensure that it operates on its breakdown curve.

9.10.3 Stabilization ratio

The degree of stabilization produced by a particular circuit can be specified as its *stabilization ratio*, which is obtained by measuring the percentage change in output voltage produced by a given percentage change in input voltage. Then

$$\text{stabilization ratio} = \frac{\%\ \text{change in input voltage}}{\%\ \text{change in output voltage}}.$$

A basic Zener diode stabilizer like fig. 9.20 usually gives stabilization ratios between 5 and 20, whilst some of the more sophisticated IC regulators give values over 1000. An alternative term which defines degree of stabilization is *line regulation*, also known as *input regulation*. Line regulation may be defined in a number of ways, one common one being

$$\text{line regulation} = \frac{\text{change in output voltage}}{\text{change in input voltage}} \times 100\%.$$

Unlike stabilization ratio, this does not take account of the fact that the input and output voltages themselves are probably different, i.e. 25 V input may give 15 V output. However, this fact does not make an order of magnitude difference to the result. It is, though, misleading to compare the change in low voltage output with the change in mains voltage in this way: this would imply wrongly that a step-down transformer by itself has a significant stabilizing effect! A typical value of line regulation for a well-stabilized supply is 0.01%. An alternative definition of line regulation is the percentage change in output voltage for a given percentage change in input voltage. If this definition is specified for a 10% change in input voltage, as it often is, then line regulation can be related to stabilization ratio (SR):

$$\text{line regulation} = \frac{10}{\text{SR}}\%.$$

In comparing the merits of power supplies, it is essential to note the means used to express stabilization because different manufacturers use different methods which may not be strictly comparable.

9.10.4 Limitations of simple Zener diode circuit

The Zener diode is not, of course, a perfect stabilizer. If we examine the breakdown characteristic carefully, as in fig. 9.21, we see that it is of finite slope. In other words, the potential difference increases slightly with increasing current through the diode. This effect is very marked at low currents, and most small Zener diodes should have at least 5 mA, and preferably 20 mA, passing through them at all times if optimum stabilization is to be obtained. To express the variation of Zener voltage with current, the term dynamic slope resistance is used where, from fig. 9.21,

$$\text{dynamic slope resistance } r_s = \frac{\Delta V}{\Delta I}.$$

The Zener diode may be considered as a battery of open-circuit e.m.f. V_Z and internal resistance r_s. It is, of course, a very unusual battery, which must always have a current supplied to it to maintain its e.m.f. This concept is used for the equivalent circuit in fig. 9.22, which is an alternative representation of the basic stabilizer of fig. 9.20.

It can be seen that the total current I drawn from the input supply splits into two currents: I_Z through the Zener diode and I_L into the load.

I is given by Ohm's law:

$$I = \frac{V_{\text{in}} - V_{\text{out}}}{R}$$

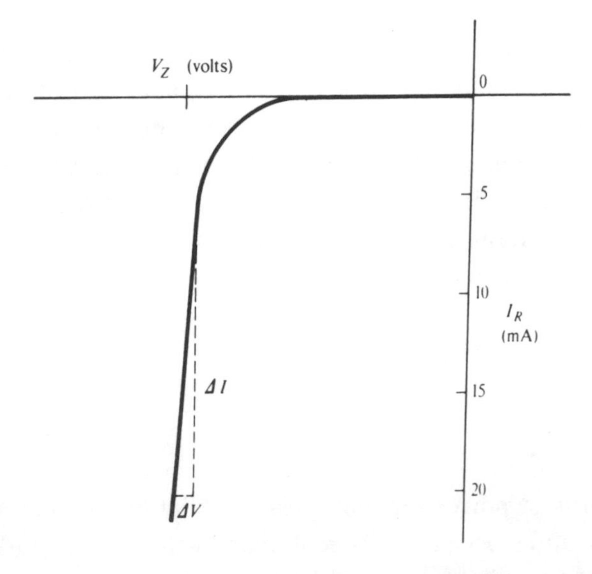

Fig. 9.21. Detailed view of typical Zener breakdown characteristic showing slope.

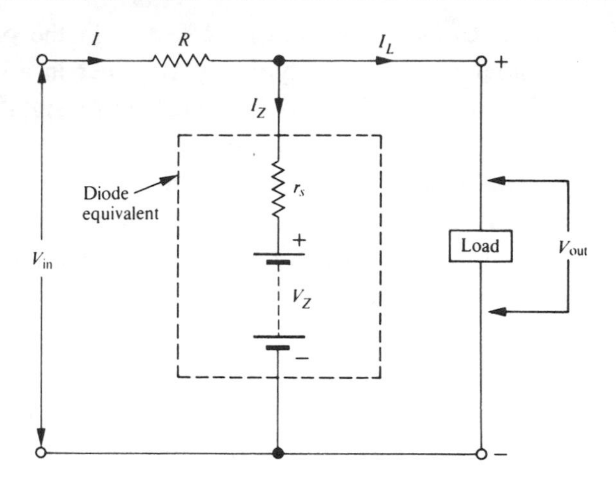

Fig. 9.22. Equivalent circuit version of Zener diode stabilizer of fig. 9.20.

and, since $V_{out} \approx V_Z$ (r_s is small compared with R),

$$I \approx \frac{V_{in} - V_Z}{R}.$$

Thus for a given V_{in}, I is virtually constant. Now, according to Kirchoff's first law,

$$I = I_L + I_Z.$$

The implication here is that, for a particular stabilizer circuit, if the load current I_L increases, the Zener current I_Z falls. The designer must be careful to ensure that, even at full load current, there is still sufficient current through the Zener diode to keep it past the knee in the breakdown characteristic. On the other hand, if the load is removed, all the current flows through the Zener diode so that $I_Z = I$. Here the diode is subjected to a high power dissipation given by

$$P_{max} \approx V_Z I_Z \text{ (r_s is once again neglected)}$$
$$\approx V_Z I.$$

The Zener diode selected must be capable of dissipating this 'no load' power without damage. The usual power rating for small diodes such as the BZY88 or BZX79 series is 400 mW, but ratings as high as 10 W are readily available. As will be seen later, however, it is usually possible to avoid the need for high Zener diode powers in stabilizer circuits.

We have so far neglected the dynamic slope resistance r_s in the current and power calculations. Where r_s does have a significant effect is in the output voltage, V_{out} in fig. 9.22. Although r_s is small, typically 10–20 Ω in a 4.7 V diode, the small changes in V_{out} resulting from its presence can be significant in an output which is supposed to be perfectly stable.

In the first place, r_s determines the load regulation since it appears as a finite output resistance. Secondly it limits the stabilization ratio. Changes in V_{in} produce corresponding changes in I and therefore I_Z; these changes in I_Z cause a varying voltage drop across r_s which changes V_{out}, given by

$$V_{out} = V_Z + I_Z r_s.$$

The simple relations established above are illustrated in the example of fig. 9.23 by calculating the answers to various questions the circuit designer might ask:

(*a*) With $V_{in} = 18$ V, what is the maximum power dissipation in the Zener diode?

Maximum diode power dissipation occurs at no load and is given by

$$P_{max} = V_Z I_Z = V_Z I,$$

where

$$V_Z = 4.7\text{V}$$

and

$$I = \frac{V_{in} - V_Z}{R}$$

$$= \frac{18 - 4.7}{1000}\text{A} = 13.3\,\text{mA}.$$

therefore

$$P_{max} = 4.7 \times 13.3\,\text{mW}$$

$$= 62.5\,\text{mW}.$$

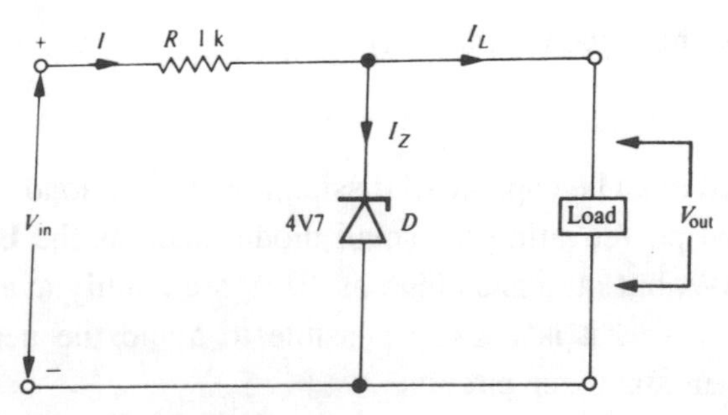

Fig. 9.23. Stabilizer circuit used in the calculation.

(*b*) With $V_{in} = 18$ V, what is the maximum load current that may be drawn?
From Kirchoff's first law,

$$\text{input current } I = I_L + I_Z.$$

With $V_{in} = 18$ V and no load we have established that $I = 13.3$ mA.

Now, to get beyond the knee in the breakdown characteristic for a small Zener diode, $I_Z \geq 5$ mA.

Hence, maximum permissible load current in this case is

$$I_L \approx 13.3 - 5 \text{ mA}$$
$$\approx 8 \text{ mA}.$$

(*c*) If the dynamic slope resistance r_s of the diode is 20 Ω, what is the change in no-load output voltage when the input voltage falls from 18 V to 9 V?

To determine the effect of r_s on output voltage, we must find the change in Zener current I_Z resulting from the input voltage reduction.

For no load and $V_{in} = 18$ V, we have already from (*a*) $I_Z = 13.3$ mA. Now for no load and $V_{in} = 9$ V,

$$I_Z = I = \frac{V_{in} - V_Z}{R}$$
$$= \frac{9 - 4.7}{1000} \text{A}$$
$$= 4.3 \text{ mA}.$$

Notice that it is convenient to neglect r_s at this stage of the calculation ($r_s \ll R$).

Therefore, for this fall in I_Z from 13.3 mA to 4.3 mA with $r_s = 20$ Ω,

$$\text{fall in } V_{out} = \frac{20(13.3 - 4.3)}{1000} \text{V}$$
$$= 0.18 \text{ V}.$$

This illustrates clearly the stabilizing action of the Zener diode. A 50% fall in input voltage causes only 0.18 V fall in output, which is equivalent to (0.18/4.7) × 100%, or approximately 4%. The stabilization ratio is thus 50/4 or 12.5.

(*d*) What is the change in output voltage when the load current increases by 5 mA?

An increase in load current of 5 mA means a decrease of Zener current by 5 mA.

Thus,

$$\text{fall in } V_{\text{out}} = \frac{5}{1000} \times r_s$$

$$= \frac{5 \times 20}{1000} \text{ V}$$

$$= 0.1 \text{ V}.$$

9.10.5 Dealing with high load currents

In part (*b*) of the example just discussed, we found that the maximum permissible load current for the Zener stabilizer of fig. 9.23 was about 8 mA, any greater load robbing the diode of current so that it could not give its nominal Zener voltage. In most cases, load currents much greater than this must be supplied. One way of achieving this is simply to reduce the value of the series resistor R so that a greater current I flows to the diode and load; care must be taken, though, that the maximum rated power of the diode is not exceeded. This line of action can involve considerable waste of power and require an expensive high-power Zener diode. Fortunately, there is a more elegant solution to the problem of high-current supplies.

We have already seen that the emitter follower can be employed to reduce the output resistance of a simple potentiometer (figs. 9.16 and 9.17). If an emitter follower is added to the simple Zener stabilizer, the available output current is increased by the current gain of the transistor. In fig. 9.24, such a circuit is shown. The BFY50 transistor is chosen because of its ability to dissipate a power of 800 mW, or 1.5 W if used with a clip-on heat sink. The voltage of Zener diode D_1 is chosen according to the output voltage required,

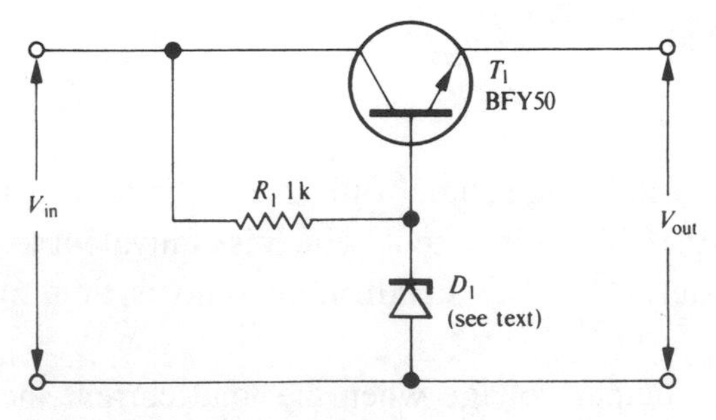

Fig. 9.24. Zener stabilizer with emitter follower.

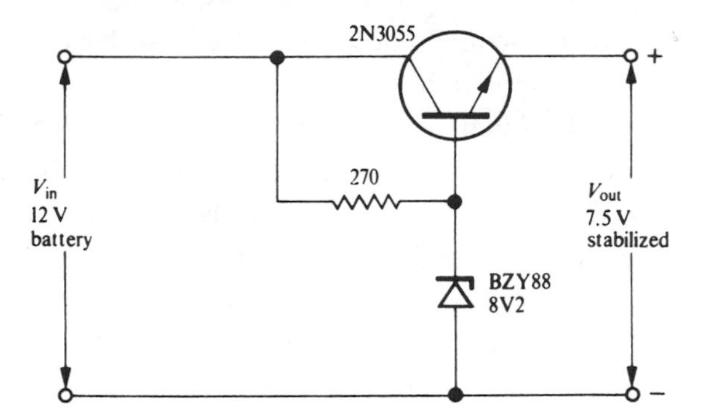

Fig. 9.25. Practical stabilizer which will operate a cassette tape recorder from a 12 V car battery.

the latter being approximately 0.7 V less than the diode voltage owing to the transistor base–emitter drop. The value of R_1 (1 kΩ) assumes a difference of about 10 V between input and output voltage, giving approximately 10 mA current in the diode. The latter can be a standard 400 mW component.

The maximum available output current is chiefly limited by the power dissipation of the BFY50; for a voltage drop of approximately 10 V the output current should be limited to 80 mA to avoid excess dissipation. Lower voltage drops will allow greater currents to be drawn.

Fig. 9.25 shows a rather higher-power circuit designed for a specific purpose, which is to provide a 7.5 V supply from a 12 V car battery to run a battery-operated radio-cassette. The 2N3055 power transistor is required to dissipate about 3 W; a heat sink is not essential for this power, but if a metal box or chassis is used to contain the unit, the 2N3055 can be conveniently mounted on it to reduce the working temperature. Output voltage can be changed by using a different Zener diode. For example substituting a BZY88 3V6 gives a 3 V output suitable for a 'Walkman'.

Great care must be taken not to short-circuit the output of this simple type of series regulator. If as a result the transistor were to fail as a short between collector and emitter, this would apply the full input voltage to the output and probably cause catastrophic damage to the circuit being supplied. Commercial circuits normally include an overvoltage sensor connected to a fast shunt thyristor *crowbar* which kills the output before damage can occur.

9.10.6 Improved stabilization using an error amplifier

The Zener diode and emitter follower combination gives much improved load regulation compared with the Zener diode alone, but its line regulation is no better. In addition, the fixed output voltage of the stabilizers so far considered may be a disadvantage in many applications.

Regulation can be improved further by incorporating an amplifier in the circuit to compare the reference voltage from the Zener diode with a given fraction of the output voltage. Such a circuit, which also has provision for adjusting the output voltage, is shown in fig. 9.26. Here, the voltage amplifier transistor T_2 has R_1 as its collector load and feeds emitter follower T_1. The emitter of T_2 is held at a constant voltage by Zener diode D_1. Potentiometer R_3 and resistor R_4 feed a given fraction of the output voltage to the base of transistor T_2. Since the emitter of T_2 is held at the Zener voltage, the circuit output voltage adjusts itself until the base of T_2 is 0.6 V (base–emitter drop) above the Zener voltage. By adjusting potentiometer R_3 and thus feeding back a different fraction, the output voltage may be varied.

Because the Zener current in fig. 9.26 is almost independent of load current, the load regulation of this type of circuit can be very good. The line regulation is still limited by Zener slope resistance and may be somewhat improved by connecting R_2 to the stabilized output instead of the unstabilized input as shown in the diagram. Further improvements can be made by replacing T_2 with a differential amplifier (e.g. a 741 integrated circuit) and using a Darlington pair in the emitter follower instead of single transistor T_1. The differential amplifier virtually eliminates voltage drift with temperature, whilst the Darlington pair further lowers the output impedance.

The error amplifier type of stabilizer can be considered simply as a high gain d.c. amplifier with a very steady input voltage, provided by the Zener

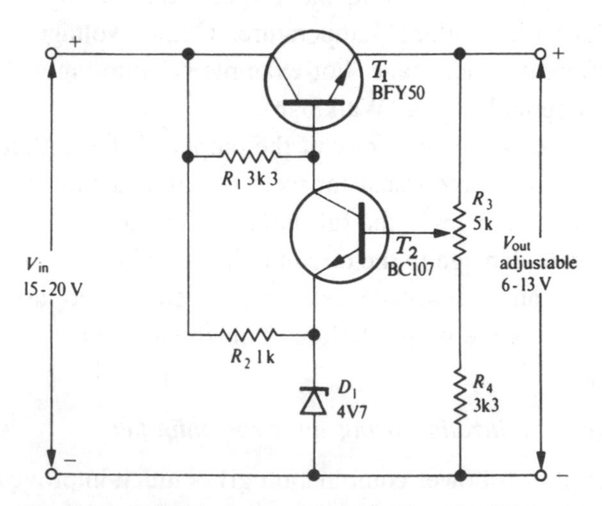

Fig. 9.26. Stabilizer incorporating an error amplifier.

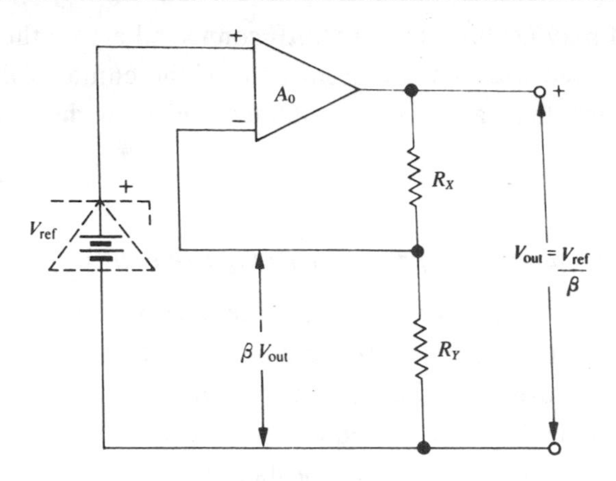

Fig. 9.27. Stabilized power supply circuit considered as an amplifier with negative feedback.

diode, and with a given negative feedback fraction (β) determined by the potential divider at the output (R_3 and R_4 in fig. 9.26). Fig. 9.27 shows the essentials of the error amplifier circuit considered in this way, where $\beta = R_Y/(R_X + R_Y)$. If A_0 is very high, then, from the feedback theory discussed in section 4.3,

$$V_{out} = \frac{V_{ref}}{\beta}$$

$$= V_{ref} \frac{(R_X + R_Y)}{R_Y}.$$

The unstabilized input simply acts as the power supply for the amplifier. Variations in that power supply should not affect V_{out} if A_0 is so high that the equations above are applicable and if V_{ref} is absolutely constant.

9.10.7 Stabilizers and ripple reduction

The use of a voltage stabilizer has a very beneficial effect in reducing the ripple on a supply. Since ripple is nothing more than a rapid periodic variation in voltage, it is subject to exactly the same stabilizing effect as longer term voltage variations. Ripple percentage is therefore reduced by a factor equal to the stabilization ratio of the stabilizer. Stabilizers of the form shown in fig. 9.27 may have their ripple reduction improved still further by connecting a large capacitor (typically 10 μF) across R_X thus feeding back the whole of the ripple of the amplifier ($\beta = 1$). The Zener diode and emitter follower combination of fig. 9.24 may usefully have 680 μF connected across the diode to reduce ripple further. Because of the current gain of the emitter follower, this is equivalent to many thousands of microfarads added to the reservoir capacitor.

9.10.8 The Zener diode as a precision voltage reference

A stabilized power supply is only as good as its reference voltage. We have already seen that, if the Zener diode is to give a constant output voltage, then, because of the finite slope resistance the current through it must be constant. Two common ways of achieving this are the use of a second diode as a pre-stabilizer and the use of a transistor as a constant-current source. A pre-stabilizer circuit is shown in fig. 9.28, where a 10 V diode is used to stabilize the supply to a 5.6 V diode. The latter therefore experiences an almost constant current irrespective of input voltage variations.

Fig. 9.29 is the Williams 'ring of two' which elegantly uses bipolar transistors as constant-current sources to supply the Zener diodes. The base of T_1 is held at 5.6 V; therefore it adjusts its emitter current so that the emitter is at $(5.6 - 0.6 = 5.0\text{ V})$; thus T_1 emitter current is given by 5.0/470 A, or approximately 10 mA. The collector current of T_1, which is approximately equal to its emitter current, supplies diode D_1 which in turn supplies the base of T_2 causing T_2 to pass a constant current of 10 mA through diode D_2. The latter diode acts as the reference and also feeds the base of T_1.

The breakdown voltage of most Zener diodes varies with temperature. Diodes designed for breakdown below 5 V depend chiefly on electron tunnelling for their action and exhibit a negative temperature coefficient, i.e. their breakdown voltage decreases with increasing temperature. At voltages above about 6 V, avalanche effect is dominant in the breakdown, and this has a positive temperature coefficient, breakdown voltage increasing with increasing temperature. There springs to mind the question of what happens between

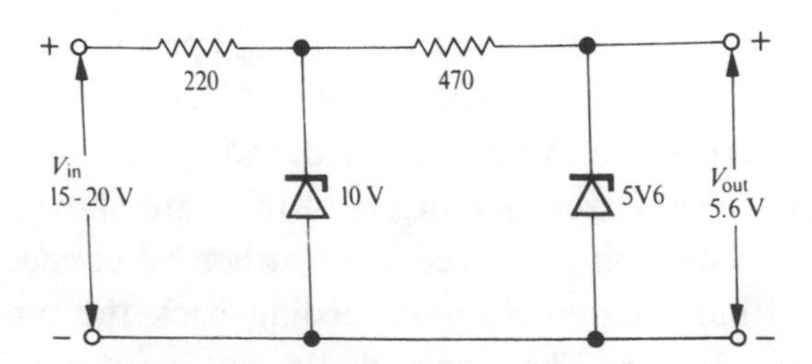

Fig. 9.28. Stable voltage reference using a pre-stabilizer.

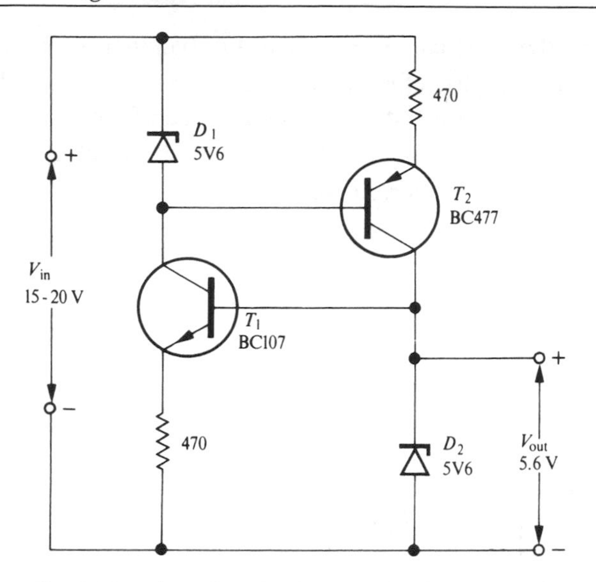

Fig. 9.29. 'Ring of two' reference circuit uses transistors as constant-current sources.

these two regions, where the breakdown is a combination of both mechanisms. The answer is that diodes with breakdowns around 5.6 V or 6.2 V can be designed with very small temperature coefficients indeed and, if used in circuits like figs. 9.28 and 9.29, can produce an e.m.f. as stable as a Weston standard cell.

9.10.9 The bandgap voltage reference

The increasing use of sophisticated battery operated electronics such as notebook computers and cellular telephones has demanded voltage stabilizer circuits which waste as little power as possible. Such circuits need a voltage reference which draws minimal current and yet gives excellent stability over a wide temperature range. The bandgap reference is used widely for this purpose. It exploits neither the Zener nor the avalanche effect but instead uses the forward p.d. (V_{BE}) of a transistor base–emitter junction. At first sight this seems improbable because of the known temperature dependence of V_{BE}.

As the junction temperature increases, the liberation of additional minority carriers results in higher reverse saturation current I_0 and a consequential reduction in the p.d. across the depletion layer. Therefore V_{BE} exhibits a negative temperature coefficient: with a constant current, V_{BE} falls as the junction gets hotter.

This temperature dependence is cancelled out in the bandgap reference by a circuit due to Widlar which exploits the fact that junctions working at different current densities have different negative temperature coefficients. Fig. 9.30 is an outline circuit of a typical bandgap reference which is being supplied with a reasonably constant external current I_{EXT}.

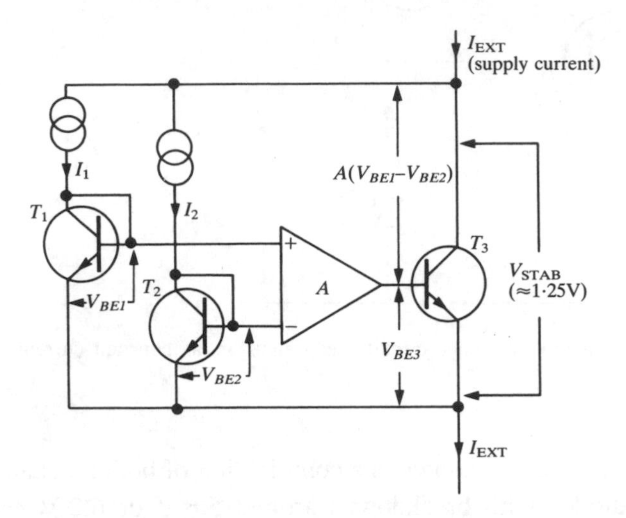

Fig. 9.30. Basic features of bandgap voltage reference.

Transistors T_1 and T_2 are supplied with different currents, I_1 and I_2 from transistor constant current sources. The resulting small difference between V_{BE1} and V_{BE2} exhibits a positive temperature coefficient and is amplified by the differential amplifier. The amplifier output voltage, $A(V_{BE1} - V_{BE2})$ is arranged to be similar in magnitude to V_{BE3} but exhibits the opposite temperature coefficient. The resulting output voltage, V_{STAB}, is the series combination of V_{BE3} and the amplifier output and is virtually constant with temperature.

For a current I_{EXT} around 1 mA, a typical bandgap reference drops a constant 1.25 V. Commercially available precision bandgap chips are normally designed to drop a voltage which is a multiple of 1.25 V, e.g. 2.5 V, 5 V etc.

9.10.10 Short-circuit protection

It is common practice for power supply units to incorporate an automatic current-limiting circuit. This is in order to avoid damage, either to the unit itself in the event of a short circuit across the terminals, or to the outside

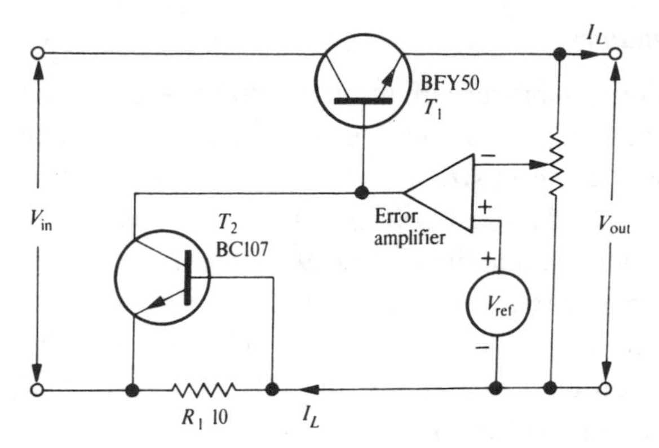

Fig. 9.31. Power supply with current-limiting circuit, R_1 and T_2.

circuit, should a fault develop. Fig. 9.31 shows a simple version of such a circuit; only the circuit elements essential to the overload protection are shown in detail, the error amplifier and reference being included in block form in the interest of clarity. The 10 Ω resistor R_1 produces a p.d. $I_L R_1$ proportional to load current I_L (its presence does not upset the regulation, because it comes before the regulator). Transistor T_2 has its base–emitter junction connected directly across R_1, so that, when the p.d. across R_1 approaches 0.6 V, T_2 begins to conduct, its collector pulling down the base of emitter follower T_1 towards earth. As the potential on T_1 base drops, so the output voltage, V_{out}, of the power supply drops. If the current continues to rise, T_2 will be saturated, the base of T_1 will be held near earth and the output voltage will be zero. In this example, $R_1 = 10\ \Omega$, so that a current of 60 mA is needed in order to drop 0.6 V to turn T_2 on; thus the current begins to limit at 60 mA in this case. Higher current limits may be obtained by decreasing R_1 and lower limits by increasing R_1.

The above circuit is an example of constant-current limiting where, as the load on the output is increased, the output current is held at its maximum rated value. Slightly more elaborate circuits can be used to give *re-entrant* or *foldback* current limiting where, if the maximum rated current is exceeded, the protection circuit trips in to reduce the output current to a safe value very much less than the maximum rating. To restore output voltage it is then sometimes necessary to disconnect the load altogether. Such an arrangement gives much better protection than the constant current circuit because circuit dissipation is very much reduced under overload conditions.

9.10.11 IC regulators

The two basic elements of a stabilizer circuit, viz. voltage reference and voltage amplifier, can easily be combined into one IC, offering the advantages of extremely good regulation, compact size and ease of use. Many such IC regulators are designed for a specific fixed voltage output, such as 5 V for logic circuits, or 15 V for operational amplifiers. Fig. 9.32 shows a circuit using a 7805 IC to give a stable 5 V supply suitable for IC logic elements. The maximum permissible load current is 1 A, and both line and load regulation 0.2%.

The 7805 should be used on a heat sink if the full rated current of 1 A is to be used. Foldback current limiting is employed, so that the short-circuit current is less than 750 mA. If the load current is under 100 mA, some design economy is possible: C_1 may be reduced to 1000 μF, the transformer may be a standard 6.3 V heater transformer, and no heat sink is needed on the regulator.

If a 12 V output is required, the 7812 may be substituted. Here, the unregulated supply needs to be at least 14.5 V. Similarly, the 7815 gives 15 V output for an input of at least 17.5 V. The maximum permissible input voltage for the 7812 and 7815 is 30 V and for the 7805, 25 V. Other output voltages are available within the 78 series, whilst the 79 series provides corresponding negative voltages.

In complex electronic installations such as telephone exchanges, where

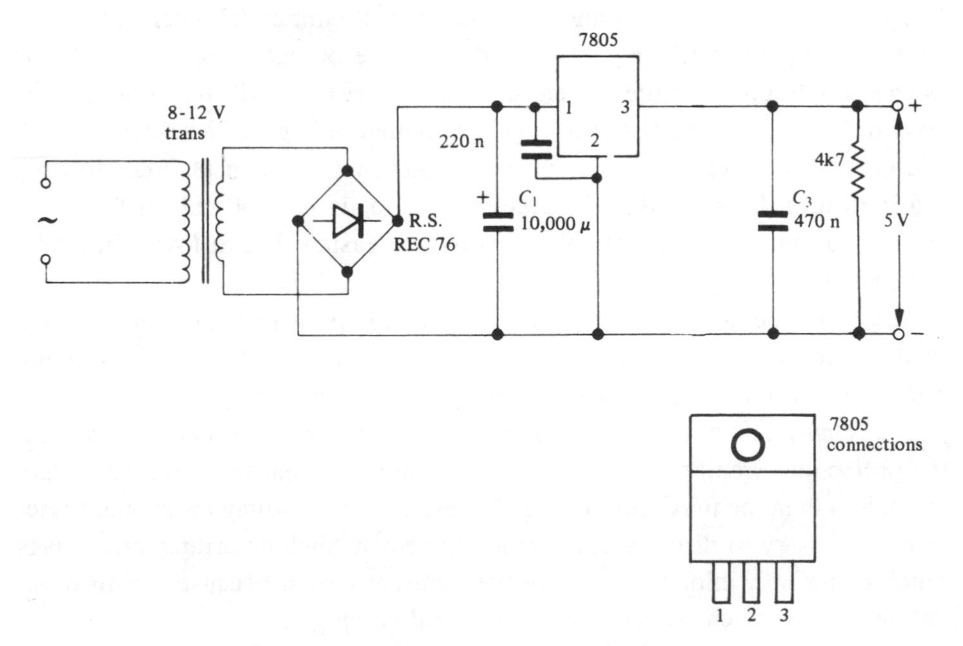

Fig. 9.32. Stabilized 5 V supply using an IC regulator.

scores or even thousands of printed circuits are involved, the distribution of power supplies presents the two problems of maintaining low supply impedance and avoiding interaction between circuits owing to the long leads involved. A common way of overcoming this problem is to mount a fixed-voltage IC regulator on each board and distribute unstabilized or roughly stabilized power so that each piece of circuitry has an individually stabilized supply. This is economically possible because the cost of IC regulators is only of the same order as that of a large electrolytic capacitor. All regulators require that manufacturers' decoupling and earthing recommendations be followed carefully. Circuit layout should keep connections to decoupling capacitors short, otherwise instability (RF oscillation) may result.

The μA723 or LM123 is a useful and long-established IC regulator with separate connections brought out from the various internal circuit sections so that many regulators of different types may be constructed. Fig. 9.33 shows a basic circuit using the μA723 to give an adjustable output voltage from 7 V to 27 V. The μA723 contains the basic stabilizer elements as follows:

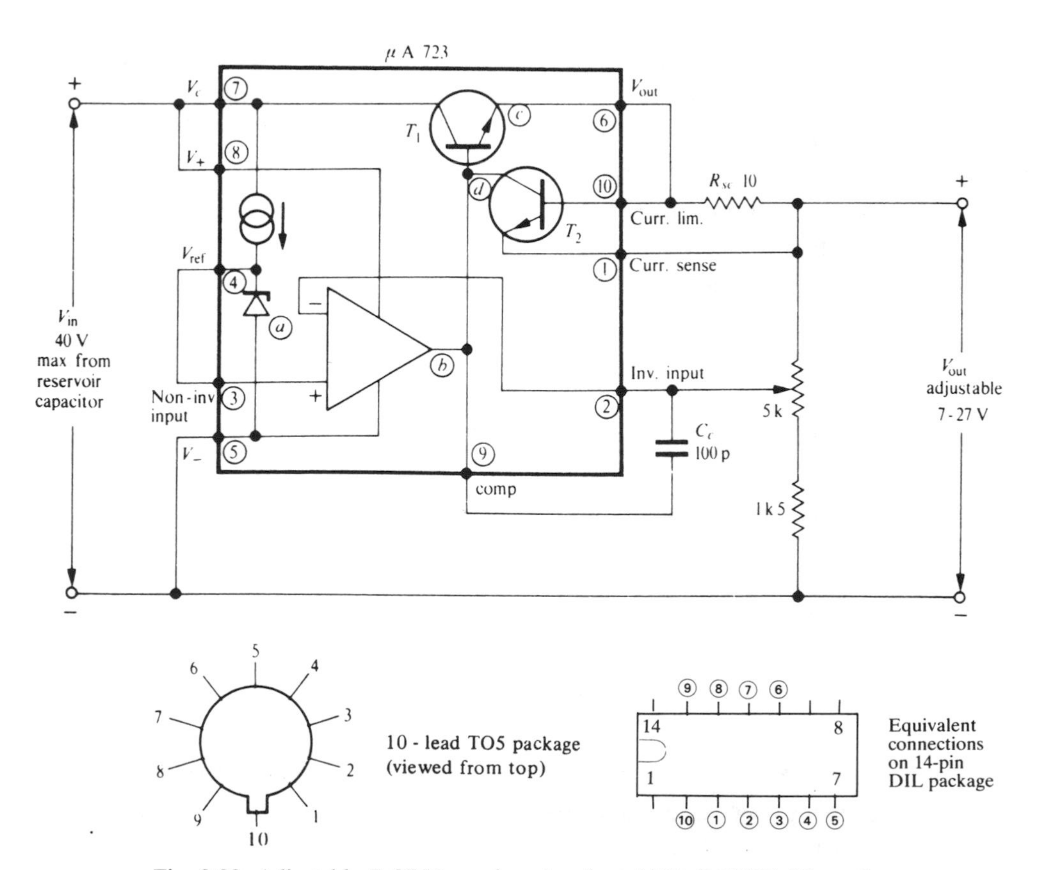

Fig. 9.33. Adjustable 7–27 V supply using the μA723 (LM123) IC regulator.

(*a*) Zener diode reference. This, as shown symbolically in fig. 9.33, is fed from a constant-current source. The diode is in fact included in a feedback amplifier which serves to lower its output resistance.

(*b*) Error amplifier. Both inverting and non-inverting inputs are brought out as external connections. The compensating capacitor, C_c, reduces high-frequency gain in order to avoid high frequency instability in what is, after all, a feedback system with a high value of β.

(*c*) Medium-power emitter follower transistor T_1. It is the emitter follower transistor which limits the power dissipation capability of the regulator. In this instance, 600 mW is the maximum permissible dissipation, but this may be increased by the use of an external transistor.

(*d*) Current limit transistor T_2. This senses the p.d. across resistor R_{sc} and begins to conduct when it reaches 0.6 V. As T_2 conducts, it short-circuits the base–emitter junction of the emitter follower, rendering it inoperative and cutting off the supply. With $R_{sc} = 10\ \Omega$ as shown, the maximum current is 60 mA. The maximum current which the IC can supply is 150 mA, but this is subject to the maximum power dissipation of 600 mW. In other words, an output current of 150 mA would only permit an input–output voltage difference of $600/150 = 4$ V, whilst a 60 mA current would allow 10 V difference before the maximum power dissipation was exceeded.

Fig. 9.34 shows how an external power transistor can be used as an auxiliary emitter follower with the μA723 voltage regulator to increase the permissible power dissipation of the circuit.

With the 2N3055 transistor specified, currents up to 2 A can be supplied,

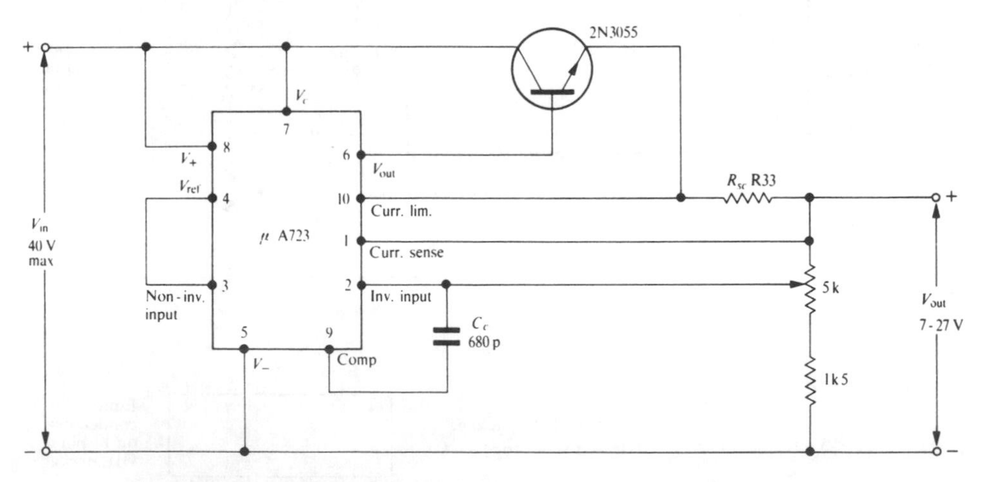

Fig. 9.34. Use of external power transistor to increase output current of μA723 regulator.

the maximum power dissipation being 30 W if a large heat sink is used on the transistor. Because of the additional high-frequency phase shift introduced by the power transistor, the compensation capacitor is increased from 100 pF to 680 pF.

9.11 Transistor heat dissipation

9.11.1 Heat sinks

In small-signal circuits, transistors rarely have to dissipate more than 100 mW of power. Heat conduction along the leads and convection from the case to the surrounding air is sufficient to avoid overheating the junction.

Transistors handling higher powers, such as the emitter followers in power supplies and the output stages of power amplifiers, require special means of heat dissipation. It is normal to use a *heat sink* with transistors fitted with heat sinks. In fig. 9.35(*a*), a corrugated clip doubles the heat loss from a TO5 transistor can such as a BFY50. The power transistor in fig. 9.35(*b*), is in the common TO3 package and is mounted on a substantial finned heat sink. Mounting a 2N3055 transistor in this way permits a 30 W dissipation; safe dissipation would be limited to 3 W without the heat sink.

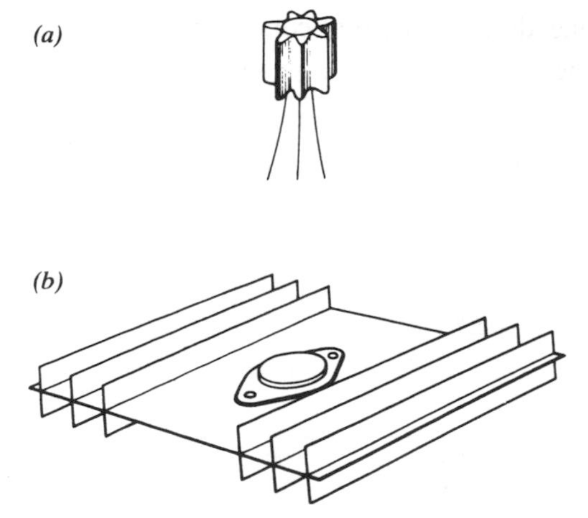

Fig. 9.35. Heat sinks.

9.11.2 Electrical insulation

A finned heat sink is usually bolted directly to an earthed metal chassis or cabinet or, in some cases, the chassis itself may serve as the heat sink. In all such instances, it must be remembered that the case of the transistor itself is normally connected to the collector and that electrical insulation is necessary between the transistor case and the heat sink. A mica or mylar washer provides insulation without significantly reducing heat conduction. Silicone grease applied to each side of the washer ensures good thermal contact.

9.11.3 Thermal resistance

The performance of a heat sink is usually expressed in terms of its thermal resistance, which takes account of the fact that the rate of conduction of heat is proportional to the difference in temperature between the heat source and its surroundings (cf. electrical resistance, where rate of flow of charge is proportional to potential difference).

As is often the case with physical terms, the units of thermal resistance (degrees centigrade per watt) give a good idea of its formal definition, which is

$$\text{thermal resistance } \theta = \frac{\text{temperature difference}}{\text{power dissipated}}$$

In other words, a finned heat sink with a thermal resistance of 3 deg C/W if called upon to dissipate 30 W, will rise in temperature to 3×30 deg C = 90 deg C above its surroundings.

A complete picture of the steady-state thermal conditions of a transistor and its surroundings is given by the thermal circuit shown in fig. 9.36. The thermal

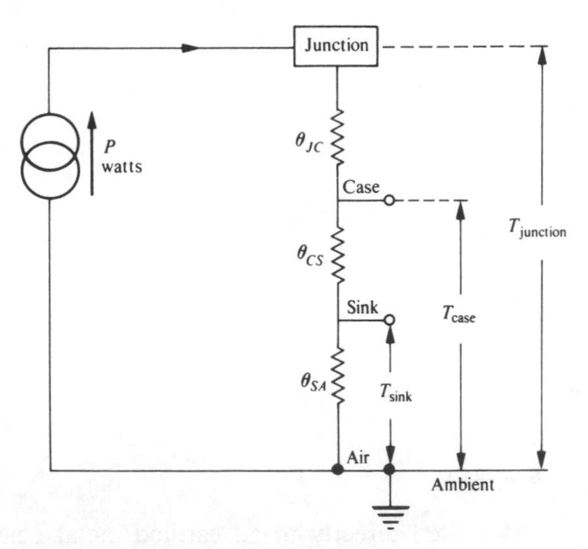

Fig. 9.36. Thermal circuit of a transistor and its surroundings.

power, P, liberated in the transistor is considered as a 'thermal current generator' which then drops temperature differences across the various thermal resistances in the system.

The maximum permissible junction temperature is usually 150 °C and the ambient temperature may be taken as 50 °C to allow for the general warmth surrounding electronic equipment.

Transistor manufacturers quote a maximum safe case temperature for their transistors (often 125 °C), in which case, θ_{JC} is eliminated from our calculations and we move one step down the ladder. Furthermore, conductivity from transistor case to heat sink is usually so good that $\theta_{CS} \ll \theta_{SA}$, so that θ_{SA}, the thermal resistance between heat sink and the air, is the dominant factor in most calculations. Knowing the watts P dissipated in the transistor, it is then easy to calculate the case temperature T_{case} assuming a 50 °C ambient:

$$T_{\text{case}} = 50 + (P \times \theta_{SA}).$$

A check with the manufacturer's literature will then indicate whether the transistor can handle the required power at this case temperature. If not, then θ_{SA} must be lowered by using a larger sink.

Large finned heat sinks for power transistors usually have a thermal resistance from 2 to 4 deg C/W which can be reduced to the region of 1 deg C/W by forced blower cooling. Small heat sinks to fit on TO5 transistor cans, on the other hand, usually average about 50 deg C/W and will increase the permissible power dissipation of a medium-power transistor such as a BFY50 or 2N3053 from 0.8 W to 1.5 W.

9.12 Switch-mode power supplies

9.12.1 d.c. to d.c. conversion

All the power supplies discussed so far have assumed a 50 Hz a.c. supply as the starting point prior to the rectifier. In some cases a d.c. supply is the starting point but it can still be useful to chop it up into a.c. and rectify to achieve the required output voltage.

A simple example is fig. 9.37 which fulfils a common requirement of generating a negative supply from a positive rail, which is of value on, say, a logic board which has only a +5 V rail but where a split-rail of ±5 V is needed for a particular circuit. Fig. 9.37 uses a 74HC14 Schmitt trigger with positive feedback using time constant R_1C_1 resulting in oscillation around 20 kHz. The

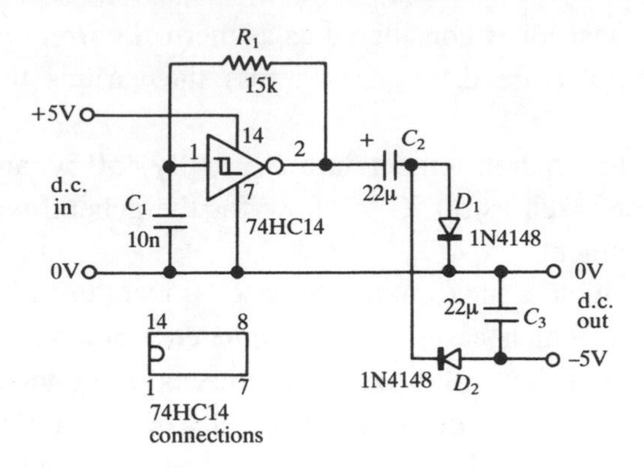

Fig. 9.37. Practical switch-mode circuit to derive –5 V supply from +5 V rail.

resulting a.c. output is then rectified, with D_1 and D_2 with C_2 and C_3 acting as a charge pump to produce approximately –5 V output. The regulation is satisfactory up to a few milliamps output.

By the use of an appropriate inductor or step-up transformer, a switch-mode circuit can achieve a d.c. to d.c. step up in voltage. An example is shown in fig. 9.38 which employs the dedicated IC type RC4190N switching regulator to provide a step up from 5 V to 15 V at 25 mA. Efficiency is 85%, line regulation 0.04% and load regulation 0.2%. Operating frequency can be from

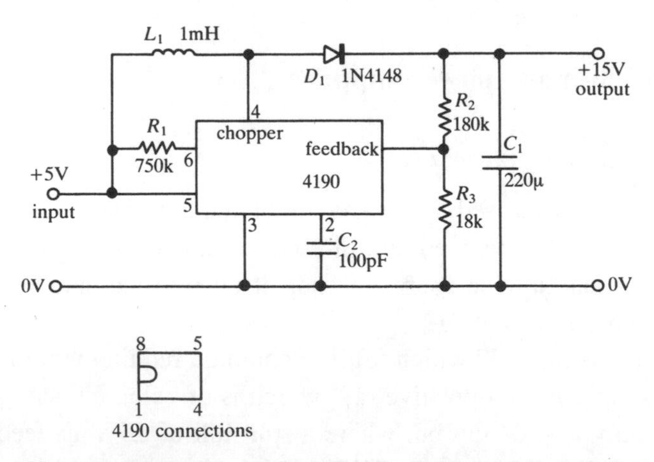

Fig. 9.38. Regulated switch-mode d.c. to d.c. step-up (boost) circuit.

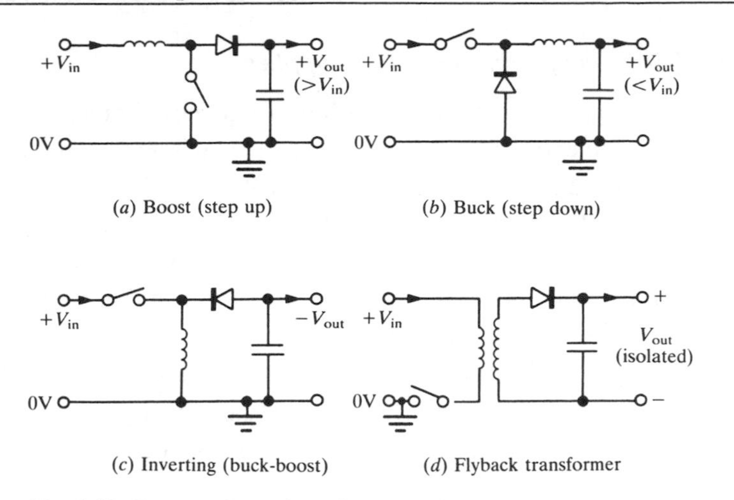

Fig. 9.39. Four configurations (topologies) of switch-mode power supplies.

100 Hz to 75 kHz: in this application 25 KHz is selected. The voltage boost is provided by the back EMF in L which is a simple 1 mH RF choke connected between voltage source and chopper. In contrast to this *boost* regulator, a *buck* or reducing circuit uses the inductor connected between chopper and output where it acts as part of the voltage step down and filtering process.

Fig. 9.39 shows four common configurations or circuit *topologies* of switch-mode power supplies. In each case it is the collapse of the inductor's magnetic field which generates the output e.m.f. The flyback transformer in (*d*), so named from its application generating E.H.T. from a TV line timebase, provides a useful isolated output.

Stabilization of a switch-mode supply is achieved in the usual way by comparing the output voltage with a reference but, instead of wasting power in a series transistor, pulse-width modulation (PWM) is employed by regulating the mark–space ratio of the pulse train fed to the inductor and rectifier, much as the thyristor circuits in the next section regulate a.c. A large range of dedicated switch mode regulator ICs can be obtained, typically operating with up to 50 V input and giving adjustable output voltages from 5 V to 40 V. Better than 80% efficiency is maintained at currents up to 10 A. As with linear regulators, the addition of external power transistors can enable higher powers to be handled.

9.12.2 Direct mains line switching supplies

By chopping the a.c. mains input directly at high frequency, immense savings in weight, size and efficiency are obtained. Direct line switching supplies typically rectify the 50 Hz a.c. mains and use high voltage switching transistors or thyristors to chop the supply into a higher frequency (usually 20 kHz–2 MHz). It is at this point that the pulse mark–space ratio is controlled to obtain the appropriate regulation. The resulting waveform can then be rectified directly to provide the stable d.c. output. In many cases, transformer isolation is still needed between mains and d.c. output but, because of the high frequency, the transformer is light and compact, usually using a ferrite core. Reservoir capacitors are also relatively small because of the very high ripple frequency.

A word of warning is appropriate here. Whilst direct line switching supplies are increasingly used in commercial equipment because of the above advantages, their practical construction lies outside the scope of our work in this book. Not only are there obvious safety hazards associated with prototyping and testing, but special suppression circuits are essential on the mains input to prevent electromagnetic interference (EMI) radiating back down the line. Protection from high voltage transient 'spikes' on the input is also essential in order to avoid catastrophic breakdown of the switching transistors.

9.13 Power control with thyristors, transistors and triacs

9.13.1 Introduction to the thyristor

No discussion of power supplies and power control is complete without reference to the *thyristor*, a semiconductor switch possessing both high breakdown voltage and very high current gain. The word thyristor is derived from the Greek *thyra*, meaning door, indicating that it can be either open or closed. Another name for the device is *silicon controlled rectifier* (SCR). As the latter name indicates, the thyristor behaves as a rectifier with the added ability of controlling the power delivered to the load.

The thyristor will not conduct until a current pulse flows in the gate circuit. Once it has been so triggered a regenerative action takes control and the device will continue to conduct until the supply voltage is removed. Fig. 9.40 illustrates this property in a simple circuit, where a variable portion of the a.c. input waveform is rectified. Conduction does not occur until a current pulse flows from gate to cathode; the phase of the gate pulses relative to the a.c.

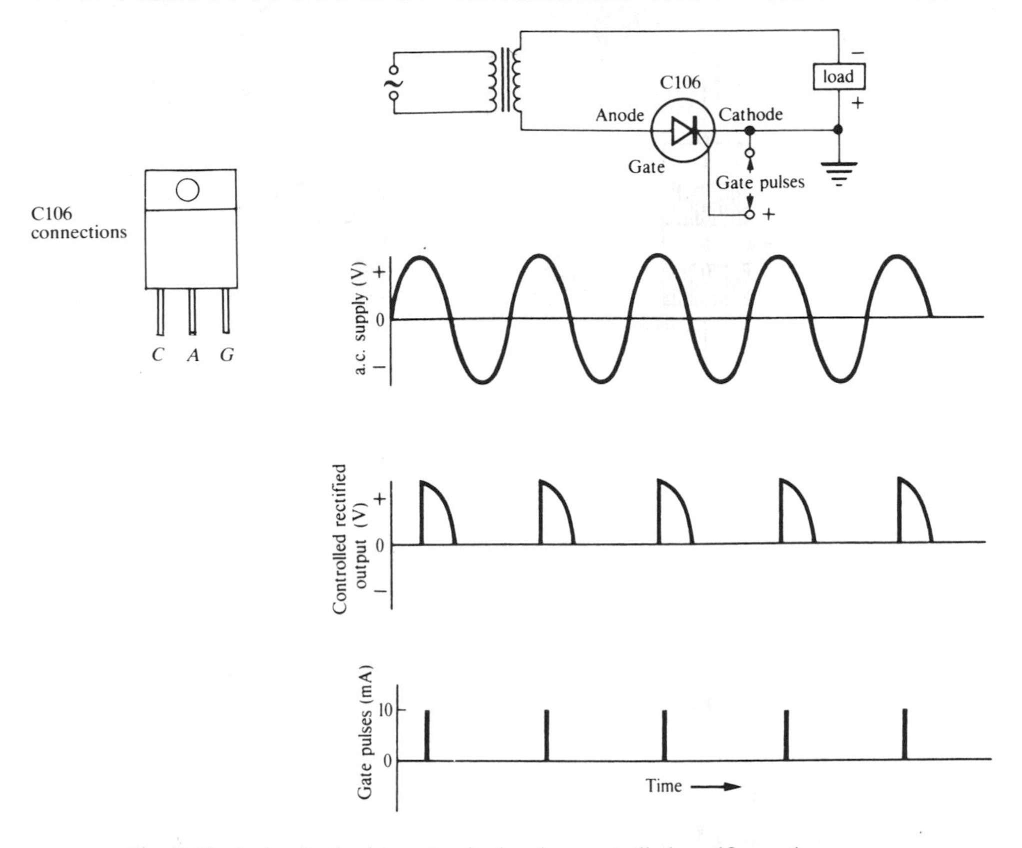

Fig. 9.40. A simple thyristor circuit showing controlled rectifier action.

supply therefore controls the fraction of the waveform transmitted by the thyristor. Conduction automatically stops at the end of each half-cycle because the input reduces to zero volts.

9.13.2 Construction and operation of the thyristor

The thyristor is a four-layer device (pnpn), shown in fig. 9.41(*a*). The circuit symbol (*b*), however, is simply that of a diode rectifier with the addition of the gate electrode.

The most useful approach to thyristor operation is to consider its four layers as two interconnected transistors, as in fig. 9.41(*c*), shown in schematic form in (*d*). The combined regenerative action of the pnp and npn transistors will now be deduced from the circuit of fig. 9.42 which shows the load current I_L, the gate current I_G, the collector current of T_1 (I_{C1}) and the collector current

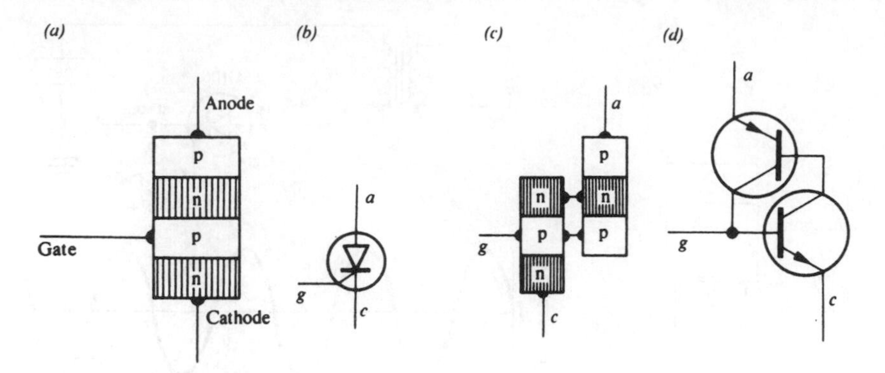

Fig. 9.41. The thyristor: (*a*) structure, (*b*) circuit symbol, (*c*) transistor equivalent, (*d*) equivalent transistor schematic.

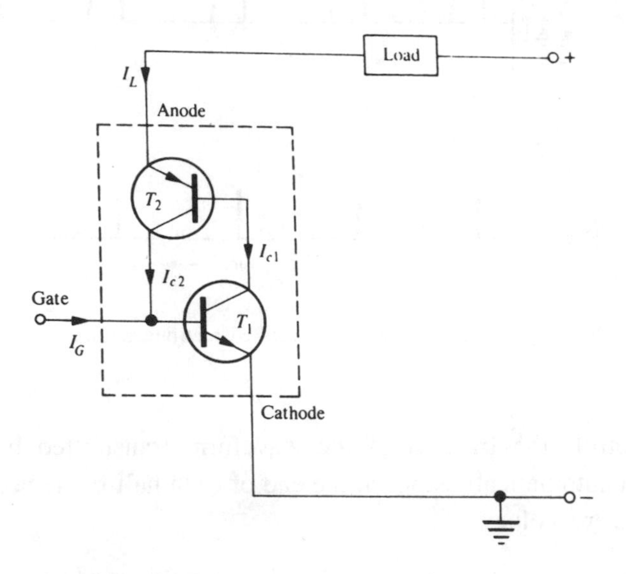

Fig. 9.42. Circuit showing thyristor action in terms of transistor equivalent.

of T_2 (I_{C2}). The d.c. common-emitter current gains of the two transistors T_1 and T_2 will be denoted h_{FE1} and h_{FE2} respectively.

From fig. 9.42,

$$\text{base current of } T_1 = I_{C2} + I_G,$$

$$\text{base current of } T_2 = I_{C1},$$

therefore

$$\text{collector current } I_{C1} = h_{FE1}(I_{C2} + I_G)$$

and

$$I_{C2} = h_{FE2} I_{C1}.$$

Substituting for I_{C2},

$$I_{C1} = h_{FE1}(h_{FE2} I_{C1} + I_G)$$

$$= h_{FE1} h_{FE2} I_{C1} + h_{FE1} I_G$$

therefore

$$I_{C1}(1 - h_{FE1} h_{FE2}) = h_{FE1} I_G,$$

i.e.

$$I_{C1} = \frac{h_{FE1} I_G}{1 - h_{FE1} h_{FE2}} \tag{9.8}$$

and

$$I_{C2} = \frac{h_{FE1} h_{FE2} I_G}{1 - h_{FE1} h_{FE2}}. \tag{9.9}$$

Now,

$$\text{load current } I_L = I_{C1} + I_{C2}$$

therefore

$$I_L = \frac{h_{FE1}(1 + h_{FE2}) I_G}{1 - h_{FE1} h_{FE2}}. \tag{9.10}$$

Clearly, as the value of $h_{FE1}h_{FE2}$ approaches unity, the load current as indicated by equation (9.10) tends to infinity: in practice, of course, it will be limited by the external load, with the thyristor simply behaving like a closed switch.

The current gain of all transistors decreases at low values of base current and the thyristor is constructed so that, when the external gate current is near zero, the product of the current gains is less than unity. As the gate current is increased to a few milliamperes, $h_{FE1}h_{FE2}$ quickly reaches unity and the thyristor switches on.

The internal feedback makes the thyristor an extremely efficient and fast switching device (typical turn-on time 1 μs). It can be used to control very high powers, since reverse-biased junctions can be made to withstand many hundreds of volts. With suitably large junctions, thousands of amps can be handled with a drop of only a volt or so across the thyristor. This combination of high breakdown voltage and high effective current gain is impossible to achieve in a power transistor: high current gain requires a thin base region, but this gives a low breakdown voltage.

9.13.3 Power control with transistors

Improvements in transistor fabrication and thermal characteristics have resulted in the increased application of power bipolar transistors and MOSFETs in switching d.c. up to 100 A or so. Such devices are, of course, free from the difficulty of turn-off exhibited by a thyristor once it is conducting d.c.: the base or gate of a transistor retains control at all times. Car ignition systems represent a common MOSFET application, where currents up to 10 A and a back e.m.f. from the ignition coil of hundreds of volts must be handled.

The *insulated gate bipolar transistor* (IGBT) is an interesting 'cross-breed' between bipolar and MOS technology and is applied as a useful medium/high power switch. Like the MOSFET, its gate draws no current from the control signal; in other respects it behaves like a bipolar power transistor. Interestingly, the IGBT has the advantage of being less likely than a MOSFET to suffer thermal runaway if switching unexpectedly high transient currents. This is because the MOSFET on resistance has a positive temperature coefficient: i.e. the hotter it gets, the more power it dissipates, leading to catastrophic failure in overload. The IGBT on the other hand has a negative temperature coefficient, a more stable situation because the volt drop across the IGBT actually falls as it gets hotter.

9.13.4 The triac and its applications

The thyristor is ideally suited to a.c. power control in all respects except one: it is only a half-wave device, which means that, even at full conduction, only half power is available. Two thyristors may be paralleled in opposition, as in fig. 9.43, to give FW operation; however, two isolated, but synchronized sources of gate control pulses are required as shown.

The most useful device for practical a.c. power control is the *bi-directional thyristor* or *triac*. As may be seen in the section diagram in fig. 9.44(*a*), the triac can be considered as two inverse-paralleled thyristors under the control of a single gate. Such is the versatility of the triac that it may be triggered

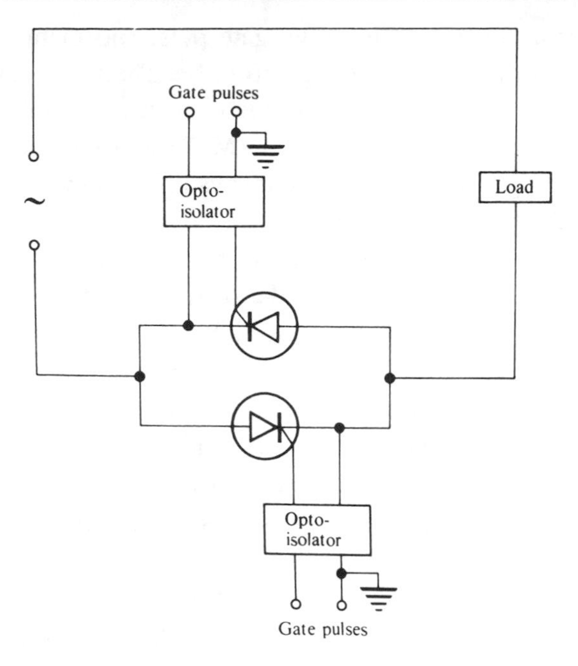

Fig. 9.43. FW control can be obtained with two thyristors. Opto-isolators are used to insulate pulse sources from mains voltage.

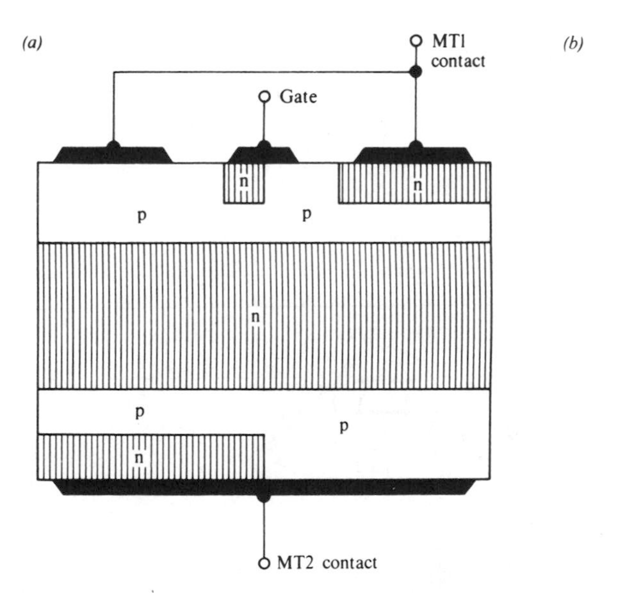

(b)

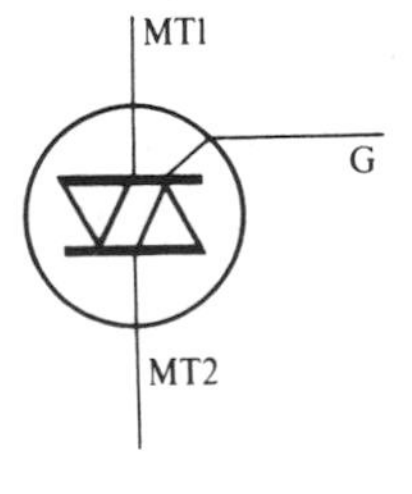

Fig. 9.44. The triac: (*a*) structure, (*b*) circuit symbol.

into conduction by either a positive or negative gate pulse, no matter what the instantaneous polarity of the a.c. supply happens to be. The terms cathode and anode cease to have meaning with the triac, the contact next to the gate being given the mundane title of main terminal 1 (MT1) and the other one main terminal 2 (MT2). The gate trigger pulse is always referred to MT1, just as it is referred to the cathode of the thyristor.

A gate current of 20 mA is normally adequate to trigger triacs up to 25 A rating, and one of the simplest applications of the device is the 'solid-state relay', where a small gate current is used to control a heavy load current (fig. 9.45). Switch SW1 can be a reed relay, a delicate thermostat, or any contact pair of 50 mA rating; circuit load current is limited only by the rating of the triac. It is useful to note that the gate resistor, R_1, only has mains voltage across it at the instant of switch-on: as soon as the triac turns on, the p.d. across R_1 falls to about one volt so that a half-watt resistor rating is ample.

A very common triac application is the lamp dimmer or motor speed control. Fig. 9.46 shows such a circuit. The timing of the gate trigger pulses is adjusted by an RC phase shifter; R_2 is the dimmer control, whilst R_1 simply limits the current in the potentiometer at its minimum resistance setting. The actual gate trigger pulses are produced by the diac or bi-directional trigger diode. The diac may be thought of as a small triac with no gate, but a low avalanche breakdown voltage of about 30 V. When the p.d. across C_1 reaches the diac breakdown level, a sudden pulse of charge from the capacitor triggers the triac.

Automatic photoelectric switching of a lamp can be readily achieved by connecting an ORP12 CdS photocell (light-dependent resistor) across C_1. The cell resistance is as high as 1 MΩ in the dark, but in daylight it drops to a few kΩ, so that the triac cannot fire and the lamp switches off. If the dimming

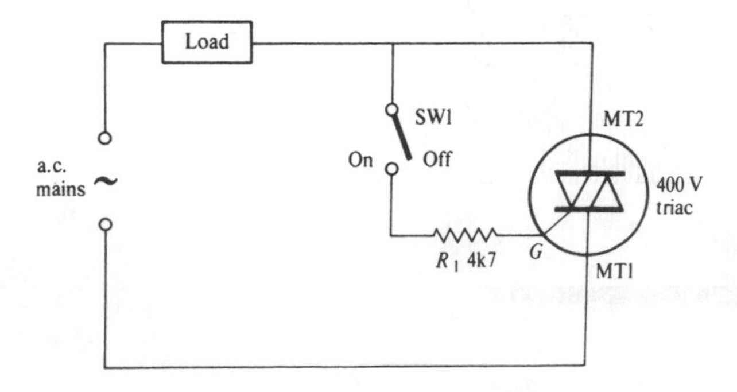

Fig. 9.45. Simple triac 'solid-state relay'.

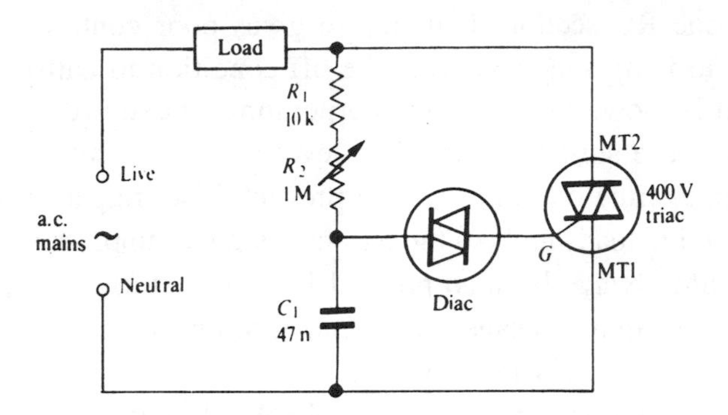

Fig. 9.46. Basic triac lamp dimmer using phase control.

facility is not required with the photoelectric switch, R_2 can be replaced by a short-circuit.

Fig. 9.47 shows the manner in which the triac controls load power by chopping off the initial part of each half-cycle. The length of the missing portion is dependent on the phase lag imposed by $R_1 + R_2$ and C_1 on the trigger pulses. The basic control circuit of fig. 9.46 cannot give more than 90° phase shift

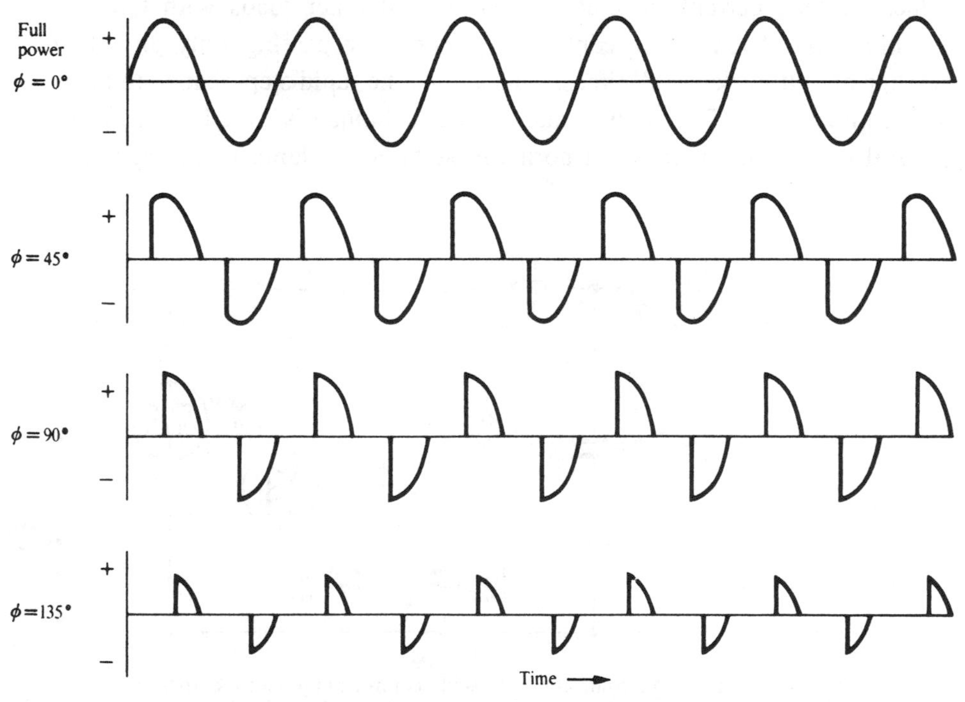

Fig. 9.47. Voltage waveforms across load of a triac dimmer as phase shift is progressively increased.

because it only uses one RC section. It therefore gives poor control at low power levels, tending to jump suddenly from the off condition to half-power.

An improved circuit is shown in fig. 9.48, incorporating an extra RC section (R_3C_3) to give greater phase shift for better low-level control. Further refinements are (*a*) the *snubber* time constant R_4C_4 to prevent false triggering from the back-e.m.f. of inductive loads and (*b*) the RF interference suppressor L_1C_1. The latter feature should always be incorporated in a 'waveform chopping' type of triac or thyristor circuit because the rapid turn-on and turn-off can generate serious mains-borne radio interference.

A very wide range of triacs and thyristors is available. As with diode rectifiers, catalogues and data sheets can be consulted in order to select a device of appropriate voltage and current rating.

Most manufacturers make a suitable diac, whilst devices combining a triac and diac are available, and are known as *quadracs*.

Fig. 9.49 shows the connections for common triac packages. If full power-handling capability is to be exploited, it is usually necessary to mount triacs on a heat sink.

The radio interference (EMI) produced by the phase-control type of triac or thyristor control becomes more difficult and expensive to suppress at high values of load current. In electric heaters and other loads with long time constants, interference-free control is possible by passing only an integral number of half-cycles at all times. This avoids the rapid step-change in current which gives rise to RF components. Such a technique is called burst firing or integral cycle control. It is not normally suitable for lamp dimming because

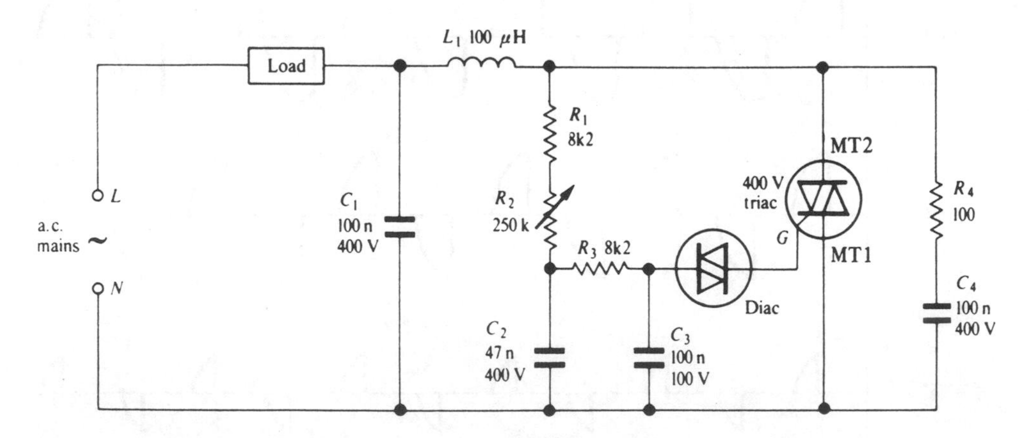

Fig. 9.48. Triac power controller with wide control range and incorporating interference suppression.

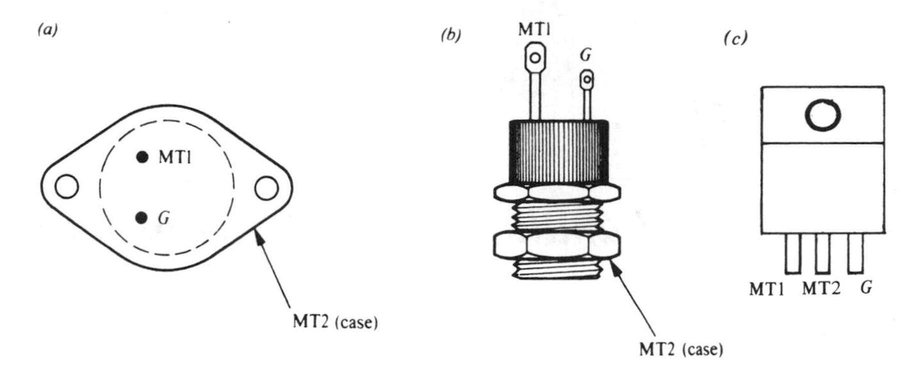

Fig. 9.49. Common triac packages: (*a*) TO66 case, (*b*) stud mounting, (*c*) TO220 plastic package.

of the flicker problem. To implement integral cycle control, zero voltage switching ICs such as the SL441 are available. These detect zero crossings in the mains waveform and provide control of triac triggering from a variable-resistance transducer such as a thermistor.

10

Pulse handling and time constants

10.1 Introduction

Our introduction to the transistor in chapter 1 used the device as a switch to turn a light bulb on and off. This is the simplest possible example of amplification, where the signal has only two voltage levels coresponding to the 'on' and 'off' conditions. Such a signal is shown in fig. 10.1; here, the two levels are zero and +5 V, the transition between the levels being virtually instantaneous so that the waveform is rectangular. Signals from radiation detectors such as the Geiger–Müller tube are of this form, whilst the enormous field of digital electronics is concerned entirely with trains of rectangular pulses. It is now common practice to transmit and record ordinary analogue signals in digital form because the signal has then only two levels and the various distortions which may occur in transmission and recording are readily corrected. This chapter is concerned with the handling of pulses and the modifications and distortions which may occur to them in various circuits.

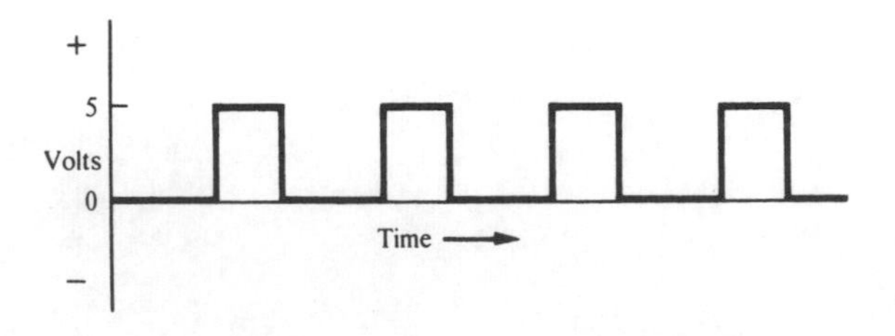

Fig. 10.1. Train of rectangular pulses.

10.2 'Squaring' a waveform

Any repetitive waveform may be converted to a rectangular wave merely by driving a simple voltage amplifier so hard that it is driven alternately into cut-off and saturation. Fig. 10.2 shows a circuit that will do this. If the input is fed with a sine wave of at least 5 V r.m.s. amplitude, the output will be a rectangular wave suitable for use with the various pulse-shaping experiments in this chapter. The positive half-cycles of the input waveform cause the transistor to saturate, taking the collector very nearly to 0 V, whilst on the negative half-cycles the transistor does not conduct, leaving the collector at +9 V. The diode from base to earth clips off negative half-cycles to avoid base–emitter reverse breakdown which would occur for inputs greater than –6 V. The output is thus a rectangular wave with an amplitude of nearly 9 V (fig. 10.3). The collector does not quite reach 0 V on the positive half-cycles: every transistor has a finite collector–emitter saturation voltage, $V_{CE(\text{sat})}$, which lies typically in the range 0.1 V to 1 V and depends on collector current and base current drive as well as on the transistor type. Transistors specifically designed for switching usually have a low $V_{CE(\text{sat})}$ so that the maximum possible output voltage swing is available. This also ensures that the transistor itself dissipates very little power: although the collector current is at its maximum under saturation conditions, the power dissipation will be negligible if $V_{CE(\text{sat})}$ is very small.

In this chapter we shall examine the various distortions which a rectangular wave may undergo owing to the effects of circuit capacitances. Since it is these same capacitances which are responsible for restricting circuit frequency

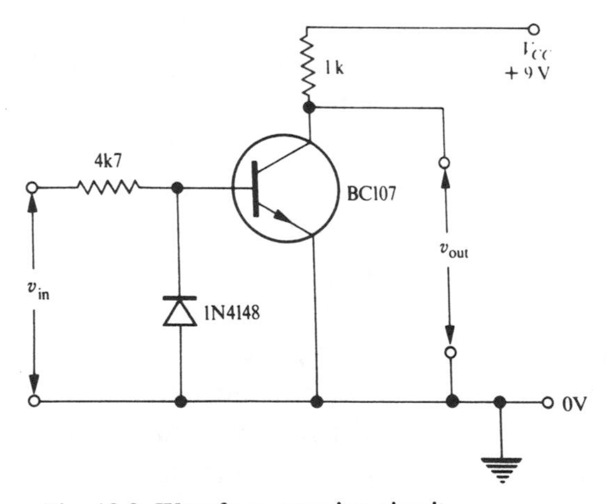

Fig. 10.2. Waveform squaring circuit.

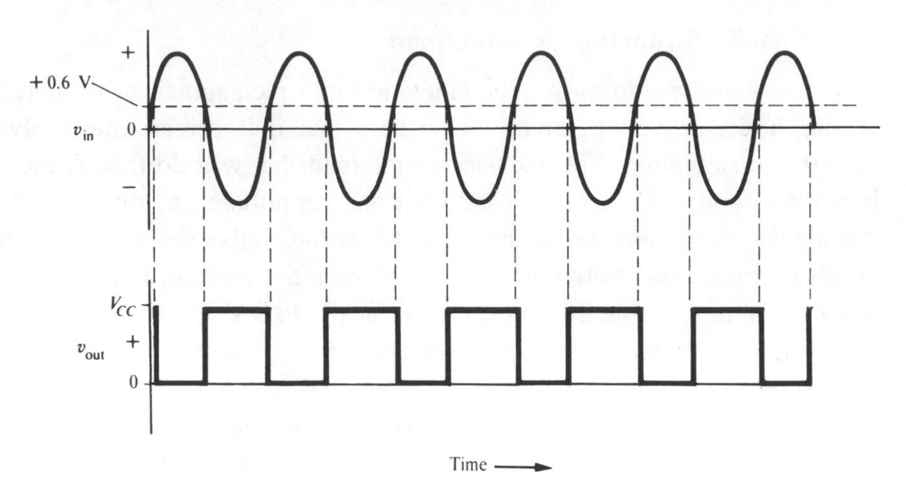

Fig. 10.3. Input and output waveforms of squaring circuit.

response, it is reasonable that there should be a relation between rectangular wave distortion and frequency response measured with a sine wave. It is appropriate to point out here that linear circuit components like capacitors and resistors can never distort the shape of a *sine* wave, though, of course, they can reduce its amplitude. It requires some sort of non-linearity, such as we have in the transistor, to distort a sine waveform. On the other hand, the shape of a 'square' wave may be completely changed merely by a resistor and capacitor.

10.3 Fourier analysis

In the discussion of distortion in chapter 4 it was pointed out that any repetitive non-sinusoidal waveform may be resolved into a number of sinusoidal components comprising a fundamental, of the same frequency f as the original waveform, plus a number of harmonics of frequencies $2f$, $3f$, $4f$ etc. A rectangular or 'square' wave contains harmonics up to very high orders, vertical edges and sharp corners implying an infinite series of harmonics. A good representation of a rectangular wave is in fact obtained if frequencies up to the tenth harmonic are present. For this reason, square waves are often used as a quick test of the frequency response of an audio amplifier, a 1 kHz square wave requiring a response up to 10 kHz for faithful reproduction.

It is possible to calculate the effect of a network on a rectangular wave by Fourier-analyzing the waveform into its individual sinusoidal components, and calculating the frequency response of the network and thus its effect on the amplitude and phase of the various components of the rectangular wave. The modified components can then be added together to synthesize the modified square wave. This technique, whilst it will give the correct result, is very cumbersome; fortunately, there is a much more direct method of assessing the effect of a resistance–capacity (RC) circuit on a rectangular wave by simply considering the charging and discharging of the capacitor(s) in the time domain.

10.4 Charging, discharging and time constants

Fig. 10.4 shows a simple RC circuit which we have already considered in terms of its effect on high-frequency response. We found in section 7.7 that it behaves as a low-pass filter, transmitting lower frequencies but attenuating high frequencies. We shall now look at the output we obtain if a d.c. voltage is suddenly applied to the input and then held steady; in other words, the input is a voltage step (fig. 10.5).

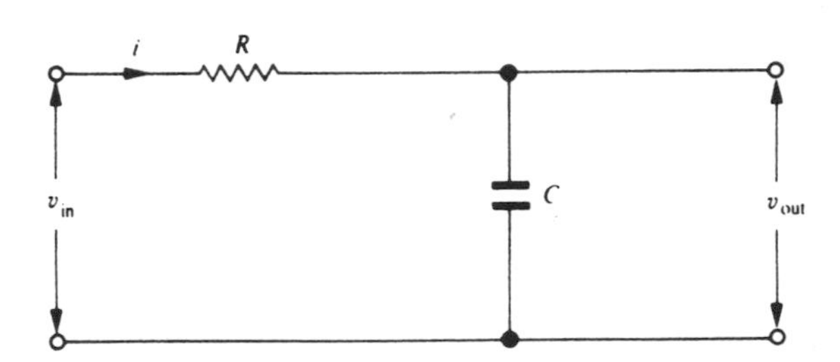

Fig. 10.4. First-order low-pass filter.

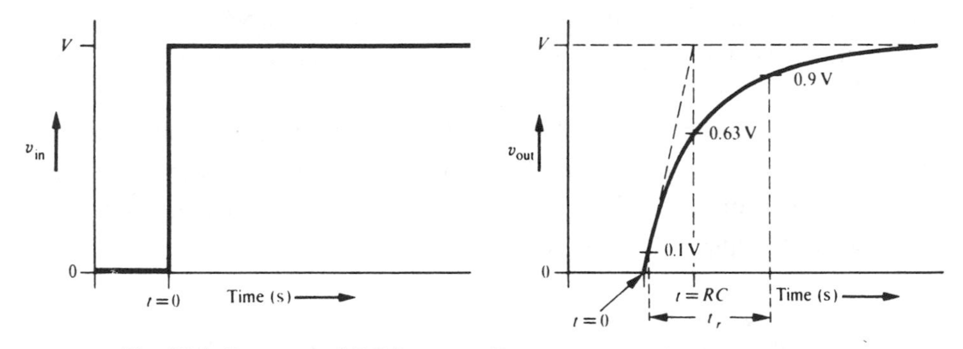

Fig. 10.5. Response of RC low-pass filter to an upward voltage step (capacitor charging).

As the voltage step, $v_{in} = V$, is applied, at a given time $t = 0$, a certain current i will flow in the circuit. If q is the charge on the capacitor at time t, then the basic capacitor equation gives

$$v_{out} = \frac{q}{C}. \tag{10.1}$$

Differentiating with respect to time,

$$\frac{dv_{out}}{dt} = \frac{1}{C}\frac{dq}{dt}. \tag{10.2}$$

But dq/dt is rate of flow of charge, which is current i. Hence,

$$i = C\frac{dv_{out}}{dt}. \tag{10.3}$$

Now p.d. across $R = V - v_{out}$. Hence, from Ohm's law,

$$V - v_{out} = iR$$

$$= RC\frac{dv_{out}}{dt}$$

or

$$RC\frac{dv_{out}}{dt} + v_{out} = V. \tag{10.4}$$

The solution of this differential equation, given that $v_{out} = 0$ at $t = 0$ is

$$v_{out} = V\left[1 - \exp\left(\frac{-t}{RC}\right)\right]. \tag{10.5}$$

The characteristic shape of this exponential charging curve is shown in fig. 10.5. Notice that the slope is initially quite steep but gradually levels off, approaching the applied voltage, v_{in}, asymptotically. It is useful to note that the shape of the curve is uniquely defined by the product RC, known as the *time constant* of the circuit. From the original differential equation, it is easy to show that the initial gradient of the charging curve is V/RC. At $t = 0$, $v_{out} = 0$, so that, from equation (10.4),

$$RC\frac{dv_{out}}{dt} = V$$

and

$$\frac{dv_{out}}{dt}=\frac{V}{RC}. \tag{10.6}$$

In other words, if v_{out} continued to rise at its initial rate it would equal V when $t=RC$. In fact however, the value of v_{out} when $t=RC$ is given by

$$v_{out}=V[1-\exp(-1)],$$

i.e.

$$\frac{v_{out}}{V}=1-\frac{1}{e}=0.63.$$

Thus, after a period of one time constant, the output is equal to 63% of the input voltage. Time constant is obtained in seconds if R and C are in the fundamental units of ohms and farads respectively. The rather more convenient units of MΩ and μF also give RC in seconds, or an answer in milliseconds is obtained using kΩ and μF.

When examining a voltage step on an oscilloscope, it is convenient to measure the *rise time* of the pulse, defined as the time taken for the voltage to rise from 10% to 90% of its final value and normally given the symbol t_r.

If t_{10} is the time to reach the 10% level, equation (10.5) gives

$$1-\exp\left(\frac{-t_{10}}{RC}\right)=0.1,$$

i.e.

$$\frac{-t_{10}}{RC}=\ln 0.9,$$

hence,

$$t_{10}=0.105RC$$

and, if t_{90} is the time to reach the 90% level,

$$1-\exp\left(\frac{-t_{90}}{RC}\right)=0.9,$$

i.e.

$$\frac{-t_{90}}{RC}=\ln 0.1,$$

hence,

$$t_{90} = 2.303RC.$$

Thus,

$$t_r = t_{90} - t_{10} \approx 2.2RC. \tag{10.7}$$

It is useful to note here that, for the simple first-order low-pass filter under consideration, the sine-wave frequency response is such that the –3 dB (half-power) point occurs at a frequency

$$f_1 = \frac{1}{2\pi RC}\,\text{Hz} \quad \text{(from equation (7.8))}$$

i.e.

$$RC = \frac{1}{2\pi f_1}. \tag{10.8}$$

Hence rise time

$$t_r = \frac{2.2}{2\pi f_1}$$

or

$$t_r \approx \frac{1}{3f_1}. \tag{10.9}$$

Thus we see directly that there is a relationship between rise time and upper cut-off frequency, a fact which was anticipated in our brief look at Fourier analysis.

If the circuit is given sufficient time to respond to the voltage step, so that v_{out} is effectively equal to V, and v_{in} is then reduced once again to zero, the capacitor discharges through the resistor, producing the exponential decay of fig. 10.6.

Once again we have a differential equation:

$$RC\frac{dv_{out}}{dt} + v_{out} = v_{in}, \tag{10.10}$$

but here $v_{in} = 0$ at $t = 0$ and $v_{out} = V$ at $t = 0$ giving the solution

$$v_{out} = V\exp\left(\frac{-t}{RC}\right). \tag{10.11}$$

When $t = RC$,

$$v_{\text{out}} = V \exp(-1)$$

$$= \frac{V}{e} \qquad (e = 2.718)$$

$$= 0.37\ V, \text{ i.e. } v_{\text{out}} \text{ has fallen by 63\% after time } RC$$

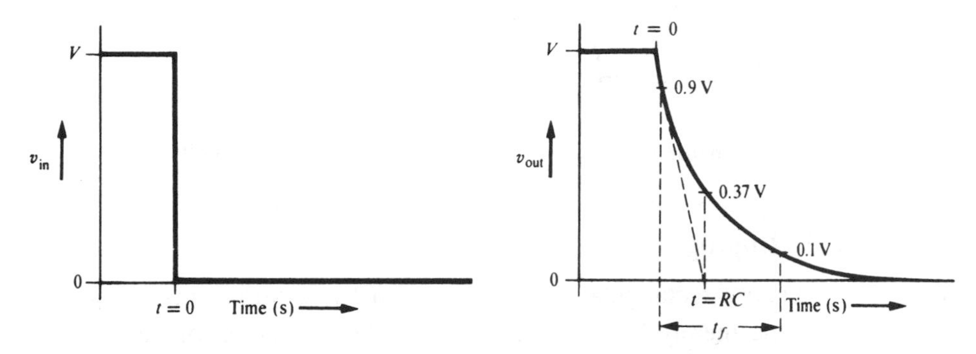

Fig. 10.6. Response of RC low-pass filter to a downward voltage step (capacitor discharging).

Fall time t_f is defined in a similar way to rise time; it is the time for the level to fall from 90% to 10% of its original level and, as with rise time, $t_f = 2.2RC$.

We can now put fig. 10.5 and fig. 10.6 together to give the response of the basic RC circuit in fig. 10.4 to a train of rectangular pulses. The result is shown in fig. 10.7, and can be experimentally observed by making $R = 100$ kΩ and $C = 1$ nF in fig. 10.4. This combination is then connected to the rectangular wave output of fig. 10.2 and the output observed on an oscilloscope. A double beam oscilloscope is useful here so that the waveforms at the input and output of the RC circuit can be compared. An input frequency in the region of 500 Hz should give a result similar to fig. 10.7. The rise time, fall time and time constant may be measured from the oscilloscope trace and compared with the theoretical values. The frequency response curve of the network can be drawn in the usual way with a sinusoidal signal, plotting gain ($20 \log_{10} v_{\text{out}}/v_{\text{in}}$ dB) against frequency. An experimental check can thus be made of the relation between upper-3 dB frequency and rise time; the exercise can be repeated for various values of R and C.

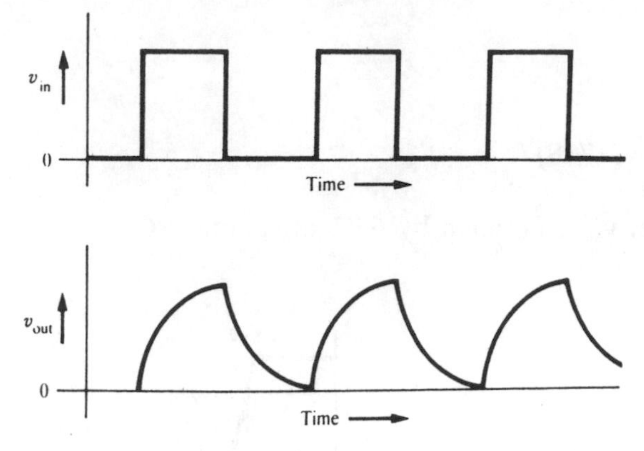

Fig. 10.7. Response of low-pass filter to a train of rectangular pulses.

10.5 Ringing

Sometimes, when a rectangular wave is fed into an amplifier, instead of the rise and fall time merely being increased, the 'ringing' effect of fig. 10.8(*a*) is observed. This indicates that the frequency response, instead of falling steadily at –6 dB per octave at high frequencies, probably rises to a slight peak before falling at a more rapid rate, as in fig. 10.8(*b*). Such a response can occur in an amplifier with negative feedback which also has excessive phase shift at high frequencies, causing the feedback to become positive.

A small amount of ringing is not usually serious, as long as the ringing frequency is higher than the maximum frequency at which the amplifier will be used (e.g. 20 kHz for an audio amplifier). Severe ringing, however, demands investigation because it may be a sign of instability in the amplifier.

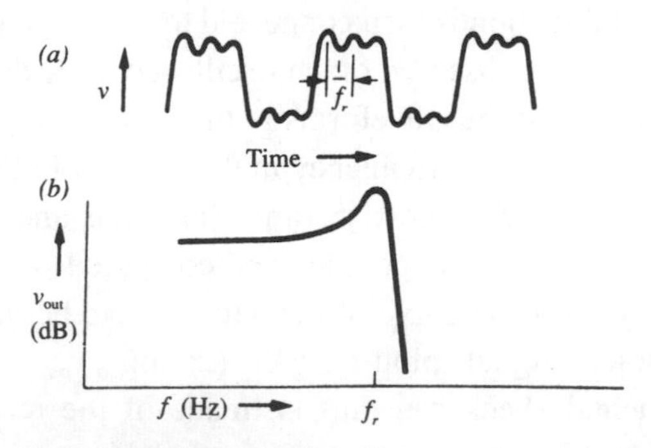

Fig. 10.8. (*a*) Ringing on a rectangular wave. (*b*) The type of frequency-response characteristic indicated by ringing.

10.6 Time constants and transistors

In fig 10.7, the rise time and fall time of the simple RC circuit are identical, the exponential fall of the discharge being simply an inverted version of the exponential rise of the charging process. This is always the case with a linear RC circuit, but is very rarely true when a transistor is used for switching and significant capacitance is present in the load.

Fig. 10.9 shows a typical example of a transistor switch with capacitance, C, included in the load. If we start with the transistor off, v_{out} will be equal to V_{CC} and C will be charged to the same voltage. Now, when the transistor turns on it appears as a very low resistance. Depending on the base current, the instantaneous collector current may be as high as 100 mA so that C discharges rapidly and the fall time is short.

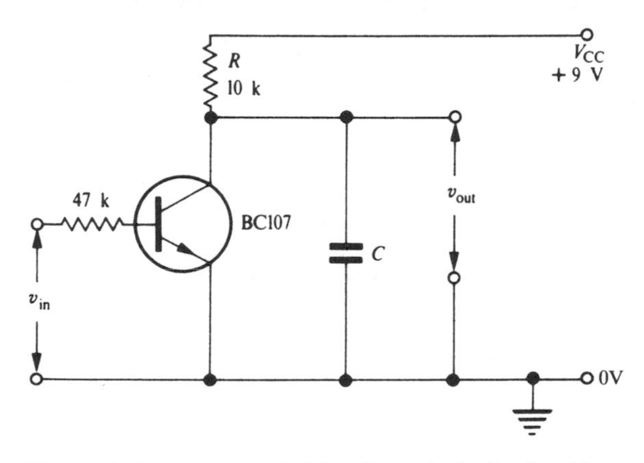

Fig. 10.9. Transistor switching (squaring) circuit with a capacitative load.

When the transistor is turned off, however, it takes no further part in the proceedings, leaving the load resistor R to charge C again with the relatively long time constant RC, giving a long rise time. Fig. 10.10 illustrates the type of waveform obtained in this case. If the circuit of fig. 10.9 is constructed with C equal to 10 nF, the effect will be clear at frequencies of about 500 Hz. A general principle in active switching circuits is that turn-on is almost invariably faster than turn-off.

Fig. 10.11 shows a method of overcoming slow turn-off using what is called *active pull-up.* This technique, common in digital logic circuits, uses in this case a pair of complementary transistors. A positive input voltage turns the npn transistor on and the pnp off, zero or negative input turning the pnp transistor on and the npn off. Thus, on both rising and falling waveform edges, there is always a transistor turning on to discharge any load capacitance, giving fast switching. A 100 Ω resistor is put in the emitter of the pnp transistor to limit the peak current flowing during switch-over, when both transistors are conducting at once; the transistors would otherwise constitute a short-circuit on the supply which could be dangerous for both transistors and supply.

Most IC logic uses active pull-up. The popular 74HC series of logic chips uses a complementary pair of MOSFETs in each output, whilst the 74LS series uses a pair of npn bipolar transistors driven in opposite phases so that one transistor turns on as the other turns off.

10.7 Coupling capacitors in pulse circuits

Coupling capacitors are very useful as a means of providing d.c. isolation between amplifier stages, and introduce no snags in the handling of normal audiofrequency a.c. signals. This is the case as long as the mean level of the

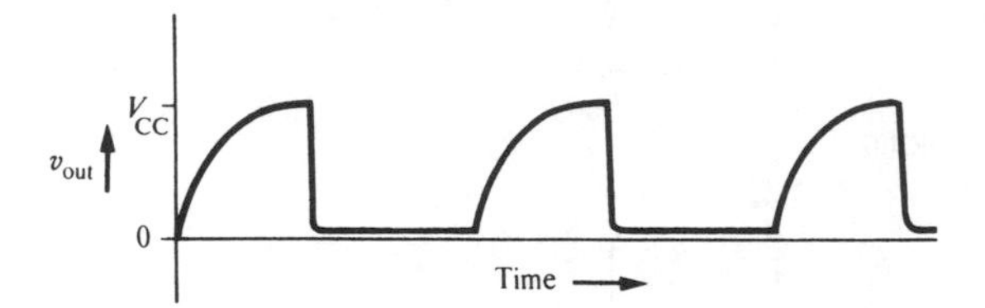

Fig. 10.10. Typical output waveform from transistor switch with capacitative load.

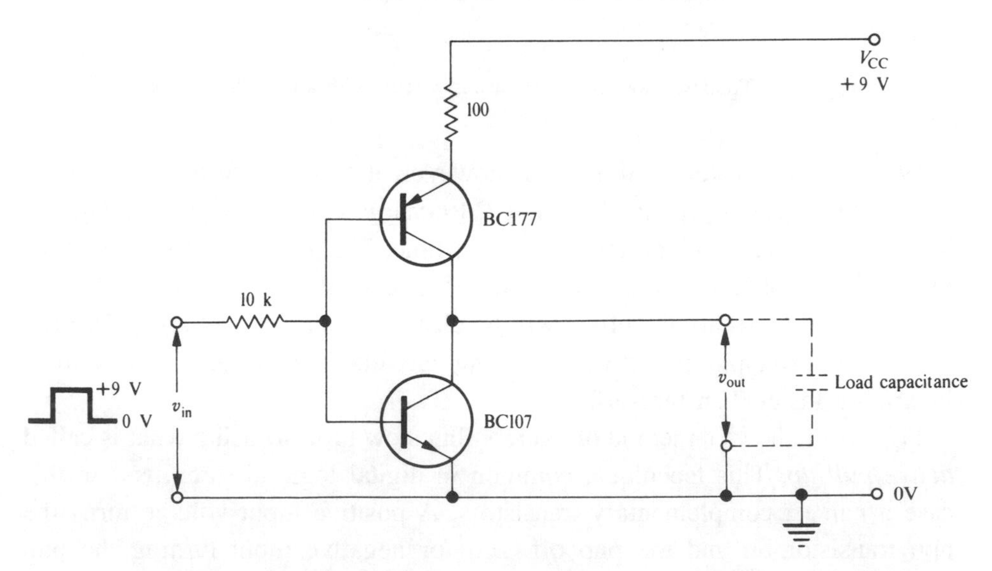

Fig. 10.11. A second transistor, giving active pull-up, avoids slow turn-off with a capacitative load.

signal is zero; i.e. that the average positive excursions are similar in size to the average negative excursions when viewed over a period of time. Most pulse trains, however, are of just one polarity, usually positive, and this can lead to unexpected results if coupling capacitors are used. Video signals, being asymmetrical, are similarly critical.

The coupling capacitor problem can be illustrated by connecting a rectangular wave signal (e.g. the output of fig. 10.2) to the input of the RC circuit of fig. 10.12, observing the output on an oscilloscope as the signal is connected to C; (the oscilloscope Y-amplifier should be switched to 'd.c.'). The output signal will be seen to move from the level of fig. 10.13(a) down to that of fig. 10.13(b). The coupling time constant in fig. 10.12 is made deliberately long so that the shift in d.c. level is observed to occur gradually over a period of seconds.

This level shift is due to the fact that the coupling capacitor charges up to a voltage equal to the mean d.c. level of the input waveform. With the usual

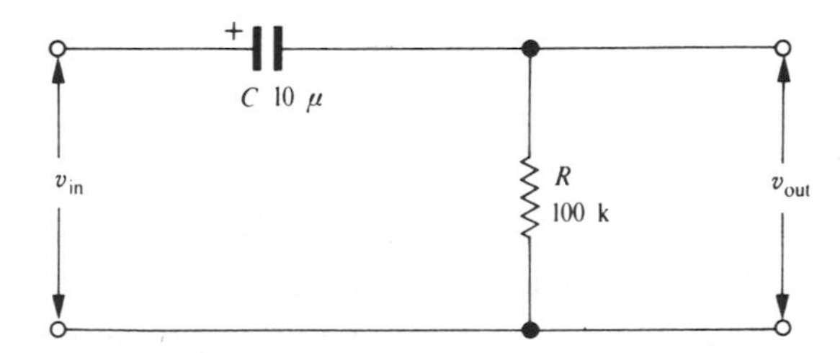

Fig. 10.12. Experimental coupling-capacitor circuit to demonstrate loss of d.c. level.

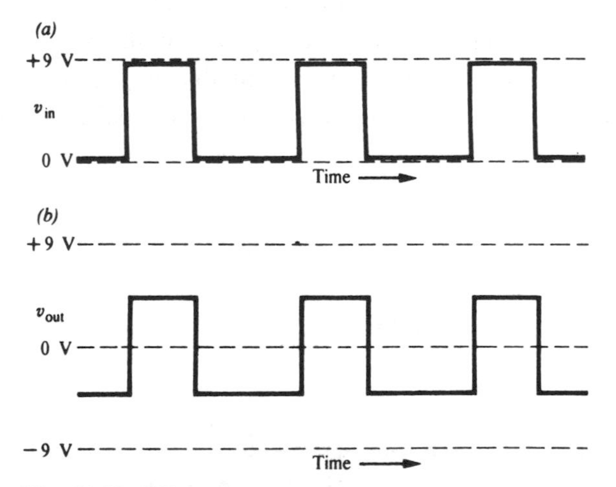

Fig. 10.13. Effect of coupling capacitor on a positive-going rectangular wave: (a) input waveform, (b) output waveform showing level shift.

type of a.c. input, the mean level averaged over several Hz is zero, but here, with a wholly positive-going pulse train, the mean level is given by the axis through the 'centre of gravity' of the waveform so that the waveform area included above the mean is equal to the area below the mean. Because waveform area is proportional to charge transferred, the output of the coupling capacitor settles so that the mean is at zero voltage, giving an equilibrium condition of zero net change in capacitor charge.

In the case illustrated in fig. 10.13, about half the pulse amplitude will be lost if the circuitry following the coupling circuit responds only to positive pulses. Furthermore, if the mark–space ratio of the input pulse train changes (i.e. the pulse duration changes in relation to the gap between pulses), the mean d.c. level will change. Thus the amplitude ratio of positive to negative half-cycles in fig. 10.13(*b*) will change. If the input positive pulses became very long compared with the gap between them, the positive half-cycle amplitude could become so small that any later circuit might fail to respond, the pulses being lost.

10.8 Clamp diode

The circuit of fig. 10.14 shows a remedy for the d.c. level shift problems associated with a coupling capacitor. The diode provides a rapid discharge path for the capacitor, 'clamping' the output to 0 V whilst the input is at zero, thus preventing a shift in d.c. level (fig. 10.15). For this reason, the clamp diode is sometimes called a *d.c. restorer*.

It is interesting to feed a sine wave into the d.c. restorer circuit. The diode will actually charge the capacitor to the peak voltage of the sine wave on negative half-cycles and 'restore' a d.c. component that was never there in the first place. The whole wave is shifted to become entirely positive-going. To

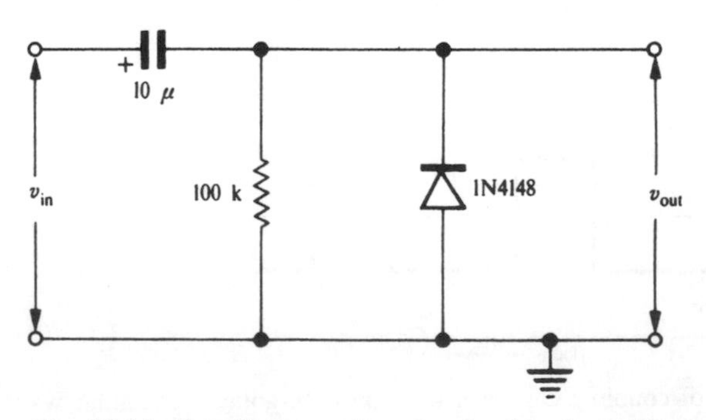

Fig. 10.14. Coupling-capacitor circuit with clamp diode to restore d.c. level.

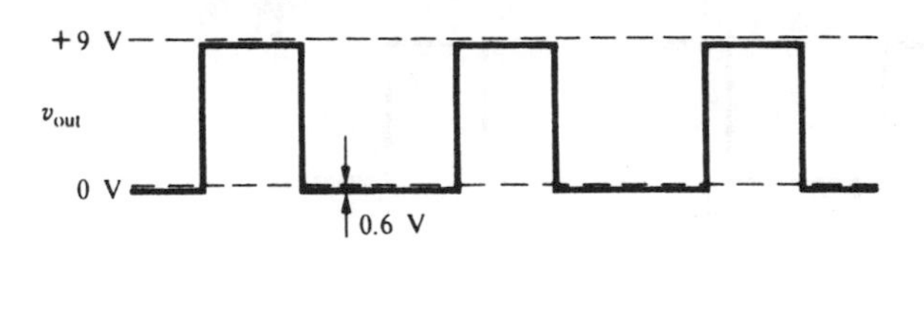

Fig. 10.15. Input and output waveforms of clamp diode circuit.

avoid distortion of the sine wave whilst the relatively heavy charging current is flowing, the d.c. restorer should be fed from a low-impedance source such as the loudspeaker output terminals of the signal generator. A small point to notice in fig. 10.15 is that the output in fact slips about 0.6 V below 0 V level owing to the forward p.d. of the silicon diode; this also occurs with a sine-wave input.

There is no reason why a clamp diode should be restricted to clamping a wave at 0 V level. If the 'earthy' end of the diode in fig. 10.14 is returned to a fixed voltage, either positive or negative, it will clamp the waveform at that level. If the diode is reversed, it has the effect of charging the capacitor so that the top of the waveform is clamped instead of the bottom.

10.9 Coupling capacitor time constant

We have so far considered the effect of a coupling capacitor with a very long time constant compared with the pulse duration. If, in fig. 10.12, C is reduced in value from 10 μF to 10 nF and a rectangular wave input applied at about 500 Hz, the effect of an inadequate coupling time constant will be clearly visible. The input and output waveforms are sketched in fig. 10.16, where the top and bottom of the rectangular wave are clearly tilted in the output waveform. The explanation is that the time constant is so short that the capacitor has time to indulge in a certain amount of charging and discharging during the period of the wave; the tilt or droop on the top and bottom of the wave is actually part of the exponential charge–discharge curve of the RC circuit.

For a small amount of tilt, it is possible to approximate the exponential

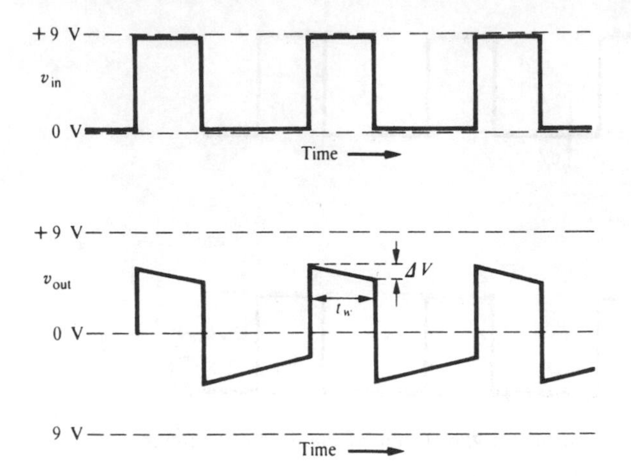

Fig. 10.16. Effect of inadequate coupling time constant on a rectangular wave.

$v = V_0 \exp(-t/RC)$ as a straight line by taking only the first two terms of its series expansion, i.e.:

$$v = V\left(1 - \frac{t}{RC}\right). \tag{10.12}$$

So, if the pulse duration is t_w, then the fractional tilt in fig. 10.16 is given by

$$\frac{\Delta V}{V} = \frac{t_w}{RC}. \tag{10.13}$$

Tilt is also related to the low-frequency –3 dB point f_1. As in section 8.2, we have $2\pi f_1 = 1/RC$, therefore, from equation (10.13),

$$\frac{\Delta V}{V} = 2\pi f_1 t_w. \tag{10.14}$$

Thus, to handle long pulses, an amplifier with an excellent low-frequency response is required. For example, using the above relation, if 1 ms pulses are required to have a tilt less than 1%, the lower –3 dB point of the amplifier frequency response must be no higher than 1.7 Hz. When this requirement is considered along with the problem of loss of d.c. level associated with a coupling capacitor, it is easy to see why d.c. coupling is preferred in pulse amplifiers.

10.10 Differentiation and integration

If, in a coupling capacitor circuit, *RC* is made very small compared with the pulse duration, then the pulse becomes so distorted as to be unrecognisable, the output becoming a *differentiated* version of the input. A pulse frequency of 500 Hz (derived from the circuit of fig. 10.2) with $C = 1$ nF and $R = 10$ kΩ (fig. 10.17) will give the wave shapes of fig. 10.18, which can be seen as the extreme case of the inadequate time constant. The capacitor initially follows the positive edge of the wave, but rapidly charges up to a p.d. of +9 V, showing only a narrow pulse at the output. When the input pulse returns to zero, on its negative-going edge, the charged capacitor takes the output right down to −9 V initially, but then rapidly discharges showing only a narrow pulse at the output. The full output pulse amplitude of ±9 V will only be seen if the input wave rise and fall times are very short compared with *RC*; sluggish edges give the capacitor time to charge before it has transmitted the full pulse amplitude.

The fact that output pulses are only obtained when the input signal is *changing* indicates why this circuit is called a differentiator. Notice that, when the rate of change of the input voltage $\mathrm{d}v_{in}/\mathrm{d}t$ is positive, a positive output pulse is obtained, and when $\mathrm{d}v_{in}/\mathrm{d}t$ is negative, a negative pulse is obtained. The theoretical input–output relation for the RC differentiator can readily be calculated. In fig. 10.17, the output voltage is simply the p.d. across *R*, i.e.

$$v_{out} = iR.$$

If q is the instantaneous charge on the capacitor,

$$i = \frac{\mathrm{d}q}{\mathrm{d}t}.$$

Now, if the period of the input waveform is very long compared with *RC*, then

$$q \approx Cv_{in},$$

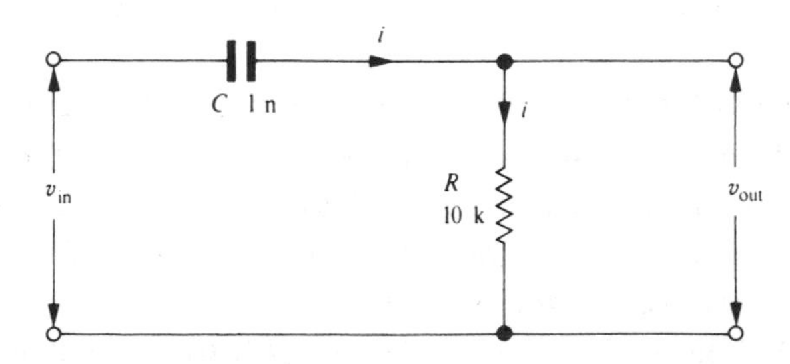

Fig. 10.17. Differentiating circuit.

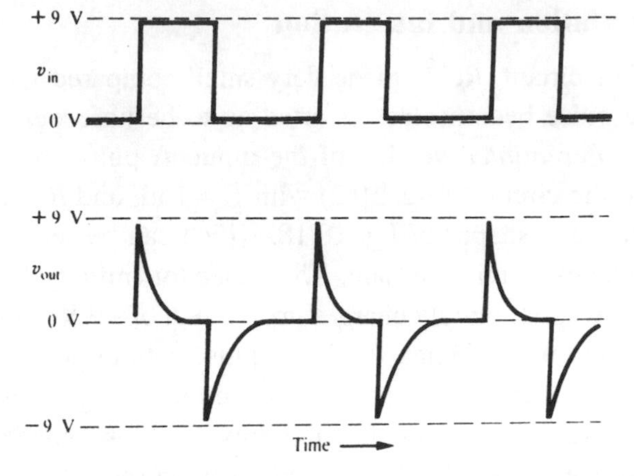

Fig. 10.18. Typical input and output waveforms of a differentiating circuit.

therefore

$$v_{out} = RC\frac{dv_{in}}{dt}. \tag{10.15}$$

If a sine wave is fed to the differentiator so that

$$v_{in} = V \sin \omega t$$

then we should expect at the differentiator output

$$v_{out} = RC\frac{dv_{in}}{dt} = \omega RCV \cos \omega t.$$

In physical terms, this means a waveshape of the same form as the input, but leading it by 90° in phase and with a peak amplitude directly proportional to frequency. These characteristics are exactly what we find experimentally and, recalling the discussion of the first-order high-pass filter in section 8.2, they indicate that we are operating on the 6 dB/octave low-frequency roll-off, giving an output amplitude directly proportional to input frequency, together with a 90° phase lead. The differentiator is thus seen as an alternative description of the high-pass filter.

If a high-pass filter can differentiate a waveform, it is reasonable to assume that a low-pass filter will *integrate*, and this is in fact the case. The circuit of fig. 10.19 will integrate tolerably well at frequencies of 500 Hz and above; the higher the frequency, the more accurate the integration. The following calculation indicates the way the integrator works.

In fig. 10.19 we shall 'freeze' all voltages and currents for a moment and let q be the instantaneous charge on C, therefore

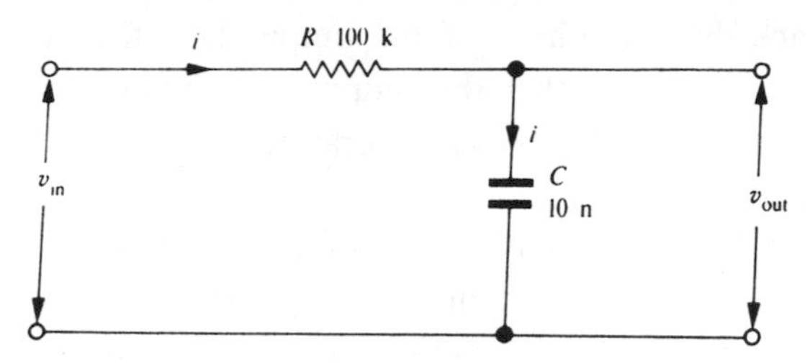

Fig. 10.19. Integrating circuit.

$$v_{out} = \frac{q}{C}. \tag{10.16}$$

Now, i is the current flowing in the resistor and capacitor (we assume the current in the output circuit to be negligible), therefore

$$\text{charge } q = \int i dt, \tag{10.17}$$

hence,

$$v_{out} = \frac{1}{C}\int i dt. \tag{10.18}$$

We now have to make the assumption that the period of the input signal is much less than time constant RC so that the signal is greatly attenuated, i.e.

$$v_{out} \ll v_{in}.$$

Then, by Ohm's law, $i \approx v_{in}/R$

$$v_{out} \approx \frac{1}{RC}\int v_{in} dt. \tag{10.19}$$

Thus we see that the low-pass RC filter acts as an integrator when the signal frequency is high enough so that the attenuation is large (say $v_{out} < v_{in}/10$). As with the differentiator, we can predict the effect of the integrator on a sinusoidal input, $v_{in} = V \sin \omega t$,

$$\begin{aligned} \text{integrated output } v_{out} &= \frac{1}{RC}\int v_{in} dt \\ &= \frac{1}{RC}\int V \sin \omega t \, dt \\ &= \frac{1}{RC}\left(\frac{-V}{\omega}\cos \omega t\right). \end{aligned} \tag{10.20}$$

This negative cosine means that the phase of the output lags the input sine function by 90°, the $1/\omega$ term means that the output amplitude is inversely proportional to frequency, indicating that we are operating on the –6 dB/octave roll-off of the low-pass filter.

The amplitude relation for the first-order low-pass filter discussed in section 7.7 can be used to assess the error in the output of an integrator. The integrator output amplitude can be rewritten from equation (10.20). If f is the signal frequency, then peak output voltage magnitude

$$|V_{\text{out}}| = \frac{|V_{\text{in}}|}{2\pi RCf}. \tag{10.21}$$

Now,

$$\text{half-power } (-3\text{ dB}) \text{ frequency } f_1 = \frac{1}{2\pi RC} \qquad \text{(from equation (7.8)).}$$

Thus, for the ideal integrator,

$$\left|\frac{V_{\text{out}}}{V_{\text{in}}}\right| = \frac{f_1}{f}.$$

In practice, for low-pass filter,

$$\left|\frac{V_{\text{out}}}{V_{\text{in}}}\right| = \frac{1}{\sqrt{(1+f^2/f_1^2)}} \qquad [(7.10)]$$

$$= \frac{f_1}{f} \cdot \frac{1}{\sqrt{(f_1^2/f^2+1)}}$$

At high frequencies, $f_1/f \ll 1$

$$\text{and } \left|\frac{V_{\text{out}}}{V_{\text{in}}}\right| \approx \frac{f_1}{f}, \text{ which approximates to the ideal case.}$$

The extent to which the term $1/(f_1^2/f^2+1)^{\frac{1}{2}}$ departs from unity thus gives the error in the integrator output amplitude. In a similar way, equation (7.12) can be used to determine the phase departure from the theoretical 90°.

With a rectangular wave input, we need to be careful in considering the effect of the integrator. Integration is simply a process of continuous summation, so that integrating a positive-going rectangular wave like fig. 10.20(*a*) will result in the output waveform of fig. 10.20(*b*); the output of a perfect integrator will simply continue climbing indefinitely as the positive areas are

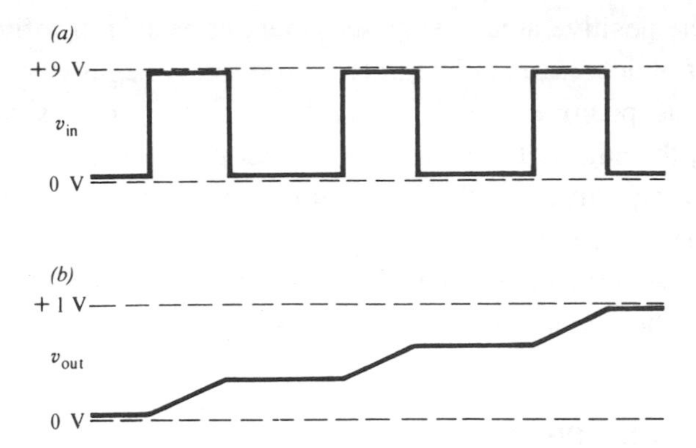

Fig. 10.20. Integration of a unidirectional rectangular wave: (*a*) positive-going input to integrator, (*b*) 'climbing' output from integrator.

summed. Notice that the positive gradient in the output corresponds to each positive-going rectangular wave, the output wave levelling off during the intervals when the input wave is at zero level.

With the simple RC integrator, the output cannot, of course, go on climbing indefinitely because, for one thing, this would violate the requirement that $v_{out} \ll v_{in}$ for accurate integration. In practice, the true action of the integrator can only be seen for the first few cycles of the input wave; thereafter the output settles down at the mean d.c. level of the input wave ($\frac{1}{2}V$ if the mark–space ratio is unity) with a superimposed triangular ripple.

A better picture of the response of the integrator to a rectangular wave may be obtained by first deliberately removing the d.c. component to obtain a wave symmetrical about zero volts. This is easily achieved by interposing the RC coupling of fig. 10.12 between the rectangular wave source and the integrator of fig. 10.19. The resulting input and output waveforms are shown in fig. 10.21. The positive halves of the input wave result in a steadily climbing

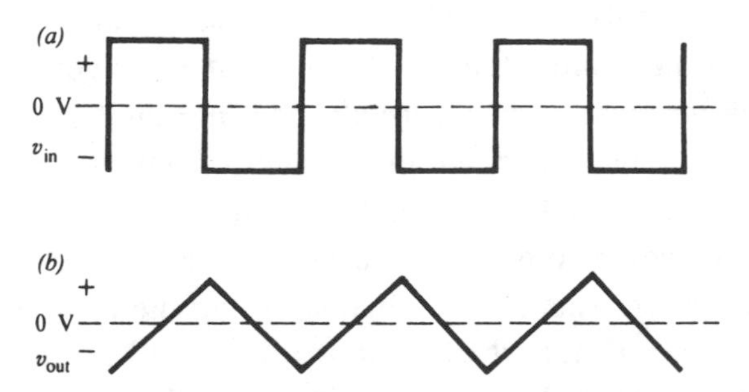

Fig. 10.21. Integration of a symmetrical rectangular wave (*a*) giving triangular output (*b*).

positive gradient as the positive area is summed, then, after each positive half comes a negative half which, as its is integrated, produces a negative gradient cancelling the previous positive one. The result is a triangular waveform centred on zero level; the edges of the wave are in fact part of the exponential charging and discharging curves of the integrating capacitor. With the time constants of fig. 10.19 and a frequency of 500 Hz or above, the condition $v_{out} \ll v_{in}$ obtains; the sections of exponential represent such a small part of the total function that they approximate to a straight line.

10.11 Electronic calculus

It is an interesting experiment to test the action of the differentiator and integrator by integrating the differentiated rectangular wave. The mathematics tells us that we should recover the original function, i.e. the rectangular wave. If the differentiating circuit of fig. 10.17 is followed by the integrator of fig. 10.19, and a rectangular wave fed into the combination, this theory can be tested by examining the output on the oscilloscope. A passable representation of the input should be obtained at a pulse frequency of about 500 Hz. Try to account for the inaccuracy of the result.

10.12 The charge pump ratemeter

A good deal of pulse circuitry is designed to handle signals from radiation detectors such as Geiger–Müller tubes, solid-state particle detectors and photomultipliers. The pulses are amplified and squared up using circuits such as fig. 10.2, often used in conjunction with the Schmitt trigger or monostable multivibrator (see chapter 12). The measurement usually required in such situations is the pulse rate, which is proportional to the intensity of radiation falling on the detector.

At low radiation levels the pulse rate may be so slow that it can be counted manually, as in the case of the familiar 'click click click' of a 'Geiger counter' with an audio output. Faster pulse rates are usually fed into digital counters, such as those discussed in chapter 13, when the number of pulses arriving in a set time is automatically displayed on a digital readout.

It is often convenient, however, to obtain an instant indication of pulse rate on an analogue (pointer) meter. This can be obtained using the *charge pump* (diode pump) ratemeter, a circuit for which is shown in fig. 10.22. The only requirement for the input pulses is that they should be positive-going and of constant peak amplitude. The wave shape does not matter.

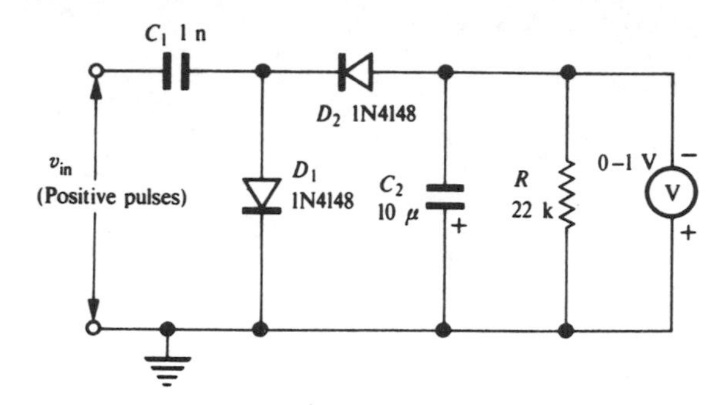

Fig. 10.22. A charge pump ratemeter.

When a positive pulse arrives, it charges the small capacitor C_1 to its peak voltage via diode D_1. In the interval between pulses, when the input is at zero volts, C_1 then quickly discharges via D_2 into the large capacitor C_2 and is then ready to receive the next pulse. Meanwhile, C_2 is continuously discharging via R. Thus, the faster the pulses arrive, the faster C_2 receives packets of charge from C_1 and the higher its p.d. will rise; i.e. the voltmeter reading is proportional to the rate of arrival of pulses at the input. This applies as long as the voltage on C_2 remains small compared with the input pulse amplitude. C_2 may be seen as a water tank from which there is a constant leakage (via R) and which is being supplied with water pumped in spurts from C_1. The level of water in the tank (output voltage) is then directly proportional to pumping rate (input frequency).

The charge pump ratemeter is commonly used as a tachometer, particularly in cars, where a source of pulses linked to the engine revolutions is readily available from the ignition system.

If the charge pump is fed with pulses of constant amplitude and frequency, and C_1 and C_2 varied, the output voltage will be directly proportional to the value of C_1 and inversely proportional to the value of C_2. A directly-reading capacitance meter is thus possible.

10.13 Pulse clipping

To obtain positive pulses of uniform height for use with the charge pump, the transistor squaring circuit of fig. 10.2 may be used with a stable power supply or, to avoid using a power supply, a Zener diode clipper may be used as in fig. 10.23(*a*). The diode begins to conduct when its Zener breakdown voltage, V_Z is exceeded, thus clipping the pulses to a uniform height equal to V_Z.

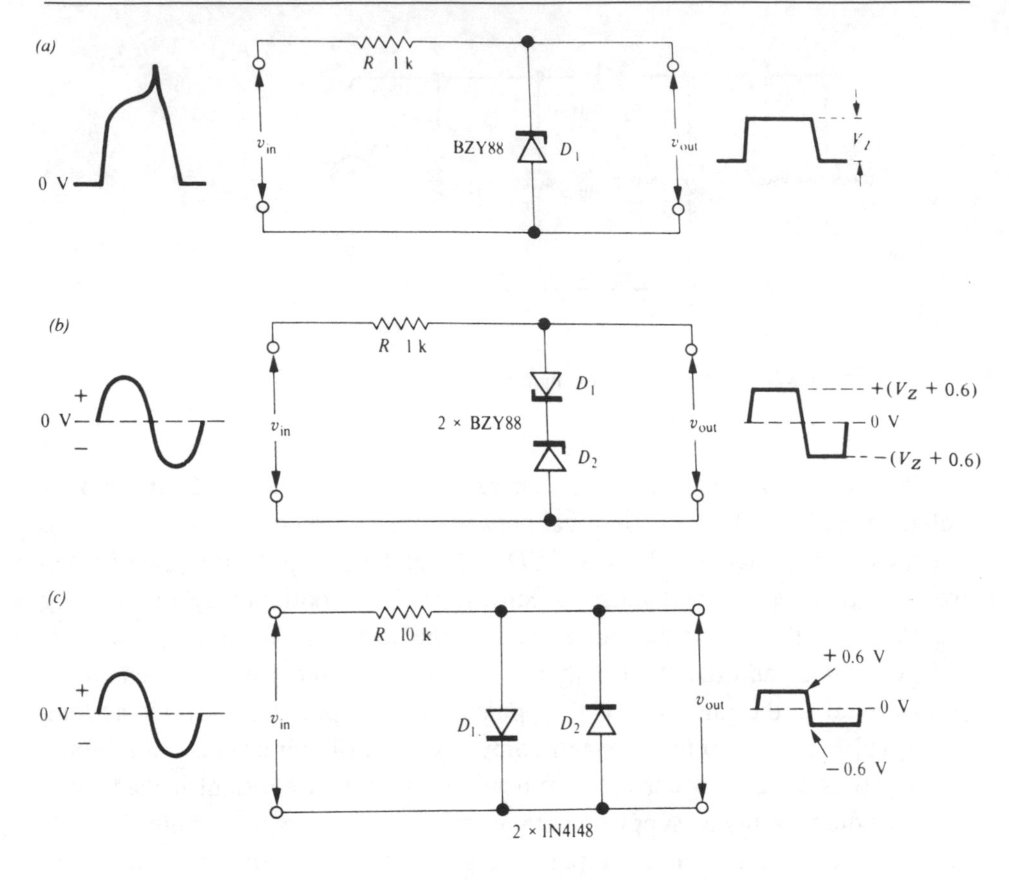

Fig. 10.23. Diode waveform clippers: (*a*) positive pulse clipper using Zener diode, (*b*) Zener diode clipper for bi-directional signals, (*c*) diodes in forward bias limit output to 1.2 V peak-to-peak.

Positive and negative pulses can be clipped by using back-to-back Zener diodes as in fig. 10.23(*b*). On positive pulses, D_1 is in its forward conduction mode and clipping occurs when the voltage across D_2 reaches V_Z; on negative input pulses, the roles of the two diodes are reversed. The clipped output is thus $\pm(V_Z + 0.6)$ volts, which allows a 0.6 V drop for the diode in forward conduction.

If positive and negative clipping is required and a clipped output of ±0.6 V is adequate, the circuit of fig. 10.23(*c*), using conventional diodes, will fill the bill. This relies on the rapid rise of the exponential forward current–voltage relationship of a silicon junction in the region of 0.6 V, the diode acting as a shunt to any higher voltage; D_1 conducts on positive pulses and D_2 on negative ones.

The clipping circuits of figs. 10.23(*b*) and 10.23(*c*) are sometimes used with analogue signals along with normal linear amplification. Here they are often called *limiters* and sometimes have a safety function to prevent damage to circuits by overloads such as static discharge and telecom line currents.

11

Integrated circuit analogue building bricks

11.1 Introduction

The integrated circuit (IC) is clearly the building brick of electronic circuits. We have already had a flavour of IC applications when looking at power supply regulators. Now we turn towards the full range of IC capabilities. A glance through a component distributor's catalogue reveals a seemingly limitless range of ICs for virtually every conceivable function. An IC may contain from a dozen transistors to a million depending on the application, together with all necessary resistors, diodes etc. The intimate thermal connection achieved by fabricating all the components on one chip of silicon generally leads to excellent stability and predictability in use.

The understanding of discrete components gained from earlier chapters will be found essential to the proper interfacing of ICs: in fact even today very few analogue circuit applications dispense totally with discrete semiconductors. However, the IC designer has today relieved the circuit designer of much of the 'donkey work'. In addition, the small size and low power consumption of ICs has made possible products like the Camcorder and hand-held GPS position-fixing satellite receiver.

This chapter deals with applications of linear ICs. They are designed to handle *analogue* signals, which carry their information in terms of amplitude and waveshape. Most audio and radio signals come into this category; they are distinct from the standard binary pulses of digital circuits which are discussed in chapter 13. To give an idea of the scope of this chapter, fig. 11.1 includes many of the basic building bricks which will be discussed and provides a quick reference to the outline circuits.

11.2 The operational amplifier

11.2.1 Simplifying assumptions

All the circuits of fig. 11.1 make use of an *operational amplifier* (op amp). The term 'operational' is generally used nowadays to describe a high-gain voltage amplifier, particularly one in IC form; the name is derived from the original use of such amplifiers in analogue computing *operations*. The characteristics of an op amp are such that the following simplifying assumptions can be made in most practical circuits:

infinite open-loop voltage gain, A_{VOL} (typically 2×10^5)
infinite input impedance (typically 2 MΩ)
zero output impedance (typically 75 Ω)

The parameter values quoted above refer to the popular 741 type IC amplifier which is used in many of the practical circuits in this book.

11.2.2 Input bias current and offset voltage

The input terminals of an op amp connect to internal transistor bases or gates which must be given some d.c. reference and be able to draw a small bias current if the amplifier is to function (there are no coupling capacitors on the chip). Input bias current in the 741 amplifier is about 100 nA. The first design consideration, therefore, is that each input of any IC amplifier must have some sort of d.c. path to earth, even if it is through a high-value resistor.

Ideally, both the inverting and non-inverting inputs should 'see' the same resistance to earth; otherwise, as fig. 11.2 shows, an effective input offset voltage will appear. We can assume that the two input bias currents are equal, i.e.

$$I_1 = I_2.$$

Hence, if $R_1 = R_2$, V_1 and V_2 will be equal and there will be zero effective differential offset voltage $(V_2 - V_1)$ at the amplifier inputs. In most circuits, the inverting input will normally have a feedback resistor R_f connected through to the output as in fig. 11.3 and therefore a proportion of the bias current will be drawn from the output through R_f. Now, if our circuit is designed correctly so that the offset voltage is zero, then the output will be at 0 V level under quiescent conditions. This means that, as far as the input bias current is

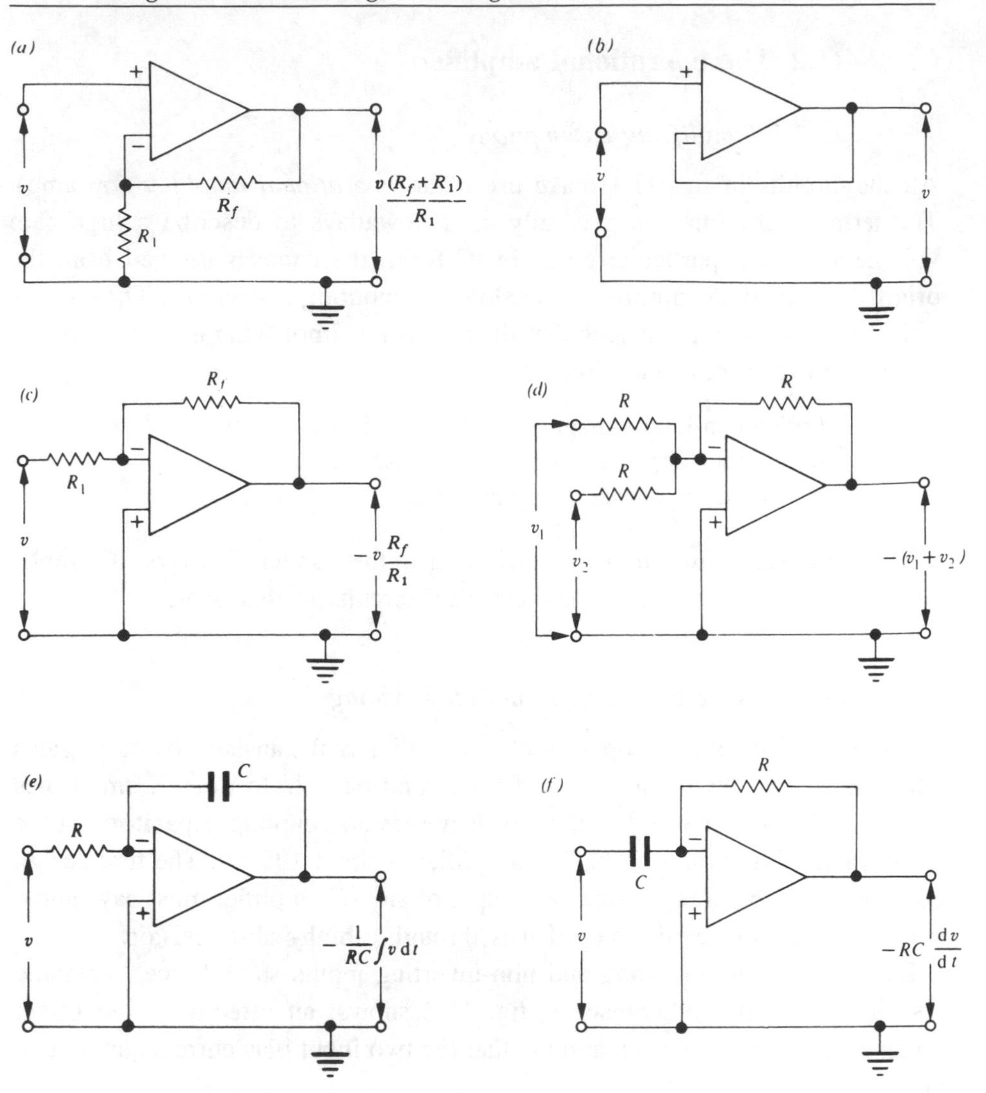

Fig. 11.1. (*a*) Non-inverting amplifier. (*b*) Voltage follower. (*c*) Inverting amplifier. (*d*) Adder. (*e*) Integrator. (*f*) Differentiator.

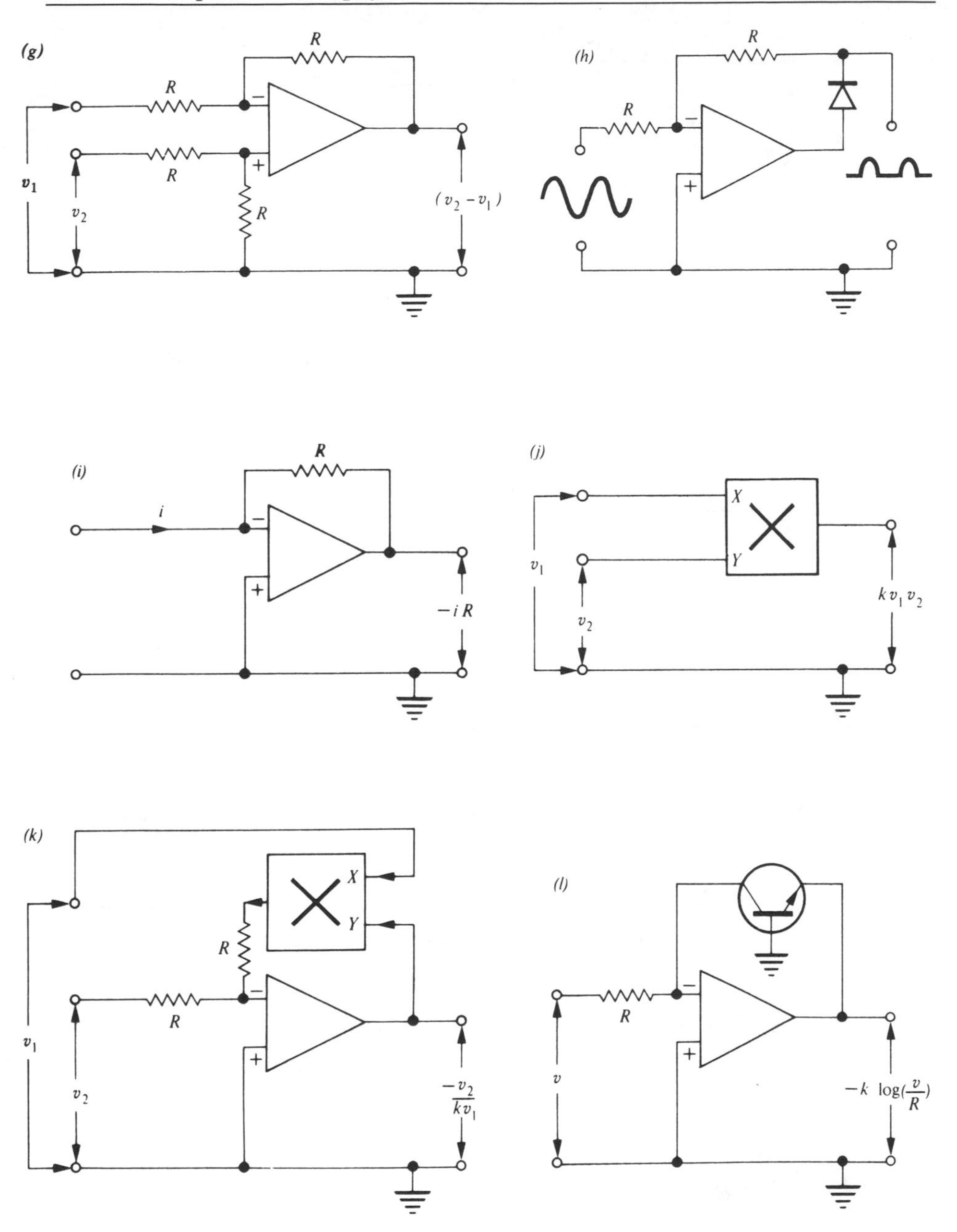

Fig. 11.1. (*g*) Differential amplifier (subtractor). (*h*) Precision rectifier. (*i*) Current to voltage converter. (*j*) Multiplier. (*k*) Divider. (*l*) Logarithmic amplifier.

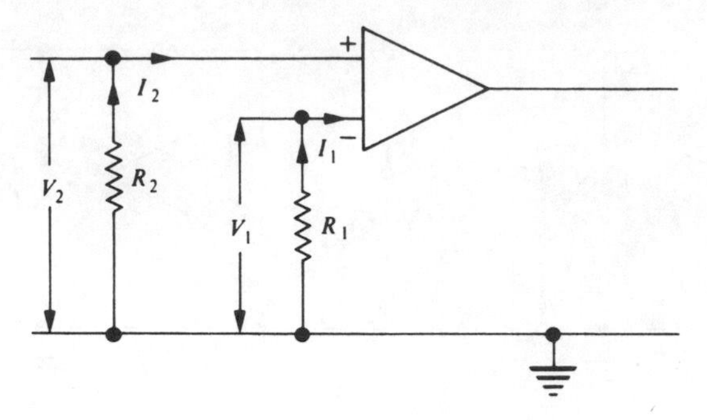

Fig. 11.2. Effect of input bias current, (I_2 and I_2) in producing differential input offset voltage $V_2 - V_1$ when inputs see different resistances to earth (R_1 and R_2).

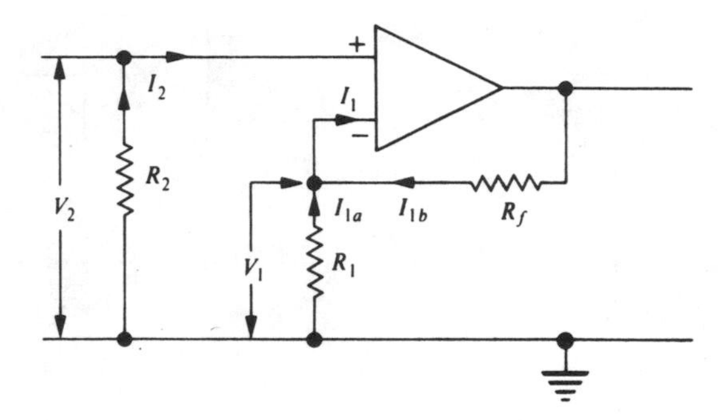

Fig. 11.3. Input offset voltage arising from a popular feedback configuration.

concerned, R_1 and R_f are effectively in parallel. For minimum offset in the circuit of fig. 11.3 we would arrange therefore that

$$\frac{1}{R_2} = \frac{1}{R_1} + \frac{1}{R_f},$$

i.e.

$$R_2 = \frac{R_1 R_f}{R_1 + R_f}.$$

It may appear that we can completely avoid any offset voltage by choosing our resistor values carefully so that both inputs see the same resistance to earth. Unfortunately, things are not quite so straightforward in practice; the internal circuitry may be slightly unbalanced so that the two bias currents are not equal.

The difference between the two input bias currents is known as the *input offset current* and typically amounts to 20 nA in the 741. It is therefore not worth bothering to match the input resistance to better than 20% or so.

Yet another factor to be considered is that, irrespective of external voltages at the inputs, the IC itself has a small inherent input offset voltage; in the 741, for example, it is about 1 mV. Bias and offset currents of the order of tens of nanoamps (10^{-9} A) sound very small until it is realized that the resistance seen by each input in a typical circuit may be some hundreds of thousands of ohms. A current of 100 nA flowing in 100 kΩ produces a potential difference of 10 mV, which may be comparable to the input signal.

More serious than the actual value of the resultant offset is the fact that the bias and offset currents vary with temperature. The offset current drift in a 741 is typically 1 nA/deg C which can result in output voltage drift.

For these reasons it is advisable to set an upper limit of 500 kΩ on the resistor value used from each input to earth on a 741. Good industrial design practice would restrict the maximum value to 50 kΩ for better d.c. stability. If higher value input resistors are required, op amps with much smaller bias currents are available, though at higher cost than the 741. The LM308 IC uses very high gain ('super-β') input transistors and achieves a typical input bias current of 7 nA and an input offset current of only 1.5 nA.

For much lower input currents, circuits with FET inputs are needed. In this category is the Texas TL071 IC which has an input bias current of 30 pA and an input offset current of 5 pA, enabling input resistances as high as 100 MΩ to be used. The use of FET inputs involves some sacrifice in inherent input offset voltage, this being typically 3 mV for the TL071 compared with the 1 mV of the 741. This is primarily due to the difficulty of producing two identically matched FETs; this fact also leads to the input offset voltage being more sensitive to temperature than in a bipolar circuit.

11.2.3 Offset null circuit

From the discussion above, it should be clear that it is virtually impossible to build an amplifier without an offset voltage at the input. In other words, even if there is no external input signal on the amplifier, it will be sitting with a few millivolts d.c. input of its own making. In a high-gain amplifier, this can be serious: if the overall voltage gain is 1000, one millivolt input offset appears as one volt at the output.

If the amplifier is intended for a.c. operation only and uses a coupling capacitor at the output, this will cut out any d.c. offset and all will be well, as long as the offset has not shifted the quiescent point so far that the available output swing is restricted. For d.c. amplification, however, we must have a means of adjusting the output voltage to zero when the input is zero. An offset null control provides this function; this is a potentiometer externally connected to the op amp. Fig. 11.4 shows the very simple connections for the offset null control on the 741, which will cancel input offset voltages of approximately ±25 mV.

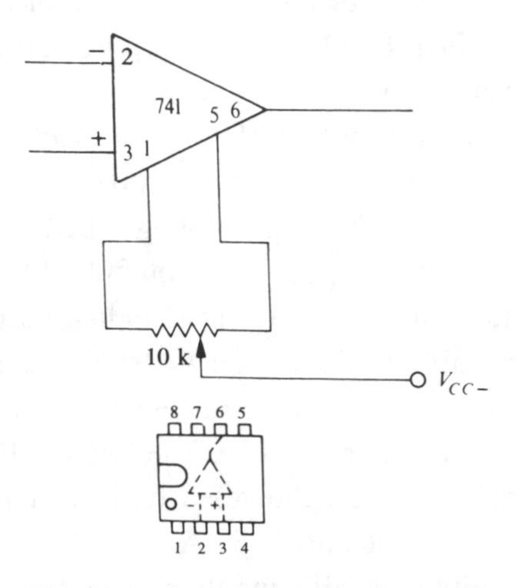

Fig. 11.4. Connection of an input offset control on a 741 amplifier (pin connection numbers refer to the 8-pin dual-in-line package shown).

11.3 Practical circuit details

In this chapter, many varieties of amplifiers and other signal processing circuits are described in practical terms so that the reader should be able to construct the circuits and experiment with them. Typical component values are given in all cases, together with a discussion of the significance of the various components so that values may be changed to suit a particular application. There is always room for experiment.

The op amp used in all the practical circuits is the 741 IC. Other chips such as FET input op amps like the TL071 can be easily substituted. Pin connections are standard except for special functions such as offset null, when a data sheet should be consulted if these are required.

Unless otherwise stated, the power supplies (V_{CC+} and V_{CC-}) are assumed to be +15 V and −15 V, connected to pins 7 and 4 respectively on the 8-pin package as shown in fig. 11.5. The common connection to the two power

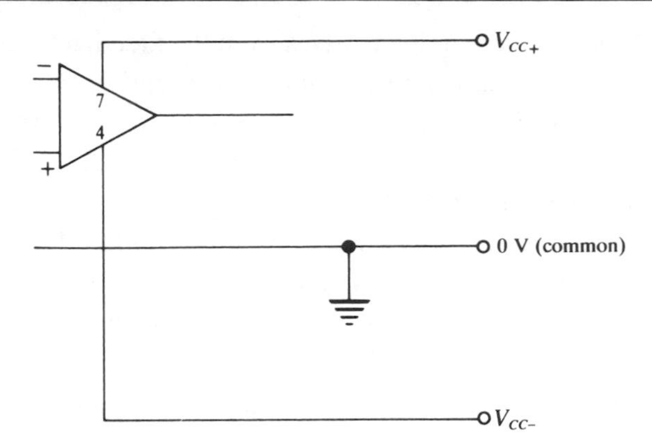

Fig. 11.5. Power supply connections on the 741 amplifier. Typical values of V_{CC+} and V_{CC-} are ±15 V. The earth (0 V) rail is connected to the 'centre tap' of the two supplies.

supplies is the earth (0 V) rail of the circuit, there being no direct earth connection to the IC. Although supplies of ±15 V are usually assumed in the quoted specifications of op amps, the 741 will work satisfactorily over a wide range of supply voltages from ±3 V to ±18 V. As might be expected, there is a variation of open-loop gain over this range, typically from 40 000 to 250 000, but this would not normally be significant with circuits using large amounts of negative feedback. More serious is the limitation on output voltage, the maximum peak-to-peak output voltage swing being limited to $(2V_{CC} - 2)$ volts.

As long as due note is taken of these limitations most of the circuits quoted will work well from +9 V and −9 V, easily obtainable from a pair of radio batteries.

11.4 The non-inverting amplifier

11.4.1 Basic d.c. amplifier

The series connection of negative feedback to provide a non-inverting amplifier was examined in section 4.8 and was employed in the thermocouple amplifier of fig. 8.12. The feedback signal is effectively subtracted from the input signal by feeding the latter into the non-inverting input of the op amp and the feedback into the inverting input.

Fig. 11.6 shows a non-inverting amplifier designed for operation down to zero frequency (d.c.) and with the input arranged for connection to a signal source of low output resistance. This latter requirement arises from the fact that the amplifier must draw its input bias current from the signal source and must therefore see a d.c. path to earth. Resistor R_x is in circuit simply to ensure that both inputs see the same resistance to earth. If the signal source output resistance is comparable with R_x, its value should be subtracted from R_x. It may be thought at first that R_x, being in series with the signal, will cause significant attenuation; fortunately this is not the case because the amplifier itself, thanks to the negative feedback, has an input impedance of at least 50 MΩ. There will thus be a negligible signal loss across the 10 kΩ or so of R_x. A signal source such as a thermocouple or a previous op amp automatically provides a negligible source resistance.

It will be noted in the table of fig. 11.6(*b*) that the circuit may be used to give unity voltage gain: the practical use of this may be questioned. The answer is that the unity-gain non-inverting amplifier is used to provide an impedance match like the emitter follower. It is called a *voltage follower*. The input

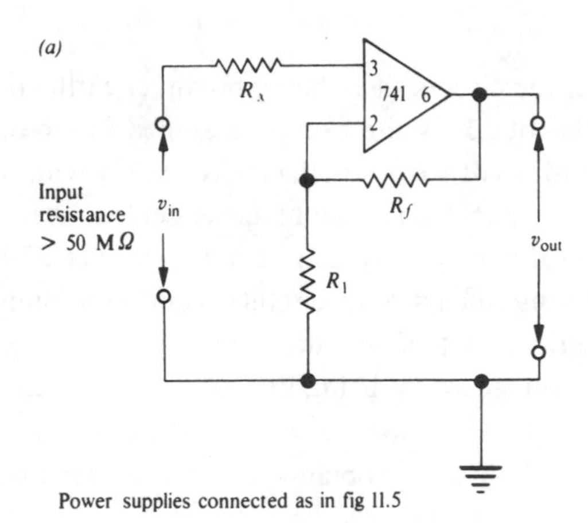

$$A_{VCL} = \frac{R_f + R_1}{R_1}$$

$$\text{for min. offset, } R_x = \frac{R_f R_1}{R_f + R_1}$$

(*b*)

A_{VCL}	R_f	R_1	R_x
1	0	∞	0
10	90 k	10 k	9 k
100	1 M	10 k	10 k
1000	10 M	10 k	10 k

Fig. 11.6. (*a*) Non-inverting amplifier. (*b*) Table of typical component values.

impedance may be many hundreds of megohms at low frequencies and the output impedance less than 1 Ω, though, owing to current limitations in the output stage, the 741 cannot feed loads of less than 2 kΩ without curtailing its available output voltage swing. Loads as low as 600 Ω can be driven by the NE5534, which also exhibits very low noise and is therefore extensively used in audio circuits. As is noted in section 5.9, low noise design requires also that resistor values be kept as low as possible.

11.4.2 Use of coupling capacitors

In many instances the amplifier will be required to handle only a.c. signals, possibly from a signal source which does not present the necessary d.c. path to ground for the input bias current. Perhaps an input coupling capacitor is needed to eliminate a d.c. component in the input signal which would otherwise overload the amplifier. In such cases, the circuit of fig. 11.6 must be modified so that it does not rely on the signal source for the input bias current path. Fig. 11.7 shows a suitable modification where R_x is connected from the

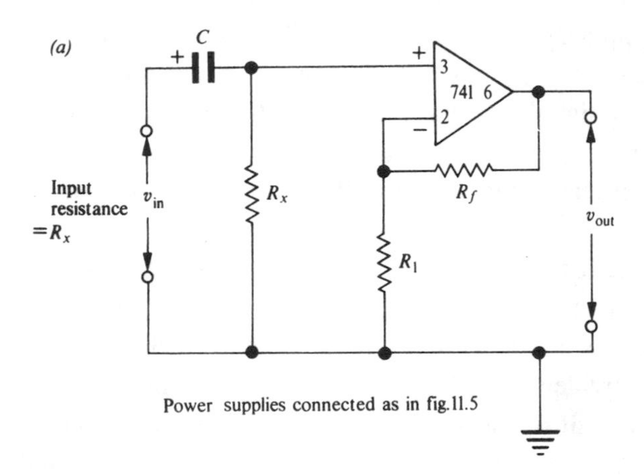

(b)

A_{VCL}	R_f	R_1	R_x	C
1	100 k	∞	100 k	1 μ
10	900 k	100 k	90 k	1 μ
100	10 M	100 k	100 k	1 μ
1000	10 M	10 k	10 k	10 μ

Fig. 11.7. (*a*) Non-inverting amplifier with a.c. coupling at input. (*b*) Table of typical component values.

non-inverting input to earth and the input fed via a coupling capacitor. The main problem with this circuit is that R_x shunts the input, so that the inherently high input impedance of the IC cannot be realized. The values of R_x in fig. 11.7 have been set as high as is consistent with a low offset voltage at the output. A coupling capacitor can usefully be connected in series with the output so that any offset is not transmitted; the offset requirements will then be less stringent.

Another technique often applied in a.c. amplifiers is to connect a coupling capacitor in series with R_1 in the feedback circuit. R_x can then be made as high as R_f without introducing a serious offset voltage. An important advantage of this arrangement is that the d.c. closed-loop gain remains at unity no matter how high the a.c. gain is set; output offset voltage is thus minimal, and temperature stability excellent.

11.4.3 a.c. amplifier with single power supply

An amplifier which is only dealing with a.c. signals and is used with coupling capacitors at both input and output can be used with a single power supply instead of the dual supplies otherwise required. The essential function of using both positive and negative supplies together on a d.c. amplifier is to ensure that, when there is no signal the input and output can sit at zero volts d.c. Coupling capacitors render this unnecessary: input and output can be arranged to sit midway between earth and V_{CC+} to allow both positive- and negative-going signal swings.

The circuit of fig. 11.8 fulfils these requirements. Capacitor C_1 in the feedback divider ensures that, as far as d.c. is concerned, the circuit acts as a unity-gain voltage follower. The non-inverting input is held midway between earth and supply by divider R_2 and R_3 so that, owing to the unity d.c. gain, the IC output will also sit at the same voltage, allowing both positive- and negative-going signals to be handled. Coupling capacitors at input and output isolate external circuits from the $V_{CC}/2$ quiescent voltage.

In the circuits of figs. 11.7 and 11.8, it is unfortunate that the presence of the input bias current prevents us from realizing the high input impedance inherent in the series feedback configuration. The input impedance of the amplifier itself with feedback may be as high as 100 MΩ, but the need to provide a d.c. path for the input bias current means that resistance to earth, and hence the overall input impedance, must be limited to a few hundred kilohms. The circuit of fig. 11.8, however, offers an opportunity to increase the effective value of the input resistor R_4 by means of *bootstrapping*, a posi-

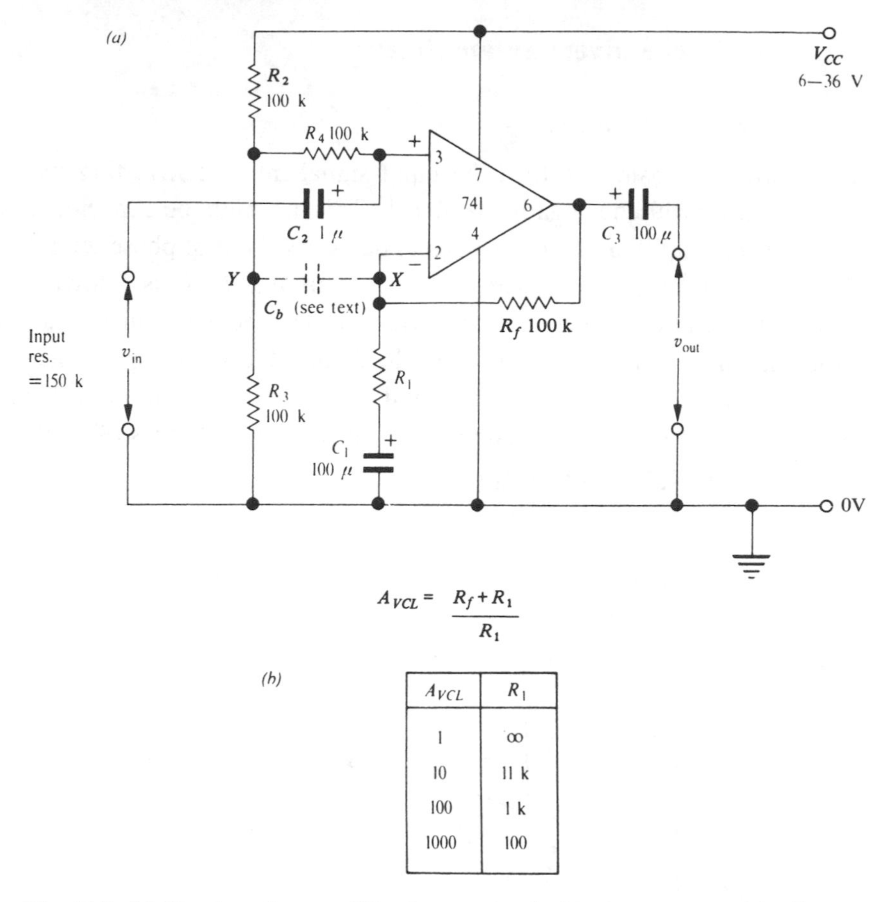

A_{VCL}	R_1
1	∞
10	11 k
100	1 k
1000	100

Fig. 11.8. (*a*) Non-inverting amplifier for a.c. signals (single power supply). (*b*) Table of values for R_1.

tive feedback technique which was discussed in connection with the emitter follower in section 5.12.6. Points X and Y can be joined by a large capacitor C_b (10–100 μF), causing the a.c. signal at X (inverting input) to appear at end Y of R_4. Owing to the almost infinite gain of the op amp, the noninverting and inverting inputs have virtually identical signals on them, the potential difference being less than a millivolt. Because of this, the bootstrapping capacitor causes both ends of R_4 to swing up and down together with the input signal; negligible current flows in the resistor, and its effective value, as far as the input signal is concerned, is increased many times.

The actual input impedance achieved depends on the ratio A_{VOL}/A_{VCL}, but is typically 300 MΩ at low frequencies, dropping to 2 MΩ at 20 kHz because of the high frequency roll-off in A_{VOL}.

11.5 The inverting amplifier

11.5.1 Introduction

It is perfectly feasible to feed the input signal into the inverting input of an op amp along with the negative feedback. This, as might be expected, produces an inverting amplifier, the output signal being 180° out of phase with the input. The circuit configuration, which is shown in fig. 11.9, is known as *shunt feedback* because the negative feedback, instead of being in series with the input signal, is in parallel with it and feeds into the same input. The study of shunt feedback is not complicated, and it is especially important because as we shall see, it forms an introduction to electronic analogues of mathematical functions (analogue computing).

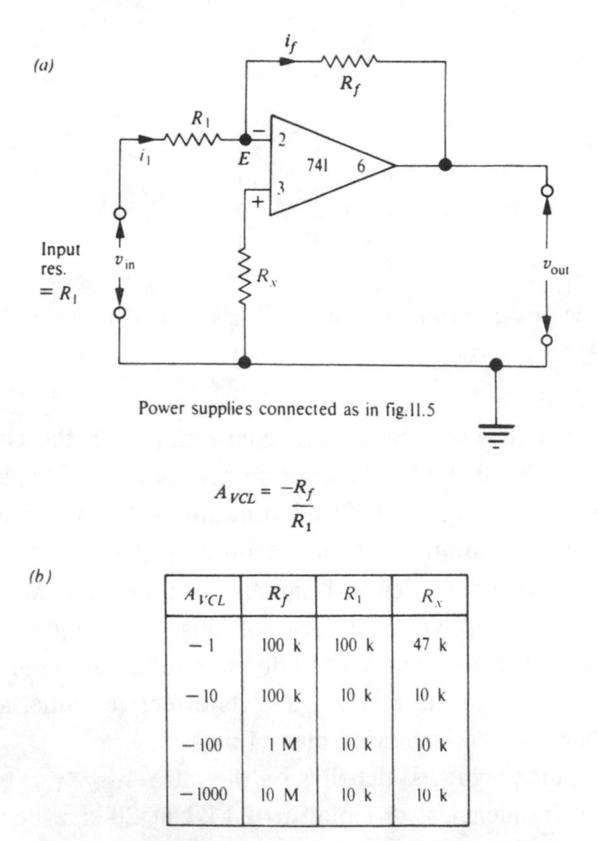

A_{VCL}	R_f	R_1	R_x
−1	100 k	100 k	47 k
−10	100 k	10 k	10 k
−100	1 M	10 k	10 k
−1000	10 M	10 k	10 k

Fig. 11.9. (*a*) Inverting amplifier. (*b*) Table of typical comonent values.

11.5.2 The virtual earth

To calculate the effect of shunt feedback, we use the three simplifying assumptions about an op amp:

open loop gain $A_{VOL} = \infty$,
input impedance $Z_{in} = \infty$,
output impedance $Z_{out} = 0$.

In fig. 11.9, the point E, where R_f and R_1 join the inverting input, is called a *virtual earth*. This is so because, if the amplifier has a voltage gain approaching infinity, there must be a negligible potential difference between the two inputs. Thus, as far as the signal is concerned, the inverting input looks virtually the same as the non-inverting input, which is earthed.

Accepting that point E is a virtual earth means that the full input voltage v_{in} appears across input resistor R_1. Thus we see immediately why the input resistance R_{in} is quoted in fig. 11.9 as being equal to R_1: R_1 is the only resistance which stands between the input terminal and the virtual earth. Signal current in R_1 is therefore given by

$$i_1 = \frac{v_{in}}{R_1}.$$

Now, considering feedback resistor R_f, which is connected between the virtual earth and v_{out}, and taking signal current i_f in the direction shown,

$$i_f = \frac{-v_{out}}{R_f}.$$

If the op amp input impedance is infinite, no signal current is drawn by the inverting input, so it follows that

$$i_f = i_1,$$

i.e.

$$\frac{-v_{out}}{R_f} = \frac{v_{in}}{R_1}$$

therefore

$$A_{VCL} = \frac{v_{out}}{v_{in}} = \frac{-R_f}{R_1}. \tag{11.1}$$

The way that the gain is dependent upon the simple ratio of two resistors makes the inverting amplifier very flexible in use. For instance, if R_f is a variable resistance box calibrated in kΩ and R_1 is made exactly 1 kΩ, the precise gain of the amplifier may be read from the scale of the variable resistance.

In analogue computing the simple inverting amplifier has two main applications. Firstly, if $R_f = R_1$ it operates as a change of sign (inverter) without changing signal amplitude. Secondly, if we require to multiply a variable (the signal) by a constant, k, the ratio R_f/R_1 is made equal to k; if the constant is positive, we must then follow up with an inverter to correct the sign.

11.5.3 a.c. operation of inverting amplifier

Coupling capacitors can be used if required with the inverting amplifier without any basic circuit modification, both op amp inputs having the necessary d.c. path to earth (via R_f and R_x). Since R_1 will now no longer be in the d.c. path for the inverting input, R_x should be increased in value to be equal to R_f if minimum offset voltage at the output is required.

11.6 The differential feedback amplifier

The circuit of fig. 11.10 employs an input arrangement which applies negative feedback and also makes available differential inputs; it is really a combination of the inverting and non-inverting configurations. Input (1) is the inverting input and (2) the non-inverting input. If input (2) is earthed and a signal fed into input (1), the circuit will be seen, both theoretically and experimentally to be equivalent to the inverting amplifier of fig. 11.9 with voltage gain $-R_f/R_1$. Interchanging inputs (1) and (2) makes the non-inverting amplifier of fig. 11.6 with gain R_f/R_1. The common-mode rejection ratio can in principle, be as high as that of the op amp itself, but in practice is limited by tolerances in the resistor values.

The resistor values in fig. 11.10 give minimum d.c. offset, but there is a considerable disparity between the input impedance at the two inputs when measured relative to earth. Looking into input (1), we see resistor R_1 followed by the virtual earth at the inverting input; the input impedance of (1) is thus equal to R_1. Looking into input (2), we see R_2 in series with R_3, the input impedance of the non-inverting input of the op amp itself being very high. Thus, in the case of the ×100 amplifier, with the resistor values suggested, input (1) will have Z_{in} of 10 kΩ whilst Z_{in} of input (2) is 1.01 MΩ. Now, one

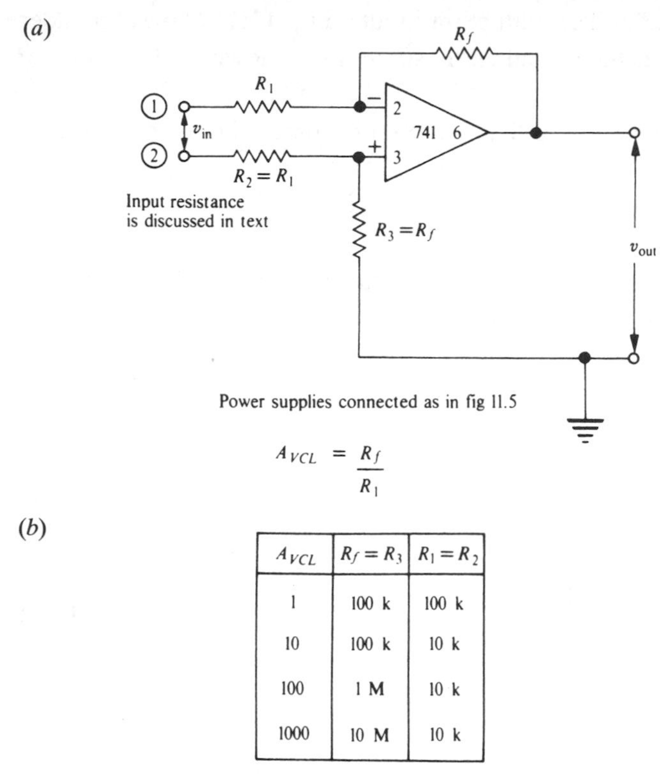

A_{VCL}	$R_f = R_3$	$R_1 = R_2$
1	100 k	100 k
10	100 k	10 k
100	1 M	10 k
1000	10 M	10 k

Fig. 11.10. (*a*) Feedback amplifier with differential inputs available. (*b*) Table of typical component values.

important application of the differential amplifier is to reject hum and noise induced into the input leads (see section 8.7). Unless the impedance of the signal source is low, such a disparity in input impedances is therefore a severe disadvantage, because the input with the higher impedance will pick up very much more hum and noise than the other. Optimum d.c. conditions can usefully be sacrificed by making $R_2 + R_3 = R_1$ and $R_2/R_3 = R_1/R_f$, thus equalizing input impedances whilst preserving high CMRR. 'Instrumentation' amplifiers, using more than one op amp can be designed to give very high CMRR values completely independent of source impedance, whilst maintaining optimum d.c. conditions. Typically, each of the two inputs is buffered by a voltage follower.

11.7 The operational adder

The concept of the virtual earth makes possible a range of circuits which perform mathematical operations very precisely. The operational adder is essentially an inverting amplifier with extra inputs. Fig. 11.11 shows an adder with two inputs. Point E is the virtual earth, sometimes known as the *summing junction*.

Equating currents and assuming that the op amp input draws no current,

$$i_f = i_1 + i_2.$$

Since point E is a virtual earth, the currents can be expressed in terms of signal voltages and resistor values thus:

$$\frac{-v_{\text{out}}}{R_f} = \frac{v_1}{R_1} + \frac{v_2}{R_2}$$

i.e.

$$v_{\text{out}} = -\left(\frac{R_f}{R_1}v_1 + \frac{R_f}{R_2}v_2\right). \tag{11.2}$$

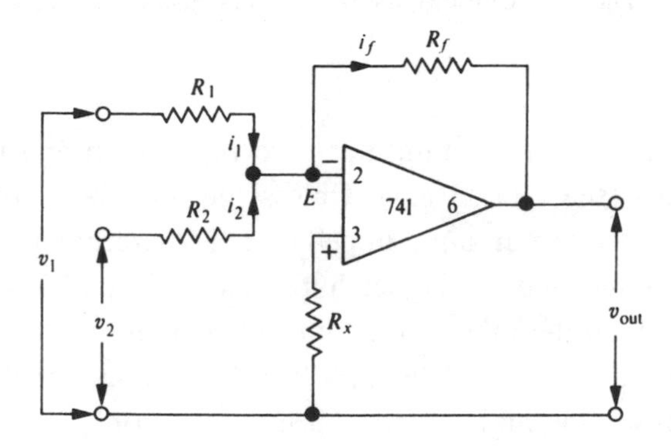

Power supplies connected as in fig.11.5

$$v_{\text{out}} = -\left(\frac{R_f}{R_1}v_1 + \frac{R_f}{R_2}v_2\right)$$

for minimum offset $\frac{1}{R_x} = \frac{1}{R_1} + \frac{1}{R_2} + \frac{1}{R_f}$

Fig. 11.11. Virtual-earth adder.

Resistor values usually lie in the range 10 kΩ to 200 kΩ, and can conveniently be chosen so that $R_1 = R_2 = R_f$, making

$$v_{\text{out}} = -(v_1 + v_2).$$

Notice that, although the magnitude of the output is equal to the sum of the inputs, there is a reversal of sign, a property of all virtual earth circuits.

A common application of the adder is the audio mixer, which is used to combine the outputs of microphones, special effects, reverberation unit, etc. in a recording studio where over fifty inputs may be mixed together to give a two-channel stereo output. On a domestic scale, the production of a 'home video' sound track usually requires the mixing of commentary, sound effects and music. An essential feature of a mixer is that fading down one input should have no effect on the gain of the others. The virtual earth acts as an isolating barrier between the inputs of the operational adder and thus provides independent operation. Fig. 11.12 shows a simple audio mixer using an adder.

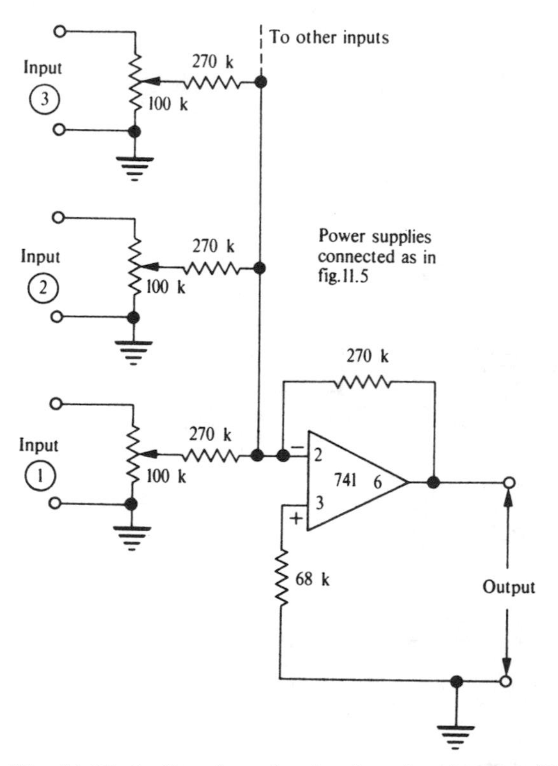

Fig. 11.12. Audio mixer circuit using virtual-earth adder (unity gain).

The feedback resistor chosen gives unity gain, but may be increased if higher gains are required. Normal procedure for use with microphones is to feed the mixer inputs via low-noise microphone pre-amplifiers. Any number of inputs may be connected to the virtual earth, as long as each has its individual gain control and input resistor. Improved HF output swing and lower noise will be obtained by replacing the 741 with the TL071.

11.8 The operational integrator

11.8.1 Basic circuit

If an op amp has a capacitor as the feedback element, it performs the mathematical operation of integration with respect to time. In other words it acts as a storage circuit which performs a summation of the input over a period of time, just as a petrol pump at a filling station records the total number of gallons delivered to a car, integrating the flow rate during the filling time.

Fig. 11.13 shows an operational integrator, sometimes known as the Miller integrator or Blumlein integrator. As before, assuming no current flowing into the op amp input,

$$i_f = i_i\,,$$

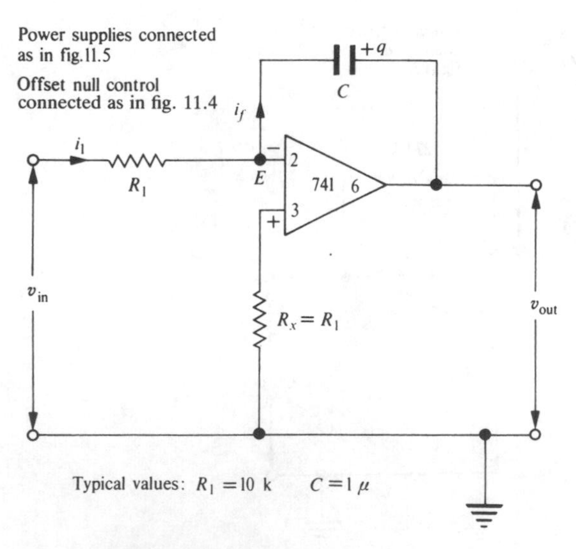

Fig. 11.13. Operational integrator.

Once again, because of the near-infinite op amp gain, point E is a virtual earth, therefore

$$i_1 = \frac{v_{in}}{R_1}.$$

If $+q$ is the charge on C at some instant, then current is rate of change of charge, i.e.

$$i_f = \frac{-\mathrm{d}q}{\mathrm{d}t}$$

therefore, since $i_f = i_i$,

$$\frac{-\mathrm{d}q}{\mathrm{d}t} = \frac{v_{in}}{R_1}.$$

Now,

$$q = Cv_{out}$$

therefore

$$\frac{\mathrm{d}q}{\mathrm{d}t} = C\frac{\mathrm{d}v_{out}}{\mathrm{d}t}$$

hence

$$-C\frac{\mathrm{d}v_{out}}{\mathrm{d}t} = \frac{v_{in}}{R_1}.$$

Integrating with respect to time,

$$v_{out} = \frac{-1}{R_1C}\int v_{in}\ \mathrm{d}t. \qquad (11.3)$$

We therefore see that the circuit of fig. 11.13 performs the operation of integration and has an effective 'gain' magnitude $1/R_1C$, where R_1 is in ohms and C in farads or, more conveniently, R_1 is in megohms and C in microfarads. For unity gain, where the output directly equals the integral of the input, the product R_1C should be equal to unity. This involves inconveniently high values of R_1 and C, particularly as electrolytic capacitors are not generally suitable because of their high leakage current and need for a polarizing voltage. In practice, values of $R_1C = 0.01$ are common (e.g. $C = 1\ \mu F$, $R_1 = 10\ k\Omega$), giving an effective gain of 100.

11.8.2 Offset in an integrator

If the circuit of fig. 11.13 is built and tested, a fault may at first be suspected because the output will almost certainly be sitting near V_{CC+} or V_{CC-}. This only means that the circuit has just integrated as far as it can go in one direction. An integrator requires very careful offset adjustment, otherwise it integrates its own input offset voltage, the output voltage gradually climbing until it reaches its limit, a volt or so below one of the supply rails. The offset control (as in fig. 11.4) should be adjusted so that the output shows no immediate tendency to creep up or down.

It is impossible, in practice, to avoid some slight offset, so that it is inevitable that any practical circuit will exhibit some drift of output voltage over a period of minutes. This is often controlled by short-circuiting the feedback capacitor C with a switch (reed relay or FET) until integration is to begin.

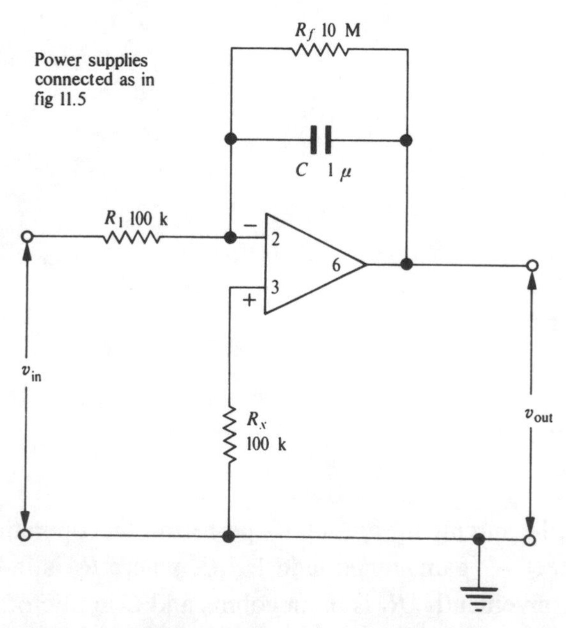

Fig. 11.14. General purpose integrator with reduced gain, and hence reduced drift.

The switch is then opened, the signal to be integrated being fed to the input; after the required integration time, the output voltage may be measured and the switch closed again to discharge the capacitor.

Sometimes integrators are required to operate continuously; one instance is when they are connected together to solve equations. At first sight one might expect such applications to be unworkable because of offset voltage. Often, however, the circuits are connected in such a way that a loop exists so that any offset voltage is ultimately fed back to the integrator input, introducing a measure of self-correction. In section 12.23, we shall see that it is possible to design an oscillator, using operational integrators, which does not require offset zero controls at all, thanks to the loop in which the integrators are connected.

If a single integrator is required to operate continuously, and offset voltage proves to be a problem, the 'tamed' circuit of fig. 11.14 may be used. Here a 10 MΩ feedback resistor R_f is added across the capacitor to reduce the open-loop gain to 100 instead of 10^5. This reduces the 'climbing offset' problem of the open-loop circuit, whilst preserving an adequately long time constant for accurate integration down to the lowest frequencies of the audio range.

11.8.3 Accuracy and integration time

A useful way of looking at the operational integrator is in terms of the simple RC integrator discussed in section 10.10, but with the effective value of the integrating capacitor increased by a factor equal to the open-loop gain of the amplifier. This result is a consequence of the Miller effect discussed in section 7.3.1. (With an amplifier of voltage gain A, the p.d. across the feedback capacitor, when seen from the input of the amplifier, appears A times smaller than it really is.) The equivalent circuit of fig. 11.15 illustrates this view of the

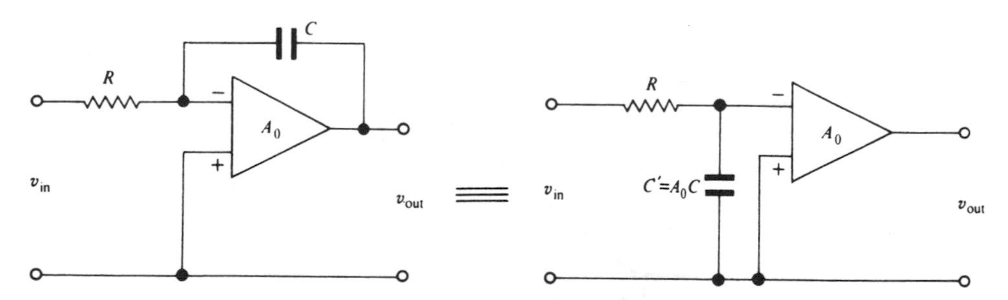

Fig. 11.15. Operational integrator with equivalent 'amplified passive' circuit, showing effective time constant.

integrator. In this example, the effective time constant, RC, is equal to A_0RC. Integrator error can then be assessed in the same way as in the passive circuits of section 10.10, but substituting A_0RC for RC in the equations. With A_0 values up to 10^5, the operational integrator facilitates precise integration over periods up to several seconds.

11.8.4 Ramp generator

The integrator is a useful source of the linear ramp waveform needed for an oscilloscope time-base sweep and for some techniques of analogue to digital conversion. If a constant d.c. voltage, $-V$, is connected to the input of an integrator then the output is given by

$$v_{out} = \frac{1}{RC}\int V\,dt$$

$$= \frac{Vt}{RC}.$$

This is a linear ramp of gradient V/RC (fig. 11.16). A repetitive rectangular waveform at the input, symmetrical about earth, produces a triangular wave output.

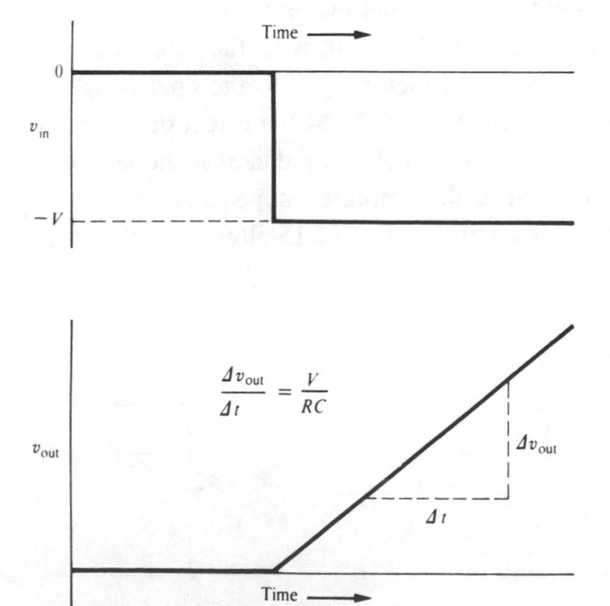

Fig. 11.16. Response of integrator to a voltage step – a linear ramp.

Fig. 11.17 shows how a FET can be used as an electronic switch to reset the integrator to zero by effectively shorting the capacitor. This can be arranged whenever required by setting the reset input to 0 V or positive thus allowing the FET to conduct (the series diode prevents the gate ever becoming forward-biased). The discharge time constant is given by the product of C and the resistance of the FET. The 2N3819 has a drain-source resistance ($r_{ds(on)}$) of approximately 300 Ω when it is turned on; special switching MOSFETs such as the ZVN 4206A are manufactured with much lower values of $r_{ds(on)}$ (typically 1 Ω). Setting the reset input to a negative voltage between −5 V and −20 V turns off the FET and allows normal integration to take place. A repetitive ramp waveform (sawtooth) can be generated by connecting the reset and signal inputs together and feeding in a negative-going rectangular wave of 5 V amplitude. The FET switch will then automatically open on each negative half-cycle and allow the integrator to give a positive ramp output. The output will reset to zero when the input returns to zero every cycle, giving the sawtooth output waveform of fig. 11.18. Such a waveform is used for an oscilloscope timebase sweep.

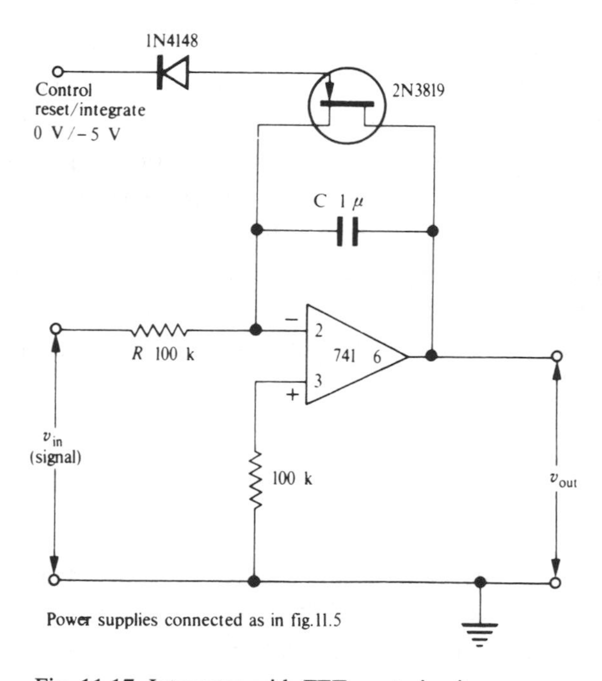

Fig. 11.17. Integrator with FET reset circuit.

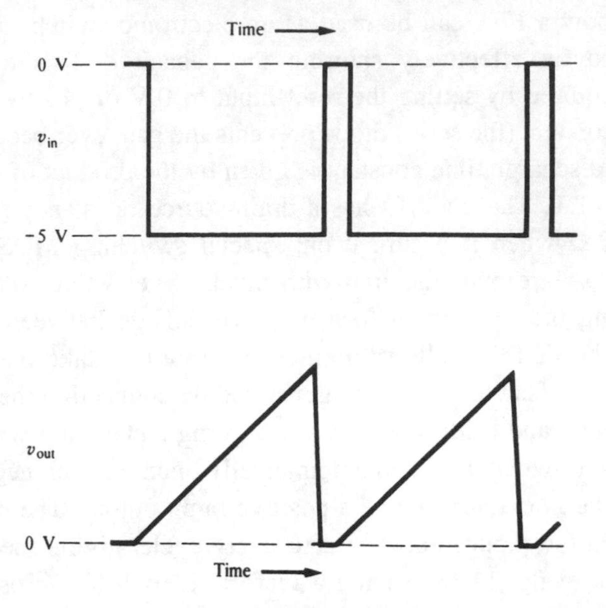

Fig. 11.18 Sawtooth output produced from rectangular wave input by integrator with automatic reset (control point coupled to input).

11.8.5 Charge-sensitive amplifier

A circuit often required in nuclear particle detection is the charge-sensitive amplifier (charge-to-voltage converter), which gives an output voltage proportional to the quantity of charge transferred to the input. The operational integrator is very useful in this application, the input resistor being removed and the input terminal connected straight to the inverting input (fig. 11.19).

Because of the high input impedance of the op amp, negligible current flows into the inverting input, so that the input current i, resulting from charge q, flows only into capacitor C. Thus q is transferred to capacitor C.

Now, point E is a virtual earth, so that the p.d. across C is v_{out}. Therefore, with charge q on C

$$v_{out} = \frac{-q}{C}. \qquad (11.4)$$

The output voltage is thus proportional to the charge which has flowed into the input. A reset arrangement such as the FET in fig. 11.17 will allow the summation of charge over a set period of time. Alternatively a resistor R can be connected across C to provide a discharge time constant. the arrival of a

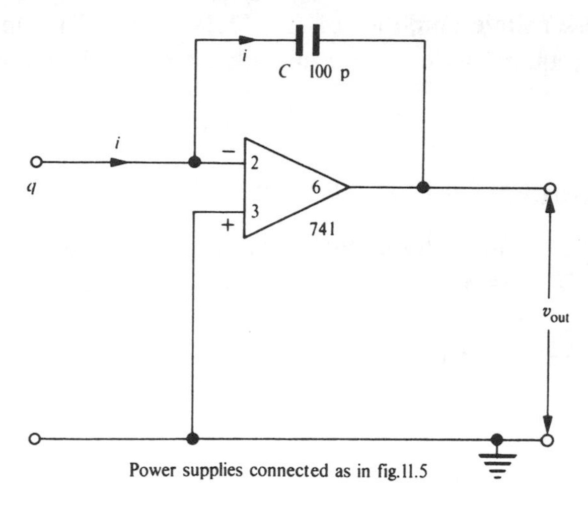

Fig. 11.19. Charge-sensitive amplifier.

charge q on the input will then give an output pulse of peak magnitude q/C which will decay with time constant RC. Care should be taken in this case that RC is short compared with the intervals between pulses, otherwise C, will not have discharged sufficiently before the next pulse arrives and the pulses will pile up one on top of the other. This can lead to limiting in the amplifier and the loss of pulses.

It may not be obvious why an op amp circuit is necessary at all in a charge-to-voltage converter; after all, a simple capacitor to earth will work perfectly well, giving $V=q/C$. The advantage of the charge-sensitive amplifier of fig. 11.19 lies in the fact that the input feeds directly to a virtual earth. Inevitably the input circuit contains stray capacitance, which is subject to variation due to changes in cable length and even changes in the internal capacitance of the particle detector. This capacitance would be added to the 'known' capacitor converting charge to voltage and could cause errors. With the virtual-earth charge-to-voltage converter, however, the low impedance of the virtual earth swamps stray capacitance. (As with the other operational integrators, the effective capacitance seen when feeding directly into the virtual earth is equal to the value of C multiplied by the open-loop gain of the amplifier.) A further advantage of the virtual-earth circuit is that the input device itself does not 'see' the output voltage pulse, since the latter appears at the opposite side of the virtual earth. If for example a solid-state particle detector were to feed a capacitor directly, it is possible that the capacitor p.d. might affect the bias voltage on the detector and upset linearity.

Although the charge-sensitive amplifier of fig. 11.19 uses a 741 op amp, some applications may require the high input impedance of a FET input such as the TL071.

11.9 The operational differentiator

Interchanging R and C in the integrator circuit produces, as might be expected, a differentiator (fig. 11.20). As before, we assume negligible current flowing into the op amp input so that $i_1 = i_f$.

Point E is a virtual earth, so that

$$i_f = \frac{-v_{\text{out}}}{R_f}$$

and

$$i_1 = C\frac{\mathrm{d}v_{\text{in}}}{\mathrm{d}t} \quad (i \text{ is rate of change of charge})$$

therefore

$$\frac{-v_{\text{out}}}{R_f} = C\frac{\mathrm{d}v_{\text{in}}}{\mathrm{d}t},$$

$$v_{\text{out}} = -R_f C\frac{\mathrm{d}v_{\text{in}}}{\mathrm{d}t}. \tag{11.5}$$

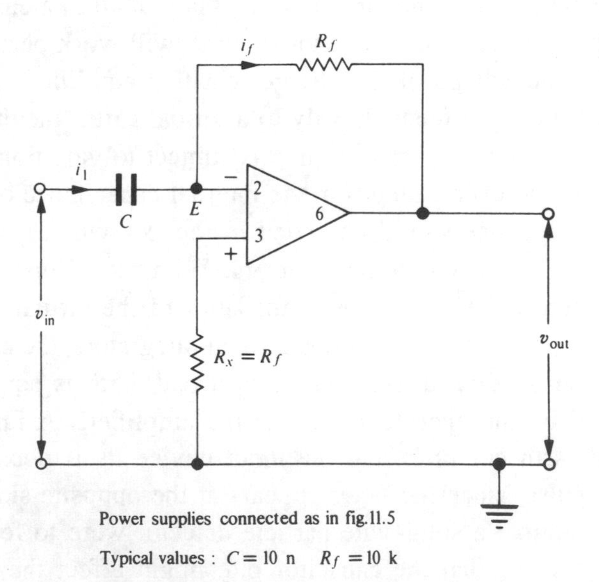

Fig. 11.20. Operational differentiator.

Thus the output voltage is directly proportional to the differential coefficient of the input voltage with respect to time. In other words, the faster the rate of change of the input, the higher the output voltage.

Because the circuit has unity d.c. gain, the connection of an offset null control is usually unnecessary.

The operational differentiator may be seen as the ultimate version of the RC differentiator in fig. 10.17. Just as the op amp in the integrator increases the effective value of the capacitor by a factor equal to the open-loop amplifier gain, so the op amp in the differentiator *decreases* the effective value of the feedback resistor, R_f, in the same proportion. This gives near-perfect differentiation, a rectangular wave input giving very sharp pulses corresponding to the edges where the input is changing (fig. 11.21).

In practice, operational differentiator circuits are troublesome because they are very susceptible to any noise in the input circuit, particularly transient interference pulses which may be picked up from electrical switching, etc. Though the actual voltage amplitude of the noise may be very low, the rate of change (dv/dt) is often very high and this results in a high spurious output from the differentiator. For this reason, the differentiator is avoided in practical

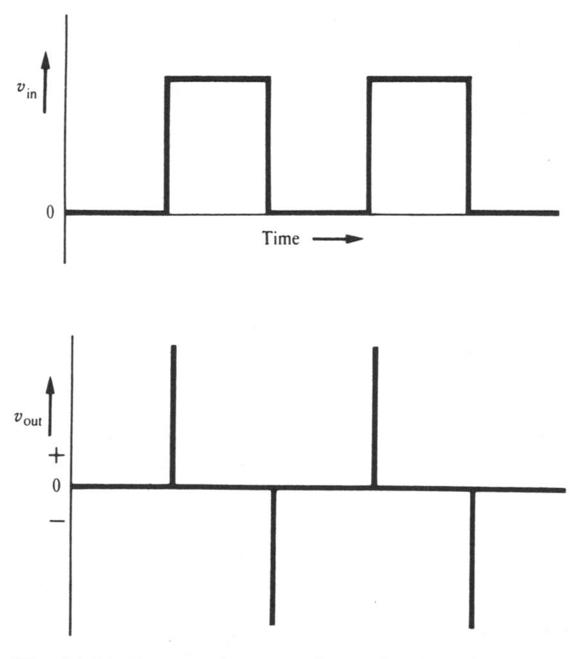

Fig. 11.21. Rectangular wave input signal, and the corresponding output obtained from an operational differentiator.

circuits whenever possible. If the circuit cannot be avoided, susceptibility to noise may be reduced by curtailing the effective amplifier gain at high frequencies. A resistor (typically 1 kΩ) is connected in series with C and a small capacitor (typically 100 pF) across R_f, the values being adjusted experimentally until a suitable compromise between noise sensitivity and differentiating performance is reached.

11.10 Current-to-voltage converter

Most electronic circuits are designed to handle voltage signals. In some cases, however, we have to deal with a current signal, such as the output of a photomultiplier or photodiode; we then want to convert the current signal into a voltage signal at the earliest opportunity. The simplest current-to-voltage converter is a single resistor (fig. 11.22(*a*)), but this has relatively high input and output impedances (equal to R). For a current input, a high input impedance

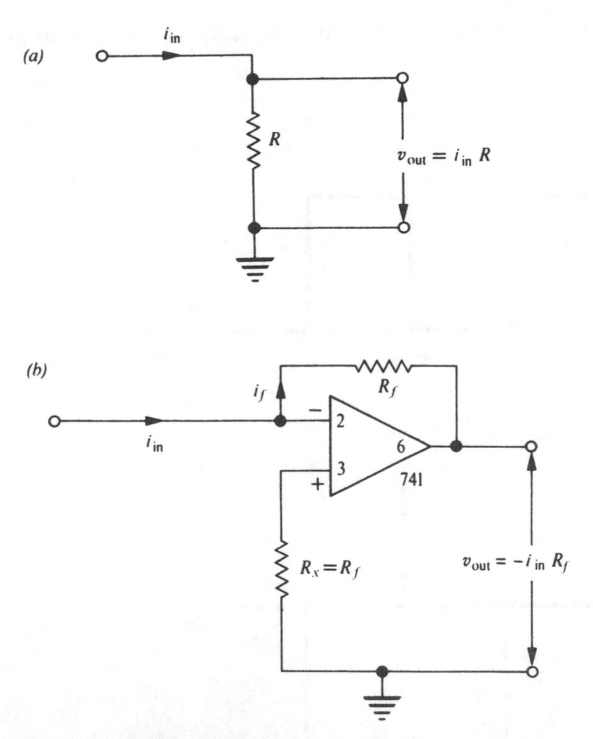

Fig. 11.22. Current-to-voltage converters: (*a*) simple resistor, (*b*) virtual-earth version.

is not required and causes degradation of high-frequency response if the input connection has significant self-capacitance. The operational current-to-voltage converter is shown in fig. 11.22(*b*), where the input current flows straight into the virtual earth. The very low impedance here means that cable capacitance has virtually no effect. In addition, the output of the circuit is also at low impedance, thanks to the op amp action.

Making the usual op amp assumptions,

$$i_f = i_{\text{in}}.$$

Since E is a virtual earth,

$$\begin{aligned} v_{\text{out}} &= -R_f i_f \\ &= -R_f i_{\text{in}}. \end{aligned} \tag{11.6}$$

11.11 Bandwidth of op amp circuits

In order to maintain stability under all conditions of feedback, in common with most op amps the 741 contains an internal compensation capacitor which rolls off the high frequencies with a first-order (–6 dB per octave) slope. In this way the maximum phase shift is limited to 90° at all frequencies where the open-loop gain is greater than unity; the feedback is thus unable to turn positive and cause instability (see sections 4.6 and 7.7).

Although this internal compensation makes for simple circuits with a minimum of external components, it does place an unnecessary restriction on bandwidth in amplifier circuits with a voltage gain greater than unity. This is because the degree of internal HF roll-off must be sufficient to maintain stability in the voltage follower mode (unity gain, 100% feedback). Stability could be maintained at higher gains with less HF roll-off, but the fixed compensation of op amps like the 741 means that bandwidth is sacrificed at gains greater than unity. The gain–bandwidth product is fixed at approximately 1 MHz (bandwidth being measured to the –3 dB point), which results in the frequency-response curves shown in fig. 11.23. As an example, a gain of 100 results in a response approximately 3 dB down at 10 kHz. This bandwidth is inadequate for most audio applications; hence, for decent quality audio circuits, the closed-loop gain of a single 741 must be limited to about 20.

The bandwidth limitations of internal compensation can be reduced by using the 748 op amp instead of the 741. The 748 is internally identical to the 741, except that the compensation capacitor is omitted and must be connected externally. A single capacitor is required, connected as in fig. 11.24, its value being determined by the amount of negative feedback in use and the bandwidth needed. Table 11.1 gives suggested minimum values of C_c for stability at different closed-loop gains.

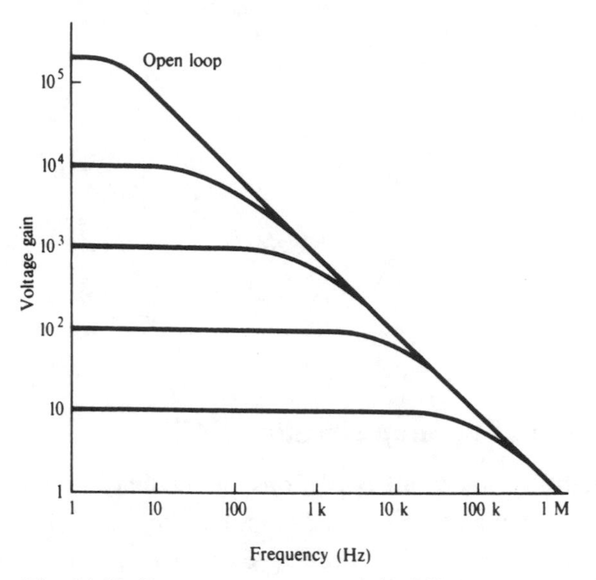

Fig. 11.23. Frequency response of the 741 op amp at various values of closed-loop gain.

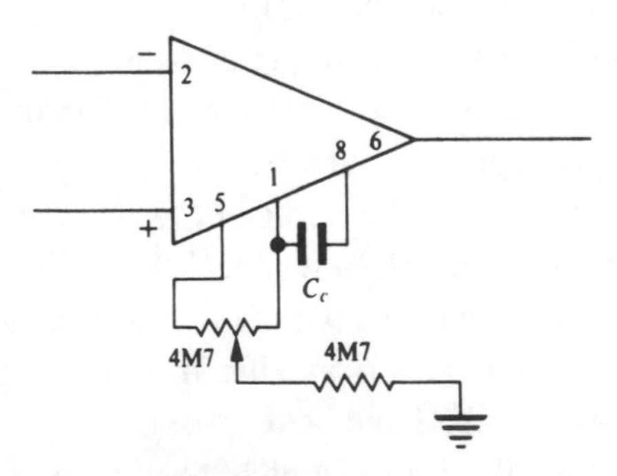

Fig. 11.24. The 748 op amp: connection of compensation capacitor C_c and offset null control. Pin connections refer to 8-pin DIL package.

Table 11.1

A_{VCL}	C_c (pF)	BW (kHz)
1	30	1000
10	5	500
100	2	200
1000	1	50

A comparison of the bandwidth figures in table 11.1 with those of the 741 in fig. 11.23 shows that the 748 is to be preferred if wide bandwidths at high gain are required. One small difference between the 748 and 741 is in the offset null control also shown in fig. 11.24. Because the control operates on a different portion of the internal circuit with the 748, a 4.7 MΩ potentiometer is required and the slider goes to earth via a 4.7 MΩ resistor instead of to negative rail.

11.12 Rise time and slew rate

Any amplifier which incorporates compensation to limit its open-loop gain at high frequencies demands careful examination if it is required to handle large-amplitude step voltages (rectangular pulses). Simple linear low-pass filter theory does not adequately predict the results.

It was shown in section 10.4 that the rise time and fall time of first-order RC low-pass filters are related simply to the bandwidth of the filter:

$$t_r \approx \frac{1}{3f_1}, \qquad [(10.9)]$$

where f_1 is the upper cut-off frequency (–3 dB point). This relationship holds good for all linear systems, but one must be cautious when applying it to op amps, because it usually only applies at low output voltages (typically <1 V). The reason for this is that the op amp circuit can run into serious non-linearity if the output voltage is asked to change (*slew*) faster than a certain rate. The basic problem is that a rapid change in output voltage means a rapid rate of charge or discharge of the compensation capacitor: this in turn demands a considerable drive current from the intermediate stages of the amplifier. In the interests of low power consumption, the available drive current from intermediate stages is limited, and this in turn imposes a limit to the output slew rate of the amplifier. Fig. 11.25 illustrates the difference between small-signal rise time and large-signal slew rate for the 741 amplifier connected as a voltage follower. Whilst the rise time for small signals is approximately 0.3 μs, as is predicted by the unity-gain bandwidth of 1 MHz, a change in output of 10 V takes approximately 20 μs because of the slew rate limit of 0.5 V/μs. It is important to realize that, unlike the rise time of a simple RC circuit, slew rate limiting is a non-linear process, because one stage of the amplifier is limiting. This means that, if the amplifier is called upon to respond to a large-amplitude signal of fast rise time, it blocks out all other signals whilst it is slewing in response to the fast signal. This phenomenon is called *slew induced distortion* (SID) or *transient intermodulation distortion* (TIM) and can cause problems if large-amplitude high-frequency signals are present in the input.

One advantage of the 748 amplifier is that, for values of closed-loop gain greater than unity, higher slew rates can be obtained with it than with the 741 because the compensation capacitor can be reduced to the minimum value

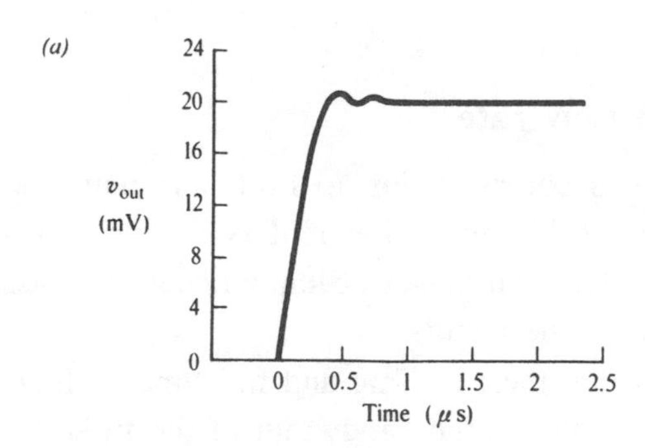

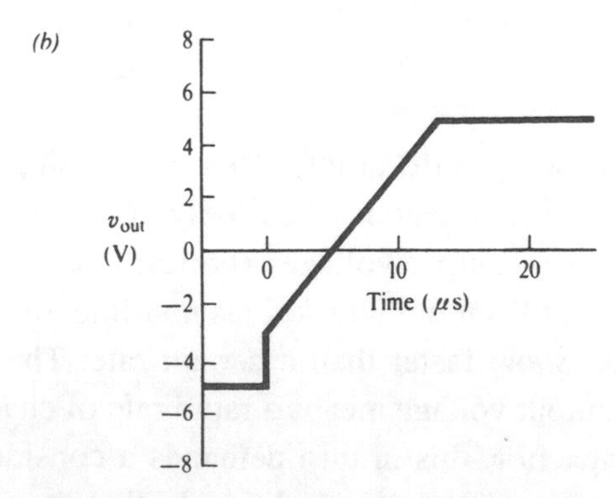

Fig. 11.25. Comparison of small-signal rise time (*a*) and large-signal slew rate (*b*) at the output of a 741 amplifier. Feedback is set for unity gain.

commensurate with stability. Where large signals of fast rise time must be handled, amplifiers such as the NE5534 or TL071 with slew rates of more than 7 V/μs are used. Such high rates are achieved by increasing the drive current available to charge capacitance within the amplifier and by minimizing high-frequency phase shift so that relatively small compensation capacitors can be used.

11.13 Power supplies

11.13.1 Supply voltage and output capability

Most of the parameters specified for the 741 and 748 assume power supply rails of +15 V and −15 V. The amplifiers will, however, work down to +3 V and −3 V and up to +18 V and −18 V. Typically, open-loop voltage gain falls from 2×10^5 at $V_{CC} = 15$ V to 4×10^4 at $V_{CC} = 3$ V. Output voltage swing capability depends, of course, on power supply voltage, and in general, the output can swing to within 1 V of the V_{CC} rail before clipping occurs. This gives the available output swing as approximately $(2V_{CC} - 2)$ V. This applies with an output load of 10 kΩ and at frequencies less than 10 kHz. Above 10 kHz, the limited slew rate of the 741 restricts available output swing; for instance, with $V_{CC} = 15$ V, the maximum peak-to-peak output is 28 V up to 10 kHz, but only 15 V at 20 kHz and 6 V at 50 kHz. Output loads of less than 10 kΩ restrict available output swing because of the current drawn, but, even with a 500 Ω load, a 20 V peak-to-peak swing is available up to 10 kHz with $V_{CC} =$ 15 V.

Where higher outputs are required, the NE5534 gives ±20 V swing into a 600 Ω load (for $V_{CC} = 22$ V) and comes in the same DIL package as the 741.

Power op amps such as the LM12 are also readily available. Packaged in a TO3-style power transistor metal can, the LM12 gives a remarkable ±10 A output current, handles ±30 V rails and can output 150 W of audio power into a 4 Ω load.

11.13.2 Power supply stability and ripple requirements

Most op amps are tolerant of power supply voltage variations. A typical power supply rejection ratio for the 741 is 30 μV/V. This figure indicates that a 1 V change in V_{CC} (or, for that matter, a 1 V a.c. ripple superimposed on V_{CC}) produces only 30 μV input to the amplifier. This figure assumes that V_{CC+} and V_{CC-} vary in step and symmetrically, which is normally the case. A typical power supply ripple of 30 mV peak-to-peak will produce only 1 μV hum at the input.

It is the differential input circuitry which gives op amps this valuable ripple rejection property. Because the ripple is common to both inputs, it is rejected along with other common-mode signals.

11.14 Active filters

11.14.1 General

There are many instances in electronics where the frequency bandwidth of a circuit must be restricted. One example is in optimizing the signal-to-noise ratio: we saw in section 5.9 that noise power is usually directly proportional to the bandwidth in Hertz. Hence, it makes sense to limit the bandwidth of the system to the minimum which will allow the wanted signal to pass unimpaired. An example here is the vintage 78 r.p.m. gramophone record which has an inherently noisy surface. If such a record is played on a wide-frequency-range hi-fi system, the music may be barely audible beneath the surface noise. However, the same record played on a pre-1950 vintage radiogram with its limited bandwidth (approximately 5 kHz) may sound quite acceptable. This is because the bandwidth is wide enough to pass the music, but sufficiently narrow to avoid unleashing the full noise power. To obtain satisfactory results on a modern wide-range system from such a record, the bandwidth must be artificially restricted by a low-pass filter which is matched to the frequency spectrum of the music.

Accurate filtering also lies at the heart of today's digital systems. As we shall see in Chapter 14, when analogue signals are sampled prior to digitising, the highest frequency must be strictly limited to half the sampling frequency, requiring in some cases a very steep low-pass filter.

11.14.2 Low-pass filters

The 'ideal' filter has a precisely defined pass-band of rectangular shape (fig. 11.26(*a*)). Such a pass-band shape is, as might be expected, unrealizable in practice. Fig. 11.26(*b*) is a more realistic aim, the HF cut-off having a finite slope; in general, the more elaborate the filter, the steeper the slope. It is usually desirable that the 'corner' should be as sharp as possible, giving a flat response in the pass-band and a rapid transition to the roll-off.

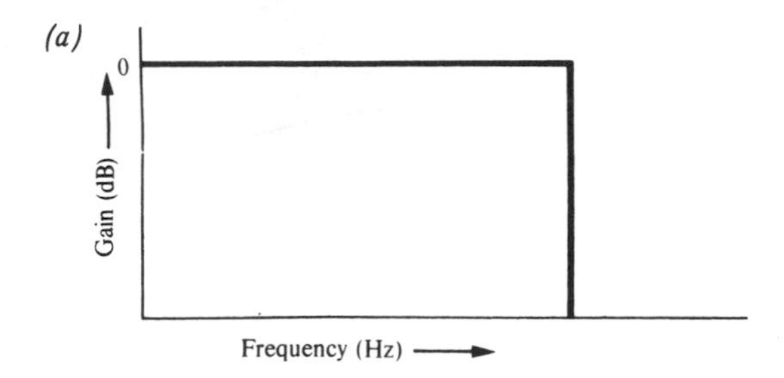

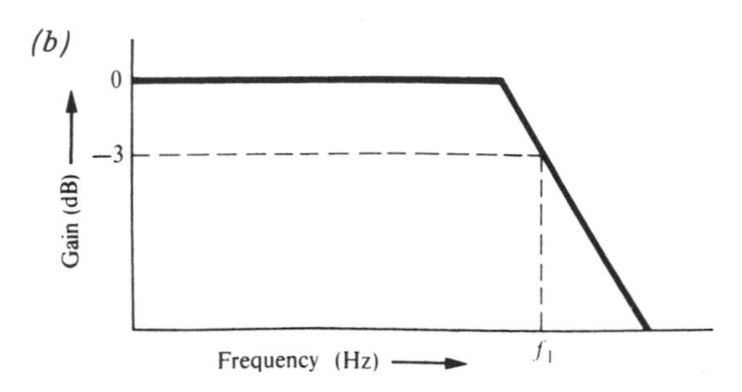

Fig. 11.26. Idealized low-pass filter responses. (*a*) Unrealizable rectangular pass band. (*b*) One step nearer reality: a flat pass-band with a finite roll-off slope. Bandwidth is usually measured to the –3 dB point (f_1).

Some practical low-pass filters are shown in Fig. 11.27 together with typical frequency response plots. In fig. 11.27(*a*), the familiar single RC filter gives a maximum slope of 6 dB per octave, together with a gradual transition from the level region to the maximum slope. The second-order filter of fig. 11.27(*b*) with two RC sections produces a faster roll-off with a maximum slope of 12 dB per octave, but the corner is still too much rounded for many applications. The introduction of inductance in fig. 11.27(*c*) brings a distinct improvement in flexibility. The presence of two reactive elements means that the maximum slope is 12 dB, whilst the damped resonance of the LC system, when both reactances are equal, can be used to sharpen up the corner of the response so that the result approximates to the ideal case of fig. 11.26(*b*). The height of the resonant peak is governed by the '*Q*' of the circuit, as we shall see shortly.

Inductors are very valuable in filter design where a precisely defined response is concerned, and yet they do have disadvantages. They are expensive, particularly if precise values of inductance are required, and they tend to

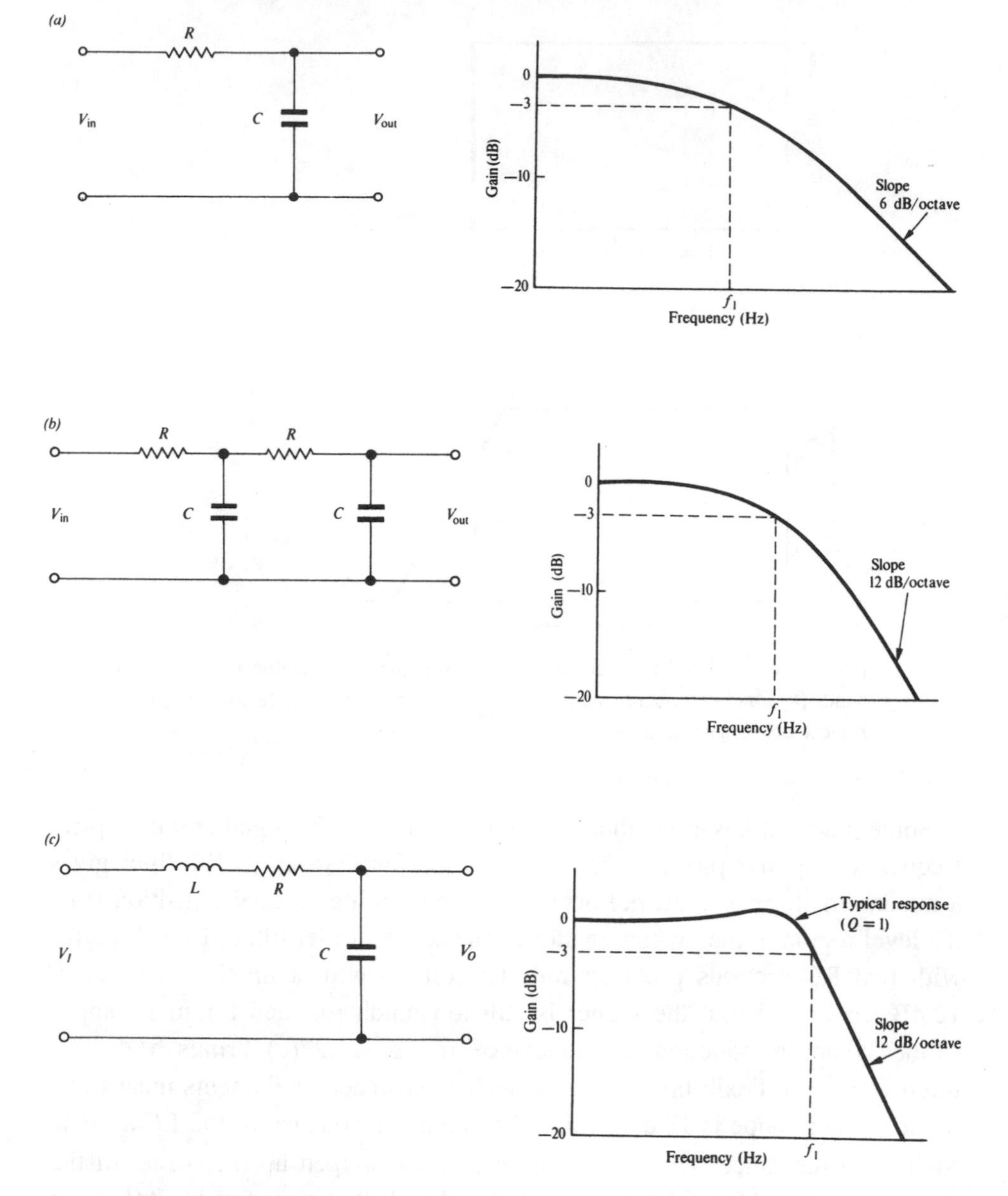

Fig. 11.27. Passive low-pass filter circuits and their frequency responses:
(*a*) single-section RC (first order),
(*b*) double-section RC (second order),
(*c*) single-section LC (second order).

pick up magnetically-induced a.c. hum from any mains transformers in the vicinity. Mu-metal screening can virtually eliminate this latter snag, but adds further to the cost. The circuit designer therefore prefers to avoid using inductors if possible.

Fortunately the op amp provides a solution here by simulating the behaviour of the LCR filter of fig. 11.27(*c*) using only resistors and capacitors. Such a filter incorporating an amplifier is known as an *active filter*. There is an enormous range of possible active filter circuits and it would take a separate book to deal with them adequately.

We shall examine one of the most useful active filters, the popular Sallen and Key type. By comparing its frequency response equation with that for the LCR filter, we shall aim to show that it can fulfil the same task.

Initially we can return to the LCR low-pass filter of fig. 11.27(*c*). Consideration of circuit impedances gives the input–output (transfer) relation

$$\frac{V_O}{V_I} = \frac{1}{1 - \omega^2 LC + j\omega CR}. \tag{11.7}$$

Resonance occurs at a frequency ω_0, where

$$\omega_0^2 LC = 1,$$

so that V_O/V_I has here its maximum value. The sharpness of the resonant peak is controlled by the circuit Q where, as usual, Q is defined as

$$Q = \frac{\omega_0 L}{R} = \frac{1}{R}\left(\frac{L}{C}\right)^{1/2}. \tag{11.8}$$

Fig. 11.28 shows the effect of two different Q-values in the LCR low-pass filter. A high Q gives a sharp peak, whilst a rounded corner is produced by a low Q (high value of R).

A Sallen and Key active filter is shown in basic low-pass form in fig. 11.29(*a*). To prove that this circuit can simulate the LCR filter, it is instructive to work through the analysis of the various voltages and currents to obtain the transfer function V_O/V_I. To facilitate analysis, the circuit is redrawn in the equivalent form of fig. 11.29(*b*); the feedback loop has been replaced by an extra generator, V_O, feeding the bottom end of C_1.

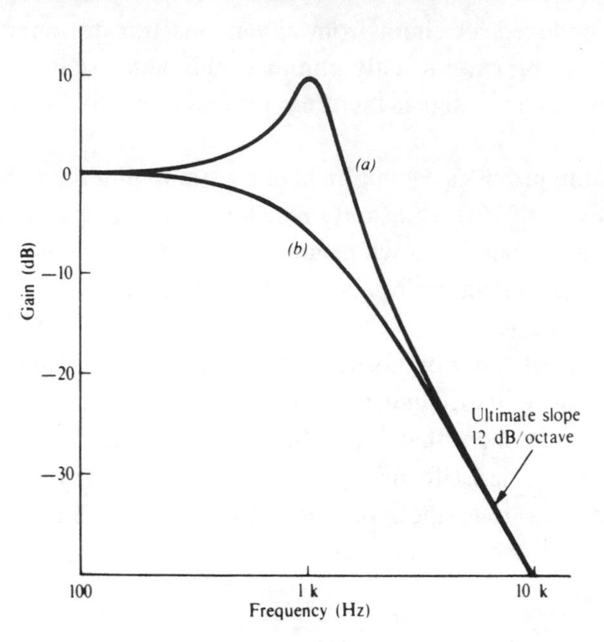

Fig. 11.28. Effect of varying circuit Q in the LCR low-pass filter of fig. 11.27(c): (a) high Q ($Q=3$), (b) low Q ($Q=0.5$).

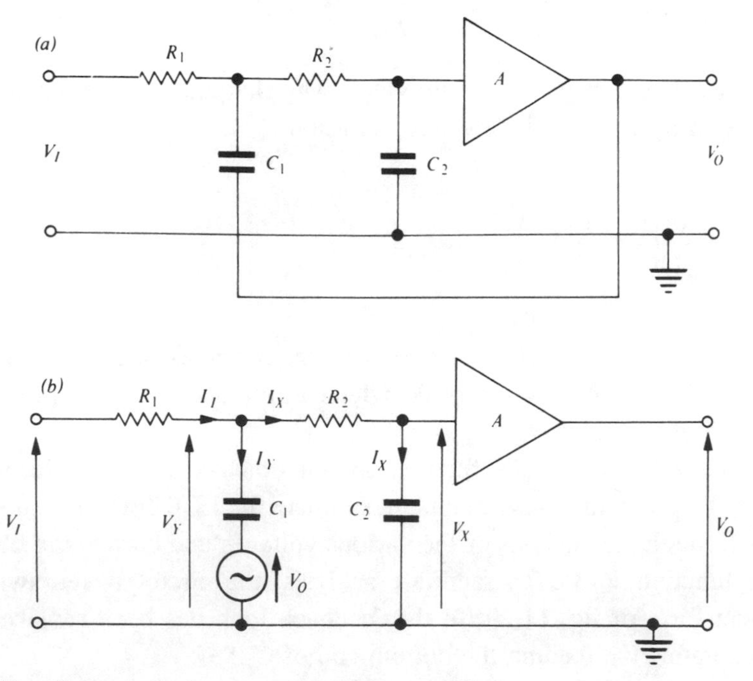

Fig. 11.29. Active low-pass filter (Sallen and Key).

Working back step by step from the output V_O, we have initially

$$V_X = \frac{V_O}{A}. \tag{11.9}$$

Now,

$$I_X = \frac{V_X}{1/j\omega C_2}$$

therefore

$$I_X = j\omega C_2 \frac{V_O}{A}. \tag{11.10}$$

Going a stage further back, V_Y can be expressed as the sum of V_X and the voltage drop across R_2, i.e.

$$\begin{aligned} V_Y &= V_X + R_2 I_X \\ &= \frac{V_O}{A}(1 + j\omega C_2 R_2). \end{aligned} \tag{11.11}$$

Now,

$$\begin{aligned} I_Y &= \frac{V_Y - V_O}{1/j\omega C_1} \\ &= j\omega C_1 \left[\frac{V_O}{A}(1 + j\omega C_2 R_2) - V_O \right] \\ &= j\omega C_1 \frac{V_O}{A}(1 - A + j\omega C_2 R_2). \end{aligned} \tag{11.12}$$

By Kirchhoff's current law, we can find the input current, I_I:

$$\begin{aligned} I_I &= I_X + I_Y \\ &= \frac{V_O}{A}[j\omega C_2 + j\omega C_1(1 - A + j\omega C_2 R_2)] \\ &= \frac{V_O}{A}[j\omega C_2 + j\omega C_1(1 - A) - \omega^2 C_1 C_2 R_2]. \end{aligned} \tag{11.13}$$

Now the input voltage V_I is given by

$$V_I = R_I I_I + V_Y$$

$$= \frac{V_O}{A}[j\omega C_2 R_1 + j\omega C_1 R_1(1-A) - \omega^2 C_1 C_2 R_1 R_2]$$

$$+ \frac{V_O}{A}[1 + j\omega C_2 R_2]$$

$$= \frac{V_O}{A}\{1 - \omega^2 C_1 C_2 R_1 R_2 + j[\omega C_1 R_1(1-A) + \omega C_2(R_1 + R_2)]\}.$$

And the transfer function V_O/V_I is given by

$$\frac{V_O}{V_I} = \frac{A}{1 - \omega^2 C_1 C_2 R_1 R_2 + j[\omega C_1 R_1(1-A) + \omega C_2(R_1 + R_2)]}. \qquad (11.14)$$

Resonance occurs when

$$\omega^2 C_1 C_2 R_1 R_2 = 1.$$

If the resonant frequency is f_0,

$$f_0 = \frac{1}{2\pi}\left(\frac{1}{C_1 C_2 R_1 R_2}\right)^{1/2}. \qquad (11.15)$$

Resonant frequency f_0 comes close to the corner of the filter response. Here the output begins to fall off, ultimately reaching a slope of 12 dB/octave. f_0 is sometimes called the *cut-off frequency* of the filter.

The rather lengthy equation for V_O/V_I can be simplified if we impose restrictions on the filter design. One such restriction is to make amplifier gain A equal to unity. In practical terms, the amplifier can then be a simple voltage follower; even a single transistor in the emitter follower mode can be used.

Setting $A = 1$ then, we have

$$\frac{V_O}{V_I} = \frac{1}{1 - \omega^2 C_1 C_2 R_1 R_2 + j\omega C_2(R_1 + R_2)}. \qquad (11.16)$$

At low frequencies (ω small) we see that the gain of the filter is unity.

Imposing further simplification, let

$$R_1 = R_2 = R_x$$

and

$$C_2 = C_x,$$
$$C_1 = nC_x,$$

where n is a constant denoting the ratio of the capacitor values. Then

$$\frac{V_O}{V_I} = \frac{1}{1 - n\omega^2 C_x^2 R_x^2 + 2j\omega C_x R_x}. \tag{11.17}$$

Now compare equation (11.17) with that for the LCR filter (fig. 11.27(*c*)), where we had

$$\frac{V_O}{V_I} = \frac{1}{1 - \omega^2 LC + j\omega CR}. \qquad [(11.7)]$$

These two equations become identical, showing that the active filter is equivalent to the LCR, if we make $C_x = C$, $nC_x R_x^2 = L$, and $2R_x = R$.

Now, with the LCR filter,

$$Q = \frac{\omega_O L}{R} = \frac{1}{R}\left(\frac{L}{C}\right)^{1/2}.$$

Therefore, in terms of the active filter,

$$Q = \frac{R_x n^{1/2}}{2R_x} = \frac{n^{1/2}}{2}.$$

A typical practical Q-value is one, giving $n = 4$; i.e. $C_1 = 4C_2$ in fig. 11.29.

Resonant frequency ('cut-off' frequency) is given by

$$f_0 = \frac{1}{2\pi(LC)^{1/2}} = \frac{1}{2\pi C_x R_x n^{1/2}}. \tag{11.18}$$

A second way of simplifying the original transfer function (equation (11.14)) is to make the R and C values equal and vary Q by changing amplifier gain A, i.e.

$$R_1 = R_2 = R_a$$

and

$$C_1 = C_2 = C_a$$

then

$$\frac{V_O}{V_I} = \frac{A}{1 - \omega^2 C_a^2 R_a^2 + j\omega C_a R_a (3 - A)}. \tag{11.19}$$

The low-frequency gain is this time equal to A. Comparison once again with the LCR section shows that the active filter is equivalent if $C_a = C$, $C_a R_a^2 = L$ and $R_a(3 - A) = R$.

We find that resonant frequency

$$f_0 = \frac{1}{2\pi C_a R_a} \tag{11.20}$$

and

$$Q = \frac{1}{3 - A}. \tag{11.21}$$

A useful value of Q is that which gives a maximum sharpness to the corner of the high-frequency roll-off without actually causing a peak in the pass-band frequency response. This is known as the *maximally flat* or *Butterworth* filter and is given by

$$Q = 1/\sqrt{2}.$$

The required value of A is calculated from equation (11.21), $A = 3 - (1/Q)$, and works out to 1.6 for the Butterworth response.

Fig. 11.30 shows a practical low-pass filter with a Butterworth response, a cut-off frequency of approximately 1 kHz, and a maximum slope of

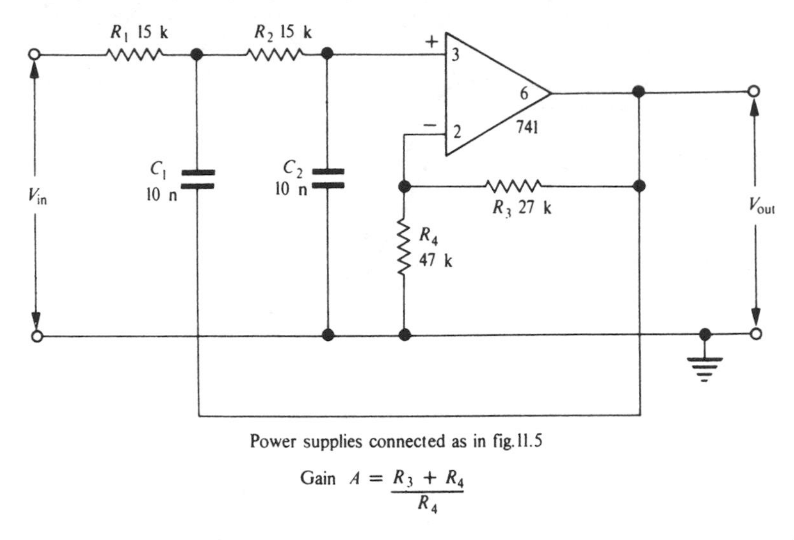

Fig. 11.30. Practical low-pass active filter, giving a Butterworth response. Values of R_1C_1 and R_2C_2 give a –3 dB point at 1 kHz. Values may be scaled for a roll-off at a different frequency.

12 dB/octave. Different cut-off frequencies can be arranged with a minimum of calculation by simply scaling the capacitor or resistor values inversely with frequency. For example, a cut-off at 100 Hz would require the capacitors to be increased by a factor of 10, to 100 nF; alternatively the 10 nF capacitors could be retained and the resistors increased to 150 kΩ. Replacing the resistors by a twin-gang potentiometer would give a continuously variable cut-off frequency.

The Q of the filter, and thus the sharpness of the roll-off 'corner' can be changed easily by varying gain A, given in the usual way by $(R_3 + R_4)/R_4$. Fig. 11.31 shows some typical frequency-response curves.

Cut-off slopes steeper than 12 dB/octave may be obtained by adding further RC sections (as in fig. 11.27(*a*)) on the input side of the filter prior to R_1. Each section increases the slope by an extra 6 dB/octave. The amplifier gain can then be experimentally adjusted to obtain the required sharpness at the corner frequency. Two or more active filters can be cascaded to obtain very steep cut-off slopes.

It is worth noting that, as gain A is increased towards three, Q increases towards infinity. Values of A above two are not generally recommended because the circuit performance is then sensitive to small variations in component values.

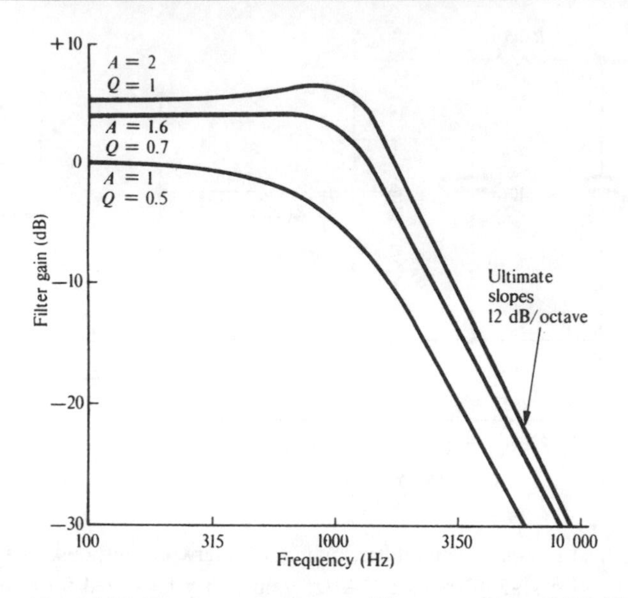

Fig. 11.31. Frequency responses of low-pass filter of fig. 11.29 with $R_1 = R_2 =$ 15 kΩ and $C_1 = C_2 = 10$ nF.

11.14.3 High-pass filters

Interchanging resistors and capacitors in the Sallen and Key circuit produces, as might be expected, a high-pass filter, shown in fig. 11.32. Following a similar analysis to the low-pass case, it is not too difficult with patience to derive the following equation for the gain:

$$\frac{V_O}{V_I} = A \bigg/ \left\{1 - \frac{1}{\omega^2 C_1 C_2 R_1 R_2} - j\left[\frac{1}{\omega C_2 R_2} + \frac{1}{\omega C_1 R_2} + \frac{1}{\omega C_1 R_1}(1-A)\right]\right\}. \quad (11.22)$$

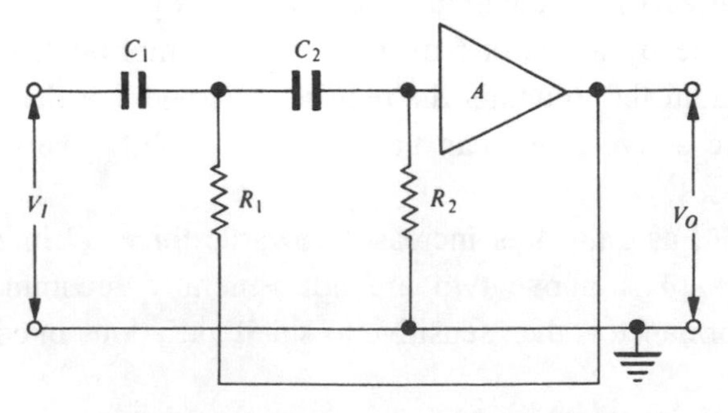

Fig. 11.32. Active high-pass filter (Sallen and Key).

And, if $C_1 = C_2 = C_a$ and $R_1 = R_2 = R_a$ this simplifies to

$$\frac{V_O}{V_I} = A \bigg/ \left[1 - \frac{1}{\omega^2 C_a^2 R_a^2} - j\frac{3-A}{\omega C_a R_a}\right]. \tag{11.23}$$

The frequency response given by this equation is plotted in fig. 11.33 with $R_1 = R_2 = 15$ kΩ and $C_1 = C_2 = 0.01$ μF and three values of gain A.

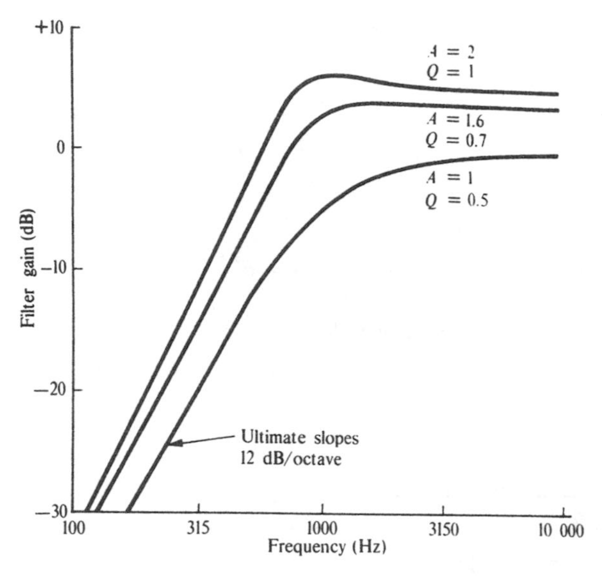

Fig. 11.33. Frequency responses of low-pass filter of fig. 11.32 with $R_1 = R_2 =$ 15 kΩ and $C_1 = C_2 = 10$ nF.

11.14.4 Band-pass filters

Band-pass filters which, as the name suggests, transmit only a certain range of frequencies, can be constructed by using separate high- and low-pass filters to define the band limits. This approach gives a nice flat top to the band-pass characteristic. Where the band required is narrow, however, some form of tuned filter is the most elegant and economical solution. This is readily achieved by using a frequency *rejection* circuit in the feedback loop of an op amp, shown in fig. 11.34. The value of R_f determines both the maximum gain at resonance ($A = -R_f/R_1$) and the Q of the circuit, the latter because of its damping effect on the tuned rejection network. Fig. 11.35 shows two frequency rejection circuits suitable for use in a feedback loop. Fig. 11.35(*a*) is the familiar parallel LC circuit with a resonant frequency f_0 given by

$$f_0 = \frac{1}{2\pi\sqrt{(LC)}}. \tag{11.24}$$

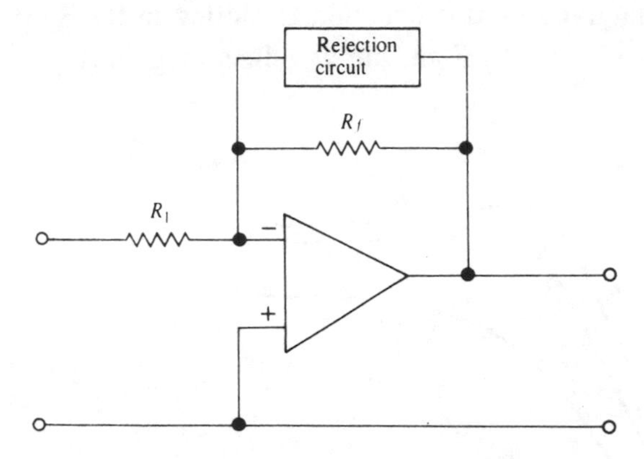

Fig. 11.34. Basic tuned band-pass active filter.

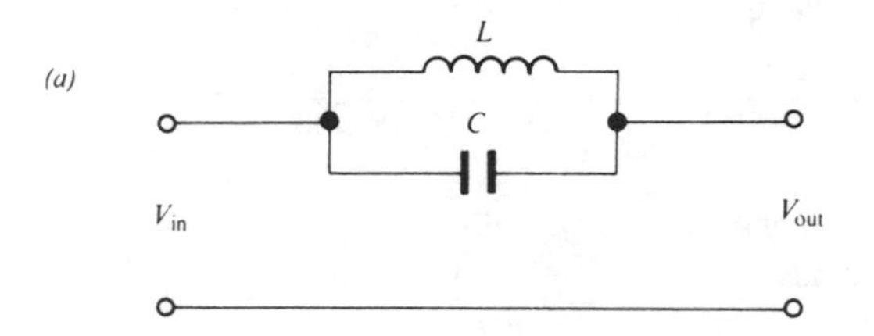

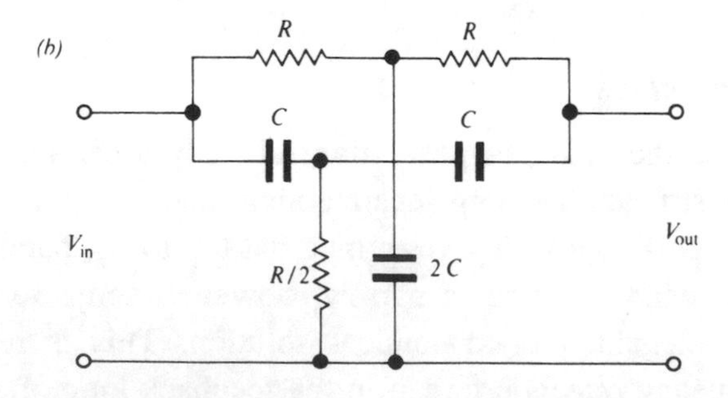

Fig. 11.35. Frequency rejection networks: (*a*) parallel LC circuit, (*b*) twin-T network.

The twin-T network of fig. 11.35(*b*) avoids the use of inductors and yet gives a sharp rejection at its resonant frequency f_0 given by

$$f_0 = \frac{1}{2\pi RC}. \tag{11.25}$$

The disadvantage of the twin-T network is that, if the resonant frequency is to be adjustable, three resistors must be varied at the same time; this is both inconvenient and expensive, requiring a triple-gang potentiometer.

The multiple-feedback filter, shown in fig. 11.36, is one of the most useful RC band-pass filters. Application of the usual op amp assumptions gives resonant frequency (pass-band centre-frequency) as

$$f_0 = \frac{1}{2\pi C}\left(\frac{R_1 + R_2}{R_1 R_2 R_3}\right)^{1/2}. \tag{11.26}$$

The bandwidth (to –3 dB points) is given by

$$\Delta f = \frac{1}{\pi C R_3}. \tag{11.27}$$

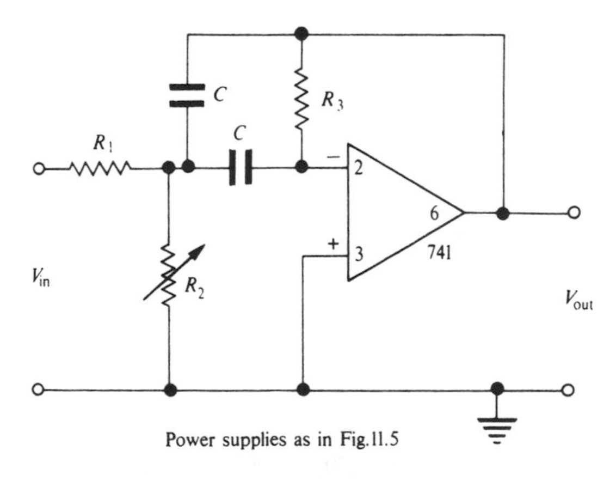

Fig. 11.36. Multiple-feedback tuned active filter.

Voltage gain at resonance is

$$A_{max} \approx \frac{R_3}{2R_1}. \tag{11.28}$$

It is clear from these equations that reducing R_2 increases f_0, but leaves Δf and A_{max} unaltered. R_2 is thus a variable tuning control. The Q of the circuit may be calculated as

$$Q = \frac{f_0}{\Delta f} = \frac{1}{2}\left[\frac{R_3(R_1 + R_2)}{R_1 R_2}\right]^{1/2}. \tag{11.29}$$

A typical frequency response for the multiple-feedback filter is shown in fig. 11.37. This is the result obtained with the following component values: $R_3 = 100\ \text{k}\Omega$, $R_1 = 47\ \text{k}\Omega$, $R_2 = 68\ \Omega$, $C = 150$ nF, giving maximum gain = 1, $Q \approx 20$ and $f_0 \approx 400$ Hz.

If two tuned filters are cascaded (the output of the first feeding the input of the second), the skirts of the pass-band are steepened considerably. With slight staggering of the resonant frequencies, a close approximation to the ideal flat-topped bandpass characteristic can be obtained (fig. 11.38).

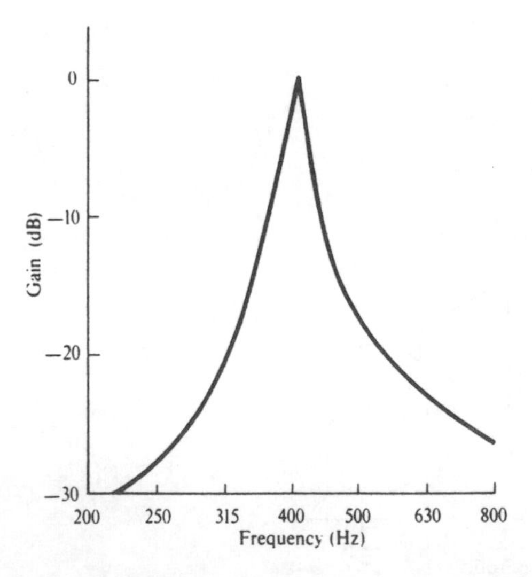

Fig. 11.37. Typical frequency response of multiple-feedback tuned active filter ($f_0 = 400$ Hz, $Q = 20$).

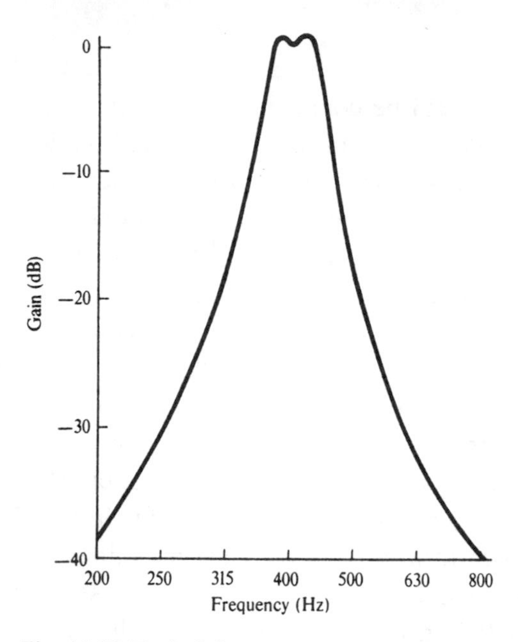

Fig. 11.38 Typical frequency response of two cascaded tuned filters with staggered resonant frequencies, $f_0 = 400$ Hz, −3 dB bandwidth = 50 Hz.

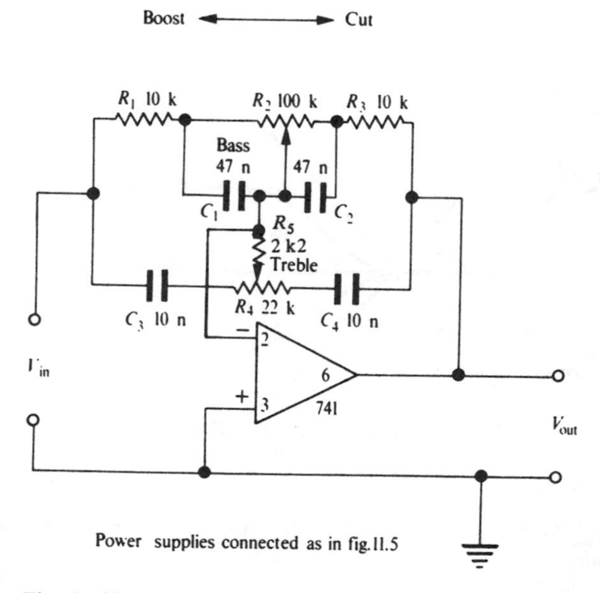

Fig. 11.39. A version of the 'Baxandall' bass and treble tone-control circuit.

11.14.5 Tone controls

No discussion of active filters would be complete without reference to what is perhaps the most common filter circuit of all: the bass and treble controls used on audio amplifiers. Fig. 11.39 shows the popular feedback tone-control circuit originated by P.J. Baxandall. A complete analysis of this circuit for the various possible settings of the bass control R_2 and the treble control R_4 would be very lengthy, but a qualitative understanding of the circuit may be quickly grasped. Considering the treble control first, capacitors C_3 and C_4 transmit only the high frequencies to R_4. Adjusting the slider of R_4 is then in effect changing the ratio of input resistor to feedback resistor on a virtual earth inverting amplifier like that of fig. 11.9, but handling only the high frequencies. R_2, the bass control, performs a similar function at the low-frequency end of the audio spectrum, being isolated from the higher frequencies by R_1, C_1 and R_3, C_2 which form a first-order low-pass filter. The effect of the circuit is to provide independent control of treble and bass boost or cut up to a maximum slope of 6 dB/octave. Mid-position on both controls corresponds to an overall gain of unity with a flat response over the audio band. Fig. 11.40 shows some typical response curves produced by various settings of the tone controls.

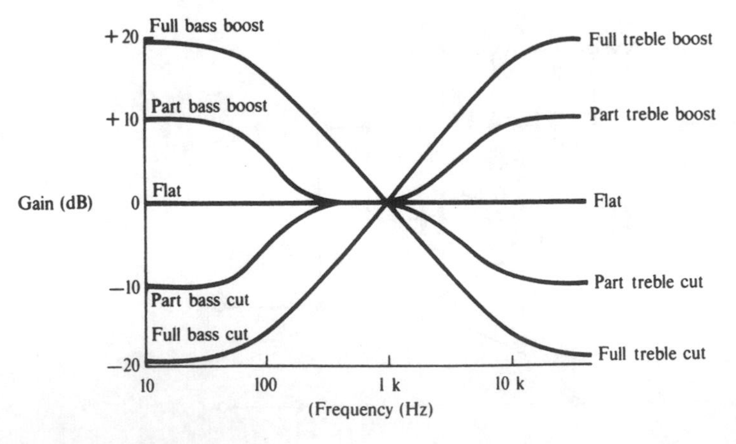

Fig. 11.40. Typical frequency-response curves obtained with a Baxandall-type tone-control circuit.

11.15 Logarithmic amplifiers

A logarithmic amplifier gives an output of the form

$$V_{out} = A \log_{10} kV_{in},$$

where A and k are constants. Such amplifiers have applications in instrumentation where a wide range of signal amplitudes must be displayed on a single meter scale; each major scale division can then represent one decade of the parameter measured.

In acoustical measurement work, sound pressure levels are usually expressed in decibels relative to a reference level of 2×10^{-5} N/m^2, which corresponds roughly with the threshold of hearing. A logarithmic amplifier connected prior to the read-out device can give readings directly in decibels. Some sound level meters cover the range 70 dB to 120 dB on a single linear scale.

This facility is also useful for measuring the reverberation time of a room, which is defined as the time taken for a sound impulse to fall in level by 60 dB. The sound level may be displayed on an oscilloscope via a logarithmic amplifier so that the exponential reverberation decay becomes a straight line, of which the slope can be easily measured. There are many such phenomena in physics which have an exponential decay with time. Such results can conveniently be displayed as a straight line decay via a logarithmic amplifier.

The basic circuit of a logarithmic amplifier is given in Fig. 11.41 which uses a transistor in the feedback loop of a conventional virtual-earth inverting amplifier. R_5 is connected in series with the amplifier output in order to limit the current which the transistor base–emitter junction can draw from the output at the higher signal levels. As was discussed in section 6.3, the emitter current of a bipolar transistor is related to the base–emitter voltage by a form of the pn junction equation

$$I_E = I_0' \exp\left(\frac{eV_{BE}}{kT}\right), \qquad [(6.2a)]$$

for $h_{FE} \gg 1$, $I_C \approx I_E$. Therefore

$$I_C = I_0' \exp\left(\frac{eV_{BE}}{kT}\right),$$

where I_0' is the reverse leakage current, e is charge on electron, k is Boltzmann's constant, T is absolute temperature. Or, taking natural logarithms,

$$\ln I_C = \ln I_0' + \frac{eV_{BE}}{kT},$$

i.e.

$$V_{BE} = \frac{kT}{e} \ln \frac{I_C}{I_0'}. \tag{11.30}$$

Now in fig. 11.41, making the usual assumptions about the virtual earth at the op amp inverting input,

$$I_F = I_I = \frac{V_I}{R_1}.$$

The basic feature of this circuit is that the feedback current I_F forms the collector current I_C of the transistor, i.e.

$$I_C = I_F = \frac{V_I}{R_1}.$$

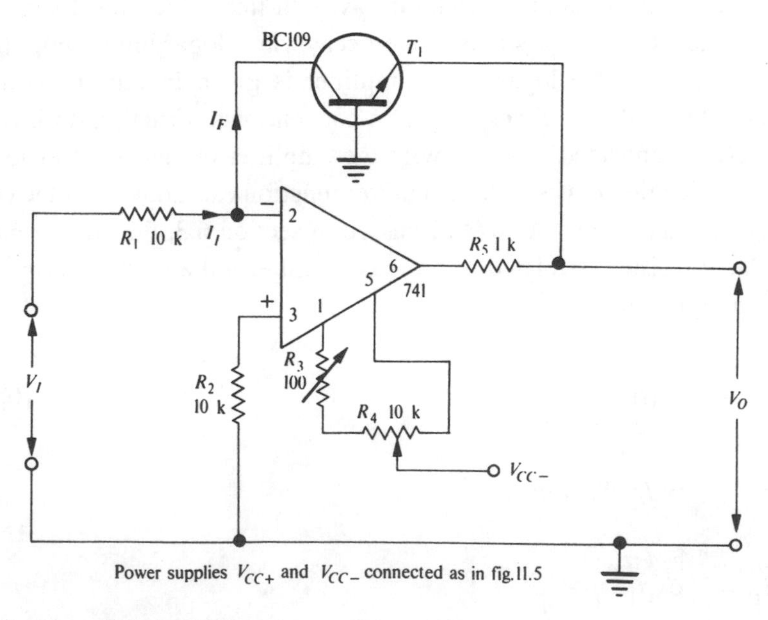

Fig. 11.41. Basic logarithmic amplifier circuit.

Now the emitter–base junction is connected directly between output and earth, therefore output voltage

$$V_O = V_{BE}$$

$$= \frac{kT}{e} \ln \frac{I_C}{I_0'} \qquad \text{(from (11.30))}$$

$$= \frac{kT}{e} \ln \frac{V_I}{R_1 I_0'}. \qquad (11.31)$$

Thus we have an output voltage proportional to the natural logarithm of the input voltage. This is readily expressed in the form of the more common logarithms to base 10 using the relation $\log_{10}x = \ln x \log_{10}e$. Therefore

$$V_O = \frac{kT}{e} \cdot \frac{\log_{10} (V_I / R_1 I_0')}{\log_{10} e}. \qquad (11.32)$$

This may be more simply stated as

$$V_O = b \log_{10} V_I + c, \qquad (11.33)$$

where b and c are constants.

Fig. 11.42 shows an input–output characteristic for a basic log amplifier, and it is clear that the output voltage swing is relatively small, changing by only about 0.3 V for four decades of input signal range. At low input voltages, the setting of the offset control R_4 becomes critical. To assist adjustment, the

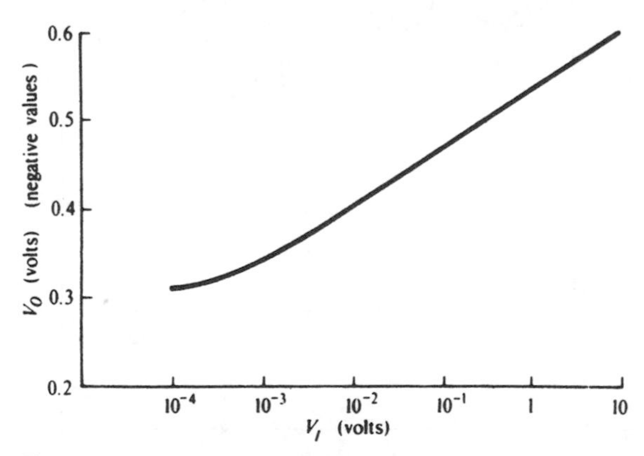

Fig. 11.42. Typical input–output characteristics of simple logarithmic amplifier.

fine control R_3 is connected in series with the normal 10 kΩ offset potentiometer. The offset controls are adjusted for zero amplifier output with the input shorted to 0 V. In this situation, the current in the feedback transistor is negligible and the amplifier is operating near its open-loop condition, making the offset adjustment very critical.

In addition to the limited output voltage range, there is another snag associated with the basic log amplifier circuit of fig 11.41: it is temperature sensitive because of the kT/e factor in the transistor equation. These two limitations may be overcome by using the more advanced circuit of fig. 11.43. Here a second stage is added, both to increase gain and to provide temperature compensation. The latter feature is achieved by the inclusion of a second base–emitter junction T_2, in opposition to the main logging transistor T_1; T_2 experiences a negligible change in operating current over the full operating range so that it does not tend to counteract the logging of T_1. It does however encounter the same temperature changes as T_1 so that resultant changes in its base–emitter p.d. compensate for the variations in T_1.

T_1 and T_2 should be mounted as close together as possible so that they are both in the same thermal environment. A 'super match' transistor pair such as the LM394N may be used here with advantage. To maintain high-frequency stability under all conditions, the two feedback loops are shunted with capacitors C_1 and C_2.

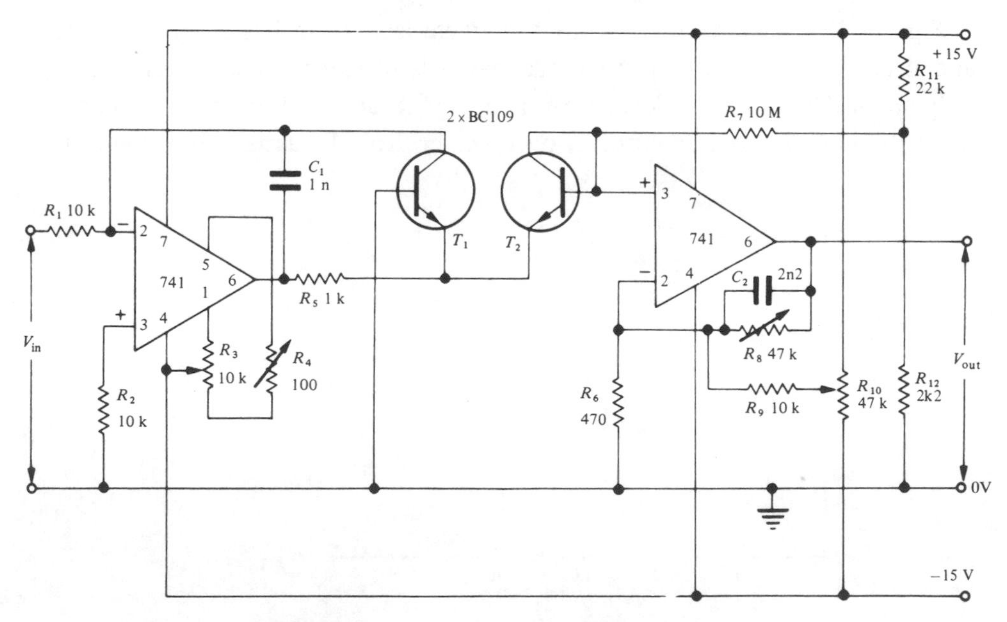

Fig. 11.43. Temperature-compensated logarithmic amplifier.

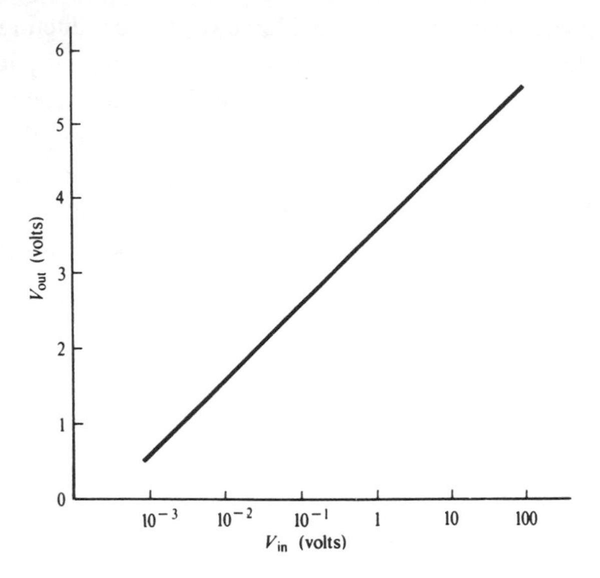

Fig. 11.44. Typical plot of V_{out} against V_{in} for the logarithmic amplifier of fig. 11.41.

Potentiometer R_8 determines the scale factor (gain) and can be readily adjusted so that a 10 dB change in input gives a convenient 1 V change in output.

The remaining three preset potentiometers are offset controls; R_3 and R_4 are adjusted exactly as in the basic circuit of fig. 11.41: with the input earthed they are adjusted so that the emitter of T_1 is as close to zero volts as possible. R_{10} is used to set the d.c. offset of the output; for most applications, the d.c. offset is adjusted so that, as the input tends to zero, the output also tends to zero. It is important to note that log zero is minus infinity and that the amplifier output will therefore, quite correctly, go 'wild' for a true zero input. It will, however, perform well down to inputs of 5 mV and up to inputs of 100 V, giving a useful dynamic range of over 80 dB. A typical plot of output against input for the temperature-compensated logarithmic amplifier is given in fig. 11.44.

11.16 Precision rectifiers

The op amp may be used to advantage where precise rectification of low a.c. voltages is required for measurement purposes. The non-linearity of the diode forward characteristic makes the direct rectification of small a.c. signals very inaccurate; indeed, a silicon diode may be considered to stop conducting once the forward e.m.f. falls below 0.4 V. This problem can be overcome by including the diode in the feedback loop of an op amp as in fig. 11.45. Here diode D_1 is the rectifier and its inclusion in the feedback loop of an amplifier has the effect of dividing its effective forward p.d. by a factor equal to the open-loop gain of the amplifier, so that the rectifier is operative down to input voltages of well under 1 mV. The second diode, D_2 and its associated feedback resistor are in the circuit simply to provide feedback on positive half-cycles of the input and thus avoid amplifier overload.

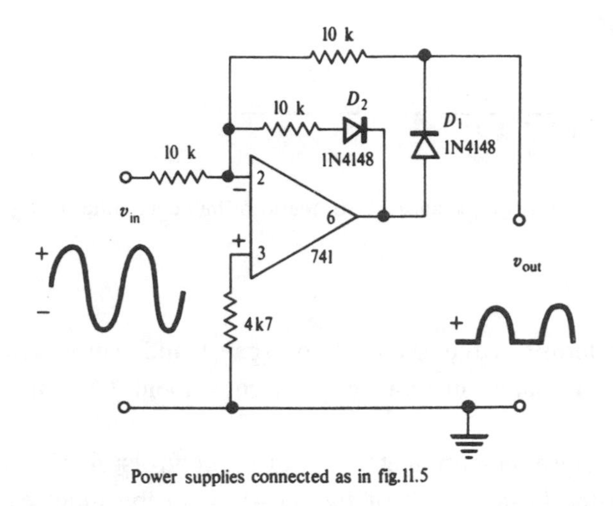

Fig. 11.45. Precision HW rectifier.

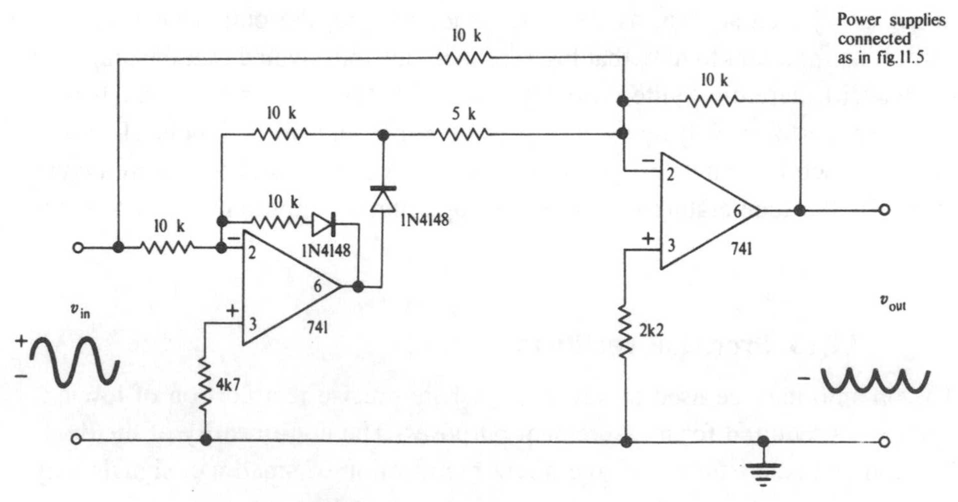

Fig. 11.46. Precision FW rectifier (absolute value circuit), $v_{out} = -|v_{in}|$.

A full-wave version of the precision rectifier is easily constructed by including an operational adder to insert the 'missing' half-cycles (fig. 11.46). The accuracy of this circuit depends on the precision of the 10 kΩ and the 5 kΩ resistors, which should preferably by of ±2% tolerance. An alternative name for fig. 11.46 is the *absolute value circuit* since, for a wide range of signal levels,

$$v_{out} = -|v_{in}|.$$

11.17 The differential comparator

11.17.1 Introduction

There are many instances in electronics when we require to make a comparison to determine whether a signal voltage is higher or lower than a certain reference level. The simplest example is a polarity detector, where the reference level used as a basis for comparison is the earth (common) rail.

Any differential amplifier may be used as a comparator, the amplifier being used in this application without negative feedback (fig. 11.47). The resulting

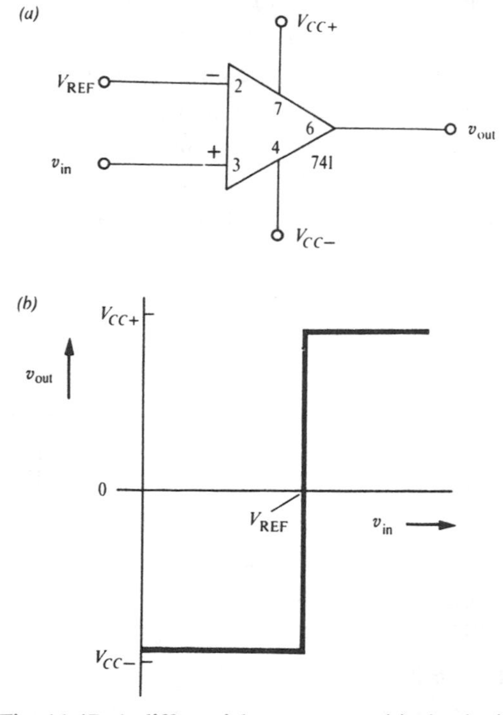

Fig. 11.47. A differential comparator: (*a*) circuit, (*b*) characteristic.

high open-loop gain means that a slowly changing input voltage, v_{in} will cause a rapid change in v_{out} as v_{in} reaches the value V_{REF}. For example, v_{in} may be the output of a photodetector which is required to switch on lights at a given stage of twilight; the value of V_{REF} will then determine the point at which switching occurs.

A general purpose op amp, such as the 741 makes a useful comparator for a great many purposes. Its main disadvantage is that its slow slew rate gives a relatively long switching time (approximately 20 μs). The 748, used without compensation, is much faster (<1 μs switching time), but this is still too slow for the reliable operation of many logic circuits. An additional problem where interfacing with logic circuits is concerned is that the output swing amplitude, from V_{CC+} to V_{CC-} is inconveniently wide.

The 311 comparator IC is specifically designed to overcome these problems and is useful where an interface is required between analogue and digital systems. Internally it has an op amp connected to an output transistor as shown in fig. 11.48. This arrangement makes it possible to operate from a single-rail supply, the +5 V supply shown giving a 0 V/+5 V step compatible with most digital logic circuits. An external 1 kΩ collector load (*pull up* resistor) is provided for the 'open collector' output transistor. Rise time of the output pulse is typically less than 50 ns.

A limitation of the simple single-rail circuit of fig. 11.48 is that neither v_{in} nor V_{REF} can go to zero or go negative without upsetting the operation of the internal op amp due to common mode failure. The cure is to take pin 4 (V_{cc-}) to a −5 V rail instead of 0 V; both inputs can then swing to within 1 V or so of either supply rail. Pin 1 remains connected to 0 V in order that the output transistor can continue to switch between 0 V and −5 V. With these dual supply rails, V_{REF} can be connected to 0 V, making the comparator into a

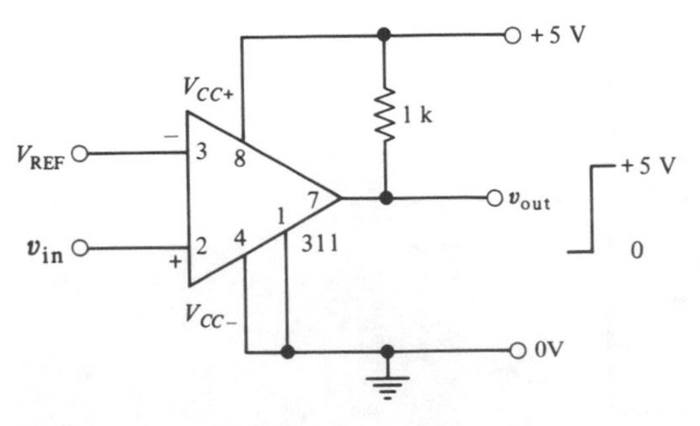

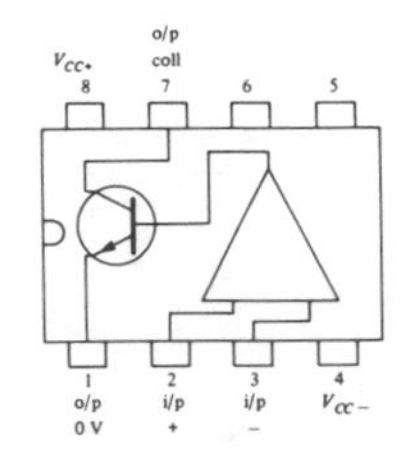

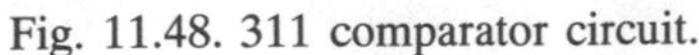
Fig. 11.48. 311 comparator circuit.

zero-crossing detector, which is useful for converting an a.c. input signal into a pulse train for digital counting.

Maximum supply rails of ±18 V on pins 8 and 4 allow the 311 to handle inputs up to ±17 V in peak value. The output transistor can continue to run independently from a +5 V supply to retain compatibility with logic circuits. Its 50 mA current handling capacity allows the direct switching of lamps and relays.

11.17.2 Hysteresis by positive feedback

Trouble from 'jitter' may be experienced in comparators, particularly if the input signal has a significant noise content. Instead of a single transition of output level, there may be rapid oscillation between logic states as the input approaches V_{REF}. Such a phenomenon creates havoc with digital counters, and can be avoided by introducing *hysteresis* into the comparator characteristic. This hysteresis or 'backlash' is the electrical analogue of a spring door latch and is introduced by means of a small amount of positive feedback.

Fig. 11.49 shows a 311 comparator with a 50 mV hysteresis incorporated. The positive feedback is, of course, applied to the non-inverting input and, in order that the feedback (and hence the hysteresis) is precisely determined, it is necessary that this input be fed from a low-resistance source. Rather than specify a low-resistance signal source, it is usually easier to arrange that the reference voltage feeds the non-inverting input. With the input signal then feeding the inverting input, the comparator assumes the inverting characteristic of fig. 11.50 which also shows the effect of hysteresis. When the comparator output changes state, the positive feedback has the effect of changing the reference voltage slightly so that a relatively large change of input signal is

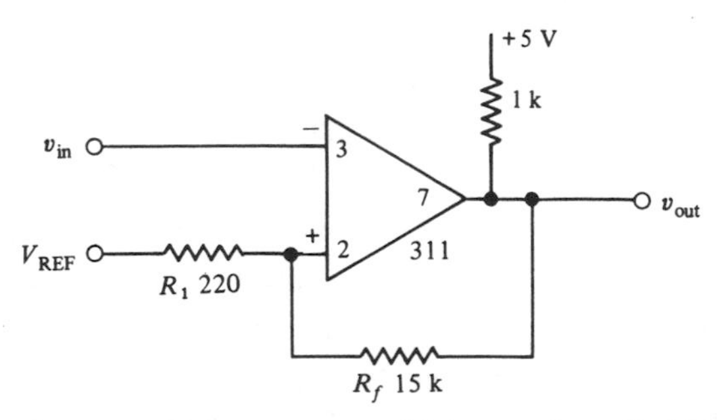

Fig. 11.49. 311 comparator with hysteresis applied (power supplies and earth rail as in fig. 11.48).

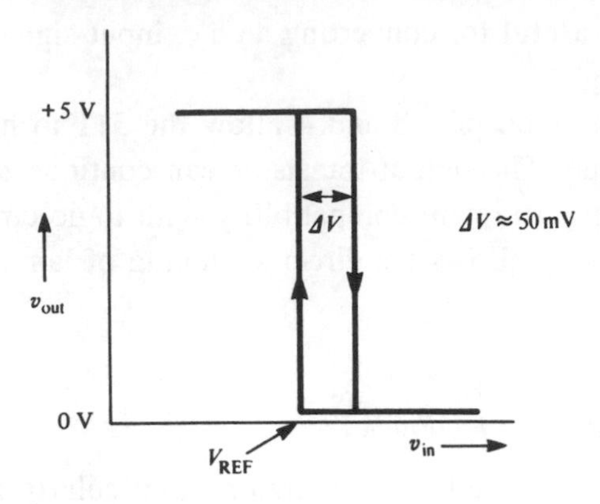

Fig. 11.50. Typical characteristic of 311 comparator with hysteresis.

then required to reverse the output state. In fig. 11.49, the hysteresis ΔV is given by

$$\Delta V = \frac{R_1}{R_1 + R_f} \cdot V_{pp}, \tag{11.34}$$

where V_{pp} is the peak-to-peak output voltage change.

11.18 Op amp data

Information on characteristics, pin connections and type numbers of op amps can be found in Appendix 3.

11.19 Analogue multiplier

In our discussion of the differential amplifier in section 8.4.2, it emerged that the differential voltage gain is directly proportional to the current in the tail of the two differential transistors. Clearly, we have here a means of electrically controlling the voltage gain of an amplifier. The tail current of a differential amplifier may be controlled independently of the amplifier input by varying the bias voltage on the tail transistor used as the constant current source (fig. 8.9). If a balanced output is taken from the two transistors, as in fig. 8.10, variation in tail current has no effect on the differential output voltage, but only changes the gain.

Now let the input voltage signal to the amplifier be V_X, then

$$V_{out} = A_V V_X,$$

where A_V is the voltage gain. Now suppose

$$A_V = kV_Y,$$

where V_Y is the bias voltage controlling the tail current, then

$$V_{out} = kV_X V_Y$$

and the circuit gives an output proportional to the product of the two input voltage signals. This is the principle of the variable-transconductance (Gilbert cell) analogue multiplier, which is available at relatively low cost in IC form, e.g. Motorola MC1495L or Analog Devices AD834.

The typical range on both X and Y inputs is ±5 V for <1% error in multiplication. Fig. 11.51 shows a practical circuit using the 1495 multiplier. The

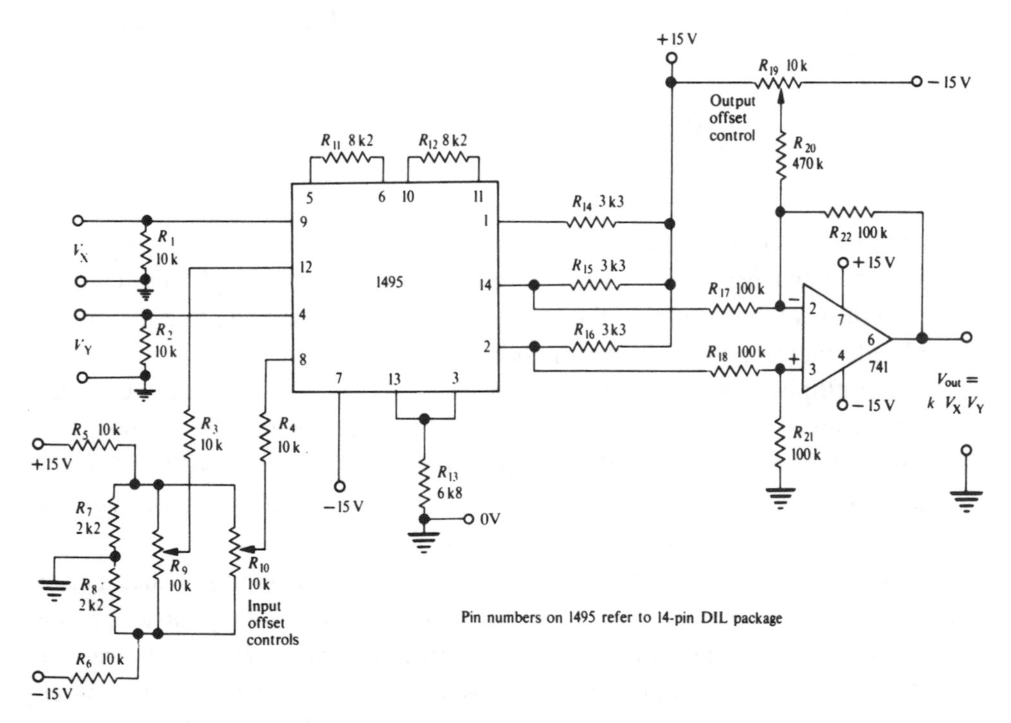

Fig. 11.51. Analogue multiplier using the 1495 IC.

output of the multiplier chip is in differential form; it is in fact sitting at a high common-mode d.c. voltage, and the 741 differential amplifier is used to obtain an output free from d.c. offset. The output offset control, R_{19}, is capable of a wide range of adjustment and is set for zero output with $V_X = V_Y = 0$. Input offset controls are also necessary, since zero output must also be obtained if only one of the inputs is zero, i.e.

$$V_X \times 0 = 0$$

and

$$V_Y \times 0 = 0.$$

The input offset controls are conveniently set using an a.c. input. A sine-wave signal of about 2 V r.m.s. is fed into the X input and nothing into Y. With an oscilloscope or millivoltmeter on the output, R_{10} is now adjusted for zero a.c. output. The a.c. signal is then fed into Y, with $V_X = 0$ and the procedure repeated using R_9 to null the offset. A useful check on the operation of the multiplier is then to feed a sine wave into both inputs and examine the output. The sine wave should be multiplied by itself and therefore squared.

Now, since

$$(V \sin \omega t)^2 = \frac{V^2}{2}(1 - \cos 2\omega t),$$

the output waveform should appear as double the input frequency but sitting with its negative peaks clamped at earth level. The squared amplitude function may be observed by doubling the input amplitude and checking that the output amplitude increases by a factor of four.

The multiplier output is of the form $V_O = kV_XV_Y$, where k is a constant which is approximately equal to 0.1 in this circuit; k may be trimmed to exactly 0.1 by varying R_{13}. With maximum values of V_X and V_Y of ±5 V for linear operation, the relatively low value of k ensures that the output voltage swing remains well within the capability of the 741.

As it is shown, the circuit will work with a variety of inputs, either a.c. or d.c. R_1 and R_2 are included to tie the inputs to earth if coupling capacitors are used. If, however, the input is of low source resistance, R_1 and R_2 can be removed; R_3 and R_4 should then be short-circuited to maintain the low source resistance at all inputs and reduce the inherent offset. It is useful to note that

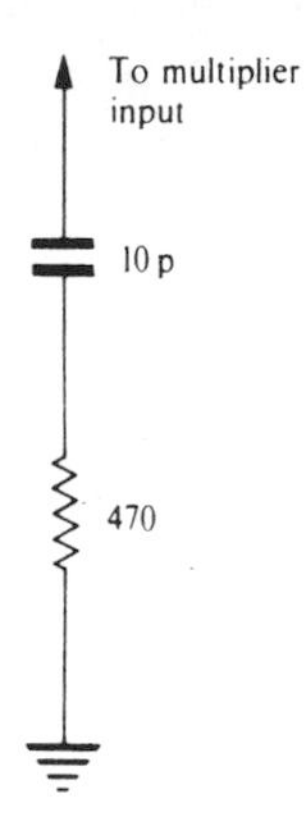

Fig. 11.52. Suppressor for connection to each multiplier input to reduce the risk of high-frequency instability.

pins 9 and 12 form one differential input and pins 4 and 8 the other differential input; these inputs can be treated in the same way as op amp inputs, taking due note of the fact that input bias current is relatively high at 3 μA, but that the input impedance is also high at 20 MΩ.

In order to avoid unwanted high-frequency oscillation, it is a useful precaution to connect a parasitic suppressor (fig. 11.52) from pin 9 to earth and pin 4 to earth. This is particularly necessary if long input leads are used, the high-frequency response of the multiplier being only 3 dB down at 3 MHz so that stray coupling can cause oscillation. The use of the series combination of capacitor and resistor reduces the Q of any stray tuned circuits at high frequencies and thus prevents oscillation.

The analogue multiplier has several intriguing applications. A circuit is sometimes required to simulate a particular equation, which might say, be of the form $V = k(lm + xy)$, where l, m, x and y are all variables and k is a fixed constant. Analogue multipliers would be used here to produce products lm, xy. The two products would then be summed in an analogue adder which may be arranged to have the appropriate gain corresponding to constant k.

The analogue multiplier makes a useful voltage-controlled attenuator if used as in fig. 11.53. A simple application of such a circuit is a remote-controlled audio-gain control: because the gain is varied by adjusting a d.c. voltage, there is no need to run the audio leads to the remote control, and the risk of induced hum is thereby avoided. An AGC (automatic gain control) facility may be designed by rectifying and smoothing the signal itself and subtracting the resulting varying d.c. level from the steady control voltage. Thus, if the input

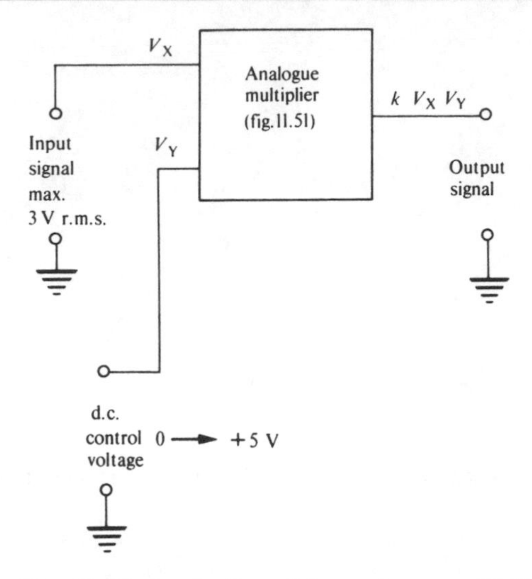

Fig. 11.53. Application of the analogue multiplier as a voltage-controlled attenuator.

signal increases, the control voltage reduces and the output remains relatively constant. Volume compressors and expanders can be made on this basis. If the control voltage is derived from a second signal, e.g. speech from a microphone, a circuit can be derived to reduce the volume of music automatically during an announcement. This is the 'ducking' device, much used by disc jockeys.

It is useful to note that if only a.c. output is required from the 1495 multiplier, the circuit of fig. 11.51 may be simplified by omitting the differential amplifier on the output. Instead, the d.c. component at the output can be eliminated by taking the output signal via a coupling capacitor straight from pin 14 or pin 2 of the 1495.

11.20 Analogue divider

Division is accomplished by including a multiplier in the feedback loop of an op amp. Fig. 11.54 shows how this may be done by using the multiplier of fig. 11.51 with an additional 741 amplifier.

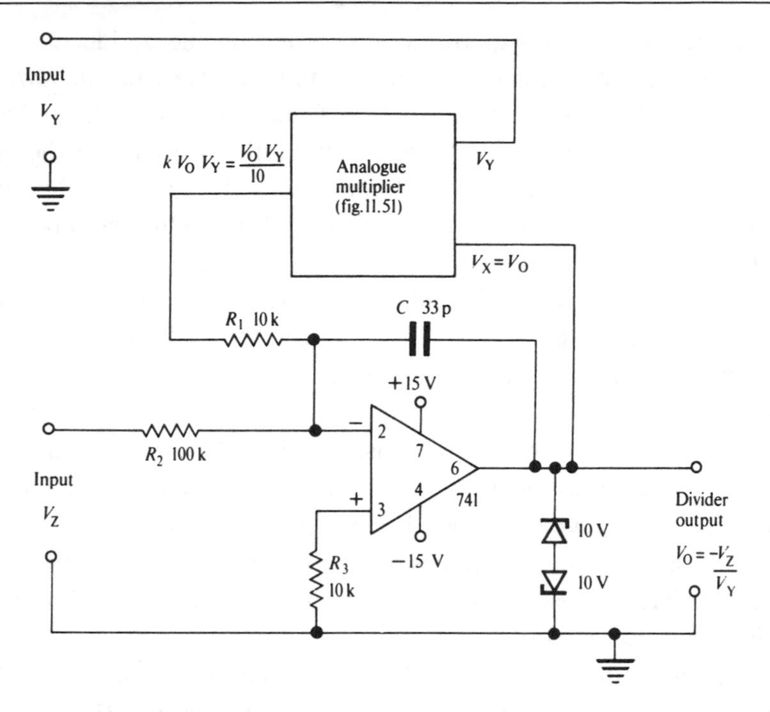

Fig. 11.54. Analogue divider.

Treating the inverting input of the 741 as a virtual earth and making the usual assumptions gives:

$$\frac{V_Z}{R_2} = \frac{-kV_Y V_O}{R_1} \tag{11.35}$$

or, rearranging,

$$V_O = \frac{-R_1}{kR_2}\frac{V_Z}{V_Y}$$

Thus, with $k = 0.1$ and, with the resistor values given, $R_1/R_2 = 0.1$, we have

$$V_O = \frac{-V_Z}{V_Y}. \tag{11.36}$$

Some difficulty may be experienced in setting up the divider unless it is realized that a zero denominator gives an infinite answer. In other words V_Y must be set to a finite value, say 1 V, before the divider is operational. Division by low denominators ($V_Y < 0.5$ V) tends to be critical because of the op amp is working at high gain and the offset controls need to be carefully adjusted if precision better than ±5% is to be maintained. The multiplier is best offset-nulled first before it is connected into the feedback loop.

Capacitor C_1 is used to provide additional high-frequency compensation to maintain stability despite the extra phase shift in the multiplier. The back-to-back Zener diodes across the output prevent *latch-up* if V_Y does go to zero. Without the diodes, if V_Y falls to zero, the 741 output aims at infinity, but stops just short of 15 V for obvious reasons. More than 12 V on the X input is, however, sufficient to render the multiplier temporarily inoperative so that it is unable to respond when V_Y does become non-zero. The latch-up condition thus persists until the power supplies are switched off and then restored. The 10 V back-to-back Zeners limit the output to approximately 10.5 V which is insufficient to cause latch-up.

The maximum value for V_Y is +5 V, and it is important to note that V_Y must be positive. V_Z on the other hand may cover a range from −25 V to +25 V.

If the multiplier X and Y inputs are connected together, then V_O is the square root of the magnitude of V_Z. In this application V_Z must be negative, and in the range 0 V to −25 V.

11.21 Analogue simulation

11.21.1 Introduction

We have already seen how adders, amplifiers and multipliers can be used to simulate various algebraic equations. These are examples of analogue computing, where voltages are used to represent physical quantities. Even the complexities of the differential equations used to describe moving mechanical systems can be readily simulated. We shall here consider one of the most common and useful simulations: the second-order differential equation describing a damped resonant system (e.g. a mass on a spring).

11.21.2 Damped resonant system (harmonic oscillator)

Consider a mass m hanging on the end of a spring of stiffness k (stiffness = increase in force/increase in length) in the presence of viscous damping described by the damping constant b and driven by some force which is a function of time $F(t)$. The differential equation for the distance x of the mass from some datum point is

$$m\frac{\mathrm{d}^2x}{\mathrm{d}t^2}+b\frac{\mathrm{d}x}{\mathrm{d}t}+kx=F(t). \tag{11.37}$$

In designing an analogue computer to solve this equation, the easiest starting point is to assume that an acceleration signal, corresponding to $\mathrm{d}^2x/\mathrm{d}t^2$, is available. The equation can then be rearranged;

$$\frac{\mathrm{d}^2x}{\mathrm{d}t^2}=-\frac{b}{m}\frac{\mathrm{d}x}{\mathrm{d}t}-\frac{k}{m}x+\frac{1}{m}F(t). \tag{11.38}$$

To simulate this equation, $\mathrm{d}^2x/\mathrm{d}t^2$ is integrated to yield $-\mathrm{d}x/\mathrm{d}t$, then a second integration produces x. To obtain the correct polarity and value of the various constants, a fraction b/m of $-\mathrm{d}x/\mathrm{d}t$ is obtained from a potentiometer on the output of the first integrator; this is inverted and fed together with the fraction k/m of x, from the second integrator output, to an operational adder. This performs its usual inversion, giving an output

$$-\left(\frac{b}{m}\frac{\mathrm{d}x}{\mathrm{d}t}+\frac{k}{m}x\right).$$

If we ignore for a moment the $F(t)$ term, which is often only an initial transient 'kick' anyway, the output represents the right-hand side of our rearranged equation. This must now be equated with $\mathrm{d}^2x/\mathrm{d}t^2$: all we need do is return the output to the input, where we assumed that $\mathrm{d}^2x/\mathrm{d}t^2$ would be available to start with. The $F(t)$ term may be conveniently added in now via an extra input resistor into the first integrator. This is shown in fig. 11.55, where the op amps are drawn in schematic form omitting the non-inverting inputs, which are earthed in the usual way. It is convenient if the gain of each integrator is arranged to be unity so that the equation constants are completely determined by the potentiometers. Now, since the equation of the integrator output is

$$v_{\mathrm{out}}=\frac{-1}{RC}\int v_{\mathrm{in}}\,\mathrm{d}t,$$

time constant RC should be one second for gain of −1. The input bias current of the 741 limits the maximum usable R to 100 kΩ, so that C should be 10 μF. Electrolytic capacitors are not suitable because there is normally no d.c. polarizing voltage present; thus a relatively bulky paper or polyester capacitor is necessary. To save space, 100 kΩ and 1μF are frequently used, giving $RC = 0.1$ and an integrator gain of −10 which must be allowed for in setting the potentiometers.

It is interesting to notice that at all times the voltage across the first integrator capacitor represents velocity and that across the second integrator capacitor represents displacement. Thus, prior to computation, the displacement and velocity may be assigned initial values by charging the integrator capacitors to predetermined voltages. This can be accomplished by the circuit of fig. 11.56, which also includes a hold facility so that the computation may be interrupted and resumed as required. Switch S_2 is closed to set the intial condition, determined by the adjustment of the potentiometer. To 'hold', S_2 and S_1 are both opened, whilst to compute, S_1 is closed and S_2 opened. In practice, the switches may be relay contacts, or FETs used as switches.

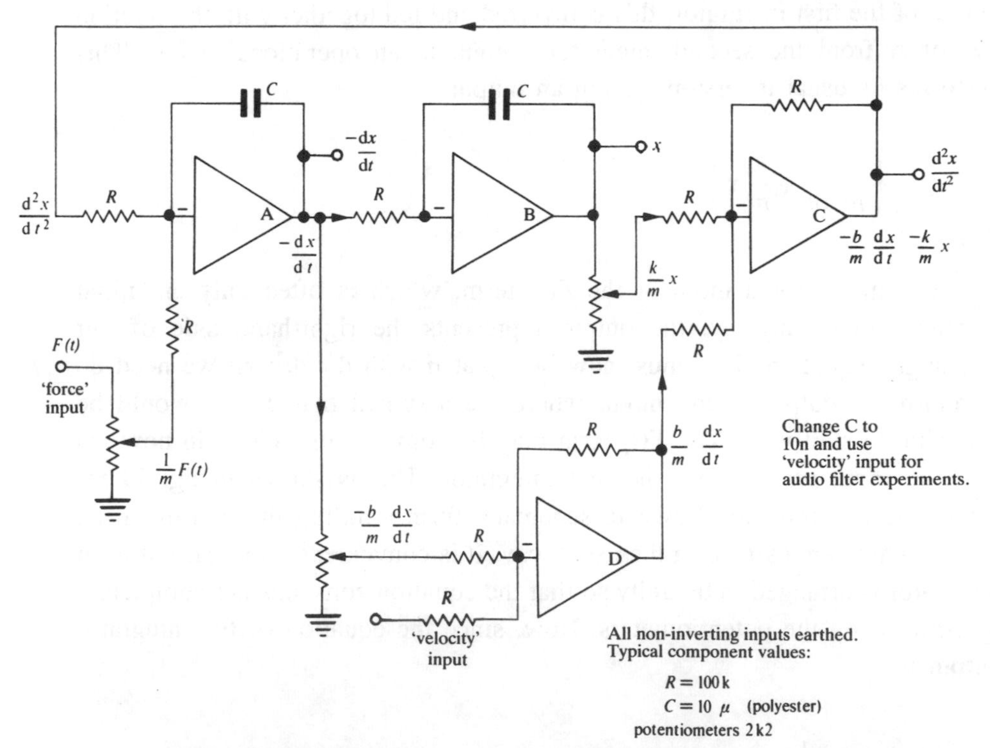

Fig. 11.55. Analogue computer to simulate a resonant system (biquadratic or state-variable filter).

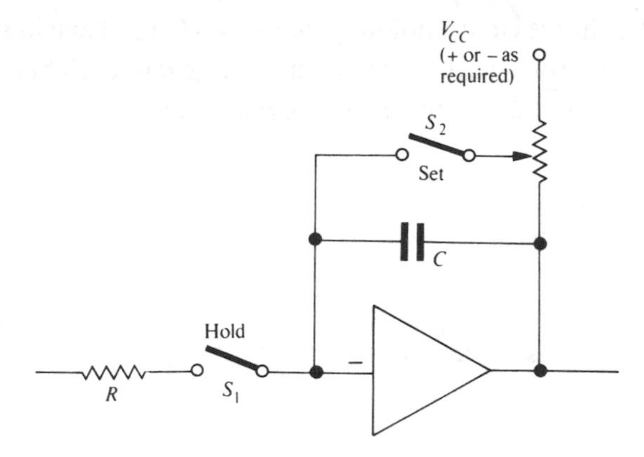

Fig. 11.56. Integrator arranged with 'set initial value' and 'hold' switches.

11.22 State-variable filter

The reader will note that, although the second-order integrator loop of fig. 11.55 has been presented as a simulation of a mechanical resonant system, it is also a further example of an active filter. Being a resonant system with controllable damping, it exhibits the properties of an LCR resonant circuit. This type of *state-variable* or *biquadratic (biquad)* circuit is widely used for audio equalization and may be investigated by setting $C = 10$ nF in fig. 11.55 and feeding an audio signal into the 'velocity' input (op amp D).

Monitoring the various amplifier outputs whilst varying signal frequency is very instructive. It is useful to experiment with various potentiometer settings. The output of A gives a peaking band-pass filter. At B a low pass characteristic is available whilst C gives a high pass output. Changing the gains of the two integrators modifies the resonant or roll-off frequency. The 'damping' potentiometer controls Q.

11.23 Switched-capacitor filter

The state-variable circuit of fig. 11.55 is an example of building a flexible active filter using integrators. Cut-off frequency is varied by effectively changing integrator gains at critical points using potentiometers. The switched-capacitor filter uses integrators where the gain (charge flow) is regulated not by a resistor but by a rapid switch.

In the integrator of fig. 11.57, switch S is responsible for charging integration capacitor C_2 by alternately charging C_1 up to the input voltage v_{in} and then discharging C_1 into C_2.

With S in position 1, C_1 charges to v_{in}, holding charge $q = C_1 v_{in}$. Then in switch position 2, C_1 discharges to virtual earth, transferring charge q to C_2. This causes v_{out} to change by a small amount Δv_{out} in one clock cycle time Δt.

$$\Delta v_{out} = -q/C_2 = -\frac{C_1}{C_2} v_{in}$$

In continuous time, this happens f_{clk} times per second giving

$$v_{out} = -f_{clk} \frac{C_1}{C_2} \int v_{in} \, dt. \tag{11.39}$$

The integrator gain is therefore a function of the clock frequency instead of an input resistor. Hence a remote-controlled state-variable filter can be built by varying the clock frequency. Also the dependence on a capacitor *ratio* rather than absolute capacitor value makes the switched-capacitor filter ideal for fabrication on an IC. Switch S is typically a pair of MOSFETS.

The disadvantage of the switched-capacitor filter can be some clock breakthrough into the signal but f_{clk} is typically very much higher than the signal frequency and a little conventional low-pass filtering on the output can eliminate the problem. It is worth noting in comparing equation (11.39) with the equation for integrator gain (11.3) that the value of R in the conventional integrator has become $1/(f_{clk}C_1)$ in the switched-capacitor version. The switched input capacitor is behaving like a resistor, in fact a charge flow controller.

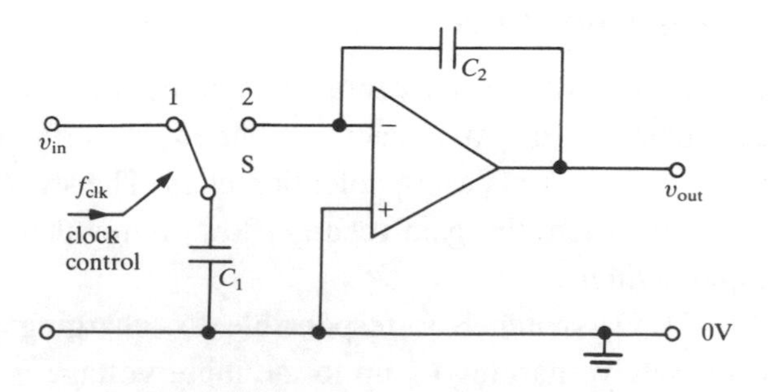

Fig. 11.57. Integrator for switched-capacitor filter.

Many switched-capacitor networks with their associated amplifiers can readily be fabricated on one chip, giving filters of very steep roll-off and closely defined bandwidth. They are commonly used in connection with analogue to digital conversion.

12

Positive feedback circuits and signal generators

12.1 Positive feedback

One of the most useful test instruments in experimental electronics is the signal generator. It is appropriate, therefore, that we should consider the methods used for signal generation.

Positive feedback is the basic factor common to signal generators. We found in the discussion of negative feedback in section 4.3 that the gain of any amplifier with feedback is given by

$$A = \frac{A_0}{1 - \beta A_0}, \qquad [(4.1)]$$

where A_0 is the gain without feedback and β is the fraction of the output fed back to the input. Now, with negative feedback, either β or A_0 is negative, so that the denominator is always greater than one. In the case of positive feedback, however, it is possible to arrive at the condition

$$1 - \beta A_0 = 0, \qquad (12.1)$$

which gives an infinite value for $A = A_0/(1 - \beta A_0)$. This implies that the amplifier produces an output signal with no input, which is the condition for oscillation. An oscillator forms the heart of every signal generator. A sine-wave generator is designed so that the condition of (12.1), often called the *Barkhausen criterion*, is satisfied at only one frequency.

12.2 Sinusoidal oscillators

12.2.1 The phase-shift oscillator

A useful starting point in the study of practical oscillator circuits is to apply positive feedback to a single-stage voltage amplifier. This is accomplished in the phase-shift oscillator of fig. 12.1.

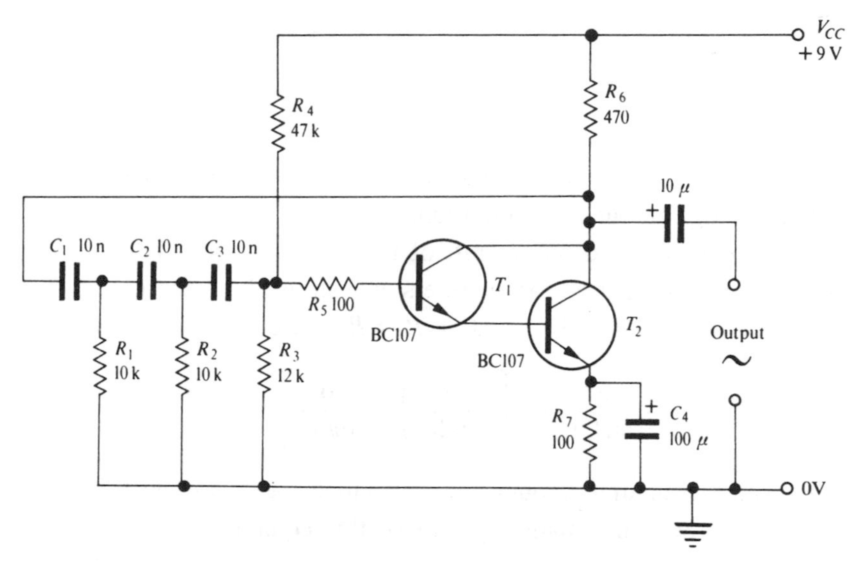

Fig. 12.1. Phase-shift oscillator (sine-wave output).

The phase-shift circuit is necessary if oscillation is to be achieved because, as we saw in section 1.6.3, the amplifier output is 180° out of phase with its input. Hence, for positive feedback, the external circuit (R_1, C_1, R_2, C_2, R_3, C_3) must shift the phase through a further 180°. This three-section phase-shifter is shown in isolation in fig. 12.2. Notice that R_3 in fig. 12.2 is made up of the parallel combination of R_3 and R_4 in fig. 12.1. The amplifier uses a Darlington pair, so that we can neglect the transistor input impedance, which is very high. It is interesting to notice that, when 180° phase shift is required, a passive network requires three RC stages if significant output is to be available. One RC stage will just manage a 90° phase shift, but the capacitor reactance must be very high compared with the resistor, giving near infinite attenuation.

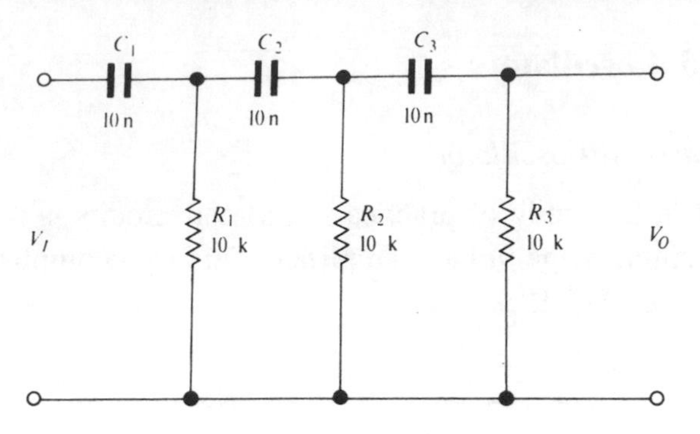

Fig. 12.2. The phase shifter in isolation.

The 100 Ω resistor R_5 in fig. 12.1 is a 'stopper' resistor to damp out unwanted high-frequency oscillation.

The attenuation and phase shift of fig. 12.2 may be calculated by the type of network analysis used for the filters in section 11.14, working back from the output of the network. If $R_1 = R_2 = R_3 = R$ and $C_1 = C_2 = C_3 = C$, the result is

$$\frac{V_I}{V_O} = 1 - \frac{5}{(\omega RC)^2} + j\left[\left(\frac{1}{\omega RC}\right)^3 - \frac{6}{\omega RC}\right] \tag{12.2}$$

To find the angular frequency ω_0 at which the phase shift of the feedback network is 180°, the imaginary part of this equation is made equal to zero, i.e.

$$\frac{1}{(\omega_0 RC)^3} = \frac{6}{\omega_0 RC}$$

therefore

$$\omega_0^2 = \frac{1}{6(RC)^2},$$

$$\omega_0 = \frac{1}{\sqrt{6}RC}$$

or

$$f_0 = \frac{1}{2\pi\sqrt{6}RC}\,\text{Hz.} \tag{12.3}$$

Substituting back for $\omega = \omega_0$ in equation (12.2) gives a value for the attenuation of the network, which leads to the feedback fraction β in equation (4.1) as follows:

$$\frac{V_I}{V_O}=1-5\times6=-29,$$

therefore

$$\beta=\frac{V_O}{V_I}=-\frac{1}{29}.$$

The negative sign indicates a 180° phase shift, whilst the numerical value shows that an amplifier gain of at least 29 times is necessary to maintain oscillation. The circuit of fig. 12.1 is very suitable for experiments. Here are some suggestions:

(*a*) Measure the frequency of oscillation by comparison with a laboratory signal generator on the oscilloscope. Better still, use a digital frequency meter. Compare your experimental result with the theoretical value.

(*b*) Examine the output waveform on the oscilloscope. Is it a good sine wave? If a dual-trace oscilloscope is available, use the output of a laboratory signal generator as a comparison. Distortion is often quite serious in a phase-shift oscillator output because the only factor available to limit the amplitude of the output waveform is the non-linearity which occurs in the transistor itself at high outputs. In addition, the rate of change with frequency of the network phase shift is rather gentle. As a result, considerable circuit gain still exists at harmonics of the basic frequency, and the distortion products are therefore transmitted.

(*c*) If a variable-voltage power supply is used, it is a convenient way of making a fine adjustment to the gain of the circuit. Reduce V_{CC} until the circuit is only just oscillating: the waveform should then be a respectable sine wave. The gain of the amplifier section can now be measured by removing C_3 and feeding a signal into the junction of R_3 and R_4 via the usual coupling capacitor of 10 μF or so. Compare the measured gain with the theoretical value of 29.

(*d*) The easiest way to change the output frequency is to change the values of C_1, C_2, C_3 (the values of R, R_2 and R_3 could be changed instead, but R_4 would have to be changed in proportion to maintain the correct bias on the amplifier). Explore the frequency range over which the circuit will oscillate. Low-frequency performance is limited by the reactance of C_4, whilst stray capacitances become significant at the high-frequency end, reducing gain and introducing unwanted phase shift.

12.2.2 The Wien bridge oscillator

The Wien bridge oscillator is the most popular audiofrequency signal generator circuit. It possesses good frequency stability and can give very low distortion; it is also easy to tune. Fig. 12.3 shows a basic Wien bridge oscillator which employs a 741 op amp as the amplifier. The frequency-selective part is the Wien network, consisting of R_1C_1 and R_2C_2, which is used to apply positive feedback to the amplifier, whilst negative feedback is applied via R_3 and the bulb.

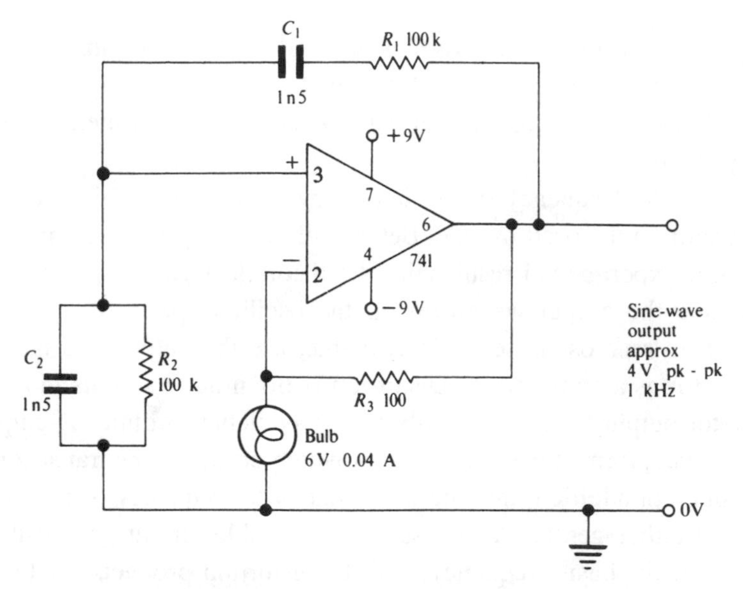

Fig. 12.3. Wien bridge oscillator using an IC amplifier.

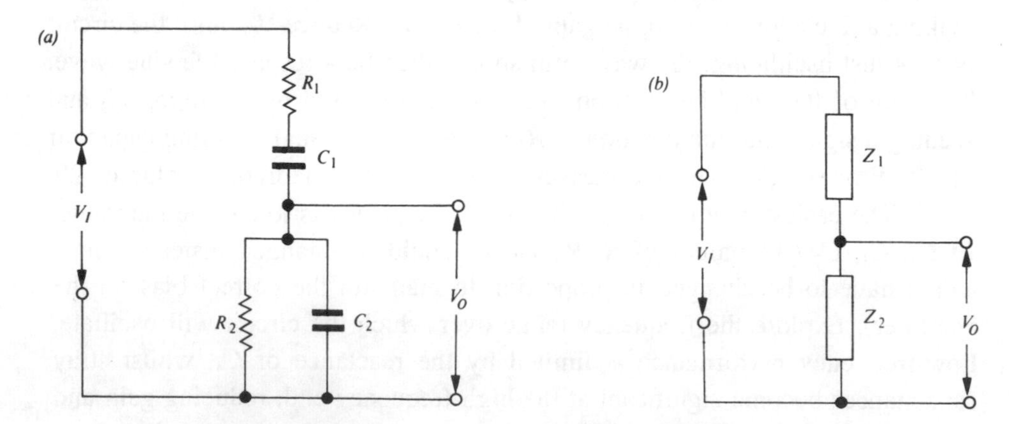

Fig. 12.4. The Wien network: (*a*) relevant components separated from oscillator, (*b*) equivalent potential divider.

It is useful first of all to look at the Wien network separately as in fig. 12.4(*a*), where it is drawn clearly in the form of a potential divider. From its equivalent form in fig. 12.4(*b*), we can write down an expression for the attenuation:

$$\frac{V_I}{V_O} = \frac{Z_1 + Z_2}{Z_2}$$

$$= 1 + \frac{Z_1}{Z_2}.$$

Now, returning to the actual Wien network,

$$Z_1 = R_1 + \frac{1}{j\omega C_1} \qquad (R_1 \text{ and } C_1 \text{ in series})$$

and

$$\frac{1}{Z_2} = \frac{1}{R_2} + j\omega C_2, \qquad (R_2 \text{ and } C_2 \text{ in parallel})$$

i.e.

$$Z_2 = \frac{R_2}{1 + j\omega C_2 R_2}$$

therefore

$$\frac{V_I}{V_O} = 1 + \frac{[R_1 + (1/j\omega C_1)](1 + j\omega C_2 R_2)}{R_2}$$

Multiplying out,

$$\frac{V_I}{V_O} = 1 + \frac{R_1}{R_2} + \frac{C_2}{C_1} + j\left(\omega R_1 C_2 - \frac{1}{\omega R_2 C_1}\right). \qquad (12.4)$$

Equation (12.4) is simplified if, as is usually the case, $R_1 = R_2 = R$ and $C_1 = C_2 = C$, then

$$\frac{V_I}{V_O} = 3 + j\left(\omega RC - \frac{1}{\omega RC}\right). \qquad (12.5)$$

The imaginary term will vanish at resonant frequency ω_0 when

$$\omega_0 RC = \frac{1}{\omega_0 RC},$$

i.e.

$$\omega_0 = \frac{1}{RC}$$

therefore

$$f_0 = \frac{1}{2\pi RC}. \tag{12.6}$$

In fig. 12.3, $RC = 1.5 \times 10^{-4}$, giving $f_0 \approx 1000$ Hz.

At f_0, the resonant frequency of the Wien network, the phase shift is zero because the imaginary term is zero and we are left with an attenuation

$$\frac{V_I}{V_O} = 3.$$

Hence, in the Wien bridge oscillator, the amplifier must have a gain $\geqslant 3$ in order to maintain oscillation. Now in section 12.2.1, we saw that the phase-shift oscillator gave a somewhat distorted waveform because the only factor limiting the output amplitude was the nonlinearity of the amplifier itself. It is easy to improve matters in the Wien bridge circuit: the relatively low gain required means that heavy negative feedback can be employed, and the amplitude stabilization is readily incorporated in the feedback loop. In fig. 12.3 the feedback divider consists of a 100 Ω resistor, R_3, and a 6 V, 0.04 A bulb. Now it is well known that the resistance of a bulb filament changes enormously as it heats up. In this circuit, as the output voltage increases, the bulb heats up; as it does so, it increases the negative feedback by increasing its resistance. When the bulb resistance reaches 50 Ω, the amplifier gain is given by $A = (100 + 50)/50 = 3$, and the theoretically stable condition is reached. With the bulb specified, this applies when the oscillator output is approximately 4 V peak-to-peak. Any tendency for the output voltage to rise further is prevented by the bulb, which would further increase its resistance and reduce the amplifier gain. Adjustment of the value of R_3 controls the output voltage, but the available swing from the 741 is limited by the fact that the resistance of the feedback loop is well below the recommended minimum value of output load. Nevertheless, the circuit gives very satisfactory results.

An alternative to the bulb as a stabilizing element is the negative-temperature-coefficient (NTC) thermistor, which decreases its resistance as temperature rises. The thermistor is connected in place of R_3, the bulb being replaced by a resistor with a value selected according to the resistance of the thermistor at the required oscillator output voltage. The latter value can be found from the manufacturers' data. Thermistors are available with relatively high resistance values, providing rather more flexibility in design than a bulb. They are the most common method of amplitude stabilization.

A third amplitude stabilizer, which is also commonly used, is the FET working as a voltage-controlled resistor. Here the oscillator output must be rectified and smoothed before being applied to the gate of the FET, a method which is more complicated than the simple thermistor, but which does give control over the stabilizing time constant. The latter can be critical if minimum amplitude 'bounce' is required together with low distortion. Too short a time constant will cause the regulation to try to smooth individual cycles at low frequencies, causing distortion, whilst too long a time constant tries the patience of the user as he waits for the amplitude to settle after changing the frequency.

In commercial signal generators, tuning of the output frequency is usually achieved by making R_1 and R_2 a dual potentiometer and switching in various pairs of capacitors for different frequency ranges.

12.2.3 *Quadrature oscillator*

The circuit of fig. 12.5 is an interesting low-distortion (<0.1%) sine-wave oscillator which provides two equal outputs 90° out of phase with each other. It is an adaptation of the biquad filter (analogue computer) of fig. 11.55 which simulates a resonant system. Basically it is two integrators connected with feedback via a unity-gain inverter. With no attenuation on potentiometers R_{1a} and R_{1b}, oscillation will occur at the frequency where each integrator has unity gain, i.e. where $1/\omega RC = 1$ (in this circuit $R_3 = R_5 = R$, and $C_1 = C_2 = C$). Potentiometers R_{1a} and R_{1b} are ganged together and give a frequency range of 11 : 1. A twin-gang linear slider potentiometer is ideal here because the matching of the two sections is much more precise than in the usual ganged rotary controls. Different pairs of capacitors can be switched in for different frequency decades.

The 4.7 MΩ resistor R_7 applies a small amount of 'negative damping' to ensure that oscillation starts quickly when the circuit is switched on. Amplitude limiting is provided by Zener diodes D_1 and D_2 and the divider R_8, R_9, which bring in positive damping when the output amplitude rises above approximately 4 V peak. Thus both output (1) and output (2) give a very steady 8 V peak-to-peak. The 90° phase difference is produced because the second integrator comes between the two outputs. If appropriate values of C_1 and C_2 are inserted, the circuit will work well at frequencies as low as

0.1 Hz. When the two outputs are connected to the X and Y inputs of the oscilloscope, a perfectly circular Lissajous figure is produced; this is a particularly effective demonstration at low frequencies, when the spot can be clearly seen in its slow circular sweep. This could form the basis of a simulated radar display.

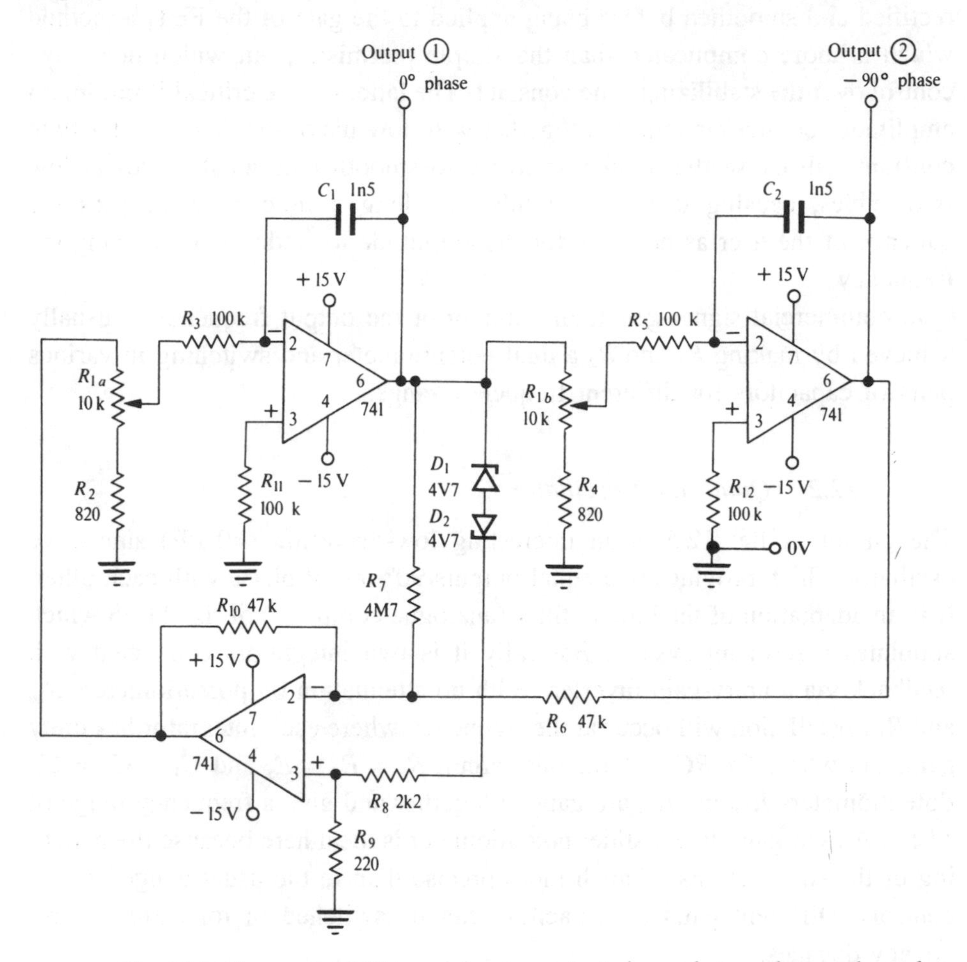

Fig. 12.5. Sine-wave oscillator giving two outputs in quadrature phase and equal amplitude.

12.2.4 LC tuned oscillator

The LC tuned circuit is perhaps the most basic frequency-selective network. It is extensively used at the higher frequencies above 50 kHz or so, but becomes rather cumbersome at the lower audiofrequencies, where the necessary inductors are bulky. The circuit of fig. 12.6 shows a simple LC oscillator

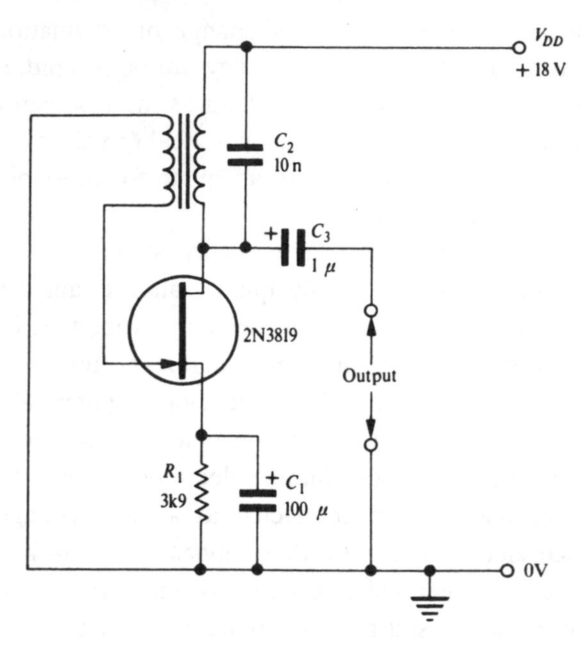

Fig. 12.6. Simple LC oscillator using a FET.

which can be used as a basis for experiments. It employs a FET amplifier stage with a transformer to provide a positive feedback coupling. The transformer primary in the drain circuit is tuned by capacitor C_2 which may be varied over a wide range. Almost any audio transformer, such as an interstage transformer or output transformer, will work in this circuit, the secondary working into the high impedance of the FET gate so that there is little damping on the parallel resonance between the primary inductance and C_2. If the circuit will not oscillate, try reversing the transformer secondary connections to ensure that the feedback is positive.

The feedback in an LC oscillator need not necessarily go to the gate (or base in the case of a bipolar transistor). It can equally well be inserted in the source, or emitter, circuit so that the stage is operating in common-gate or common-base mode. Mixer-oscillator stages in the common superhet type of radio can operate this way, the base of the transistor then being available to take the incoming signal to be 'mixed' with the oscillator output.

12.3 Crystal oscillators

The frequency stability of an oscillator is determined chiefly by the Q of the resonant circuit, no matter whether it be an LC circuit or an RC circuit such as a Wien bridge. If the Q is large, the rate of change of attenuation with frequency and rate of change of phase with frequency are both rapid, so that the circuit is reluctant to change its frequency as long as the resonant circuit components remain constant in value. The highest practical Q-values with LC circuits are of the order of hundreds, which are adequate for most purposes but are not suitable for frequency standards.

The crystal oscillator is widely used where a very stable frequency is required. Certain crystalline materials, notably quartz, and certain ceramics such as lead zirconium titanate (PZT) exhibit piezoelectric properties, i.e. they deform mechanically when subjected to an electric field. The inverse is also true: when the crystal is stressed mechanically, it develops a potential difference between opposing faces. As a result of this piezoelectric property, a thin plate of quartz or ceramic material with conducting electrodes evaporated onto its surface vibrates mechanically when connected to an a.c. voltage. The vibrations in turn produce electrical signals; these appear as a 'back e.m.f.', which determines the effective electrical impedance of the crystal. Vibration amplitude is a maximum at the crystal resonant frequency. Internal losses in the crystal are very small and Q-values as high as 100 000 can be achieved. Resonant frequencies ranging from 10 kHz to many megahertz are possible, depending on the size and shape of the crystal.

Detailed study of crystal behaviour reveals that its characteristics are a combination of those of a series-resonant and a parallel-resonant circuit. Over most of the frequency range its impedance is capacitative except near resonance, when it drops to zero at the 'series' resonance then flies off to infinity at the 'parallel' resonance. The parallel resonant frequency is always slightly higher than the series resonant frequency, but for most purposes the two may be considered to be the same. Because of this dual behaviour at resonance, a crystal can be used in a wide variety of ways in a feedback loop to produce an oscillator. One useful circuit which produces a rectangular-wave output suitable for driving logic and microprocessors is shown in fig. 12.7. It uses the inverters in the 74LS04 digital IC as amplifiers.

As well as serving as the resonant element in oscillators, crystal and ceramic devices are also used extensively as bandpass filters. They are particularly common in the intermediate-frequency amplifiers of v.h.f. radios, where a precisely-defined bandwidth must be maintained for good performance.

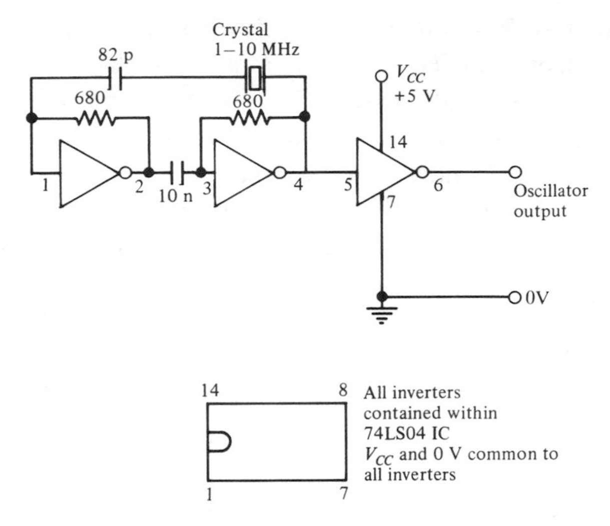

Fig. 12.7. Crystal-controlled oscillator for driving digital circuits. Uses TTL inverters as amplifiers.

12.4 The bistable multivibrator

Consider the two-transistor circuit of fig. 12.8. It is essentially two transistor switches connected to one another. Suppose that when the supply is turned on, T_1 turns on first. When saturated, its collector will be within 0.3 V of earth. This means that no current can flow through R_3 into the base of T_2, because it takes approximately 0.6 V to obtain significant conduction in the silicon junction. T_2, therefore, will remain off. Notice that, as long as T_2 is

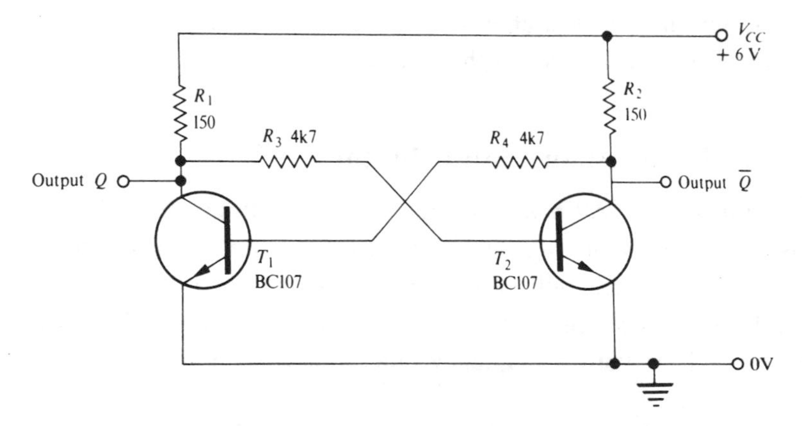

Fig. 12.8. A bistable multivibrator. A clear visual indication of state is obtained if R_1 and R_2 are replaced with bulbs (6 V, 0.04 A).

off, its collector is up at +6 V so that current is flowing via R_4 into the base of T_1, maintaining the status quo. The circuit is in a stable state.

Now let us momentarily short the base of T_1 to earth, starving it of base current. Its collector current will fall to zero and its collector voltage rise to +6 V. T_1 has turned off, but T_2 now receives a base current flowing in R_3, which switches it on. T_2 saturates, with its collector near zero volts, thus preventing any current flowing into T_1 base even when the short is removed. Once more the circuit is in a stable state, but this time T_2 is on.

As the name suggests, the bistable multivibrator has two stable states. It acts as an electrical memory, remembering which transistor was last triggered off. As we shall see in the next chapter, it is a basic building brick in digital circuits, being employed in counters and memories. In digital circuits it has earned the name 'flip-flop'.

For experimental work, the collector load resistors in fig. 12.8 may be replaced by 6 V, 0.04 A bulbs. The transistor which is on will then light its bulb, giving a clear indication of the state of the circuit. The state may be switched either by earthing the base of the transistor which is on or by earthing the collector of the transistor which is off.

The value of the base resistors R_3 and R_4, whilst not at all critical, must be low enough to bottom the transistors in use. For reliable saturation,

$$\frac{R_3}{R_1} < h_{FE},$$

where h_{FE} is the current gain of the transistor. In this case, $R_3/R_1 = 4700/150 \approx 30$, so that the bistable should work with $h_{FE} \geqslant 30$. In general, the higher the base current pushed into a saturated transistor, the lower $V_{CE(\mathrm{sat})}$ becomes, which is a desirable state of affairs because it reduces transistor power dissipation and ensures that the 'off' transistor remains off.

12.5 The astable multivibrator

The circuit of fig. 12.9 has much in common with the bistable of fig. 12.8, but uses capacitor-coupling instead of resistor-coupling between the transistors. The effect of this will be clearly seen if the circuit is constructed using 6 V, 0.04 A bulbs as the 150 Ω collector loads. The transistors switch alternately on and off with a frequency determined by the circuit time constants C_1R_3 and C_2R_4.

In order to look more closely at the operation of the astable multivibrator, let us assume that T_1 has just switched off, its collector heading for +6 V; meanwhile, T_2 has just turned on. What determines the duration of this quasi-

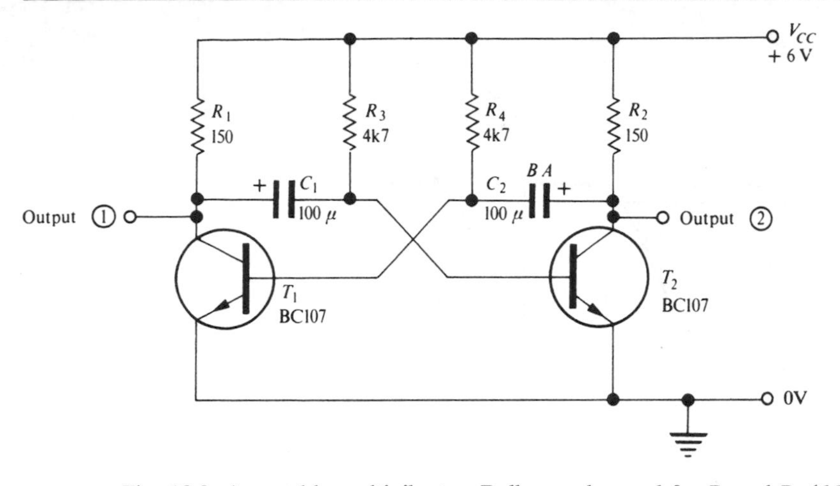

Fig. 12.9. An astable multivibrator. Bulbs can be used for R_1 and R_2 (6 V, 0.04 A).

stable state? Consider the potential difference across the plates of capacitor C_2: plate A, connected to T_2 collector, has just been at +6 V, whilst plate B, connected to T_1 base, has just been at +0.6 V because T_1 was conducting. C_2 therefore has a 5.4 V p.d. across its plates: this represents a considerable charge which the capacitor is loath to release. So, at the instant when T_2 switched on, plate A went down to earth with T_2 collector and, momentarily, the 5.4 V p.d. between plates A and B was maintained. Plate B therefore was pulled down to −5.4 V, turning off T_1 by taking its base negative. C_2 then begins to charge in the opposite direction via R_4, which is connected to +6 V. Thus, the base of T_1 is now aiming at +6 V with a time constant R_4C_2. But it never gets there: as soon as it reaches 0.6 V T_1 turns on, starting the whole process again, C_1 taking the base of T_2 to −5.4 V and charging via R_3. Fig. 12.10 shows a plot of T_1 base voltage during the cycle. Here, the interruption of the exponential charge of C_2 via R_4 can be seen; there is in fact a slight overshoot as a positive pulse is driven into T_1 base by T_2 switching off.

Neglecting the fact that the charging period does not *quite* start from $-V_{CC}$, the charging equation is of the familiar exponential form

$$v = 2V_{CC}\left[1 - \exp\left(\frac{-t}{C_2R_4}\right)\right]. \tag{12.7}$$

Notice the factor of 2 which shows that the curve is aiming at $+V_{CC}$ from its $-V_{CC}$ start. Rearranging,

$$\exp\left(\frac{-t}{C_2R_4}\right) = \frac{2V_{CC} - v}{2V_{CC}}.$$

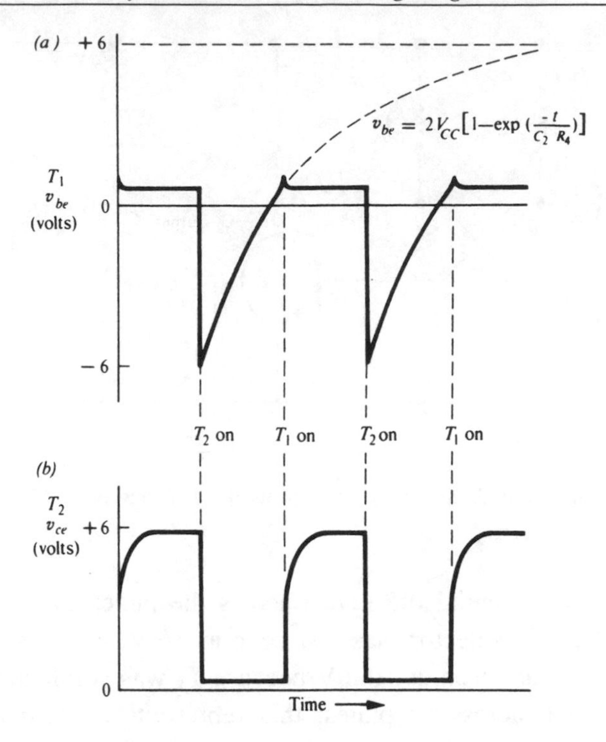

Fig. 12.10. Waveforms in an astable multivibrator: (*a*) T_1 base voltage, (*b*) T_2 collector voltage.

Now, to find the time interval for which T_1 is off, we must extract t from the exponential by taking natural logs

$$\frac{-t}{C_2R_4} = \ln\left(\frac{2V_{CC} - v}{2V_{CC}}\right).$$

Now, over the time interval in question voltage change $v \approx V_{CC}$, therefore

$$\frac{-t}{C_2R_4} \approx \ln\left(\tfrac{1}{2}\right),$$

therefore

$$t \approx C_2R_4 \ln 2 \approx 0.7\, C_2R_4. \tag{12.8}$$

An identical relationship is, of course, obtained for the other half-cycle: $t' \approx 0.7C_1R_3$.

Notice that the mark–space ratio of the output waveform can be varied by selecting different time constants for the two transistors. Care is necessary, though, in selecting time constants to ensure that the base resistors are still sufficiently low to saturate the transistors.

The output waveform of the astable is shown in fig. 12.10(*b*); although definitely rectangular it has a very slow rising edge. This is because, as a transistor turns off and the collector rises toward V_{CC}, the capacitor behaves as an enormous 'stray' capacitance and has to be charged via the collector load. (It is this charging current which produces the positive overshoot in the base voltage waveform of fig. 12.10(*a*).) Fortunately, this deficiency is readily corrected by introducing diodes to isolate the collector from the capacitor during the rising edge of the wave. This technique is used in the circuit of fig. 12.11 which is designed to give a clean square wave at approximately 1 kHz. When C_2 turns T_1 base negative, T_1 collector is able to rise up to V_{CC} almost instantaneously because diode D_1 becomes reverse biased. R_5 is now responsible for charging C_1 back up to the rail. When T_1 turns back on, D_1 conducts and the usual sequence of operations is able to take place.

The name 'multivibrator' used for this family of circuits may appear a little obscure, but is derived from the fact that the output of the astable multivibrator contains many frequencies: the fundamental frequency, as calculated above, plus a full range of harmonics, normally extending beyond the thousandth. It is interesting to note that, if the mark–space ratio of the wave is *exactly* unity, the even harmonics will be absent. A circuit such as fig. 12.11 makes a useful signal injector for circuit testing on a.m. radio work. Injecting the signal into the audio stages should produce the 1 kHz tone with its audible harmonics.

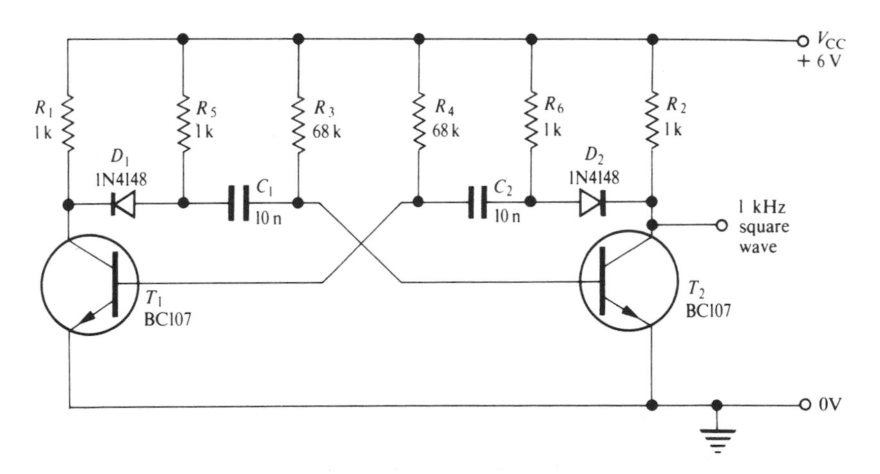

Fig. 12.11. Astable multivibrator with diodes to shorten rise time.

A response will also be obtained from the RF and IF stages if the signal is injected there, the harmonics extending well up into these higher frequency bands. By working with a signal injector backwards from the output stage of a faulty radio, the defective stage may be identified when the signal is no longer heard.

12.6 The monostable multivibrator

Fig. 12.12 is a cross between the bistable and astable, having one d.c. coupling and one capacitative coupling. The result is that the monostable has only one stable state, with T_2 on and T_1 off, but may be triggered into a *quasi-stable* state with T_1 on and T_2 off, where it will remain for a period given by $t \approx C_1 R_3 \ln 2$, as with the astable.

Triggering the monostable into its second state is readily achieved by *momentarily* shorting the base of T_2 to earth. Alternatively, electrical triggering is possible with a negative-going pulse coupled to T_2 base. The output pulse is best taken from T_2 collector, because the rising edge at T_1 collector suffers from the retarding effect of C_1.

Applications of the monostable include the production of pulses of a desired width, and the provision of an adjustable time delay between successive events (e.g. between firing a gun and photographing the bullet in flight). Although the discrete circuit discussed here is useful for illustrating the principles of

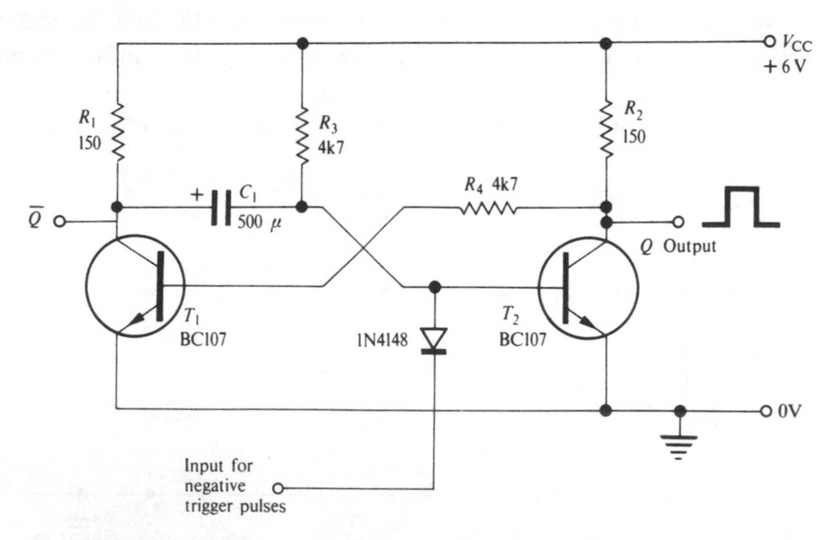

Fig. 12.12. A monostable multivibrator. Bulbs can be used for R_1 and R_2 (6 V, 0.04 A).

operation of the monostable, the IC versions are usually preferred for practical applications, generally offering greater versatility and stability and totally reliable triggering (see section 13.15).

12.7 The binary counter

A bistable multivibrator can be triggered from an astable or from another bistable by using the trigger circuit of fig. 12.13. This circuit is specially designed to cope with the low frequency of the astable of fig. 12.9. A bulb in the collector of T_2 indicates the state of the flip-flop. The slow pulses from the astable or from another bistable are fed into the trigger input and differentiated by C_1 and R_5, and C_2 and R_6. The *falling* edge of the input pulse thus gives a *negative* trigger pulse, which is steered to whichever transistor happens to be on and kicks it off by momentarily reverse-biasing the base–emitter junction.

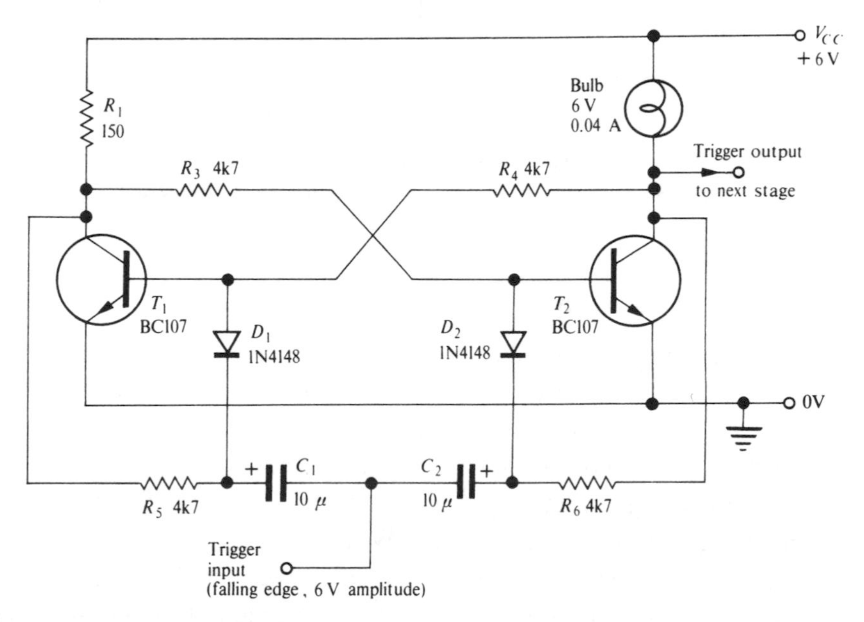

Fig. 12.13. Bistable with trigger input suitable for binary counting. Values of C_1 and C_2 may be experimentally adjusted for most reliable operation.

Because the triggered bistable only changes state on the falling edges of the trigger pulses, it switches at half the rate of the preceding stage. This is made clear in the waveform diagrams in fig. 12.14 which represent an astable triggering a bistable which in turn triggers a second bistable. This divide-by-two function is the basis of binary counting, and an examination of the outputs of the two bistables in fig. 12.14 shows that they are actually counting the number of complete cycles in the astable waveform: if 6 V is taken as equivalent to 1 and 0 V as equivalent to zero the sequence is 00, 01, 10, 11. Two flip-flops can, of course, only count up to 3, but four flip-flops can count up to 15. The use of ICs has revolutionized electronic counting, and a more detailed study is in the next chapter (section 13.12).

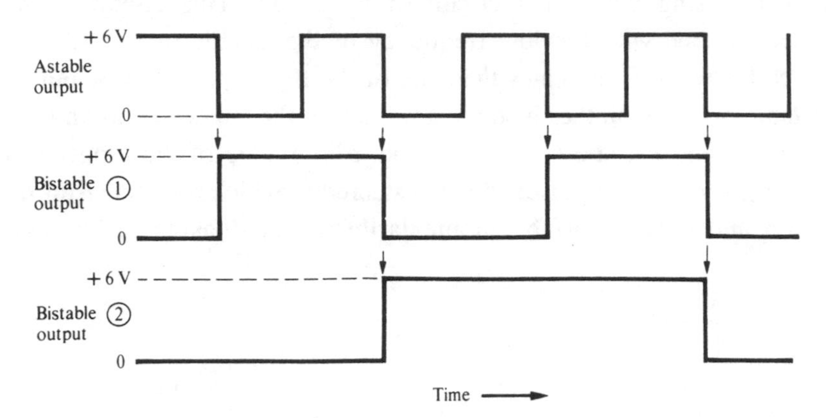

Fig. 12.14. Output waveforms from an astable triggering a bistable which in turn triggers a second bistable. Arrows show the instants of triggering.

12.8 The Schmitt trigger

The Schmitt trigger is a bistable circuit whose state is controlled by the voltage level at the input. It is used as a level-detecting circuit and also as a means of converting a sine wave into a square wave with fast rise and fall times. A typical discrete-component Schmitt trigger is shown in fig. 12.15.

With zero voltage on the input, T_1 is off and T_2 is therefore saturated, base current being supplied via R_1 and R_2. With negligible p.d. between collector and emitter of T_2, the 9 V supply is shared between R_3 in the emitter and R_4 in the collector. R_3 will drop 1.6 V and R_4 7.4 V, the p.d.s being in proportion to the resistor values, and totalling 9 V. T_1, therefore, has an emitter potential of 1.6 V, and T_2 collector is sitting at a similar level. As the input voltage is increased, T_1 begins to conduct when $v_{\text{in}} \approx 2$ V, causing a voltage drop in its collector load R_1 and robbing T_2 of its base current. As T_2 emitter current falls, so the drop across the common emitter load R_3 falls, turning on T_1 even faster. This regenerative positive feedback continues until T_1 is saturated and T_2 is off, the output then being at +9 V.

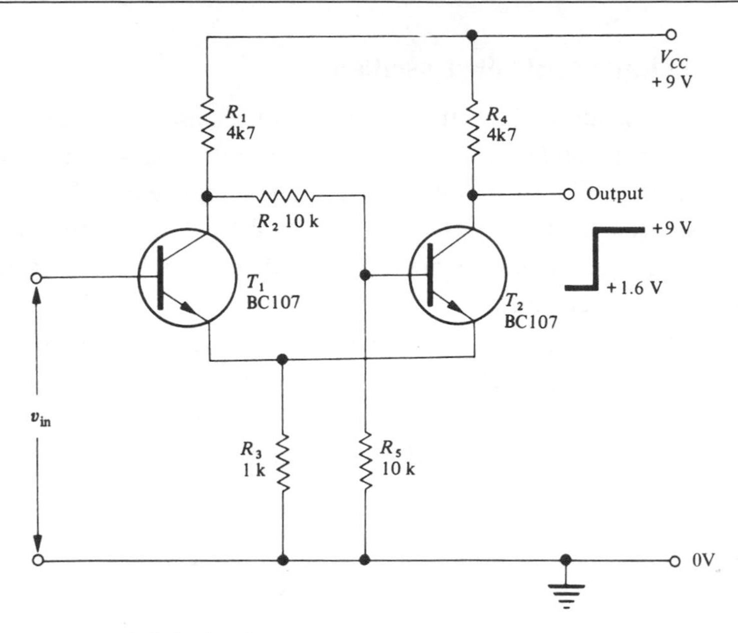

Fig. 12.15. Schmitt trigger.

If v_{in} is now reduced, T_1 collector current gradually decreases. There is, however, an additional factor this time: emitter follower action comes into play and the emitter voltage of T_1 falls along with the input voltage until T_1 collector voltage is high enough to begin to turn on T_2. When T_2 begins to draw emitter current, it initiates a rapid regenerative action: the extra voltage drop in R_3 helps to turn off T_1 by raising its emitter potential. As T_1 turns off fully its collector rises towards supply rail, turning T_2 fully on.

The emitter follower action, which only comes into play as v_{in} is reduced, causes hysteresis in the Schmitt trigger: its switch-off level is lower than its switch-on level. Hysteresis in the circuit shown is small, at approximately 0.6 V, but can be modified by changing the ratio of $R_1 : R_4$ and $R_2 : R_5$.

The base of T_1 does, of course, draw some current when the circuit is triggered. If this is undesirable, an *n*-channel FET such as a 2N3819 can be directly substituted for T_1, giving the circuit the high input impedance normally associated with a FET.

The Schmitt trigger is an ideal circuit for interfacing slowly-changing signals with logic circuits, which require fast rise and fall times. In such applications, the 7413 IC Schmitt trigger is convenient; its use is further discussed in section 13.14.

12.9 Voltage-controlled oscillator

There are many instances where the frequency of an oscillator needs to be remotely controlled by an internal voltage. Music synthesisers and frequency modulated (FM) transmitters are two examples. A circuit example of such a *voltage-controlled oscillator* (VCO) is shown in fig. 12.16 which shows an astable with the time constant resistors brought out to an external control line, $V_{control}$. Varying $V_{control}$ changes the 'aiming' voltage as C_1 and C_2 discharge through R_1 and R_2 and hence speeds up their discharge as $V_{control}$ is increased. It is worth working out a formula relating frequency and $V_{control}$. An LC oscillator can also be made into a VCO, by the use of varactor diodes in the tuned circuit, a technique widely used in RF oscillators but giving a more limited span of control.

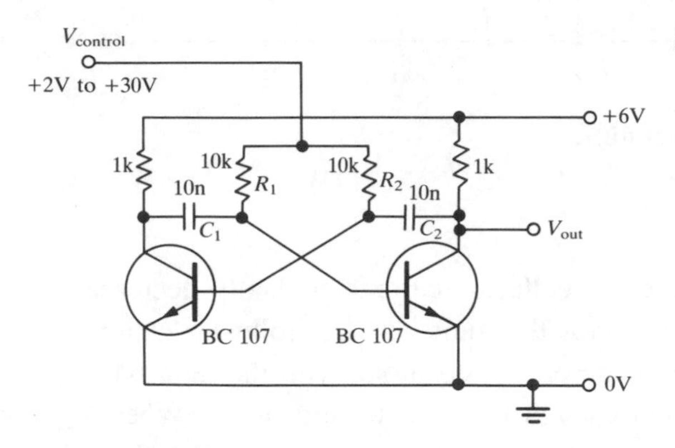

Fig. 12.16. Simple voltage-controlled oscillator using astable multivibrator.

12.10 Phase-locked loop

12.10.1 Basic outline

Sometimes it is a requirement to lock the output frequency of an oscillator to that of an incoming signal. We do it every time we lock an oscilloscope timebase to get a stable waveform on the screen. The phase-locked loop (PLL) automatically locks the frequency and phase of a VCO to an external signal. Applications include FM demodulation and the detection and regeneration of signals buried in noise. There is usually a PLL on the receiving end of a TV remote control or a fibre optic cable. A car radio uses one PLL for frequency-synthesised tuning and a second one for FM demodulation. The PLL is shown in outline in fig. 12.17. The phase comparator compares the frequency and

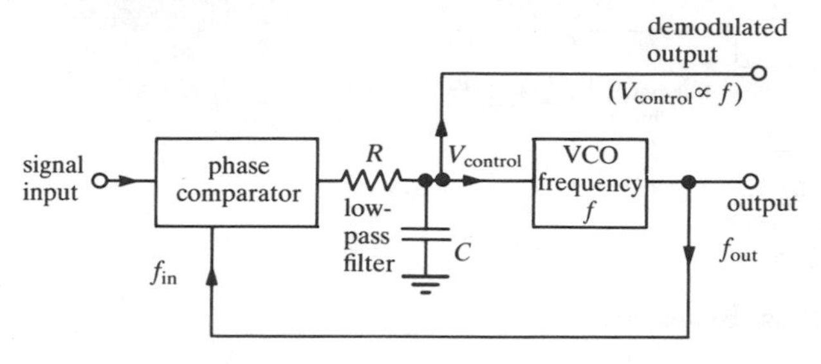

Fig. 12.17. Basic circuit of phase-locked loop (PLL).

phase of the oscillator signal (f_{out}) with that of the input (f_{in}) and generates a phase difference signal which is low-pass filtered. The filter output, $V_{control}$ then controls the oscillation frequency of the VCO until the frequency and phase difference becomes zero, at which point the PLL is *locked* to the input frequency.

If f_{out} is initially below f_{in}, the phase comparator produces a positive $V_{control}$ which raises f_{out} until it matches f_{in}, and vice versa if f_{in} starts below f_{out}. If the VCO is designed with a linear voltage/frequency relationship, then $V_{control}$ is directly proportion to f_{in} and can itself be used as the demodulated output of an FM input signal. Virtually all FM radios work this way.

12.10.2 Phase comparator

The phase comparator in the PLL is worth a closer look. It is basically a multiplier. In fig. 12.18 the multiplier is shown with the two inputs:

$$v_{in} = V\cos\omega t$$

$$v_{comp} = V\cos(\omega t + \phi)$$

Then the product at the output is

$$v_{out} = v_{in} \times v_{comp}$$

$$= V^2[\cos\omega t \times \cos(\omega t + \phi)]$$

and by trig expansion

$$v_{out} = \frac{V^2}{2}[\cos(2\omega t + \phi) + \cos\phi] \tag{12.9}$$

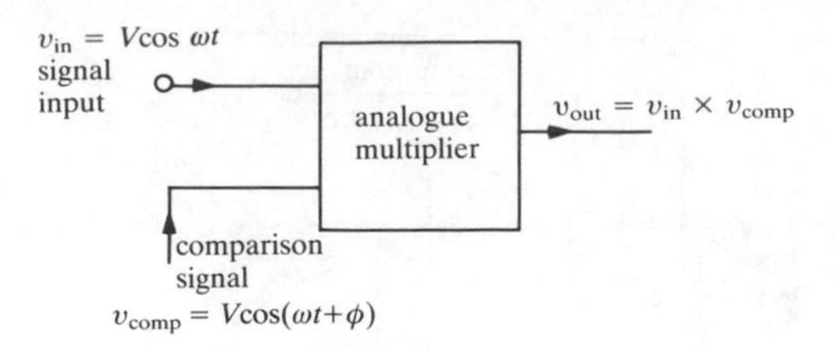

Fig. 12.18. Phase comparator.

The $\cos(2\omega t + \phi)$ term is a high frequency ripple which is removed in the low-pass filter, leaving the output of the phase comparator as:

$$v_{out} = \frac{V^2}{2}\cos\phi, \tag{12.10}$$

providing a suitable phase-dependent control for the VCO.

The PLL normally operates with phase shift $\phi = 90°$, i.e. $\cos\phi = 0$. At this point an increase in phase difference ϕ produces a negative correction voltage to the VCO and a decrease in ϕ a positive one. The PLL will remain locked to the input over the range $\phi = \pm 90°$. Once lock is lost, the oscillator starts to free run and the input signal must approach this free running frequency (it must get within the bandwidth of the low-pass filter) before lock is recaptured (acquired).

A long time constant in the low-pass filter gives a good noise rejection and a stable output frequency but reduces the acquisition range and increases acquisition time. In practice the multiplier in the PLL can be replaced by a chopper circuit which simply modulates the input signal with a square wave from the VCO. All the additional high frequency components arising from the square wave are disposed of in the low-pass filter.

12.10.3 Practical phase-locked loop

The NE565 chip shown in fig. 12.19 is a single chip PLL which contains phase comparator, VCO and provision for the low-pass filter using an external capacitor C_2. Capacitor C_1 sets the centre (free run) frequency of the VCO together with R_1. RF stability of the VCO is maintained by the 1 nF capacitor C_3.

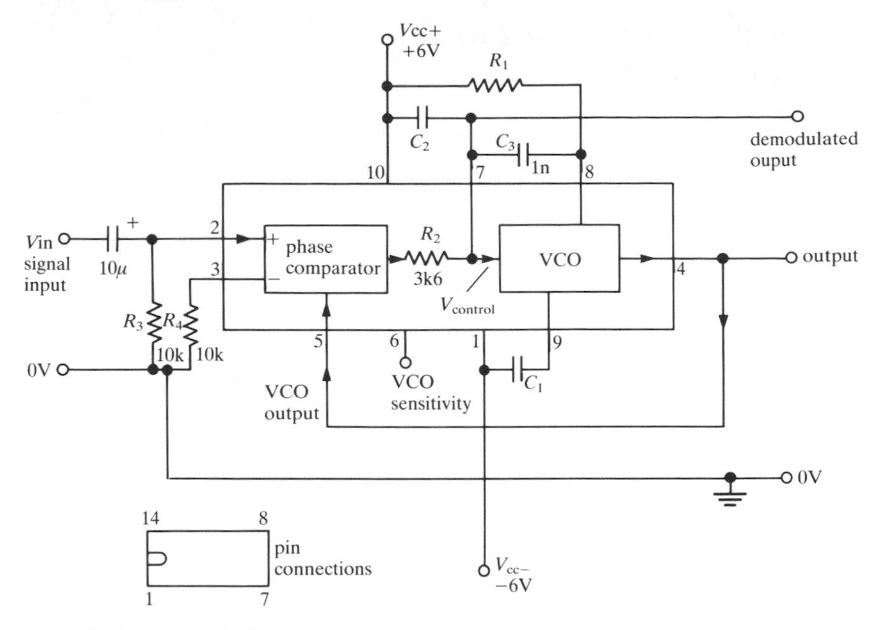

Fig. 12.19. Practical phase-locked loop using NE565 chip.

Centre frequency f_0 is given by the formula

$$f_0 = \frac{1.2}{4R_1C_1} \text{ kHz} \tag{12.11}$$

(R in KΩ; C in μF).

R_1 must be in the range 2–20 kΩ (optimum 4 kΩ). The maximum value of f_0 is 500 kHz.

The lock range f_L is very wide, being $\pm 8f_0/V_{CC}$ Hz where $V_{CC} = V_{CC+} + V_{CC-}$. With 6 V supplies shown, lock range is approximately ±60% of f_0. Connecting a resistor between pin 6 and the VCO input pin 7 reduces the sensitivity of the VCO and hence reduces lock range. At audiofrequencies a good starting value for filter capacitor C_2 is 100 nF. It is instructive to vary the internal time constant using C_2 and find out the effect on capture range f_C. In theory,

$$f_C \approx \pm \frac{1}{2\pi}\sqrt{\frac{2\pi f_L}{R_2C_2}} \tag{12.12}$$

where $R_2 = 3600\ \Omega$ (internal resistor)
C_2 is in farads.

If the input signal is frequency modulated, pin 7 (VCO control voltage) provides a demodulated output. Further low-pass filtering may be needed to remove the residual carrier.

13

Digital logic circuits

13.1 The digital world

At the heart of today's computer-based electronic systems lies the elementary transistor on/off switch. The simplicity and reliability of semiconductor switching have led circuit design into the digital world where the signals represent numbers, and circuit functions are expressed by logic and arithmetic. Even applications involving analogue inputs and outputs, such as audio recording, are rapidly becoming digital, by using analogue to digital and digital to analogue converters to convert signal voltages into numbers and back from numbers to voltages again, thereby achieving greater fidelity in the recording process.

We shall see as this chapter unfolds that the binary numbering system, working to a base of 2 instead of the base of 10 used in the familiar decimal system, is ideally suited to electronic implementation. The simple 'ons' and 'offs' of electronic switches represent the noughts and ones of binary numbers. Digital electronics is therefore free from many of the critical features of analogue circuits such as distortion and drift, the basic gate circuits being essentially simple. This simplicity has the advantage that several million such circuits can be packed onto one 'chip' in an integrated circuit microcomputer.

These final two chapters lead from the electronic fundamentals so far described into the world of microcomputers and the vital software which controls them. We begin by examining the way that logical operations can be carried out using ordinary switches and then see how simple transistor circuits can expand into arithmetical adders, memories, counters, timers and finally the computer itself.

13.2 Logic functions and gates

Logical operations are not at all complicated – we encounter them every day.

The three most basic logic functions are called *OR*, *AND* and *NOT*. Here are three examples of their application shown in fig. 13.1:

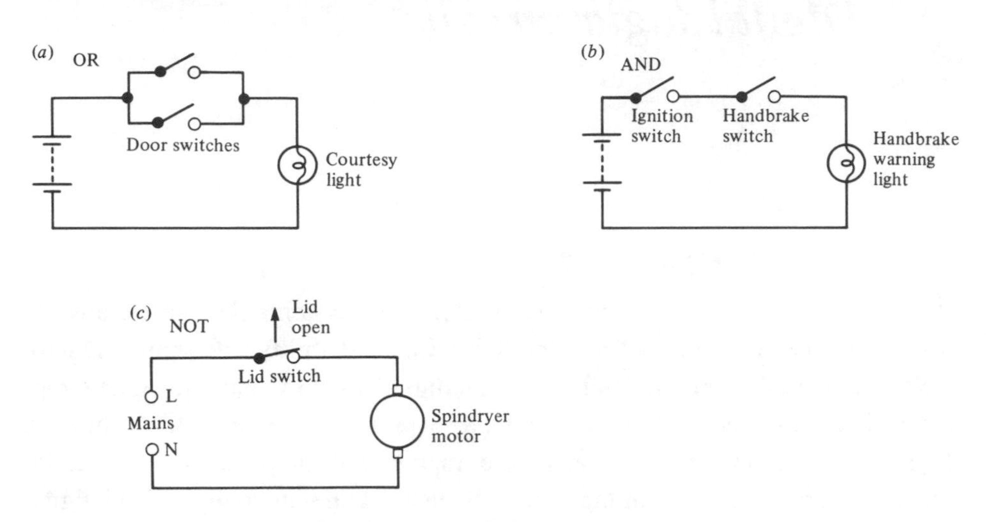

Fig. 13.1. Logic functions using switches.

(*a*) A car interior light illuminates when the nearside door is opened *OR* the offside door is opened – or if both are opened (fig. 13.1(*a*)).

(*b*) A car handbrake warning light illuminates when the handbrake is on *AND* the ignition switch is also on (fig. 13.1(*b*)).

(*c*) A spin dryer will only run when the lid is *NOT* open (fig. 13.1(*c*)).

We can readily interpret these common functions as logic *gates*, which accept electrical inputs and provide an output according to their logical function. Such logic gates normally include a simple amplifier so that they can be 'cascaded' freely in more complex networks with each gate driving one or more others. As we recognised at the very start of the book, a simple amplifier is the electromagnetic relay: our first logic gates are therefore designed using relays – indeed, for many years, this was the only type of digital circuitry used in industrial controllers and telephone exchanges.

The voltages applied to the gate inputs are arranged to be either ON (high) or OFF (low). High corresponds to logic 1 (sometimes called TRUE) and low to logic 0 (FALSE). In fig. 13.2, logic 1 is represented by +12 V and logic 0 is 0 V. In the OR gate (fig. 13.2(*a*)) +12 V applied to *A OR B* energizes the

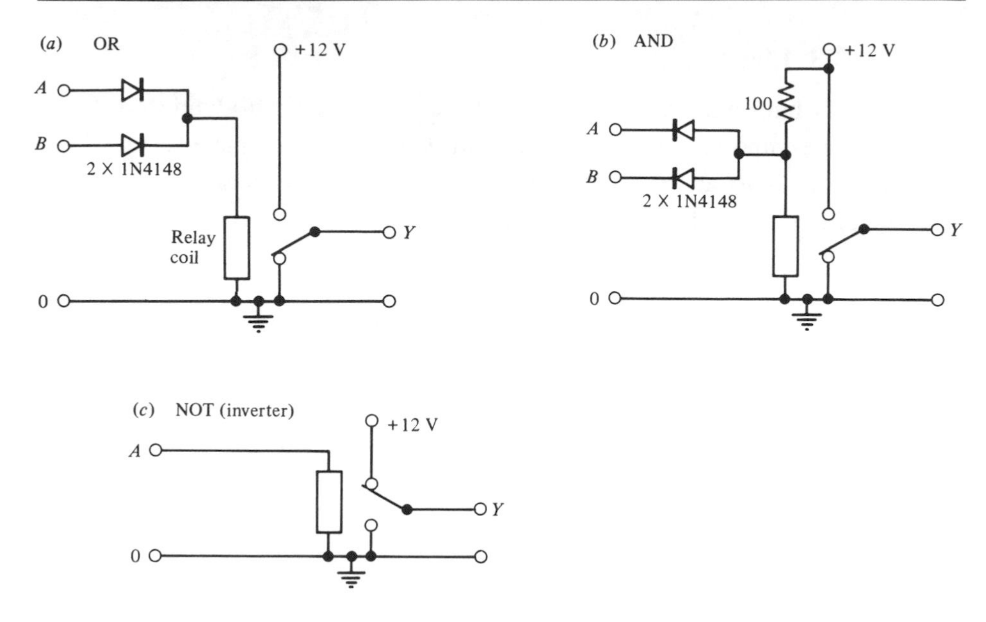

Fig. 13.2. Basic gates (relays shown in 'power off' state).

relay and raises output Y to +12 V. Fig. 13.2(*b*) shows an AND gate where the relay energizes immediately on power-up raising output Y to +12 V as long as both *A AND B* are at +12 V. If either of the inputs is earthed (0 V), its diode conducts, short-circuiting the relay coil, and output Y drops to 0 V. This example emphasizes the important point that logic 0 on an input must actually be connected to 0 V and not just left floating or the AND gate will not work; this general principle must be observed with electronic gates too.

The NOT function or *inverter* completes our trio of basic gates (fig. 13.2(*c*)). The normally closed contact is connected to +12 V until the relay is energized by raising A to +12 V (logic 1), making output Y fall to logic 0. Thus Y is high when A is NOT high.

It is with these three logical functions OR, AND and NOT that all digital systems, computers and microprocessors are built. Fig. 13.3 shows their commonly used symbols.

Fig. 13.3. Symbols for basic gates.

13.3 Electronic gates

Whilst gates built with relays are a useful experimental introduction to logic, and still have important uses in some control systems, they have serious limitations of size, cost and speed when it comes to more elaborate applications. As we might expect, the transistor provides the answer, especially as millions

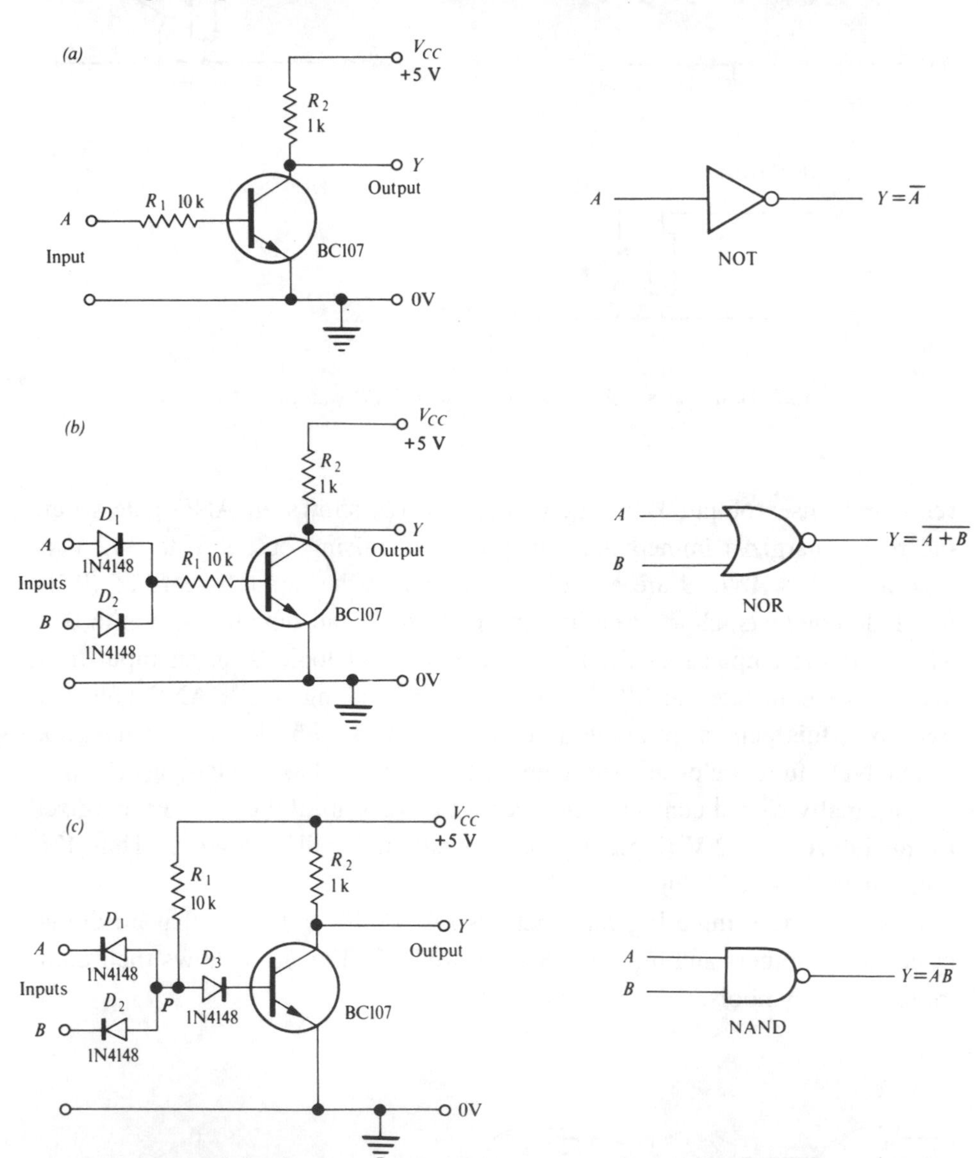

Fig. 13.4. Three discrete gates: (*a*) NOT (inverter), (*b*) NOR (OR-NOT), (*c*) NAND (AND-NOT).

of them can be included in one IC chip. Simple electronic gates automatically include the NOT function because of the signal polarity inversion inherent in the common-emitter switch. Fig. 13.4 shows basic circuits for NOT, NOR (OR–NOT) and NAND (AND–NOT) together with their symbols.

The gates are easy to construct. When testing them, it is useful to have some means of quickly checking the output to see whether a logic 0 or logic 1 is present. An oscilloscope or voltmeter can be used, but the clearest indication is given by the simple lamp-driver circuit of fig. 13.5. The bulb or LED lights brightly for inputs greater than 2 V, so that a logic 1 in virtually all systems will operate the indicator. On the other hand, inputs of less than 1 V, corresponding to logic 0, will not light the bulb at all. In these circuits, logic 1 is equivalent to +5 V (supply rail) and logic 0 is equivalent to 0 V. This convention, with logic 1 high and logic 0 low, is the normal one. The opposite convention, negative logic, is occasionally used, with logic 1 low and logic 0 high, but is rarely found in practice. The +5 V rail is used in most logic systems; in these simple examples a 4.5 V–6 V battery is suitable, but IC logic supplies must usually be within 5% of the nominal +5 V.

In the circuit of fig. 13.4(*a*), a logic 1 (+5 V) input will saturate the transistor and take the output voltage virtually to zero (logic 0). Thus, if there is logic 1 on the input, there is NOT logic 1 at the output and the circuit operates as an inverter. Its output may be expressed in the shorthand of *Boolean algebra*, the mathematics of logic, as

$$Y = \overline{A}$$

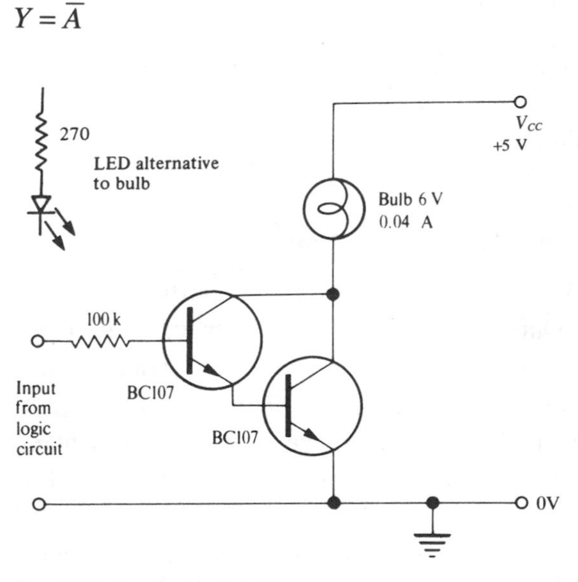

Fig. 13.5. Logic-level indicator.

where A is the input logical state. The bar indicates the complement (negative) of the variable.

In the NOR circuit of fig. 13.4(*b*) the two diodes D_1 and D_2 provide two inputs, each with independent control over the transistor. Even if one input is earthed (logic 0) the other input will turn on the transistor if it is connected to +5 V (logic 1). Thus, if A OR B is connected to logic 1 there is NOT logic 1 on the output. This OR–NOT function is abbreviated to NOR and written in the form

$$Y = \overline{A + B}$$

where the addition sign means OR and the bar means NOT as in fig. 13.4(*a*).

In fig 13.4(*c*), if the inputs are left floating or connected to logic 1, the transistor is saturated by the base current flowing in R_1 and the output is at logic 0. If, however, either of the inputs is taken to logic 0 (0 V), the base current is bypassed via D_1 or D_2. Notice that the presence of D_3 in the base circuit means that point P must be at about +1.2 V for the transistor to conduct. If either input is earthed, P is held down to +0.6 V by D_1 or D_2 and the transistor is turned off. To summarise, if A AND B are at logic 1, the output will NOT be at logic 1. Any other combination of input logic will give a logic 1 at the output. Thus we have the AND–NOT function, abbreviated to NAND and written

$$Y = \overline{A \, . \, B} \quad \text{or} \quad Y = \overline{AB}.$$

The common product function means AND in Boolean algebra. Notice that either the NOR or NAND gate may be used as a NOT merely by linking the inputs together and using them as a single input.

13.4 Input and output conditions

The circuits of fig. 13.4 once again emphasize the importance of never leaving a logic input floating. Quite apart from the problem of stray signal pickup which might be present, the apparent logic level of a floating input might not be the expected one. For instance, in figs. 13.4(*a*) and (*b*), an input must be physically connected to the +5 V rail to take it to logic 1, whereas in fig. 13.4(*c*) an input considers itself to be at logic 1 unless actually connected to earth.

Any circuit feeding (*a*) or (*b*) must be able to act as a current source to supply the base current, whilst circuit (*c*) requires to feed *out* current from an input held at logic 0, therefore any gate feeding (*c*) must be able to 'sink' this current. These current demands made on a gate output restrict the number of inputs which may be connected to any one output without significant changes occurring in its voltage level. This number is known as the *fan-out* of the output and is usually at least 10. The term *fan-in* may also be met: this refers to the number of inputs available on a gate. Circuits (*b*) and (*c*) have a fan-in of 2, but this may be increased simply by adding more diodes: the gates are said to be *expandable*.

13.5 Circuit classification

Extensive use is made of abbreviation in the logic field. Initials are commonly used to describe different types of logic circuits. We have already encompassed two types in the discrete circuits of fig. 13.4. Fig. 13.4(*a*) uses only resistors and a transistor, so it is described as resistor–transistor logic (RTL) whilst fig. 13.4(*b*) and (*c*) use diodes on the input to supplement the transistor and are termed diode–transistor logic (DTL). Both RTL and DTL are today obsolete, development having progressed via transistor–transistor logic (TTL), which will be discussed shortly. Most logic ICs today use FETs (CMOS), where the high input impedance and low on-resistance offer advantages.

13.6 Truth tables

The Boolean algebra representation is one convenient way of describing the function of a logic gate. Another method, which is invaluable when designing logic circuits, is the truth table or function table. This simply describes all possible combinations of input and output states for a gate or system. Truth tables for the NOT, NOR and NAND gates are drawn up in fig. 13.6.

NOT

Input	Output
A	*Y*
0	1
1	0

NOR

Inputs		Output
A	*B*	*Y*
0	0	1
1	0	0
0	1	0
1	1	0

NAND

Inputs		Output
A	*B*	*Y*
0	0	1
1	0	1
0	1	1
1	1	0

Fig. 13.6. Truth tables for basic logic gates. Logic 1 and logic 0 are often written H (high) and L (low), respectively.

13.7 Simple combinations of gates

A NAND gate and a NOT gate can be connected together to form an AND gate, shown in fig. 13.7 together with its truth table. The cancellation of the negation is shown on the symbol by the removal of the circle on the output. Similarly, a NOT on the output of a NOR gate gives an OR function. The result of unfamiliar combinations of gates may be found by working through truth tables. For instance, consider the combination of fig. 13.8(*a*), where a NAND gate has a NOT on each input. Is the result AND or something different? The truth table of fig. 13.8(*b*) shows the result. This is not an AND function, but is an OR, being the inverse of the NOR table in fig. 13.6. This 'experiment' is an illustration of De Morgan's theorem which states: *To obtain the complement of a Boolean function, complement each variable and exchange AND for OR*. In shorthand form,

$$\overline{A+B}=\overline{A}\,.\,\overline{B}$$

It is instructive to build the circuit of fig. 13.8(*a*) using the discrete-component gates of fig. 13.4; the result can be checked against the truth tables by using the logic-level tester (fig. 13.5).

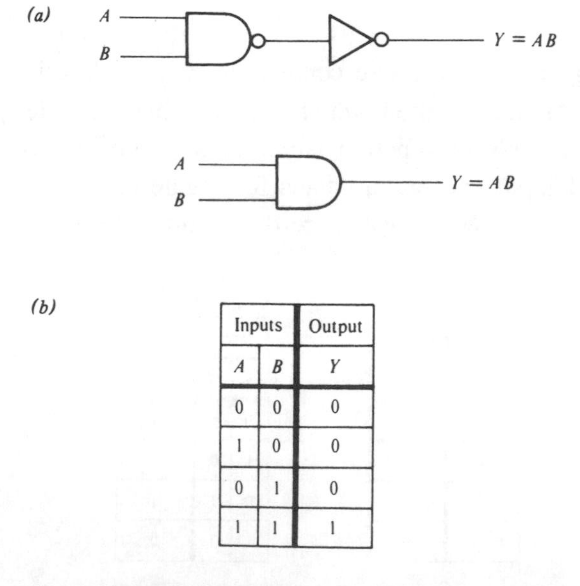

Inputs		Output
A	B	Y
0	0	0
1	0	0
0	1	0
1	1	1

Fig. 13.7. Following NAND with NOT gives an AND gate: (*a*) circuit with equivalent AND symbol, (*b*) AND truth table.

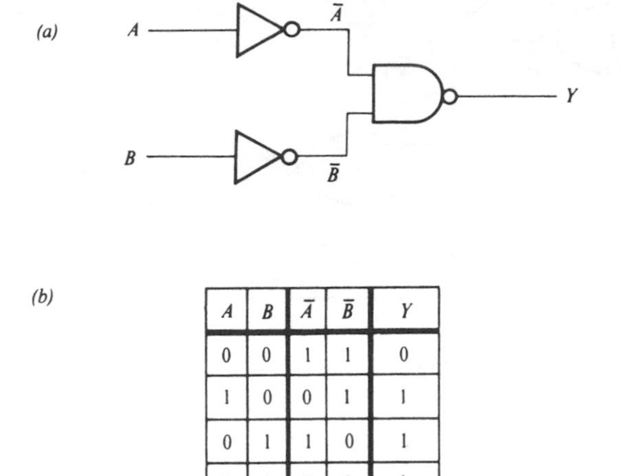

A	B	$\bar{A}$	$\bar{B}$	Y
0	0	1	1	0
1	0	0	1	1
0	1	1	0	1
1	1	0	0	1

Fig. 13.8. (*a*) NAND gate with negated inputs, (*b*) corresponding truth table.

13.8 Addition of binary numbers

The addition of numbers lies at the heart of the arithmetic unit of a computer or calculator. There is nothing complicated about the construction of binary adders; in fact, one can be built up from the discrete gates of fig. 13.4.

Fig. 13.9 shows a *half-adder*, which takes two binary digits, usually called *bits*, as its inputs and produces the bit sum and any necessary carry digit. For example, using normal binary arithmetic notation,

$$1+0=1$$

giving a bit sum of 1 and zero carry, whilst

$$1+1=10$$

giving a bit sum of zero and a carry of 1. Note that the base of 2 makes 1 the maximum allowed digit.

If the 'carry' output is not used, the half-adder is known as an *exclusive OR*, *not equivalent* or *anti-coincidence* gate. These names arise from the fact that the output is zero whenever both inputs are at the same logic level, but goes to 1 whenever they differ. The normal OR gate considered previously is called an *inclusive* OR, where a logic 1 output is obtained if one *or both* inputs go to 1. Fig. 13.10 shows the symbols which distinguish the exclusive OR (EXOR) and inclusive OR, together with their truth tables.

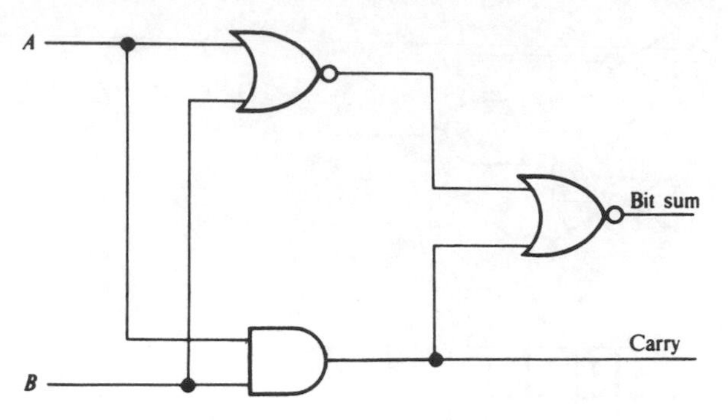

Fig. 13.9. Half-adder (exclusive OR).

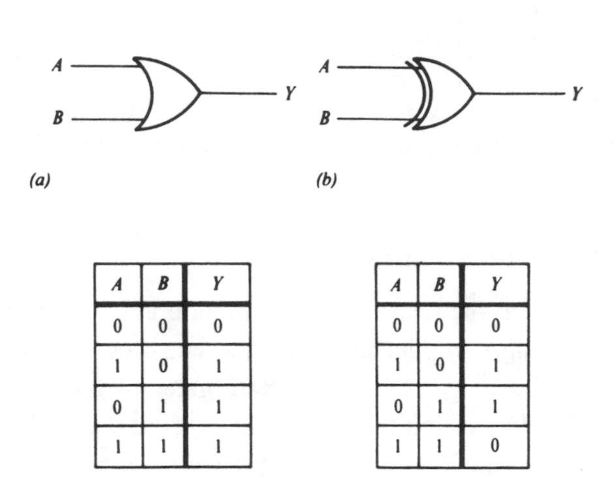

A	B	Y
0	0	0
1	0	1
0	1	1
1	1	1

A	B	Y
0	0	0
1	0	1
0	1	1
1	1	0

Fig. 13.10. Symbols and truth tables for (*a*) inclusive OR gate, (*b*) exclusive OR gate.

For complete binary addition, the *full adder* is required: this accepts two binary digits plus a carry bit and provides an output of the bit sum and a further carry bit. One way of constructing a full adder is from two half-adders and an OR gate, as in fig. 13.11. Fig. 13.12 shows the truth table for a full adder; this may be checked by starting from the truth tables of the individual gates.

The addition of two multi-bit binary numbers can be accomplished with a parallel chain of adders, as shown in fig. 13.13, where one 4-bit number $A_4A_3A_2A_1$ is added to another number $B_4B_3B_2B_1$, A_1 and B_1 being the least significant (2^0) bit in each case. Parallel addition simply means that all the digits are presented together rather than in a pulse train. The latter is known as serial addition and is much slower than parallel addition. In fig. 13.13 all

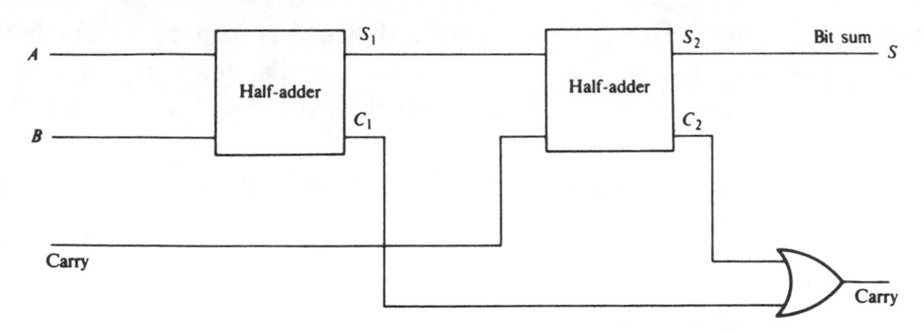

Fig. 13.11. A full adder.

Inputs			Outputs	
A	B	Carry	Carry	S
0	0	0	0	0
1	0	0	0	1
0	1	0	0	1
1	1	0	1	0
0	0	1	0	1
1	0	1	1	0
0	1	1	1	0
1	1	1	1	1

Fig. 13.12. Truth table for full adder.

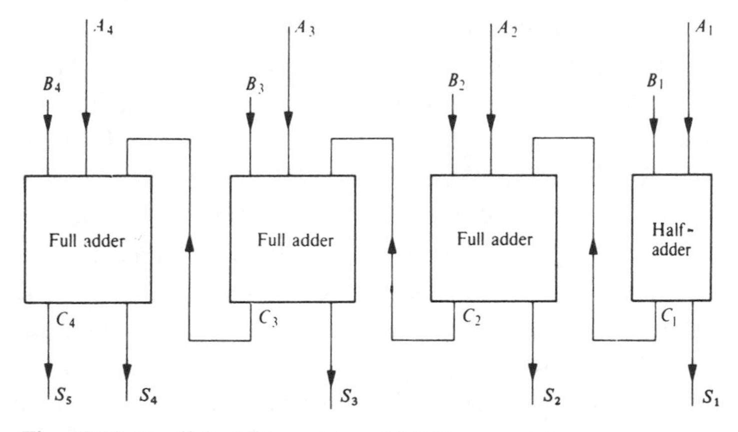

Fig. 13.13. Parallel addition of two 4-bit binary numbers.

adders except the least significant must be full adders in order to handle the carry digit from the previous stage. Notice too that the carry digit from the most significant adder is brought out as the most significant bit (MSB) of the answer (S_5).

Binary arithmetic is discussed more fully in the next chapter as a prelude to microcomputers.

13.9 Integrated circuit logic

13.9.1 Introduction

It is quite feasible to construct a single full adder from the discrete component gates of fig. 13.4, and the reader is encouraged to do this. However, the prospect of, say, a discrete-component 4-bit adder becomes somewhat daunting. We can, though, retain a sense of perspective by recalling that the first electronic digital computers, such as 'Colossus', built in Britain during the Second World War, relied on thermionic valves for their logic gates. Apart from the extreme bulk of such machines, the heat dissipation problem from thousands of valves was enormous. Fortunately, ICs provide us with inexpensive compact logic gates which have the additional bonus of being extremely fast in operation.

13.9.2 TTL

The first IC logic to be widely used began with a development of the obsolete DTL gate of fig. 13.4(*c*). The diodes at the input were fabricated in the form of a special multiple-emitter transistor, giving the circuit its name of transistor–transistor logic (TTL) which is still in use in various forms today. Fig. 13.14 shows the circuit diagram of one NAND gate from the 7400 IC. Although the output circuit is slightly more elaborate than in fig. 13.4(*c*), it can be seen how T_1 is substituted instead of diodes D_1, D_2, D_3. This use of a transistor is not just a manufacturing convenience, it also dramatically improves performance. When either input is held at logic 0 (earth), transistor action occurs in T_1, so that it momentarily transmits a heavy collector current down to earth, rapidly discharging the current carriers stored in the base of T_2. Thus T_2 is switched off quickly, giving rapid operations of the gate in about 10 ns.

At this stage of our discussion, mention of operating speed of transistors may seem at first a little irrelevant since their switching time seems infinitesimal compared with our earlier relay examples. However, today's processing

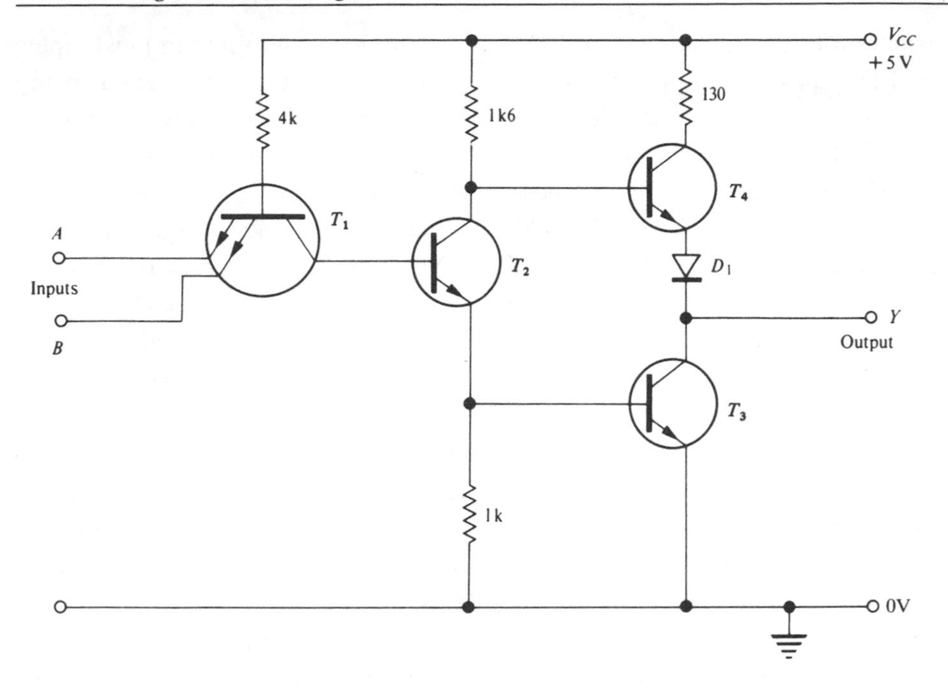

Fig. 13.14. TTL NAND gate.

systems have very quickly reached a level of complexity where the number of gate switchings required to compute an answer or sort out data is measured in millions or even millions of millions, so that operating speed (gate *propagation time*) has become a vital criterion for the designer.

The 5 V supply rail is common to all TTL circuits. For correct operation, its value is critical (between 4.75 V and 5.25 V) and it must never rise above 7 V under any circumstances, otherwise certain reverse-biased junctions run into avalanche breakdown, pass excess current and destroy the chip.

The output stage of the TTL gate uses two transistors, T_3 and T_4, in a 'totem pole' arrangement. T_4 provides *active pull-up* when the output goes to logic 1 and thus greatly decreases switching time by rapidly discharging any capacitance on the output.

Each 'standard' TTL input draws a current of 40 μA when held in the logic 1 state and feeds out 1.6 mA in the logic 0 state. Every 'standard' gate output circuit is capable of supplying at least 400 μA and sinking at least 16 mA. It therefore has a fan-out of 10, being able to feed ten 'standard' inputs. Special higher power *buffer* gates are also available. For instance, the 7437 is identical to the 7400, with four 2-input NAND gates, except that its fan-out is 30.

The 40 μA/(−1.6) mA current requirements of a 'standard' TTL input are usually referred to as one input load. Whilst these ratings apply to most inputs on TTL gates, certain inputs on flip-flops and counters may represent as many as four input loads. Care should be taken in such cases to check with the manufacturer's data that the previous gate is not exceeding its fan-out.

So-called 'standard' TTL is nowadays obsolete, having been replaced by low-power Schottky TTL (74LS00 etc.) which only takes a quarter of the current requirements of 'standard' TTL. Normally, the lower currents would involve a reduction in operating speed because internal capacitances then charge and discharge more slowly, but the device fabrication reduces those effective internal capacitances by incorporating Schottky diodes with their low forward p.d. to prevent saturation of the transistors and maintain a quick 10 ns gate switching time. Similar techniques used with 'standard' device currents (74S00 etc.) achieve extra speed, typically 3 ns per gate, whilst advanced oxide-isolated manufacturing methods give the best of both worlds by combining this 3 ns speed with a low-power requirement in advanced low-power Schottky devices (74 ALS00 etc. and 74F00 etc.).

In most commercial applications today, TTL and LSTTL have themselves been superseded by the CMOS chips described below. For our experiments here however, they do retain advantages in some applications. TTL chips handle enough output current to operate LEDs, and in some cases relays, directly. Both TTL and LSTTL have a useful variety of medium scale integration (MSI) counter and latch chips in the range, which are ideal for experimenting.

Pin-out information on a number of the most popular TTL ICs is listed in Appendix 4.

13.9.3 CMOS

In our study of transistors we recognized the benefits of the high input impedance of the FET. This is applied to advantage in the high speed complementary MOSFET CMOS family of gates, the 74HC series. The quiescent current drawn by a CMOS gate is typically less than 1 μA, compared with 400 μA for LSTTL, and the high input impedance virtually eliminates loading problems, leading to infinite fan-out capability at low speed. At normal fast operating speeds above 10 MHz, however, input capacitance must be considered; supply current also increases due to the rapid charging and discharging of capacitances and becomes comparable with LSTTL towards the 40 MHz maximum operating frequency.

Fig. 13.15 shows the basic CMOS inverter with its simple complementary

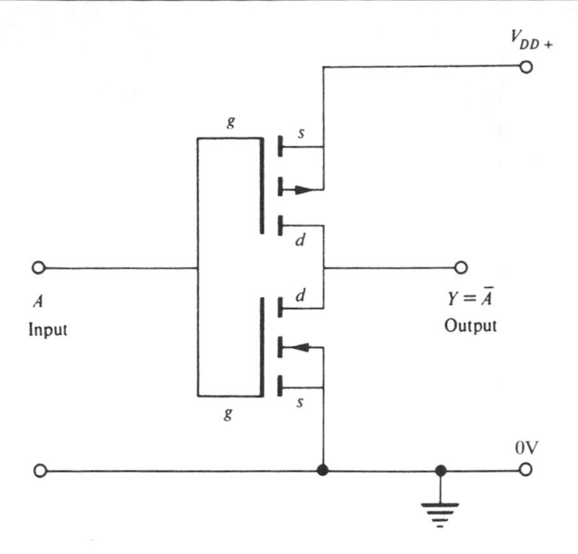

Fig. 13.15. Basic CMOS inverter.

'totem pole' form using enhancement-mode MOSFETs. Under quiescent conditions, whether the input is high or low, one MOSFET is switched off, so that the quiescent current is extremely low.

Fig. 13.16 shows how the CMOS design is extended to provide a 2-input NAND gate (e.g. 74C00) and 2-input NOR gate (74C02).

The original introduction of CMOS into digital logic was in the 4000 series of chips which exhibited long gate propagation delays of the order of 100 ns per gate, compared with 10 ns of TTL and LSTTL. Whilst this might seem a disadvantage, it was fast enough for many logic applications and had the advantage of good noise immunity because high frequency interference was ignored. For this reason, many 4000 series CMOS logic circuits are still to be found in control applications.

Today, however, 74HC series CMOS is an industry standard, using silicon gate devices of 2–3 μm dimensions to give the 10 ns gate propagation time of TTL and operating readily up to 40 MHz. For even higher speed, the Advanced CMOS 74AC series has a 5 ns gate propagation time and works up to 120 MHz clock frequency. The 74LV series is available for 3.3 volt battery operation, coping with typical supply voltage variations from 3.6 V to 2.7 V. There is now a general trend towards 3.3 V logic operation because it allows tiny device geometry, as small as 0.5 μm, without risk of internal voltage breakdown.

As with all MOSFETs and ICs of small internal geometry, CMOS chips

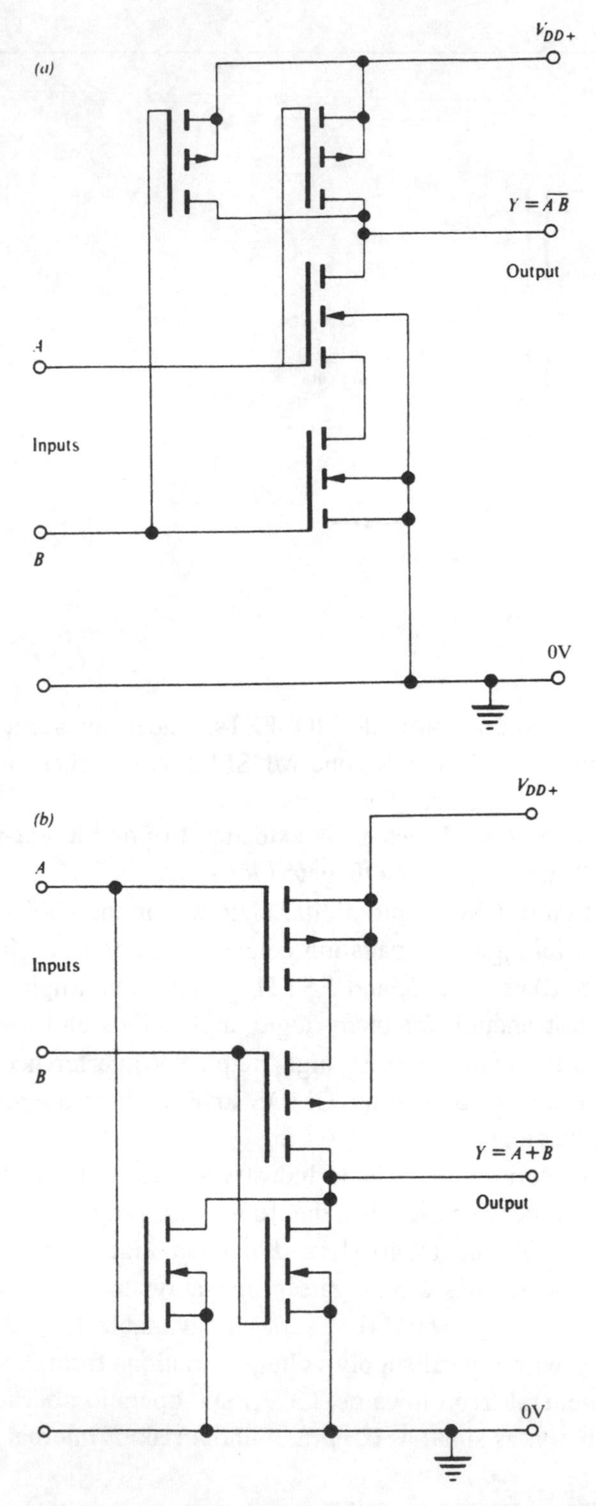

Fig. 13.16. (*a*) Basic CMOS NAND gate. (*b*) Basic CMOS NOR gate.

are easily damaged by electrostatic discharge (ESD) during handling. Protection diodes are normally included to reduce this hazard, but IC leads should preferably be kept shorted with wire or conductive plastic until wiring is finished. Metal benches, soldering iron – and even the operator – should be securely earthed. The protection diodes themselves are susceptible to damage if the input voltage to a gate is permitted to rise higher than V_{DD}. Care is therefore needed when disconnecting the power supply to CMOS circuits: any inputs must be disconnected first.

13.10 Sequential logic – flip-flops and memories

All the logic circuits discussed so far have simply responded to a particular static combination of inputs to provide the required output (e.g. two binary numbers at the adder inputs provide a sum output). As we examine logic applications, we quickly recognize that these *combinational* or *combinatorial logic* circuits are very limited when used alone. We soon find a need for some kind of memory so that we can implement a sequence of operations. This brings us to the realm of *sequential logic*. For example, once we have added together two binary numbers, we may wish to add a third number to the sum. This requires a memory to hold the sum and then return it to one adder input for the third number to be added. The alternative of more and more adders as we add in more and more numbers would be a very unmanageable system indeed! Already the principles of a computing system are emerging: memories can hold not only data but also the set of instructions for the whole sequence or *program*.

13.10.1 Introduction to flip-flops

One of the most common memory elements is the bistable multivibrator or flip-flop discussed in its discrete-component form in section 12.4 and repeated here in fig. 13.17 together with its equivalent in logic gates. This rudimentary pair of cross-coupled NOT gates has the limitation that it can only be made to change state by crudely short-circuiting to earth whichever output happens to be at logic 1.

Control is obtained if the NOT gates are replaced by 2-input NOR gates, providing the set and reset inputs of the *RS* flip-flop (fig. 13.18). If S and R are both held at logic 0, the NOR gates simply function as inverters like the simple circuit of fig. 13.17 and the flip-flop holds its state indefinitely. If now S goes high whilst R remains low, output $\overline{Q}$ is forced low because its NOR

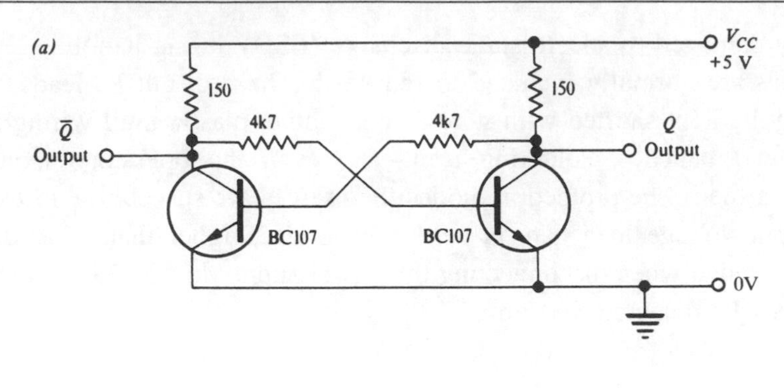

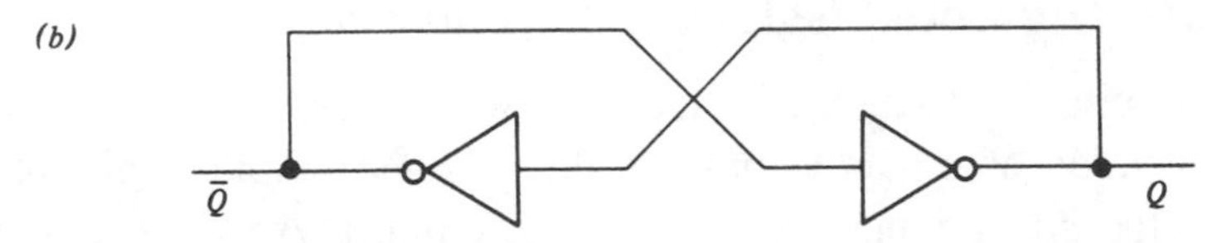

Fig. 13.17. Flip-flop (bistable multivibrator): (*a*) discrete component circuit diagram, (*b*) equivalent in logic gates.

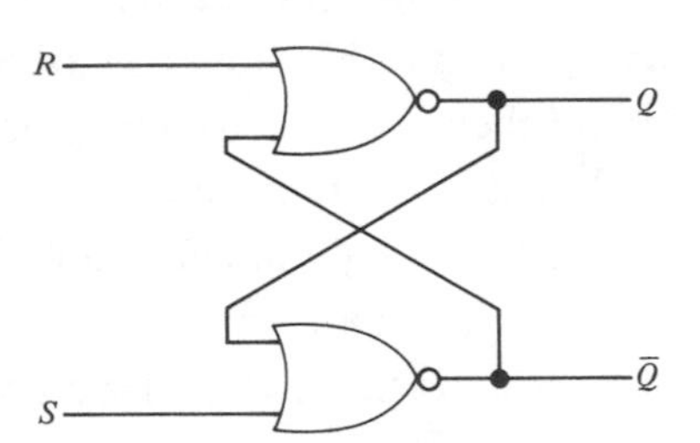

Fig. 13.18. Set-reset (*RS*) flip-flop.

gate now has an input at logic 1. Meanwhile, output Q is forced high because both its NOR gate inputs are low. If both S and R once more return to logic 0, Q remains at 1 and $\overline{Q}$ at 0, thus remembering indefinitely the fact that input S rose to logic 1. Similarly, if R goes momentarily high whilst S remains low, output $\overline{Q}$ is forced up to 1 and Q down to 0; once again this state will be held indefinitely as long as S and R remain at logic 0. Because its outputs are directly forced or 'jammed' into the required condition, the *RS* bistable is sometimes referred to as *jam-loaded.*

If both R and S rise to 1 together, the NOR gates force both outputs down to 0. This condition cannot be memorized once R and S return to 0 because the cross-coupling then requires the outputs to be different. In practice, the stored condition will depend on whether R or S reached 0 first. Because of its random dependence on timing, the condition $R = S = 1$ results in *indeterminate* stored data and should never be followed by $R = S = 0$. In most applications, $R = S = 1$ is regarded as a prohibited condition; the actual output will depend on the gate arrangement used to implement the flip-flop.

The *RS* flip-flop just described is 'active high' – the set and reset inputs must be taken to logic 1 to have their specified effect. 'Active low' circuits are also common: fig. 13.19 shows one made from two NAND gates. The two inputs are now marked $\overline{S}$ and $\overline{R}$ because they have their defined effect when taken to logic 0 instead of logic 1 (e.g. Q goes high when S is low, because its NAND gate inputs are no longer at 1). The flip-flop is shown in a useful and very common practical application as a switch 'debouncer'. In experimental logic circuits, input pulses are often derived from closing a switch or simply touching a wire on a relevant contact. If such a contact closure is used directly as an input to a sequential circuit, unpredictable results may be obtained because contact 'bounce' means that the circuit will close and open several times before the switch settles (bouncing time is usually of the order of a few milliseconds). The debounce circuit works by using the very first contact closure to trigger the *RS* bistable which then remains in stable condition despite subsequent contact bounces.

In logic diagrams, flip-flops are conveniently shown as rectangles with the various inputs and outputs marked. Fig. 13.20 shows the symbol and truth table for the *RS* flip-flop (active high). Strictly speaking, the name truth table is reserved for combinational logic, and what we show here is the *transition table* as it shows the output immediately after any transition caused by the inputs.

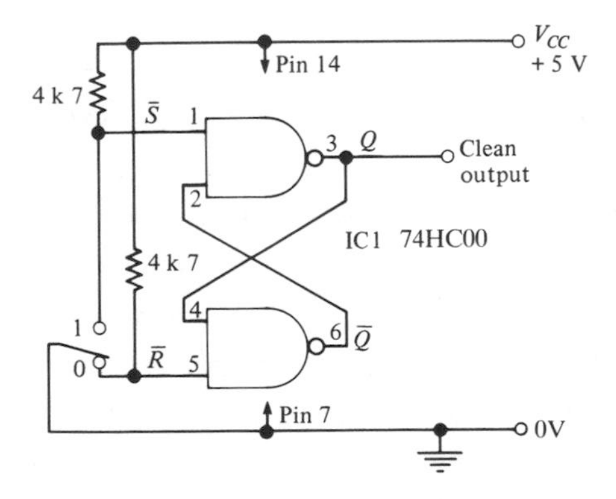

Fig. 13.19. 'Active low' *RS* flip-flop applied as a switch debouncer.

Combinations of flip-flops and gates can be devised to implement all sorts of complicated sequences of output states. Such circuits are known as *state machines* and are useful in simple industrial controllers.

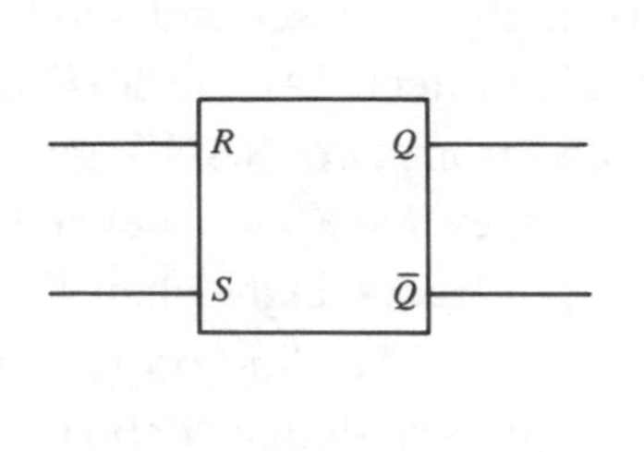

Inputs		Outputs	
R	S	Q	$\overline{Q}$
0	0	Q stored	$\overline{Q}$ stored
1	0	0	1
0	1	1	0
1	1	0 (indeterminate if stored)	0

Fig. 13.20. *RS* flip-flop symbol and truth (transition) table.

13.10.2 Clocked RS flip-flop

An essential feature of most sequential logic systems is that operations should occur in a timed regular order. This is normally achieved by a regular train of clock pulses which controls the sequence of events rather as an orchestra conductor maintains the beat with his baton, ensuring that all the musicians keep time together as they play the music.

Fig. 13.21 shows an *RS* flip-flop arranged so that it can change state only when the clock pulse input (CP) receives a logic 1 pulse. Whilst the CP input is at 0, each AND gate has an input at 0; thus, both S' and R' on the basic flip-flop are held at 0 and the Q and $\overline{Q}$ outputs therefore cannot change. As soon as CP goes to logic 1, however, the AND gates each have a high input, so that the logic levels at the external S and R inputs are transmitted through to the flip-flop, which will now therefore set its state according to the *RS* truth table in fig. 13.20. Thus, the clocked *RS* flip-flop cannot respond to the S and R inputs until the clock pulse arrives. However, notice that any change in the S and R inputs whilst the clock input is high will directly change the Q and $\overline{Q}$ outputs immediately. This flip-flop is therefore called a *transparent latch* because the outputs can effectively 'see' the inputs for the whole of the duration of the logic 1 clock pulse. A 'snapshot' of the S and R inputs at their current states is effectively stored or latched at Q and $\overline{Q}$ at the instant that the clock returns to logic 0.

The additional *preset* and *clear* inputs, normally held at logic 0, give

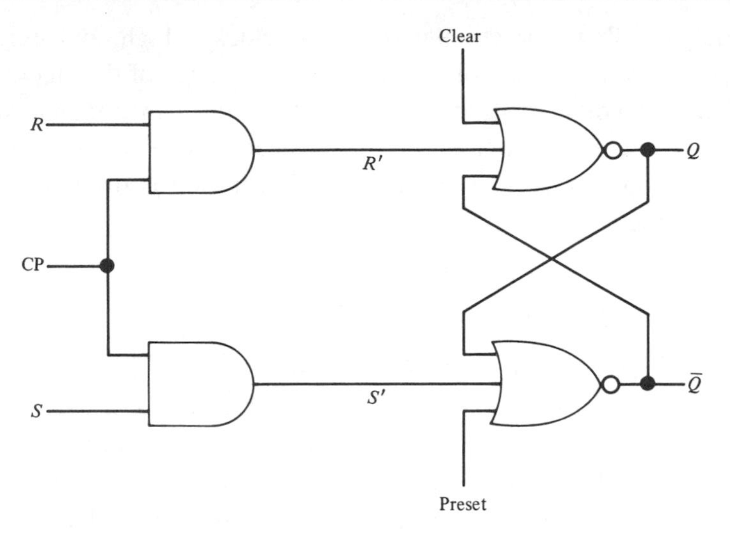

Fig. 13.21. Clocked *RS* flip-flop.

immediate access to the internal set and reset functions via the extra inputs on the NOR gates. A useful feature on any clocked flip-flop, they enable the initial state to be set up independently of the clock by momentarily raising the relevant input to logic 1.

13.10.3 D-type flip-flop

The *D*, or data, flip-flop is a clocked *RS* flip-flop operated from just one input. This has the advantage that *S* and *R* cannot simultaneously be set to 1 and give an indeterminate stored output. It is wired as in fig. 13.22, where the rectangle represents a clocked *RS* flip-flop. The output is held until the clock input goes from 0 to 1, when whatever logic level is on the *D* input is transferred to the *Q* output. In the form shown, this too is a transparent latch, so

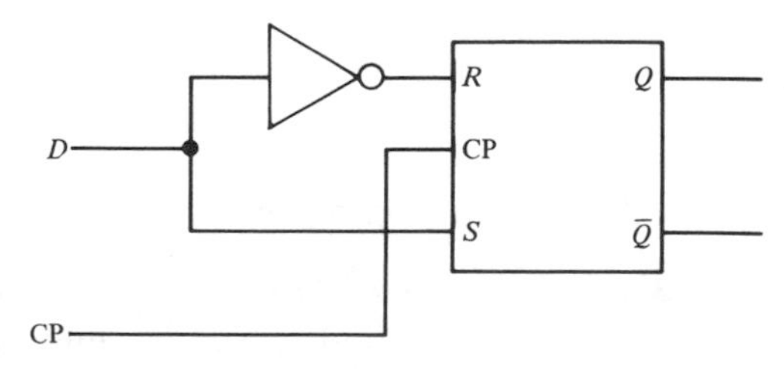

Fig. 13.22. *D*-type flip-flop (bistable latch).

that the Q output follows the D input whilst the clock is high, and stores the logic level present at the instant of the negative-going edge of the clock pulse. This useful little memory finds particular application in instruments with a digital read-out, where the output must be held steady long enough to be read; a circuit application of the 74LS75 quadruple D flip-flop is discussed in the section on counting.

13.10.4 JK flip-flop

A very versatile development of the *RS* flip-flop is the *JK* flip-flop shown in fig. 13.23(*a*). Here, unlike the *D*-type, the two inputs are still available, the indeterminate $S=R=1$ state being avoided by gating each input with the opposite output. By convention, the gated 'set' input is known as J and the gated 'reset' input as K. The inputs are enabled as soon as the clock (CP) goes high and disabled again when CP goes low. The transition table in fig. 13.21(*b*) shows a similarity to the *RS* flip-flop except that $J=K=1$ is no longer a prohibited condition but gives rise to the useful *toggle* condition where the outputs reverse their state every time the clock goes high. Thus, Q goes high on alternate clock pulses, making the toggling flip-flop a ÷2 frequency divider, with an obvious application in binary counting.

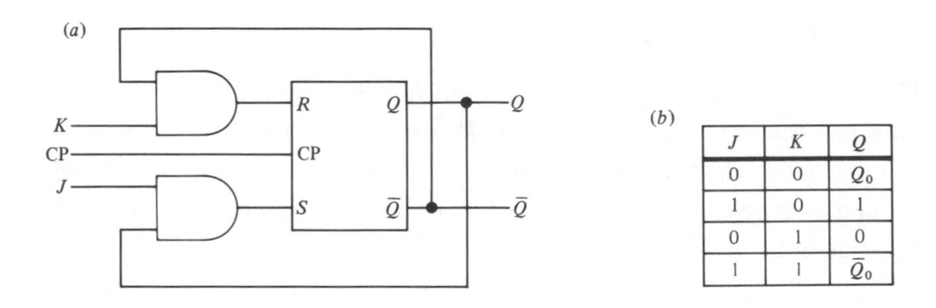

J	K	Q
0	0	Q_0
1	0	1
0	1	0
1	1	$\bar{Q}_0$

Fig. 13.23. *JK* flip-flop: (*a*) circuit showing gating of R and S inputs, (*b*) transition table (CP high).

13.10.5 Master–slave and edge-triggered flip-flops

As with the *RS*- and *D*-types, the basic *JK* flip-flop is transparent to any change on the inputs whilst CP is high. For example, if J and K both go high during the clock pulse when they were not both high at the positive transition, the output will toggle right away without waiting for the next CP transition. To be of general application, the *JK* flip-flop must exhibit two timing criteria:

(*a*) Output transitions must only be permitted immediately after CP transitions.

(*b*) The *J* and *K* inputs should be 'read' a defined time before the output transition and should be locked out whilst the outputs are actually changing.

These timing characteristics avoid uncertainty in the outputs during fast changes in the inputs and enable a chain of flip-flops to be connected as a shift register or counter, where the inputs are normally received from the outputs of the previous flip-flop. Transparent latches fail to work in this situation because the original input pulse just races through the chain on the first clock pulse without being stored in any of the flip-flops.

Fig. 13.24 shows one way of avoiding the race condition: two *JK* flip-flops are used in a *master–slave* configuration. During the time that CP is high, FF1 responds to inputs *J*1 and *K*1. As CP goes low, CP2 goes high via the inverter, thus enabling FF2 and transferring the condition set in 'master' FF1 into 'slave' FF2 and hence to main outputs Q and $\overline{Q}$. At the same time, FF1 is now locked out whilst CP is low, isolating the *J* and *K* inputs and preventing a race condition. Thus, our timing criteria are fulfilled, with the inputs 'read' on the positive clock edge and the outputs only changing on the negative clock edge.

Whilst such master–slave flip-flops are very widely used, a short word of warning is appropriate. Although the outputs can only change on the negative

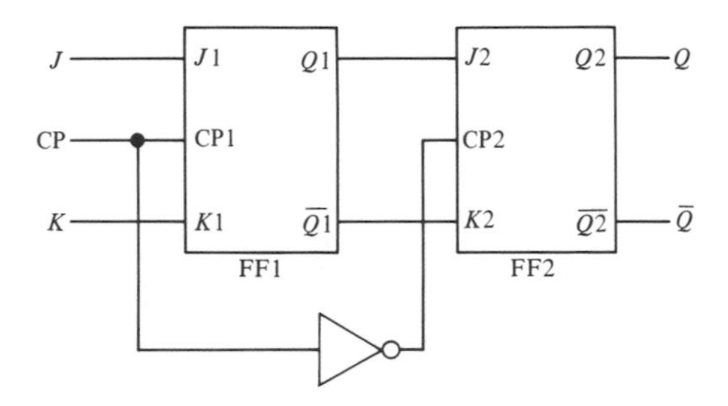

Fig. 13.24. Master–slave *JK* flip-flop.

clock edge, the actual output obtained then can sometimes surprise the unwary designer. This is because FF1 is responsive to J and K during the whole time that CP is high, so that any 'ones' occurring on J and K during that interval are transmitted as far as $Q1$ and $\overline{Q1}$, and will then change the main outputs on the negative clock edge even though J and K may have been innocently sitting low during both positive and negative clock transitions. It is therefore vital that J and K should remain stable whilst CP is high if operation is to be predictable. This type of flip-flop, exemplified by the 7473, 7476 and 74107, is termed *pulse triggered* and is for obvious reasons also called a *ones-catching* master–slave. Risk of unpredictable operation is minimized by using a clock with a narrow positive pulse.

Most current flip-flops such as 74LS versions include a further enhancement to make them purely edge triggered. This is usually arranged as negative-edge triggering, where the inputs are not read until the start of the negative clock transition, the outputs being changed later on during the same negative transition. Fig. 13.25 shows how the negative clock edge can be detected and converted to a short positive pulse by means of a simple *RS* flip-flop and NOR gate. CP is connected to the reset input and, via an inverter, to the set input. The output pulse comes from a NOR gate with one of its two inputs derived from input CP itself and the other from the Q output of the flip-flop. At the negative transition of CP, the output NOR momentarily has logic 0 simultaneously on both its inputs – one directly from CP and the other from Q due to the finite transition time in the *RS* flip-flop before it goes high in response to the CP transition. The resulting positive output pulse may then be used as the CP input of a master–slave flip-flop.

This useful edge detector circuit may be regarded as a sophisticated version

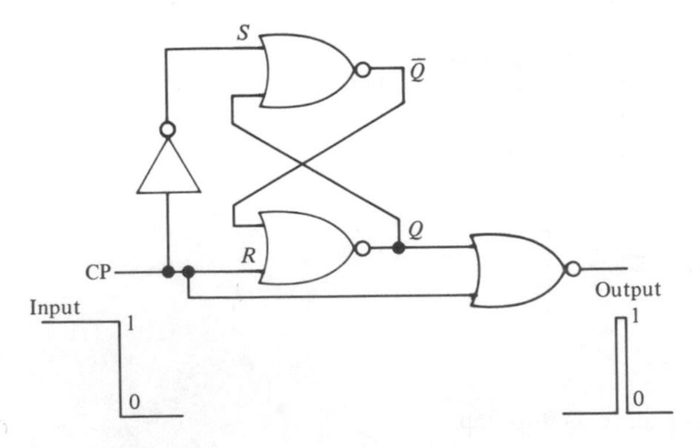

Fig. 13.25. Clock pulse negative-edge detector.

of the simple CR differentiating network used to trigger the discrete bistable flip-flop in fig. 12.13. It does, however, have the important advantage that it is purely *level* sensitive, the pulse duration and height being independent of input pulse rise and fall times. In a typical edge-triggered flip-flop, the J and K inputs only need to be stable for a short *set-up time* (typically 20 ns) before the CP transition. Some flip-flops also have a *hold time* requirement of up to 5 ns after the CP transition before J and K are allowed to change again.

13.11 Registers

13.11.1 Memory registers

We have seen how a D flip-flop stores a single bit on the clock transition. It is only a small step to connect eight such flip-flops together in parallel to a common clock to provide temporary storage for a whole byte (8 bits) of data. Fig. 13.26 shows such a parallel register which is a vital part of any computer system, being the equivalent of the scratch pad used for jotting down intermediate results in hand calculations.

13.11.2 The shift register

A string of JK flip-flops can be cascaded to store a series of digits. Such an arrangement, the *shift register*, is shown in fig. 13.27. It takes its name from the fact that it takes in one new digit for each clock pulse, *shifting* the existing digits along one stage to accommodate the new one. It is a *first-in first-out* (FIFO) register.

A study of fig. 13.27 will reveal how the shift register works. The reset line is first set to logic 1, then returned to 0. Now, suppose that the input data is initially at 1 and that a train of clock pulses is arriving at the clock input. In this condition, FF1 has its J input high and its K input low, so that Q_1 goes high after the first clock pulse. Meanwhile, suppose that the input data returns to 0 and stays there. The second clock pulse finds FF2 with J high, so transfers this logic 1 to Q_2. At the same instant, FF1 now has a logic 0 on J, so that Q_1 is taken low on this second clock pulse; if the input data remains at 0, Q_1 will stay low with every clock pulse. The logic 1 bit, however, is moved one stage along with each clock pulse so that, after four pulses, it arrives at Q_4. Altogether, 4 bits of input data have now been stored. Further clock pulses cause this data to be lost, and more new data to be stored.

In the shift register shown in fig. 13.27, the stored data is available in *parallel* form if required, access being provided to Q_1, Q_2, Q_3 and Q_4 outputs. Such an arrangement is known as a serial-in parallel-out register: the data must be 'clocked in' sequentially (serial form) via the single input, and is then available in parallel form at the register outputs. Such serial to parallel conversion is a very common application, being used, for example, to convert the recorded program digits from a computer disk into parallel form for entry into main memory.

If each flip-flop is provided with a separate preset input in addition to the common clear, the data can be entered in parallel form via these inputs. With the data thus 'jam-loaded', the clock pulse train can feed the digits out of Q_4 in serial form. This is a parallel-in serial-out register, a device which is often

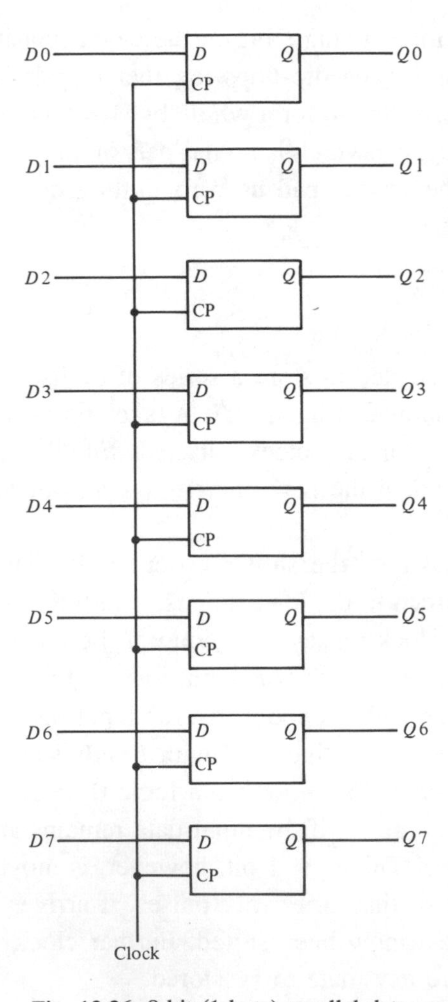

Fig. 13.26. 8-bit (1 byte) parallel data register.

used to convert the parallel output of a microprocessor, where digits are present simultaneously on typically sixteen output lines, to serial form to feed out on a single pair of wires to a network or a modem. The *UART* (universal asynchronous receiver–transmitter) is a popular dual-purpose combined parallel to serial and serial to parallel data converter containing the necessary shift registers, control logic and line drive circuitry in one IC chip.

In fig. 13.27, if the Q_4 output is connected back to the data input, then data, which may be entered in parallel form via the preset inputs, need never overflow the register, but will simply circulate. This is known as an *end around* shift register or *ring counter*. A decimal counter can be constructed using ten flip-flops numbered 0 to 9 connected as a ring counter. Initially, flip-flop 0 is set high and the rest are reset to zero. The pulses to be counted are then fed into the clock input so that, at each pulse, the flip-flop which is set moves one stage along. After nine pulses, flip-flop 9 is set and one more pulse restores the initial state to zero. A connection from flip-flop 9 to the clock input of another ring counter will record the tens, and yet another ring counter could record the hundreds. Despite the apparent elegance of this scheme, it is almost always more convenient to count in binary form, and then to decode the output into decimal form.

Another application of the ring counter is as the replacement for the distributor in a car electronic ignition system. Instead of a mechanical cam opening and closing contact points to produce the ignition spark, clock pulses are generated by an optical or magnetic sensor on the engine flywheel. These are used to shift a logic 1 round a ring counter which has one stage for each engine cylinder. The phase of the clock pulses can be precisely adjusted so that the logic 1 arrives at each stage at exactly the right time to fire the mixture. Correct ignition timing is thus readily set and, furthermore, once set, it should

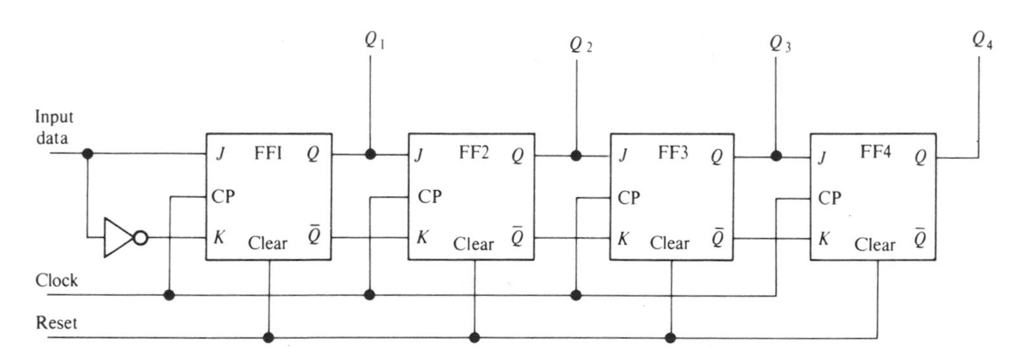

Fig. 13.27. Interconnection of *JK* flip-flops to make at 4-bit shift register.

never change because there is no mechanical wear involved in such an electronic distributor.

The shift register circuit of fig. 13.27 may be used as the basis for experiments with all types of shift registers and ring counters. The 74LS76 is recommended for the *JK* flip-flop: each package contains two negative-edge-triggered flip-flops with separate clear and preset inputs. Pin connections are shown in Appendix 4.

13.12 Binary counting

13.12.1 Introduction

One of the most important aspects of digital electronics is the function of counting. With the exception of the ring counter just discussed, counting is normally carried out in binary, since we are dealing all the time with two-state circuits. Conversion of the binary output into decimal form is easily carried out with a single IC decoder.

The basic element in most binary counters is the master–slave or edge-triggered flip-flop set to toggle (reverse its state) on each clock pulse. Fig. 13.28 shows four *JK* flip-flops arranged in sequence as a 4-bit binary counter. All *J* and *K* inputs are raised to logic 1 so that the flip-flops are in their toggle condition. All outputs can be set initially to zero by momentarily raising the common reset input to logic 1. The circuit is then ready to count input pulses. The binary count is available on outputs *A* to *D*, and it is important to notice that output *A*, although shown on the left-hand side, is in fact the least signifi-

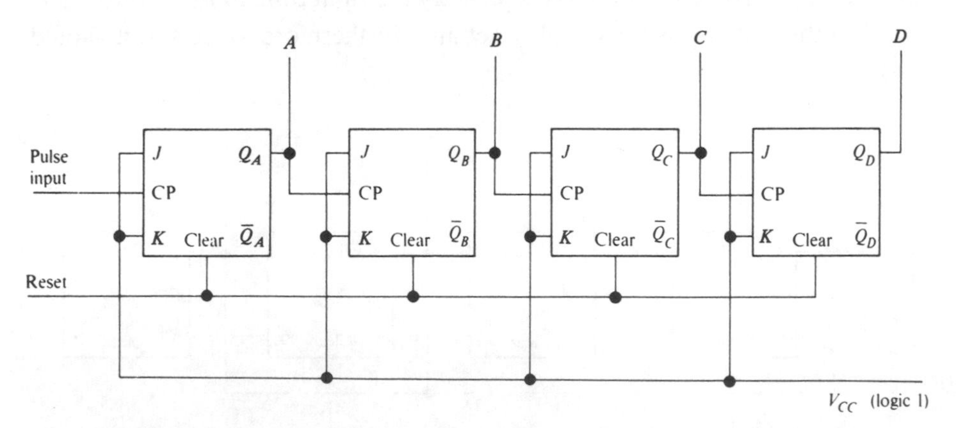

Fig. 13.28. Interconnection of *JK* flip-flops to make a 4-bit binary counter.

Number of input pulses	*D*	*C*	*B*	*A*
0	0	0	0	0
1	0	0	0	1
2	0	0	1	0
3	0	0	1	1
4	0	1	0	0
5	0	1	0	1
6	0	1	1	0
7	0	1	1	1
8	1	0	0	0
9	1	0	0	1
10	1	0	1	0
11	1	0	1	1
12	1	1	0	0
13	1	1	0	1
14	1	1	1	0
15	1	1	1	1

Fig. 13.29. Outputs of 4-bit binary counter.

cant bit (LSB). The numerical count should therefore be read in the order *DCBA*, and the table in fig. 13.29 confirms this.

The 74LS93 IC is a 4-bit binary counter; its connections and internal circuit arrangement are shown in fig. 13.30. Notice that flip-flop *A* is separated from the other three in order to allow independent ÷2 and ÷8 counting. Input *A* and output *A* together give a ÷2 count, whilst input *B* and outputs *B*, *C* and *D* give a ÷8 count. All the flip-flops have a common reset line. For normal 4-bit binary counting, output *A* is connected to input *B*.

Input and output waveforms for the 4-bit binary counter are shown in fig. 13.31, where it is seen that all output waveforms have a unity mark–space ratio and that the frequency is halved by each flip-flop. It is worth mentioning in passing that, if a true square wave of unity mark–space ratio is required for general testing purposes, the easiest way of achieving complete symmetry is to generate a repetitive pulse at twice the frequency and then use a flip-flop to divide by two, guaranteeing exactly unity mark–space ratio.

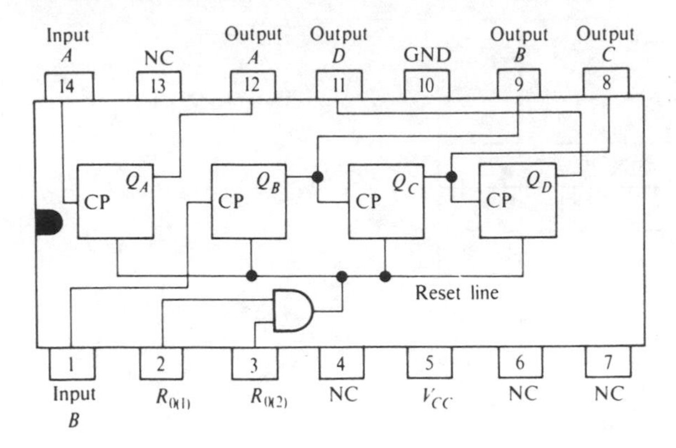

Fig. 13.30. 72LS93 TTL 4-bit binary counter: connections and circuit arrangement. NC indicates no connection to pin. GND is ground (earth) connection.

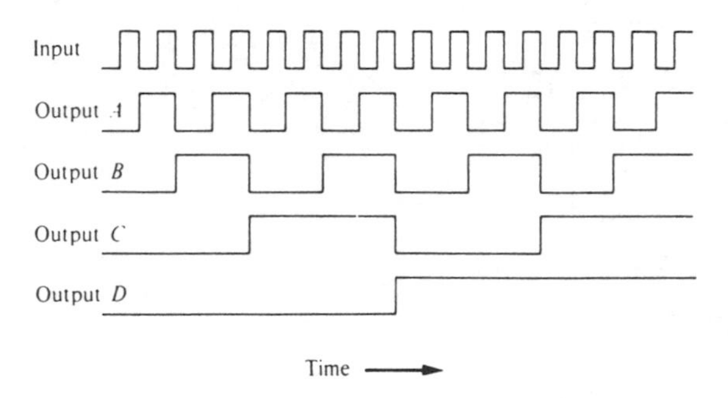

Fig. 13.31. Input and output voltage waveforms for the 4-bit binary counter.

13.12.2 Reset inputs and counting sequence

All four outputs of the 74LS93 can be reset to zero simultaneously by means of the two reset inputs $R_0(1)$ and $R_0(2)$. Both reset inputs must be at logic 1 together to reset the flip-flops. To permit counting, at least one reset input should be earthed (logic 0).

The availability of two resets makes the 74LS93 a very flexible device. By linking the reset inputs to various output combinations, a variety of different counting sequences can be generated. Reference to the count-sequence truth table of fig. 13.29 will show how the counter can be automatically reset at a specific number. For instance, if $R_0(1)$ is connected to output D and $R_0(2)$ to output B, the counter operates normally until the instant that the output reads 1010, when it resets to zero. Fig. 13.32 gives a table of reset connections for different counting sequences. In applying these 'feedback to reset' circuits,

Function	Counting sequence	$R_{0(1)}$ to	$R_{0(2)}$ to
÷ 3	0 1 2	A	B
÷ 4	0 1 2 3	V_{CC}	C
÷ 5	0 1 2 3 4	A	C
÷ 6	0 1 2 3 4 5	B	C
÷ 9	0 1 2 3 4 5 6 7 8	A	D
÷ 10	0 1 2 3 4 5 6 7 8 9	D	B
÷ 12	0 1 2 3 4 5 6 7 8 9 10 11	D	C

÷ 2 and ÷ 8 are available separately

Fig. 13.32. Different counting sequences available with the 74LS93 4-bit binary counter.

the designer must note that the reset takes a finite time (approximately 40 ns), so that a momentary 'spike' or 'glitch' will appear at the outputs immediately prior to resetting to zero. This does not matter in applications such as directly fed displays, but can cause problems if an output is used as a source of clock pulses for another circuit.

Such glitches can be a common source of unpredicted or erratic performance in digital hardware; they can often arise because of transit time differences which result in timing 'races' of pulses through different parts of the circuit. Different production samples can then give completely different results depending on which pulse 'wins the race'.

13.12.3 The BCD counter

In many counting applications, the output is ultimately required in decimal form so that it can be understood by the human operator. It is therefore often convenient to divide the counting flip-flops into groups of four, each group being arranged to give a ÷10 count. In this way, the individual digits of the decimal answer are readily available, each being separately expressed in binary form. This code is known as binary-coded decimal (BCD).

For example, the decimal number 2901 would be expressed in BCD code as follows:

0010 1001 0000 0001.

The keyboard data inputs on computers normally work on BCD code too: for example, the popular 'ASCII' code sends out character by character the numbers and letters typed on the keyboard, with the letters having defined binary codes of their own. This is further discussed later in connection with alphanumeric displays (see fig. 13.43).

We have already seen how a 74LS93 ÷16 counter can be arranged to reset after binary 9. This is one way of making a BCD counter, but, in addition to the 'glitch before reset' problem, it means that the reset inputs are then unavailable for any other purpose unless extra OR gates are added. It is more convenient to use a purpose-built counter, such as the 74LS90. Its pin connections are shown in fig. 13.33.

In addition to the two reset to zero inputs, the 74LS90 features a pair of *reset to nine* inputs marked $R_9(1)$ and $R_9(2)$. This facility is used in a form of BCD subtraction known as ten's complement. $R_9(1)$ and $R_9(2)$ are normally earthed when not required, but it is worth noting that they can be used to give a ÷7 function which is otherwise unavailable. For this ÷7 facility, outputs *B* and *C* are coupled to the R_9 inputs, this connection setting the counter to 9 immediately the count of 6 (0110) is reached. This gives the counting sequence 0, 1, 2, 3, 4, 5, 9 which, although of no use for direct decimal decoding, can be used for dividing the input frequency by 7 if the output is taken from pin *D*.

As with the 74LS93, the *A* flip-flop in the 74LS90 is separate from the other three, so that independent ÷2 and ÷5 functions are available. For normal ÷10 (BCD) counting, output *A* is joined to input *BD*. The separate ÷2 counter can be used to overcome the disadvantage of a BCD counter when used simply for ÷10 frequency division. Although the waveform at the *D* output is one-tenth of the input frequency, it is asymmetrical, having a mark–space ratio of 1 : 4 (*D* is low on counts 01234567 and high only on 8 and 9). If a ten-times frequency division with a unity mark–space ratio is required, the ÷5 counter should precede the ÷2 counter. In other words, the *D* output should be connected to the *A* input and the input pulses fed to the *BD* input. The ÷10 square wave is then obtained from output *A*.

It is relevant to note here the fact that the *BD* input represents four input

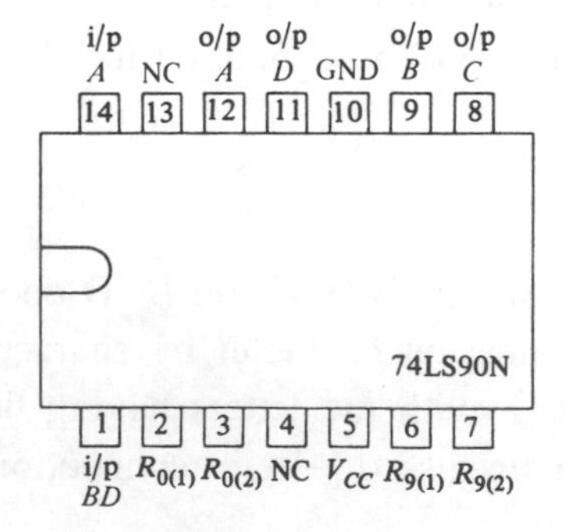

Fig. 13.33. Connection diagram for 74LS90 TTL decade counter.

loads, and care should be taken not to exceed permissible fan-out if one gate is used to feed several such inputs.

13.12.4 Cascading BCD counters

A string of BCD counters such as in fig. 13.34 is used to count units, tens, hundreds etc. The D output of the first counter feeds the A input of the second and so on. As the units counter reaches 9 and returns to 0, the D output goes from high to low, and the tens counter responds to the falling edge, recording the carry digit each time.

A variety of division ratios greater than 10 may be achieved with two counters in cascade using feedback to the reset inputs. One example is the circuit of fig. 13.35, which gives a ÷24 function, suitable for the hours counter in a digital clock. Reset occurs when the output reaches 0010 0100.

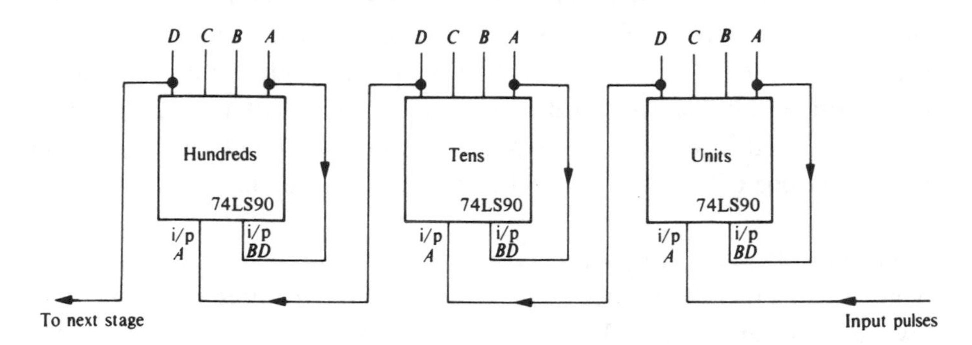

Fig. 13.34. Cascaded BCD counters.

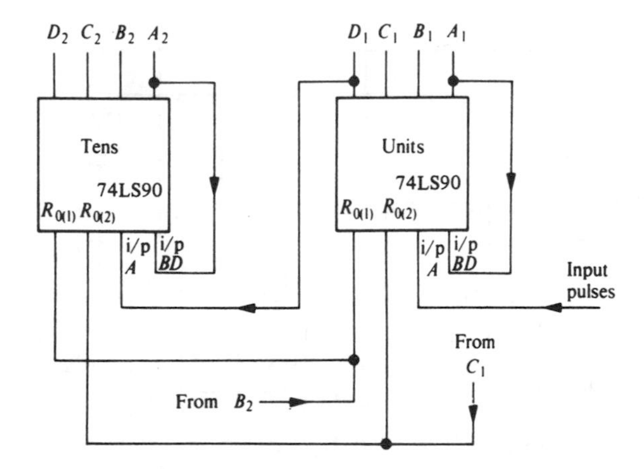

Fig. 13.35. Use of reset inputs on two decade counters to make a ÷24 (hours) counter.

13.12.5 Synchronous counting

All the counters considered so far have consisted of cascaded toggling flip-flops with each CP input fed from the previous output. This simple 'ripple-through' arrangement suffices for many basic counting operations, but it does exhibit disorderly timing, with the number transitions occurring in sequence, owing to gate transition times, instead of together. It is therefore known as an *asynchronous* counter because the clock cannot trigger all the flip-flops simultaneously. The typical propagation delay through four stages is about 70 ns, and thus can result in spurious glitches if decoded counter outputs are used as a source of clock pulses for other circuits.

The *synchronous* counter overcomes the problem by using a little intermediate logic to ensure simultaneous transitions and is shown in fig. 13.36. The first (least significant bit A) flip-flop has $J = K = 1$ so that it changes state with each clock pulse. The second (B) flip-flop has $J = K = A$ so that it toggles only when $A = 1$. The third (C) flip-flop can toggle only when B and A are 1, whilst the fourth (D) needs C, B and A to be 1 before it can toggle. Following through this transition table, we find that it is identical to that of the asynchronous counter except that all the toggles occur together, all the clock inputs being linked to the one CP input instead of being cascaded from the previous flip-flop output.

The 74HC191 is a useful general-purpose synchronous binary counter. It counts either up or down according to the logic level presented to the up/down control input. A further facility is the ability to jam-load a specific

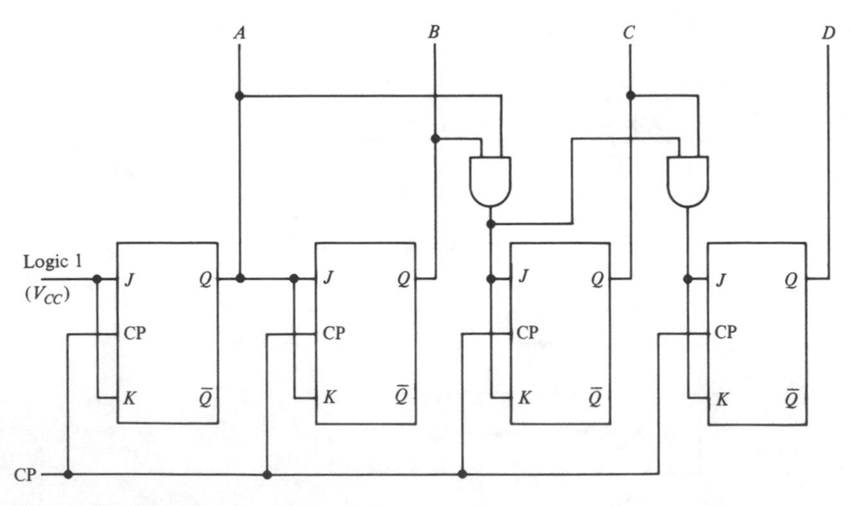

Fig. 13.36. Synchronous 4-bit counter.

number into the counter as its starting point, this being done via four data pins which access the data inputs on the four flip-flops.

13.13 Decoders and displays

13.13.1 Direct decoding – decimal and hexadecimal

Having acquired the means of counting pulses, the next logical step is to display the resulting number. This can of course be done directly in binary form using a lamp or LED with transistor drive (fig. 13.5) on every output. Where several outputs are to be displayed together as a binary number, the circuit of fig. 13.37 is economical. It uses the 7404 IC, which contains six inverters, to drive six LEDs directly. Each 'standard' TTL output is able to sink 16 mA in its 'low' state, thus handling the LED current (approximately 10 mA) comfortably when the appropriate input goes high.

Whilst binary numbers are very convenient and economical for electronic handling, with their straight ON/OFF nature, the strings of ones and zeros are cumbersome for the human brain to manipulate. The most generally

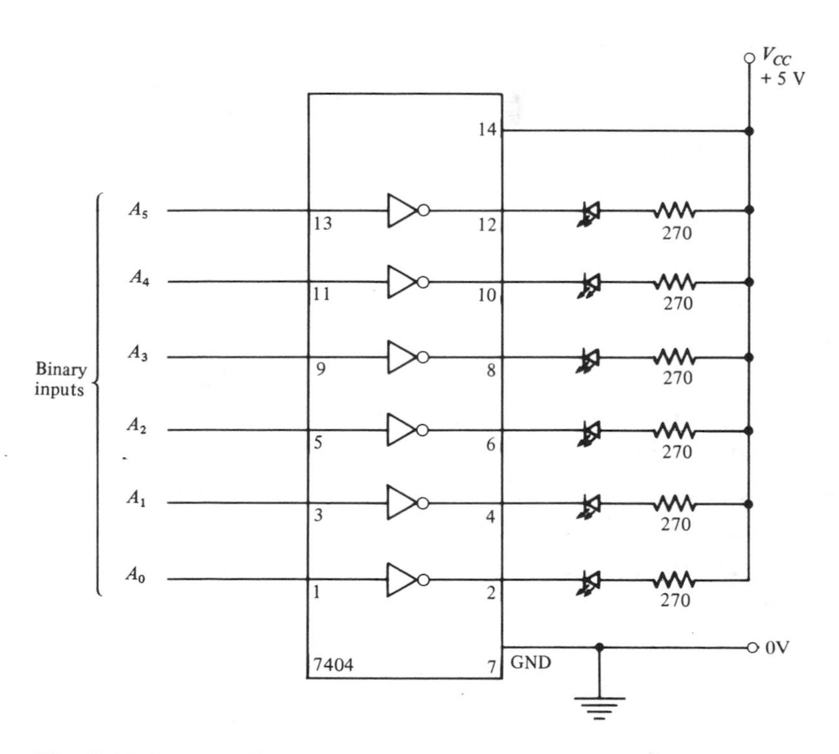

Fig. 13.37. Inverter chip used as LED driver to display 6-bit binary number.

Table 13.1

Binary	Decimal	Hexadecimal
0000	0	0
0001	1	1
0010	2	2
0011	3	3
0100	4	4
0101	5	5
0110	6	6
0111	7	7
1000	8	8
1001	9	9
1010	10	A
1011	11	B
1100	12	C
1101	13	D
1110	14	E
1111	15	F

convenient representation of a binary number uses *hexadecimal* numbers, working on a base of 16 instead of the base 10 of the decimal system. The use of base 16 means that a hexadecimal ('hex') number represents a 4-bit binary number as a single digit. The only unfamiliar aspect of 'hex' numbering is that extra characters have to be invented to correspond with the decimal numbers 10, 11, 12, 13, 14 and 15, so that a carry digit is not needed until decimal 16 (10 hex). The extra characters use the first six letters of the alphabet as shown in table 13.1.

A long binary number may be simply split into groups of 4 bits and each group converted into its corresponding hex form. The alternative of decimal representation requires each bit to be given its value as the corresponding power of 2 and then the results added – a much lengthier job.

The convenience of hexadecimal representation is demonstrated by taking a 16-bit binary number, typical of those used in computers, together with its hex equivalent:

$$0111101001011111 \equiv 7A5F \text{ (hex)}.$$

To avoid risk of confusion with decimal numbers, hex numbers are normally labelled in a shorthand form. Our hex number is here shown labelled in five common ways:

$$\&H7A5F \quad \&7A5F \quad 7A5Fh \quad \$7A5F \quad 7A5F_{16}.$$

Decoding 4-bit groups into hex can be implemented by the 74154 IC shown in fig. 13.38 connected to 16 LEDs labelled 0–F in order to display the hex

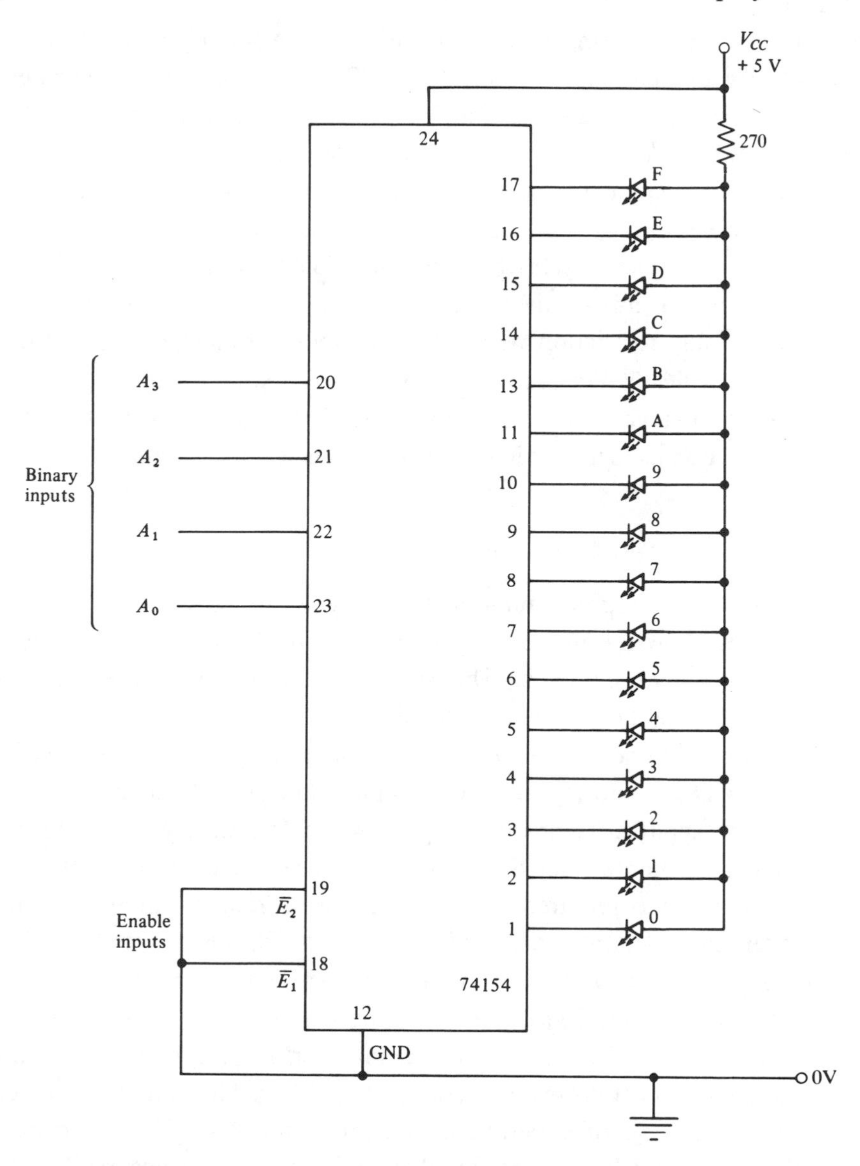

Fig. 13.38. Binary to hexadecimal conversion using 74154 4-line to 16-line decoder chip.

equivalent of the binary input number $A_3A_2A_1A_0$. Outputs are normally high and go low when the corresponding binary number is on the input, thus passing current through the corresponding LED. 'Enable' inputs $\overline{E_2}$ and $\overline{E_1}$ gate the decoder and may be operated from clock pulses to prevent 'glitches', which may have arisen because of counting and decoding mis-timing, from appearing on the outputs thus avoiding false outputs if further logic is driven from the 74154.

For systems operating in BCD form where only the numbers 0–9 (binary 0000 to 1001) need to be decoded, the 7442 provides a compact decoder which fits into a 16-pin IC package. The 7445 is a high-current version of the 7442, having *open collector* outputs instead of the usual totem pole TTL circuit. Unusually, its output transistors are freed from the 5 V limitation of the TTL circuit and can handle up to 30 V, with a maximum 'ON' current rating of 80 mA, making them useful for lamp and solenoid driving. The 74LS145 is a low-power version of the 7445, which retains the 80 mA output current rating, the only restriction being the maximum voltage rating of the output transistors, which is reduced from 30 V to 15 V. Note that these open collector chips are still subject to the maximum +5 V V_{CC} rating on the chip itself – it is only the output loads which may be connected to a higher voltage.

13.13.2 Seven-segment displays and decoders

By far the most popular numerical display is the seven-segment type. Fig. 13.39 shows how the numerals from 0 to 9 are displayed by various combinations of seven segments. The segments are often light-emitting diodes (LEDs) using gallium arsenide phosphide.

The most common type of seven-segment indicator is the *liquid-crystal* display (LCD). Liquid crystals are not light emitters, but become opaque when an e.m.f. is applied. Their operating current is extremely low, which makes them ideal for digital clocks and watches where a continuous display with battery operation is required. They do, of course, require ambient illumination to render the indication visible, but this is no disadvantage for clocks and watches. If necessary, liquid-crystal displays can incorporate front or rear illumination for operation in poor lighting conditions. Commercial LCD displays are normally designed with integral interface chips to connect direct to computer data and address buses (see section 14.4). They are therefore difficult to use for the kind of experiments we are doing here. We are hence using readily available LED 7-segment displays, which interface readily to logic chips.

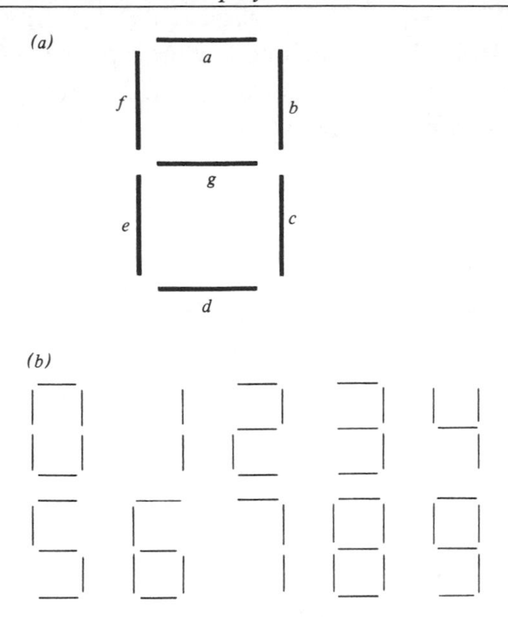

Fig. 13.39. Seven-segment numerical indicator: (*a*) segment identification, (*b*) numerical displays.

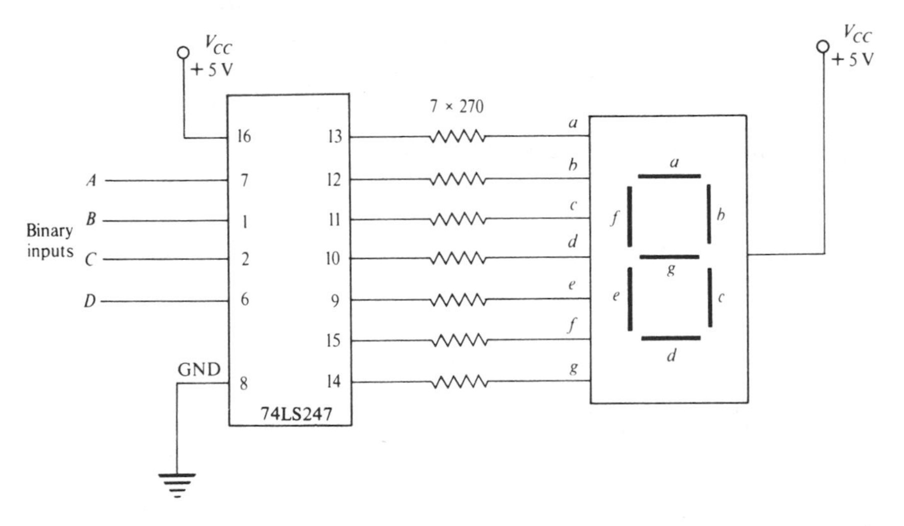

Fig. 13.40. 74LS247 BCD to seven-segment decoder driving a LED indicator. Pin numbers refer to 16-pin DIL package.

The decoder/driver circuit in fig. 13.40 employs the 74LS247 chip. Each LED segment has a 270 Ω series resistor for current limiting: a common resistor is impracticable owing to the varying load, which depends on how many segments are illuminated. The indicator may be blanked by supplying

1111 as a binary input to the 74LS247, or by pulling low the separate blanking input (pin 4). The 7447 decoder/driver is also available; it is directly interchangeable with the 74LS247 except that it omits the top and bottom 'tails' on digits 6 and 9.

13.13.3 The 74LS75 bistable latch

When a numerical display and decoder are fed directly from a counter, a clear indication is only obtained at low pulse rates (<2 Hz). Counting rates much faster than this result in a rapidly changing display which dissolves into a multi-digit blur.

The problem can be overcome by connecting a temporary store between the counter and decoder. The binary count may then be captured by the store in response to the appropriate instruction and held for display on the indicator whilst the counter continues to count the next batch of pulses. The chip usually used here is the 74LS75 quadruple bistable latch (*D* flip-flop), which stores four binary digits for as long as necessary (fig. 13.41).

Digits at the data inputs (*D*) are transferred to the outputs (*Q*) when the clock input is at logic 1, the *Q* outputs following the input data as long as the clock remains high. When the clock goes low (logic 0) the digits that were present at the data input at the instant of the clock transition are retained at the *Q* output until the clock goes high again.

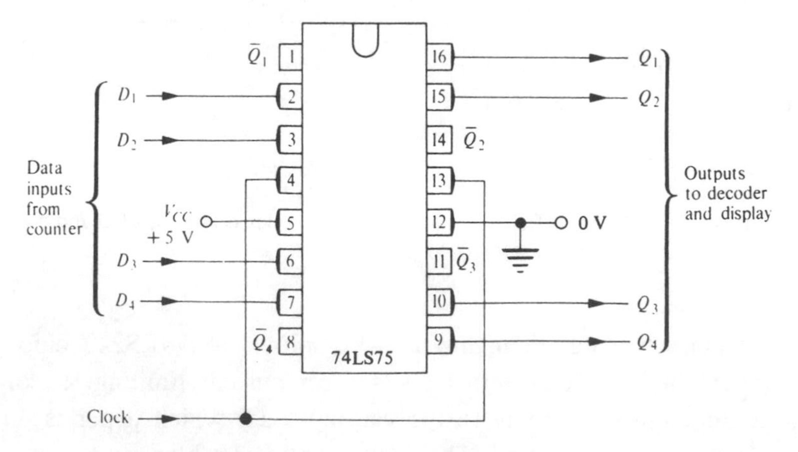

Fig. 13.41. Applications of the 74LS75 bistable latch.

13.13.4 Multiplexed displays

A multiple digit display would at first sight seem an unwieldy circuit with a large number of connections to the indicator segments, together with a separate decoder chip for each digit. To avoid this problem, with its consequential cost, most such displays are multiplexed with all the corresponding segments connected in parallel, each digit having its own common anode or cathode connection. In this way, the output of the seven-segment decoder can be arranged to illuminate whichever of the digits is required. A multi-digit number is shown by sequentially illuminating successive displays; persistence of vision gives the illusion of simultaneously displayed digits. Fig. 13.42 shows the connections to a typical 4-digit common anode display (RS 587–024). The switching transistors (BC177 or similar) switch on the anode supply to each digit in turn, each being activated by a logic 0 pulse synchronized with the arrival of the appropriate binary number at the common decoder. Four separate seven-segment displays can, of course, be used with the cathodes connected together.

Displays with larger numbers of digits can be conveniently scanned by connecting the digit enable lines of fig. 13.42 to the outputs of a decoder such as the 74154 (use fig. 13.38 with the outputs connected to the enable lines of fig. 13.42 instead of to LEDs). A 4-bit binary count on the inputs will then scan the digits as fast as required. Notice that each display digit is now individually

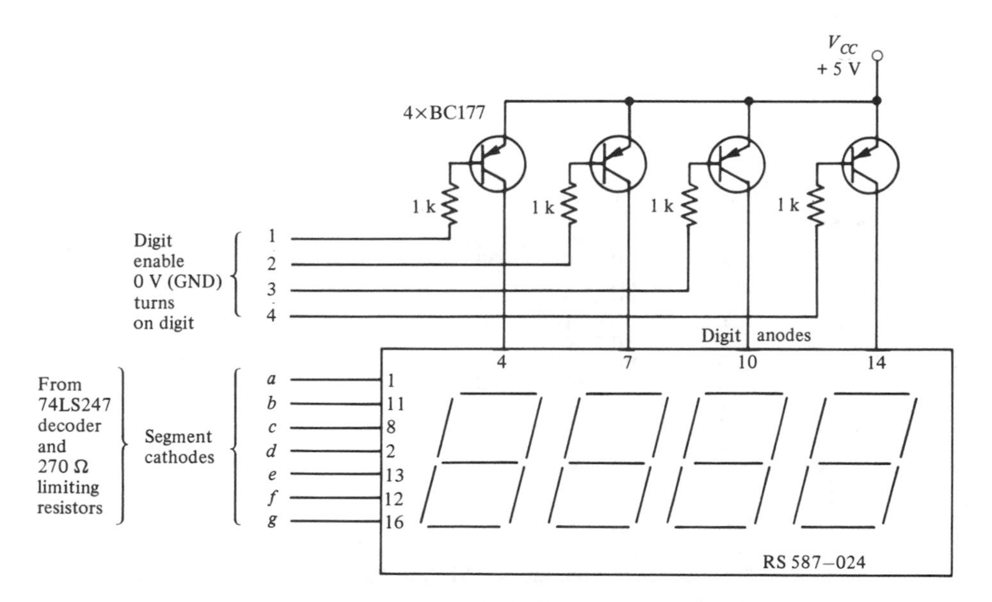

Fig. 13.42. 4-digit-multiplexed display.

addressed by inputting its appropriate binary number on the 74154. This addressing principle for activating a particular device out of a number of devices all connected in parallel on one set of data lines is fundamental to the computer system *bus*, which we shall explore in the next chapter.

Displays often use a high degree of integration to include in a single device decoders, drivers, latches and demultiplexing circuits as well as the LED segments themselves. One such popular package is the TSM2416 (RS 585–208), which contains four 16-segment 'starburst' displays capable of showing the full alphabet as well as numbers and a variety of other characters. Its built-in memory and addressing circuits allow data for each character to be loaded whenever required, quite independently of the LED multiplex timing. The DLR2416 (RS 589–317) is a similar display but uses a 5×7 matrix of LED dots to form the characters. Standard binary codes are used for letters as well as for numbers; they form part of the internationally agreed *ASCII* 7-bit code used in virtually all microcomputers (fig. 13.43).

ASCII	Character	ASCII	Character
010 0000	space	011 0000	0
010 0001	!	011 0001	1
010 0010	"	011 0010	2
010 0011	#	011 0011	3
010 0100	$	011 0100	4
010 0101	%	011 0101	5
010 0110	&	011 0110	6
010 0111	'	011 0111	7
010 1000	(	011 1000	8
010 1001	)	011 1001	9
010 1010	*	011 1010	:
010 1011	+	011 1011	;
010 1100	,	011 1100	<
010 1101	-	011 1101	=
010 1110	.	011 1110	>
010 1111	/	011 1111	?

ASCII	Character	ASCII	Character
100 0000	@	101 0000	P
100 0001	A	101 0001	Q
100 0010	B	101 0010	R
100 0011	C	101 0011	S
100 0100	D	101 0100	T
100 0101	E	101 0101	U
100 0110	F	101 0110	V
100 0111	G	101 0111	W
100 1000	H	101 1000	X
100 1001	I	101 1001	Y
100 1010	J	101 1010	Z
100 1011	K	101 1011	[
100 1100	L	101 1100	\
100 1101	M	101 1101	]
100 1110	N	101 1110	^
100 1111	O	101 1111	_

Fig. 13.43. Main characters from the ASCII 7-bit binary code (used on 2416 alphanumeric display/decoder chip).

13.14 The Schmitt trigger 7413

For reliable operation, all gates and counters require input pulses with very fast rise and fall times.

In the intermediate region between logic 0 and logic 1, the gate is in fact biased as a high-gain amplifier, and it is then that it is most vulnerable to noise and instability. A fast rise and fall to the input waveform means that the gate spends only a short time in this critical region and reduces the likelihood of spurious signals. In practice, optimum noise immunity requires logic edges of less than 50 ns duration. Interfacing with slow pulses can clearly present a problem. One solution is the use of a comparator in conjunction with positive feedback to give hysteresis (section 11.17.2). An easier solution is most cases is, however, provided by the 7413 Schmitt trigger and its LS and HC derivatives.

The connection diagram in fig. 13.44 shows that the 7413 chip contains two identical Schmitt trigger circuits: logically each circuit functions as a 4-input NAND gate but, because of the positive feedback action, the input threshold levels are different for positive- and negative-going signals. A typical rising-edge threshold is 1.7 V and a falling-edge threshold 0.9 V. This hysteresis of approximately 0.8 V provides a fast, jitter-free output waveform even with slowly changing input signals. Once the gate begins to change its output, the positive feedback ensures that the logic state reversal is rapidly completed without interference from noise at the input.

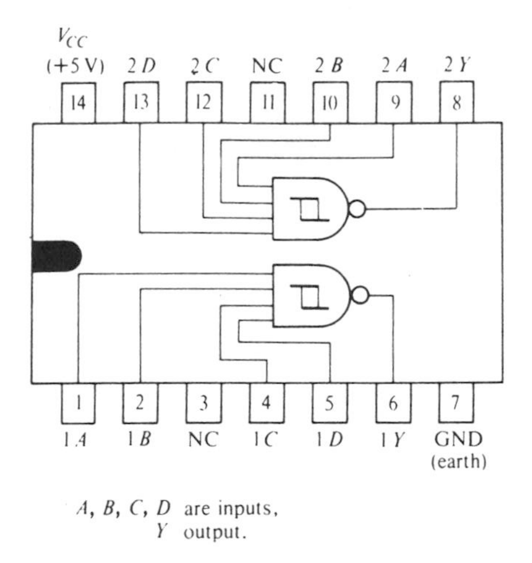

Fig. 13.44. Pin connections to 7413 dual Schmitt trigger.

High-frequency noise is often present on signals encountered in an industrial environment. If these signals are to be used with digital logic, the noise must be removed: this can be achieved by interposing a low-pass filter and then reshaping the pulses into TTL-compatible form with the 7413 (fig. 13.45). The optimum filter capacitor value will depend on the nature of the noise and the pulse repetition frequency; some experiment may be necessary here.

Feedback via an RC network applied to the 7413 Schmitt trigger makes a versatile wide-frequency-range pulse generator (fig. 13.46). It has, of course, a TTL-compatible output and can be used as a clock generator in any simple logic circuit. Output frequency with the 390 Ω feedback resistor shown is given by the approximate relation

$$f \approx \frac{2000}{C} \text{Hz}$$

where C is in μF. The output can be gated by separating off one or more of the four inputs: connection to 0 V disables the generator.

In order to satisfy both input and output requirements of the 7413, the permissible range of the feedback resistor value is limited, but it may be varied from 330 Ω to 470 Ω to act as a fine frequency control. The 74HC13 allows a much wider range of resistor values, up to several megohms.

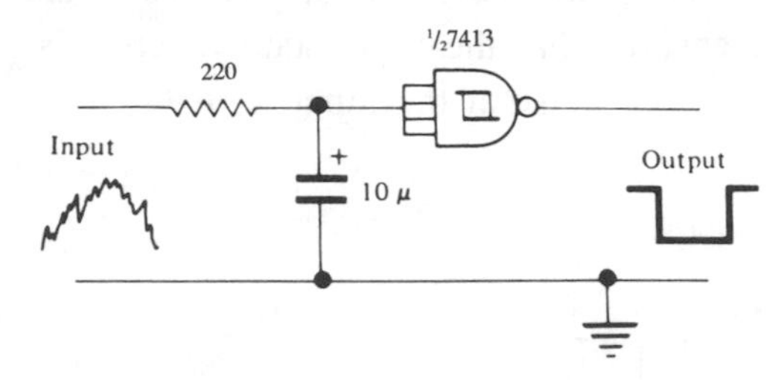

Fig. 13.45. Use of Schmitt trigger to interface IC logic with noisy signals.

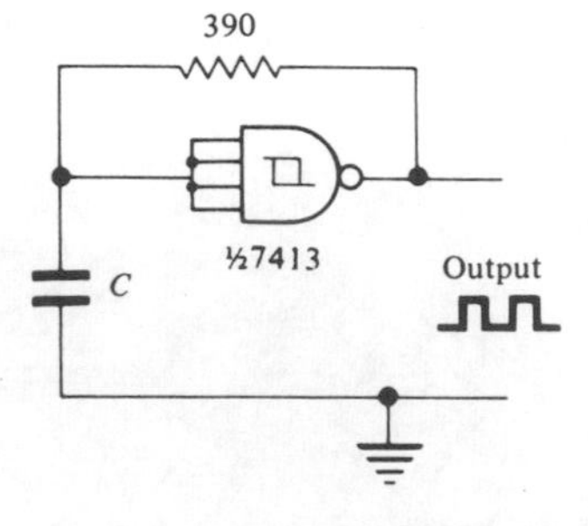

Fig. 13.46. Simple pulse generator for logic circuits.

13.15 Monostables and timers

13.15.1 General

There are many applications for electronic time delays. A short pulse of, say, 10 μs duration may need to be 'stretched' to 30 ms duration in order to operate a relay. A process timer might be triggered by a *start* pulse and be required to supply a *stop* pulse after a fixed time interval. These functions are readily performed by monostable and timer ICs.

13.15.2 The 74121 monostable

The 74121 is a TTL version of the monostable multivibrator discussed in section 12.6. On receipt of an input trigger pulse, the Q output changes from low to high for a time determined by an externally connected time constant. Fig. 13.47 shows the external connections to the 74121. There are three inputs: $A1$ and $A2$ operate as a NAND function, triggering on a falling edge, whilst input B operates the monostable via a Schmitt trigger working on a rising edge.

To use the NAND facility of $A1$ and $A2$, all three inputs must initially be held at logic 1; changing either or both of $A1$ and $A2$ to logic 0 will then trigger the monostable. Input B may thus be used here as an overriding inhibit facility: no triggering can be obtained if input B is held at logic 0. The Schmitt trigger incorporated in input B can be very useful for reliable triggering from

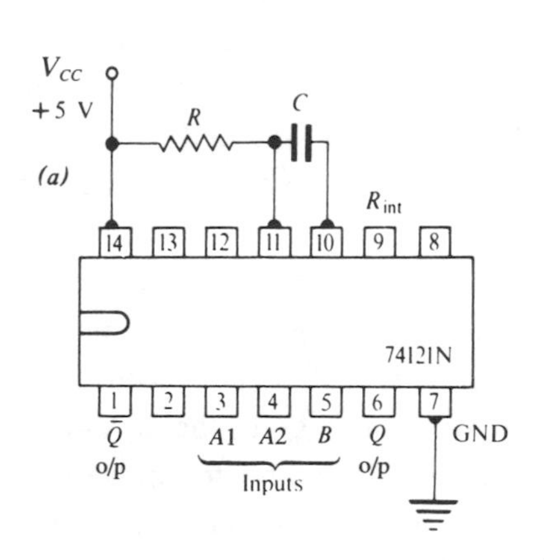

(b)

Input conditions			Output mode
$A1$	$A2$	B	
1	1	Any	Inhibit
Any	Any	0	Inhibit
↓	1	1	Trigger
1	↓	1	Trigger
↓	↓	1	Trigger
0	1	↑	Trigger
1	0	↑	Trigger
0	0	↑	Trigger

Fig. 13.47. The 74121 TTL monostable: (*a*) IC connections, (*b*) function table. Arrows refer to direction of input pulse edge.

slow pulse edges. It is used by ensuring that either $A1$ or $A2$ is at logic 0 and feeding a positive-going trigger pulse to input B; holding both $A1$ and $A2$ to logic 1 inhibits triggering. The table in fig. 13.47 summarizes the trigger functions; the arrow direction shows which pulse edge is operative.

The pulse duration is given by the same relationship as the discrete monostable, i.e.

$$t = RC \ln 2$$

$$\approx 0.7\,RC. \qquad [(12.8)]$$

Values of R from 1.4 kΩ to 40 kΩ are permissible, whilst C may be selected from the range 10 pF to 1000 μF, giving output pulse lengths up to 28 seconds.

Fig. 13.48 shows three common applications of the 74121. In (*a*), the pulse stretcher, a short input pulse triggers input B on its rising edge. The corresponding pulse on the Q output is then of duration 0.7 RC. Any further input pulse arriving before the output resumes logic 0 will be ignored: the 74121 is non-retriggerable. A retriggering facility is available on the 74122 monostable, which can provide very long output pulses by retriggering just before $t =$ 0.7 RC. The 74122, and its dual version 74123, can be used to convert a group of short pulses into a continuous high logic level by making t just longer than the pulse period. The 74HC123 dual retriggerable monostable, being a CMOS device, has the advantage of not imposing an upper limit on the value of timing resistor R.

Fig. 13.48(*b*) uses two 74121 monostables to delay the arrival of a pulse. The input pulse rising edge triggers IC1 for t_1 seconds, at the end of which period its Q output goes low. This delayed falling edge triggers IC2 via the A inputs so that an output pulse of length t_2 is generated. The appropriate choice of R_2 and C_2 can give an output pulse identical to the input pulse, but delayed in time by t_1. With no external timing components (R_{int} connected to V_{CC}) an output pulse of typically 30 ns is obtained, which may be used as a counter reset signal.

The monostable is another of those useful interface circuits between a manual switch or keyboard and a digital circuit. Fig. 13.48(*c*) may be used here, the B input being connected to earth via 330 Ω and with a normally open switch between B and +5 V. Closing the switch gives the necessary rising edge to trigger the monostable: any raggedness in the switch pulse is ignored by the combination of the Schmitt trigger action and the non-retriggerable operation. Output pulse width should be set to at least 20 ms to avoid unwanted retriggering by contact bounce.

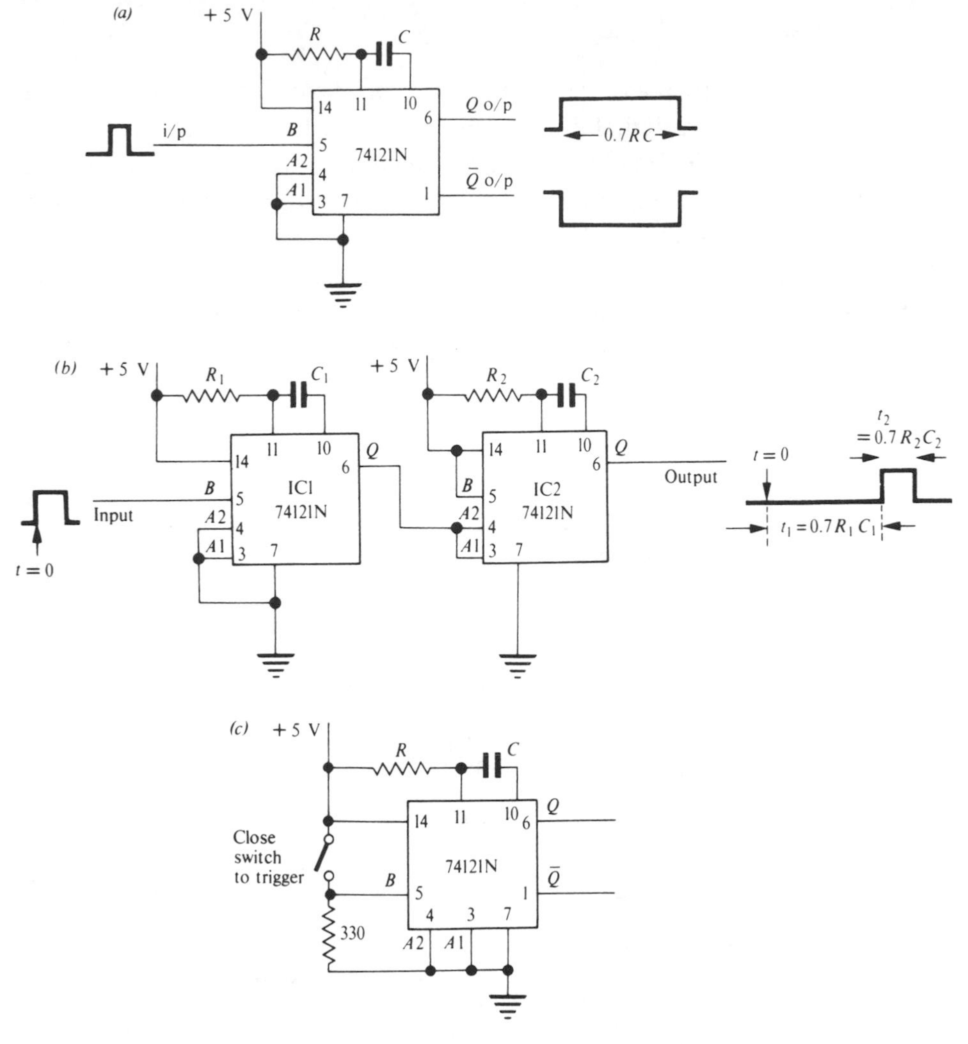

Fig. 13.48. Some applications of the 74121 monostable: (*a*) pulse stretcher, (*b*) pulse delay, (*c*) manual triggering.

Whilst the monostable can be a useful design tool and can provide many a 'short cut' solution to timing problems, its general use in professional digital products is not to be encouraged. Its dependence on 'analogue' characteristics for its timing accuracy can cause problems in critically timed systems; for example, a chip replaced during repair may be found to give different timing from the original and upset the whole circuit. Digital systems are preferably designed to run from a common synchronizing clock and reliance on monostables for critical timing should be avoided.

13.15.3 The 555 timer

The 555, produced by many semiconductor firms, is an analogue timer which works well with many digital circuits. When operated from a +5 V supply the 555 is compatible with IC logic. However, the device will operate safely from a supply as high as +15 V, and the available output current (sink or source) is 200 mA, so that a relay or bulb can be readily driven. Fig. 13.49(*a*) shows the 8-pin DIL version of the 555 with the various connections, the significance of the latter being indicated in the block diagram of fig. 13.49(*b*), shown with the external timing resistor R_T and capacitor C_T in position.

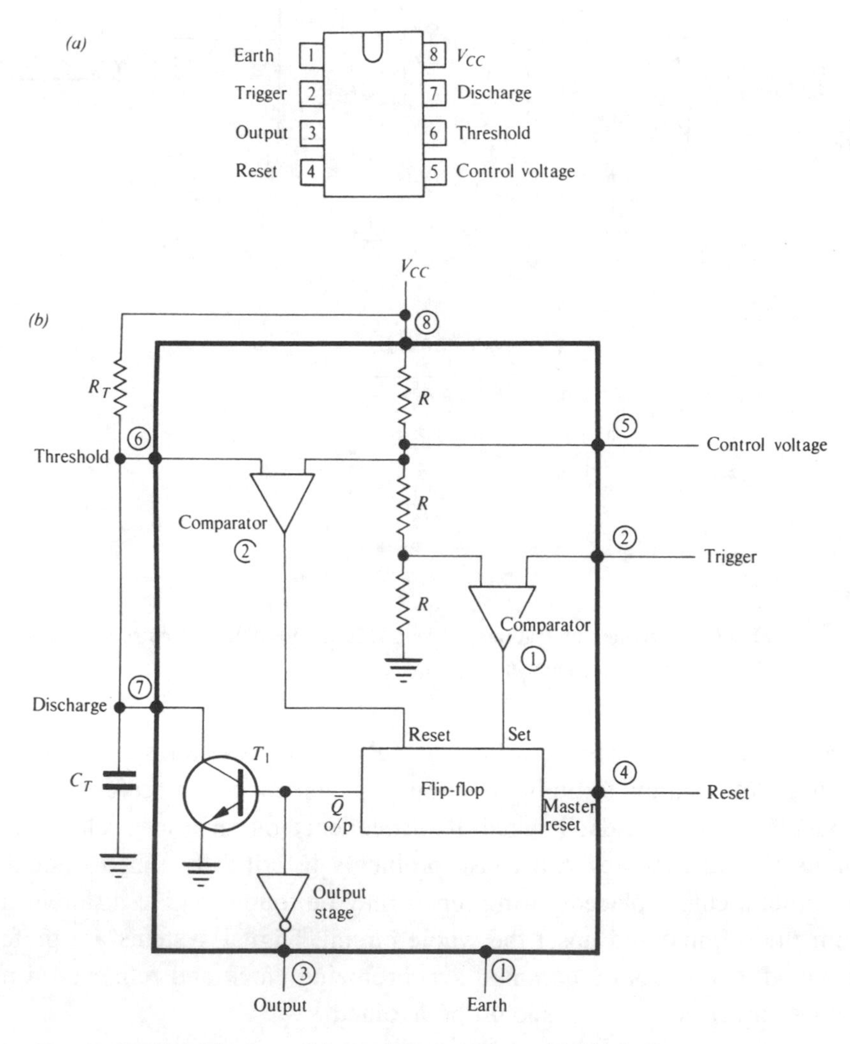

Fig. 13.49. The 555 timer: (*a*) IC pin connections, (*b*) block diagram.

The basis of the circuit is a flip-flop with set and reset inputs controlled by comparators. The quiescent condition is with the $\overline{Q}$ flip-flop output high so that the timing capacitor C_T is shorted by the transistor T_1; the output is then low because of inversion in the output stage.

The trigger is normally held high by the internal circuit, and operates on a falling input edge: when the trigger input falls to $V_{CC}/3$, comparator 1 changes state and sets the flip-flop. The $\overline{Q}$ output goes low, sending the output terminal high and also turning off T_1, thus allowing timing capacitor C_T to begin charging through resistor R_T. Charging continues until the p.d. across C_T rises to $V_{CC}/3$, when comparator 2 resets the flip-flop: T_1 is then turned back on, discharging C_T ready for the next trigger pulse. If, however, the trigger input is still held below $V_{CC}/3$ at the end of the timing cycle, the flip-flop will not reset and the output will remain high until the trigger is released. Pin 4 is an overriding reset which can be momentarily earthed to reset the device at any point in the timing cycle. The timed period can be calculated from the usual charging equation for an *RC* time constant:

$$V = V_0\left(1 - \exp\frac{-t}{CR}\right).$$

Here $V_0 = V_{CC}$, and we require the value of t which gives $V = \frac{2}{3}V_{CC}$:

$$\frac{2}{3} = 1 - \exp\frac{-t}{C_T R_T},$$

$$\exp\frac{-t}{C_T R_T} = \frac{1}{3}$$

therefore

$$t = C_T R_T \ln 3$$

or

$$t = 1.1\, C_T R_T.$$

Hence, the timed period of the 555 is 10% greater than the $R_T C_T$ time constant. The maximum value of R_T is about 10 MΩ, limited by internal leakage current. Whilst there is no theoretical limit to C_T, the time constant of large electrolytic capacitors is limited anyway by internal leakage. Reliable delays of about one hour are obtainable, given by $C_T = 1000$ μF and $R_T = 3.3$ MΩ. The 7555 CMOS version of this chip allows still higher impedances, with R_T up to 100 MΩ and timing periods of many hours.

Fig. 13.50(*a*) shows the 555 operated by pushbutton switches and controlling a relay. Here the relay is energized when the start switch SW1 is momentarily closed to reset the timer. In addition to the usual damping diode across the relay coil, the series diode in the output avoids negative back e.m.f. reaching the timer, since this can upset its operation. The 4.7 kΩ resistor in the trigger circuit limits input current, and should also be included if a TTL

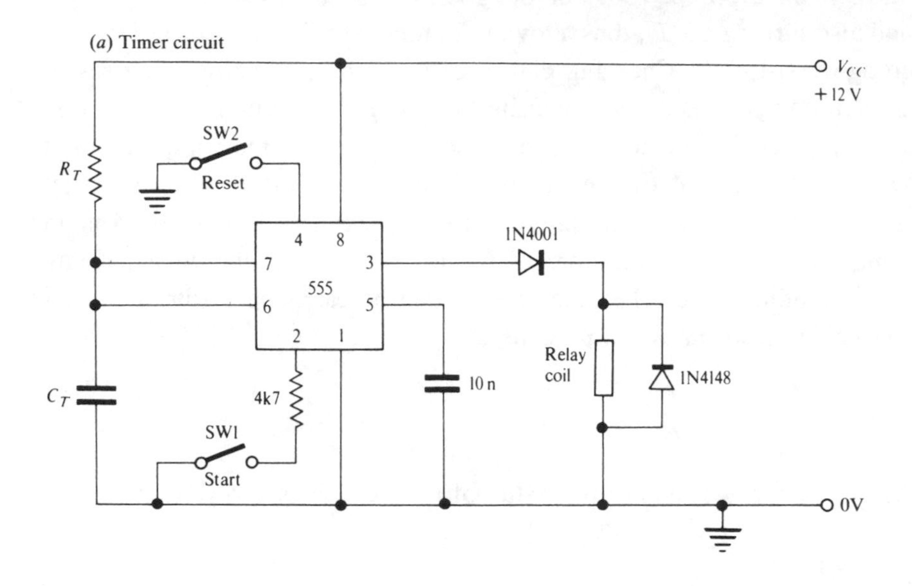

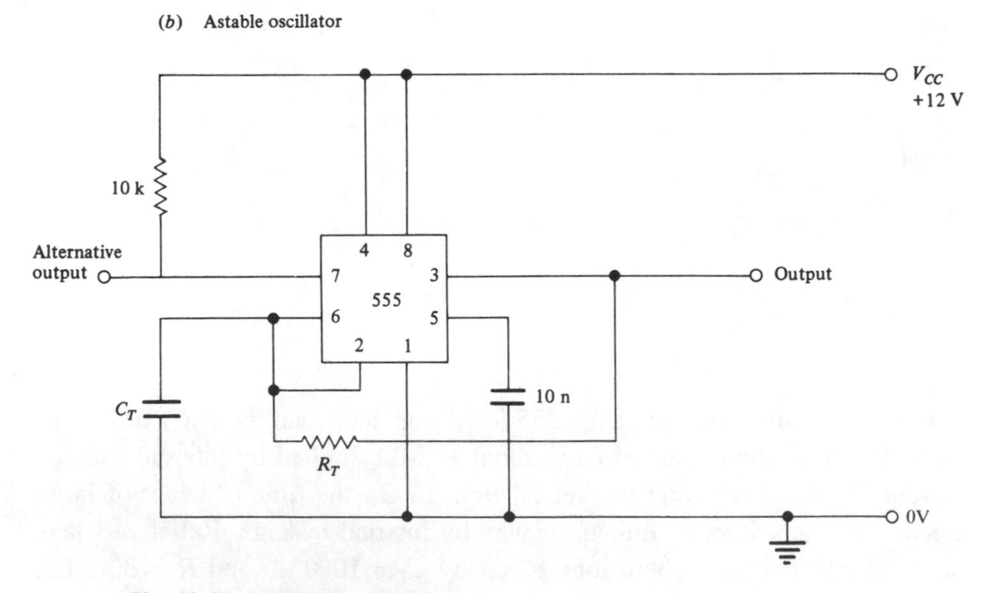

Fig. 13.50. Applications of 555 timer.

drive is used. It is normal to decouple pin 5 with 10 nF as shown: it reduces noise on the reference input of comparator 2.

A useful astable oscillator is shown in fig. 13.50(*b*) and has applications from lamp flashers to tone generators. Timing resistor R_T is fed from output pin 3 instead of V_{CC}. C_T is disconnected from the discharge transistor on pin 7 and connected to pin 2 trigger input instead. As before, C_T charges to $2V_{CC}/3$, at which point the flip-flop resets, switching the output to 0 V. Now, instead of C_T being suddenly discharged by T_1, its discharge takes place via R_T until it reaches $V_{CC}/3$, when the trigger input sets the flip-flop again. The result is a symmetrical 50% duty cycle square wave with a frequency

$$f = \frac{1}{1.4 C_T R_T}.$$

555 timers can be operated in cascade to obtain complicated sequences. Fig. 13.51 shows how a 1 nF capacitor is used to trigger the input from the previous output. It is convenient in this type of application to use the 556, which is a 4-pin IC containing two identical 555 timers.

Even more compact are the 553 and 554 containing no less than four timers in one 16-pin package.

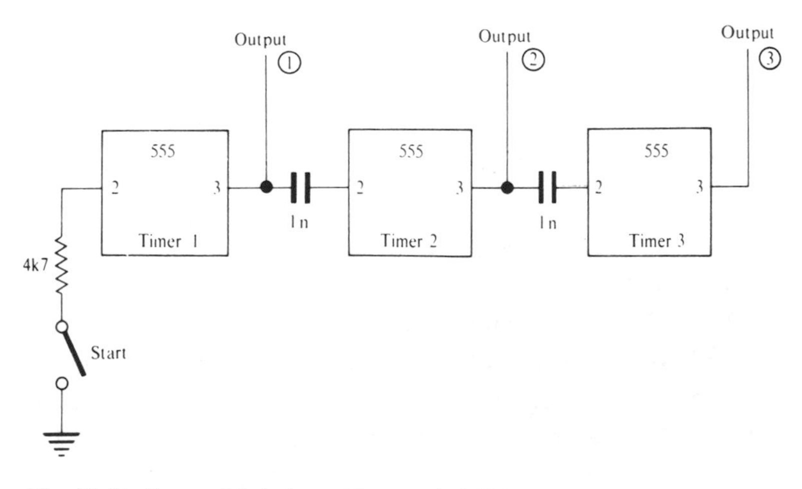

Fig. 13.51. Sequential timing with cascaded 555s.

13.16 Data multiplexers

A vital activity in digital systems is that of ensuring that data is switched to the right places. For example, if we need to add together the digits on the outputs of two registers we need to ensure that those register outputs are connected to the adder inputs for that particular operation. At other times, we may need the adder inputs to operate on some entirely different signals such as two counter outputs. Such a *data selection* function is the electronic equivalent of a multi-way switch and is known as *multiplexing* (often abbreviated to *MUX*).

Fig. 13.52 shows how simple gates can be used to select one of two inputs on to a single data line. When select line S is low, D_0 is selected through to line Y because its AND gate is enabled, D_1 being selected instead when S goes high.

As the name suggests, *demultiplexing* is the opposite operation, used to switch a single data output to several different places. Once again, a two-way operation can be implemented with simple AND gates (fig. 13.53). Data input D is sent to Y_0 with select line S low and to Y_1 with S high.

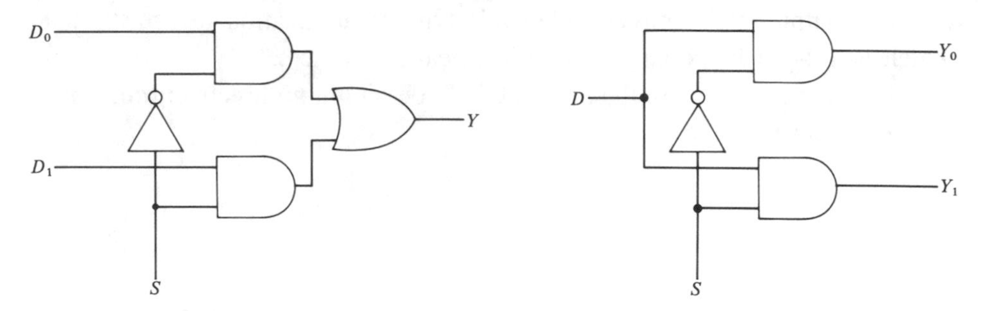

Fig. 13.52. 2-way data selector (multiplexer).

Fig. 13.53. 2-way demultiplexer.

Multiplexing or demultiplexing usually involves more than two data sources or destinations, and special chips are made for the purpose. Fig. 13.54 illustrates the 74HC151 8 into 1 multiplexer, which is the logical implementation of a single pole 8-way switch where the switch position is determined by the binary number on the select inputs $S_2S_1S_0$ and the output taken from the wiper. The single output Y carries data from whichever input is selected from D_0 to D_7. The complement of Y, $\overline{Y}$, is also available. If none of the inputs is required then the enable input $\overline{E}$, shown here connected to 0 V, is taken high.

Fig. 13.55 shows how eight outputs can be demultiplexed from a single line using the 74HC138, which is like a single pole 8-way switch with the input connected to the wiper. The bar symbol on data input $\overline{E}_1$ indicates that it is *active low*, this inverting characteristic being cancelled out by outputs Y_0 to Y_7 also being negated. The net result is no inversion through the demulti-

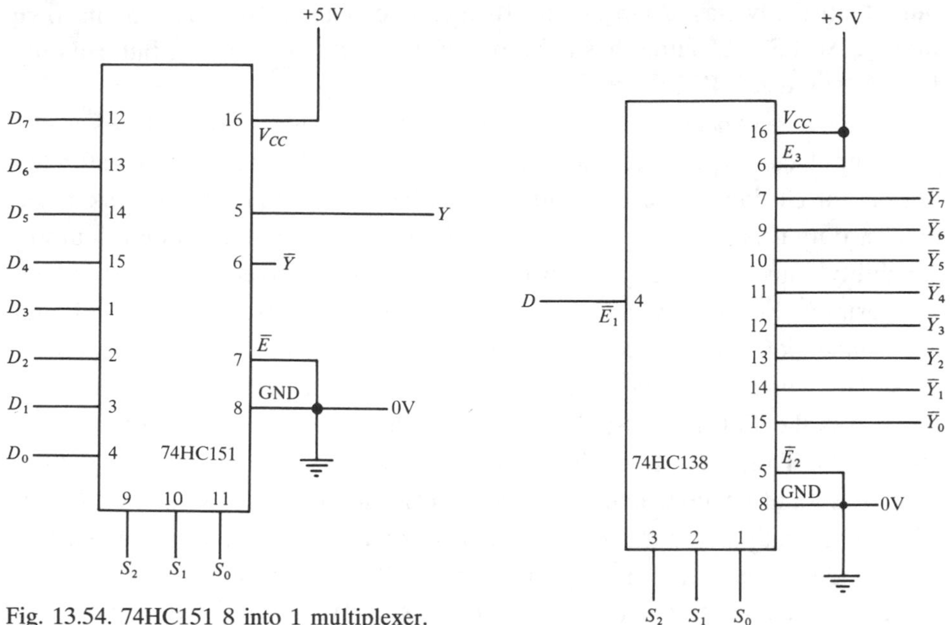

Fig. 13.54. 74HC151 8 into 1 multiplexer.

Fig. 13.55. 74HC138 1 of 8 demultiplexer.

plexer, but it should be noted that unselected outputs sit high instead of the low that would be expected from a simple rotary switch. Additional enable inputs $\overline{E}_2$ (active low) and E_3 (active high) are also available.

It will be noticed that the demultiplexing process is very similar to decoding binary numbers, and reference back to fig. 13.38 will indicate that the 74154 can be used as a 1-of-16 demultiplexer by removing the 0 V link to $\overline{E}_1$ and using it as a data input.

13.17 Interconnection of logic chips

13.17.1 General precautions

Logic circuit design demands attention to both theory and practice. Not only must the resulting system of logic gates, arithmetic functions, counters, etc. carry out the required process in principle, but all the chips must interface properly with one another. In other words, each gate must be capable of feeding the next gates in line. In some circuits it is necessary to think how many gate inputs we can connect to one output before it overloads. This digital example of impedance matching is described by the gate *fan-out*. One TTL output normally has a fan-out of 10, i.e. it can feed 10 inputs of its own family. An LSTTL output has a fan-out of 24 for its own family, but can only feed 6 'standard' TTL inputs.

As might be expected, the 74HC00 series, with its FET inputs, draws negligible input current at d.c., and hence any output exhibits almost unlimited fan-out for HC inputs. Care should however be taken with the total capacitance that an output is asked to drive at high frequencies. Its 4 mA output current capability imposes an upper capacitance limit for a given frequency which the designer can readily calculate. Each HC input represents about 4 pF and interconnection capacitance, for example of a PCB track, should be included.

An HC output can drive up to 10 LSTTL inputs and 2 TTL inputs. Surprisingly, it is the TTL and LSTTL outputs which need a little help when driving HC inputs. To get the transition voltage to the right level, a 4.7 KΩ *pull-up* resistor should be connected from the TTL output to V_{CC} as shown in fig. 13.56. On multiple outputs, a *resistor network* containing 4, 8 or more resistors in a single in-line (SIL) package is a neat pull-up solution.

The pull-up resistor can be avoided by specifying the 74HCT family of CMOS chips which is designed for direct connection to TTL. It is also useful to note, if using the LV 3.3 volt logic, that the 74LVT family uses hybrid CMOS/TTL circuitry enabling interfacing to and from 5 volt TTL.

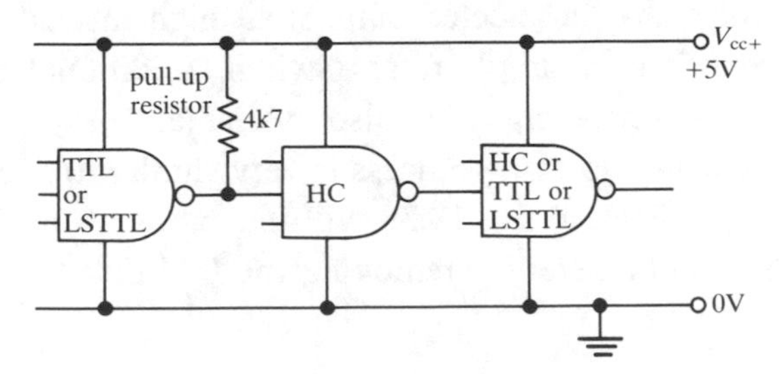

Fig. 13.56. Interconnection of TTL and HC logic families.

13.17.2 Gate connections and operation

(*a*) Data pulse rise and fall times should be <50 ns for maximum noise immunity. Use a 7413 or 74HC13 Schmitt trigger to interface slow-edged waveforms.

(*b*) Work well within the fan-out capability of gate outputs, otherwise logic levels get too close to indeterminate state. Particular care is needed on clock lines which are feeding many gates. The 7437 NAND buffer can be useful here; it is able to feed as many as 60 LS inputs comfortably.

(*c*) Width of data pulses should be at least 30 ns for reliable gate and counter operation.

(*d*) 400 mm is the maximum length of wire which should be used on the output of a flip-flop or counter (100 mm with 74AC logic). Delayed pulse reflections occurring in longer connections can upset flip-flop operation. Strictly speaking, this criterion should be adopted for all gate outputs, although longer lengths are usually satisfactory if run close to an earthed chassis (ground plane). Lines of 500 mm or more should be run with an adjacent earth wire in twisted pair or coaxial form, but ensure that sufficient drive current is available to cope with the line capacitance. Line termination resistors can also be useful for avoiding reflections and hence multiple pulse transitions. Experiments with values from 150 Ω–470 Ω to 0 V and to V_{CC} can often prove worthwhile.

(*e*) Unused inputs of NOR gates should be paralleled with the inputs in use or, if this will cause fan-out problems, they can be tied to earth.

(*f*) Unused inputs of AND and NAND gates should be paralleled with the inputs in use or tied to V_{CC}. It is a useful precaution to connect a 1 kΩ resistor in series with any such inputs tied to V_{CC}. In the event of a voltage surge or spike on the supply, the resistor will limit input breakdown current to a safe value, avoiding destruction of the gate. This precaution also applies to preset and clear inputs of flip-flops and counters. Up to 25 inputs can be connected to one 1 kΩ resistor.

(*g*) The completed circuit should be electrically screened by mounting in a metal box. Diecast boxes are useful for small assemblies, whilst steel or aluminium instrument cases will house larger logic systems.

13.17.3 Power supplies

(*a*) The specified power supply for TTL is 5 V ± 0.25 V. 74HC and 74AC CMOS chips cope with 2 V to 6 V supply range, but must run from 5 V if interfaced to TTL. Ripple should be <5% peak-to-peak. These requirements are readily met by an IC fixed-voltage regulator such as the L005, 7805 or LM309 (see fig. 9.32). Remember to allow sufficient current output – even modest logic systems can draw several hundred milliamps. As a rough guide, most LS basic gates draw from 2 mA to 3 mA per IC package, whilst LS counters and registers typically draw 10–30 mA per package.

(*b*) The fast rise and fall times of logic waveforms involve rapid charging and discharging of stray capacitance, drawing transient currents from the power supply. There is also slight 'conduction overlap' in the two transistors in the totem pole output stage so that a momentary current surge occurs at the logic transition point. Such surges must not be permitted to propagate along the 0 V and power supply lines because the resulting noise could interfere with gate operation elsewhere. The solution is to decouple the 5 V supply to earth at every IC package, using a 100 nF ceramic capacitor, which presents a very low impedance at high frequencies. The current surges are thus supplied by the local decoupling capacitor rather than the main power supply line. In order to break up the regular and potentially high-Q resonant ladder network of decoupling capacitors, it is also good practice to insert a 22 μF electrolytic capacitor every 10 chips or so.

Multi-layer PCBs are frequently used. These provide separate V_{CC} and 0 V (ground) planes to achieve 'stiff' low impedance power feeds.

(*c*) Electrical noise often arrives via the mains supply. It is advisable to interpose a proprietary mains suppression filter where the mains enters the power supply screening case. These filters incorporate special low-capacitance inductors together with ceramic capacitors to give rejection of some 30 dB over a typical frequency range of 500 kHz to 100 MHz. Today's stringent electromagnetic compatibility (EMC) standards make mains filters necessary for most equipment.

13.18 Emitter-coupled logic

We have seen that high speeds in logic circuits can be achieved by preventing the switching transistors entering saturation, with its inherent slow recovery due to carrier storage effects. Schottky TTL avoids saturation by 'bypassing' critical junctions with fast diodes of low forward p.d. An alternative and even more effective, though power-hungry, approach uses a variant of the long-tailed pair differential amplifier where switching simply involves transfer of current from one transistor in the differential pair to the other. This *emitter-coupled* logic (ECL) also achieves the low output impedance necessary for driving fast pulses into input capacitance by feeding the differential amplifier into an emitter follower stage. Clock frequencies of up to 3 GHz are possible.

13.19 Gate arrays

Many of the circuits in this chapter have been relatively simple applications of logic in themselves, but they provide an indication of the way that logic systems can quickly build up to a complex level. The reader will soon find that use of the various counters, displays, registers, arithmetic and logic building blocks in measurement and control experiments can lead to some quite impressive systems, all built from the basic elements described.

The next stage of circuit development, to be described in chapter 14, is the use of a microcomputer with appropriate software which, perhaps surprisingly, provides a great simplification in sophisticated logic systems. All microcomputer systems, however, also need a miscellany of *random logic* in addition to the few main microprocessor, memory and interface chips. These miscellaneous requirements range from data registers and counters to basic NAND gates and inverters; they are sometimes called 'TTL glue' because of their role in 'holding' the more sophisticated elements together.

Electronic design for quantity production aims for the minimum number of chips, thus reducing costs of both assembly and testing. It is therefore an advantage to collect together the 'TTL glue' in an *application-specific IC* (*ASIC*), reducing chip count by 100 or so. Clearly, the precise contents of such an ASIC depend upon the specific circuit, so that they cannot usually be economically designed from scratch as a standard component. The answer lies in the *semi-custom* IC which is a largely 'predesigned' array of general-purpose gates or gate-combinations (standard cells). The final stage of the design, the mask defining the interconnections of gates or cells, is then determined by the user's requirement. In this way, the circuit designer can have his own special chip designed and manufactured from a gate array for a small fraction of the cost and time required for a totally custom IC.

Gate arrays and cell arrays are normally made in CMOS, with some types even allowing the inclusion of linear circuit elements. Different manufacturers choose their own terms for the chips, e.g. PLA: programmable logic array and ULA: uncommitted logic array.

ASIC design is facilitated by Computer Aided Design (CAD) systems which enable the schematic circuit diagram to be drawn on screen and checked for its correct functionality before the gate interconnection *netlist* is supplied to the ASIC manufacturer. Special software design tools such as *VHDL* are increasingly exploited in order to ensure that a logical and self-consistent approach is adopted to logic hardware design. Unlike traditional interactive prototyping approaches to circuit design, CAD systems enable the whole system to be modelled on the computer so that it is right first time when committed to silicon chip fabrication.

13.20 Programmable logic devices

Increasingly, small ASICs are being superseded in small quantity applications by user-programmable gate arrays known as *programmable logic devices* (PLD) or *field programmable gate arrays* (FPGA).

PAL (programmable array logic) is one form, also known as IFL (integrated fuse logic). This is programmed by the user, being manufactured to include ready-made gate interconnections in the form of titanium–tungsten fuses. These fuses are then 'blown' by the user as required (similar to PROM programming described in the next chapter), thus leaving only the interconnections needed for the application in question.

FPGAs use a clever circuit whereby each fuse is associated with an inverter, which actually enables the gate when the fuse is blown (known as an anti-fuse). Thus the circuit can be readily synthesized from scratch. Today's FPGA chips achieve up to 20 000 gate capacity on one chip and can exceed clock speeds of 100 MHz by the use of 0.5 μm CMOS device dimensions. Erasable programmable logic devices (EPLDs) are available either as UV or electrically erasable devices and are useful for small quantity prototyping but lack the speed of the FPGA.

13.21 Switching analogue signals with CMOS

Unlike the path from emitter to collector of a bipolar transistor, the channel of a MOSFET contains no pn junctions. This means that CMOS techniques can be used to make a symmetrical switch which can be interposed in a circuit path and will pass signals in either direction. Fig. 13.57 shows such a CMOS *bilateral switch.* The switch 'contacts' may be opened or closed at will by the application of the appropriate logic level to the control input. The 74HC4016 chip contains four such switches.

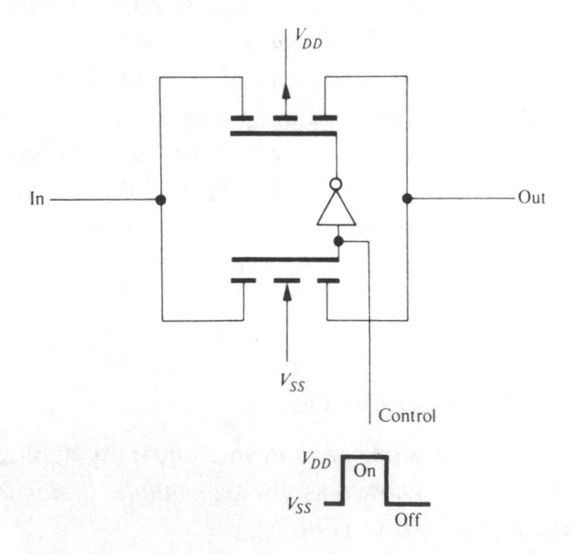

Fig. 13.57. CMOS bilateral switch. Note the inverter between the gates.

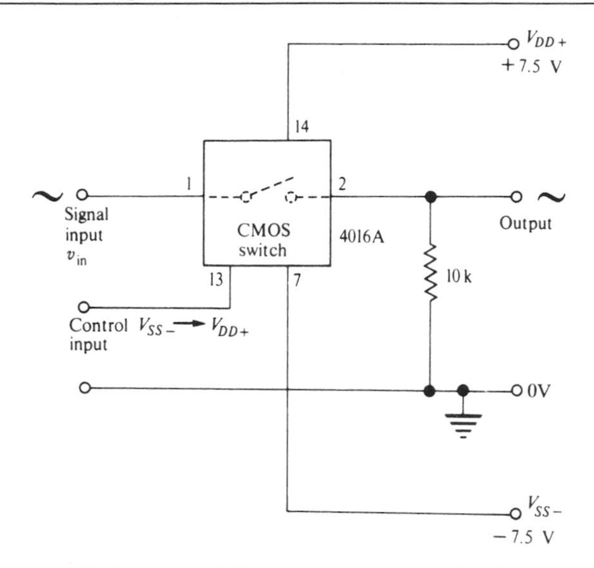

Fig. 13.58. CMOS bilateral switch with split-rail power supplies for a.c. signal operation $V_{SS} < V_{in} < V_{DD}$. IC pin numbers refer to 74HC4016 DIL package, which contains 4 switches.

For operation with normal positive logic signals, V_{SS} is the common line (0 V), and the circuit employs the usual single supply rail V_{DD+} (+3 V to +15 V). The switch is on (closed) when the control input is held at V_{DD+} and off with the control earthed. In the on condition, the switch typically exhibits a resistance of 20 Ω, rising to 100 MΩ in the off condition.

One of the most interesting features of the CMOS bilateral switch is its ability to transmit analogue signals faithfully. With a single supply rail, transmission is limited to positive-going signals, but operation with normal a.c. signals can be achieved with split rail power supplies, V_{DD+} and V_{SS-}. The signal earth is the supply centre point. The switch is on with the control input at V_{DD+} and off with the control at V_{SS-}. Fig. 13.58 shows a bilateral switch connected for operation with a.c. signals. The maximum peak signal amplitude which can be switched is $\pm V_{DD}$ (normally ±7.5 V). With appropriate drive circuitry the switch operates at frequencies up to 100 MHz.

CMOS bilateral switches find application wherever analogue signals are controlled by digital signals. Control of audio signals by microcomputer in broadcast and recording studios is one such application. Many other specialized analogue switching chips are available, such as the popular DG508 8 into 1 multiplexer/demultiplexer.

14

Microcomputer circuits and applications

14.1 Computer operations

The word *computer* is one we naturally associate with highly complicated calculations carried out at a speed immensely faster than human thought. Indeed, the speed of today's 'supercomputer' is limited only by the velocity of light itself, which ultimately determines how fast the binary logic signals can travel round the circuits. In this chapter, we shall see that a computer is much more than a fast calculator. In addition to performing arithmetic, most computer operations are dominated by the rather dull but vital business of shifting, sorting and matching data. Whether the end result is the diagnosis of a patient's medical condition, a foreign language translation or even a video game missile shooting across the screen, most of the impressive results achieved in this computer age are achieved simply by moving numbers around as required and comparing them with one another.

Although actual arithmetic is a much less dominant function in most computers than is popularly believed, the *arithmetic and logic unit* (ALU) lies at the centre of every computer. This is an arrangement of gates capable of carrying out the logical functions discussed in the last chapter together with elementary arithmetic. It is appropriate first therefore to look further at electronic arithmetic circuits.

14.2 Electronic arithmetic

14.2.1 Addition

In chapter 13, simple combinational logic experiments led quickly to the addition of two binary numbers in the full adder (section 13.8). In this chapter, we are able to extend the study of electronic implementation of arithmetic to include not only addition but also the other main operations of subtraction, multiplication and division.

Fig. 14.1 shows a practical addition circuit using the 74LS83A single chip which adds two 4-bit numbers plus incoming carry. The input numbers, $A_3A_2A_1A_0$ and $B_3B_2B_1B_0$, are provided by binary encoded 'thumbwheel' or rotary switches; these may be ten-position BCD-type with numbers 0–9 or hexadecimal with 16 positions and the numbers 0–F (hex). The latter, although less familiar than the decimal version, have the advantage of exploiting the full 4-bit range. They also provide practice in the hexadecimal system, which is helpful shorthand for handling binary numbers and is therefore used frequently in this chapter. The output 5-bit binary number, $S_4S_3S_2S_1S_0$, includes the output carry and can be displayed directly using a 7404 inverter chip to drive LEDs to represent the bits (fig. 13.35).

For a more familiar output, a restricted range of output numbers (0–9) can be displayed by decoding $S_3S_2S_1$ and S_0 into a BCD 7-segment display (fig. 13.38). A hexadecimal display covering the full range (0–1F) could con-

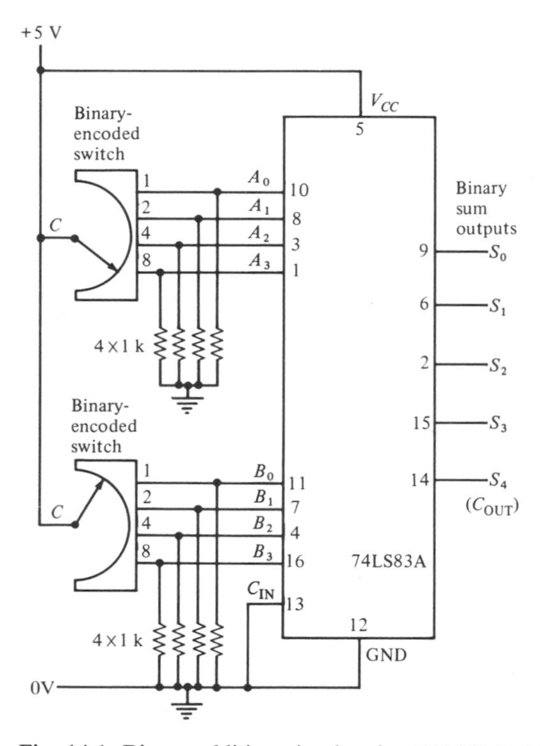

Fig. 14.1. Binary addition circuit using 74LS83A chip. Binary sum may be displayed using direct LED drivers or decoder/drivers.

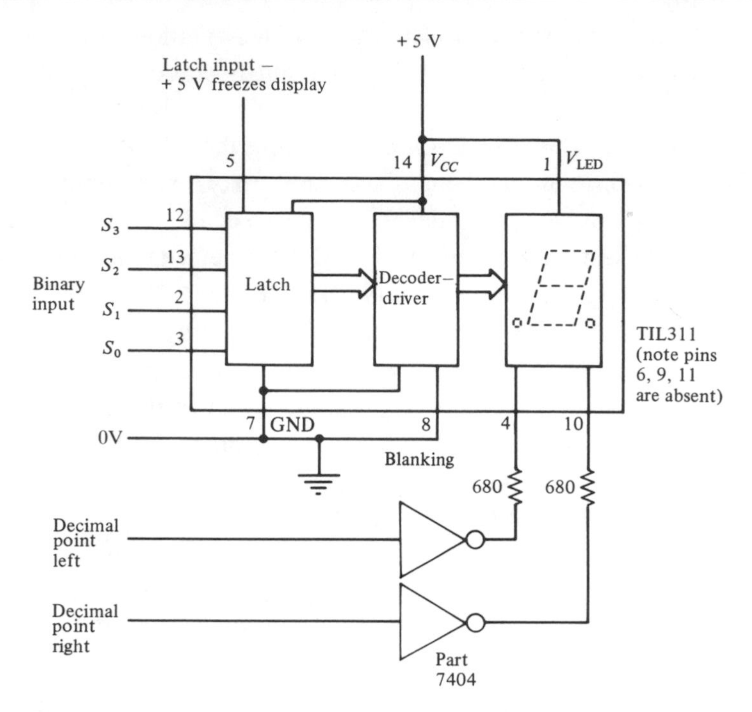

Fig. 14.2. Application of TIL311 to produce hexadecimal character from binary input.

veniently use the 4–16 line decoder of fig. 13.36 supplemented by a single 'carry' LED driven by an inverter. A rather more costly, though elegant, hex display is provided by the Texas TIL311 (RS 586–734) IC, which contains a decoder, latch and an integrated dot-format LED display capable of indicating 0–9 and A–F in response to the 4-bit binary input. Fig. 14.2 shows the circuit required for this chip to give a hexadecimal display from a binary input. Multiple digit displays are readily built up by allocating the binary lines in groups of four to separate TIL 311 chips, beginning with the least significant bit.

14.2.2 Subtraction

It is only a small step from addition to subtraction, since the latter operation really just involves making one number negative and then adding. A negative value presents no problem for analogue circuits, where it is simply a negative voltage, but digital circuits cannot inherently represent ordinary negative signals. With only the levels +5 V and 0 V (logic 1 and logic 0) we have no equivalent of the negative sign. Fortunately, the rules of mathematics are very adaptable and, as long as we are dealing only with a limited range of numbers, we can represent both positive and negative numbers without using a negative sign. The principle relies on the fact that a negative number is simply the number which has to be added to a positive number of the same magnitude to get to zero. The key to avoiding the need for a negative sign is first to define the range of numbers that we are working with. Just suppose, as an example, we are limited to the digits 0–10, then we can define positive numbers by counting upwards as usual, i.e. 1, 2, 3 etc., and 'negative' numbers by counting downwards from 10, i.e. 9, 8, 7 etc. No negative sign is then needed. What is more, if we rule that 10 is effectively identical to zero (just forget the carry one), then the rules for negative numbers come right.

Fig. 4.13 represents the 'circular' form of this *ten's complement* convention: it may be helpful to think of it as a continuous analogue meter scale. The '+/–' equivalent numbers are shown in brackets. So long as we ignore the carry digit in any arithmetic, then all the calculations work out correctly. What about 5 which lands on the positive/negative boundary – is it considered positive or negative? We can actually decide either, as long as we stick to the convention once decided. For our purpose, we shall call it negative, therefore the maximum positive number must stop short at 4.999 recurring. Any negative number of magnitude greater than 5 or any positive number greater than 4.999 recurring is said to *overflow* because it runs off one end of the scale only to reappear with the opposite polarity at the other end.

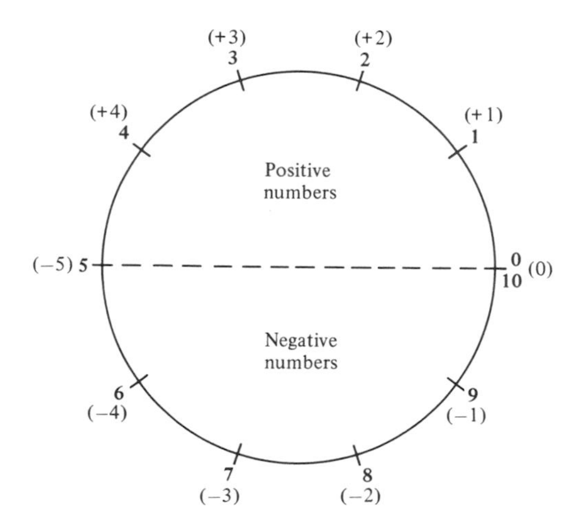

Fig. 14.3. 'Ten's complement' convention for representing negative numbers without a minus sign. Equivalents in +/– convention are shown in brackets.

The examples in Table 14.1 show how the complement representation works:

Table 14.1

+/– convention	Ten's complement convention
4 + (–1) = 3	4 + 9 = 13
1 + 3 = 4	1 + 3 = 4
4 + (–3) = 1	4 + 7 = 11
2 + (–5) = –3	2 + 5 = 7
–3 + 3 = 0	7 + 3 = 10

Everything works out in our new convention as long as we ignore any carry digit and we do not stray outside the boundaries of our defined scale from +4.999 recurring to –5.000 (+/– convention) to avoid overflow.

All this discussion is leading up to the expression of positive and negative numbers in binary form so that we can handle them with digital electronics. Here, instead of ten's complement we use *two's complement* convention. Fig. 14.4 shows how a 4-bit binary number can represent the decimal range +7 to –8 instead of 0–15 as we have considered it previously. Of course, we must be consistent in our rules so that we never try simultaneously to use 4 bits to represent the positive numbers 0–15 and also in two's complement form for +7 to –8. If in the latter case we need to represent numbers larger than +7 then we need more bits.

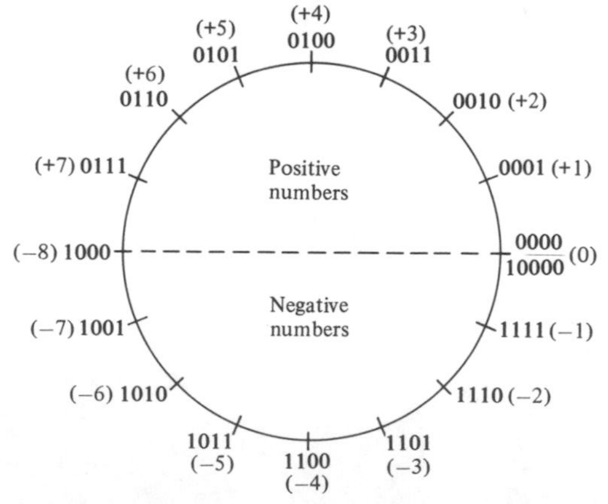

Fig. 14.4. 4-bit binary two's complement convention. Equivalent numbers in +/– decimal convention are shown in brackets.

In fig. 14.4, an important and rather interesting feature of two's complement convention can be seen: all the positive numbers begin with 0 and all the negative numbers with 1. This immediately gives quick identification of sign; in fact the most significant bit is often known as the 'sign bit'. It is, however, important to remember that two's complement is a cleverer system than merely attaching a 1 or 0 to a number to indicate its sign. We can express very simply the three rules for converting a positive binary number into its negative equivalent in two's complement form:

(1) Express the positive number in the specified number of bits.
(2) Complement every bit (turn 0 to 1 and 1 to 0).
(3) Add 1.

The reason for adding 1 will become clear from fig. 14.4, where we see in our 4-bit example that zero (0000) is equivalent not to 1111 but to 10000, with the carry 'one' ignored.

The following examples demonstrate the working of two's complement binary arithmetic. As a guide, the +/– convention decimal equivalents are given alongside:

0110	6
+1101	–3
~~1~~0011	3
0010	2
+1110	–2
~~1~~0000	0
0111	7
+1111	–1
~~1~~0110	6
1010	–6
+1110	–2
~~1~~1000	–8

The adder of fig. 14.1 is readily extended to provide subtraction as well by interposing the exclusive OR gates of fig. 14.5 to complement the bits on binary input *B*. The two's complement form is completed by adding a 1 via the C_{IN} carry input of the 7483. By switching the exclusive OR inputs and

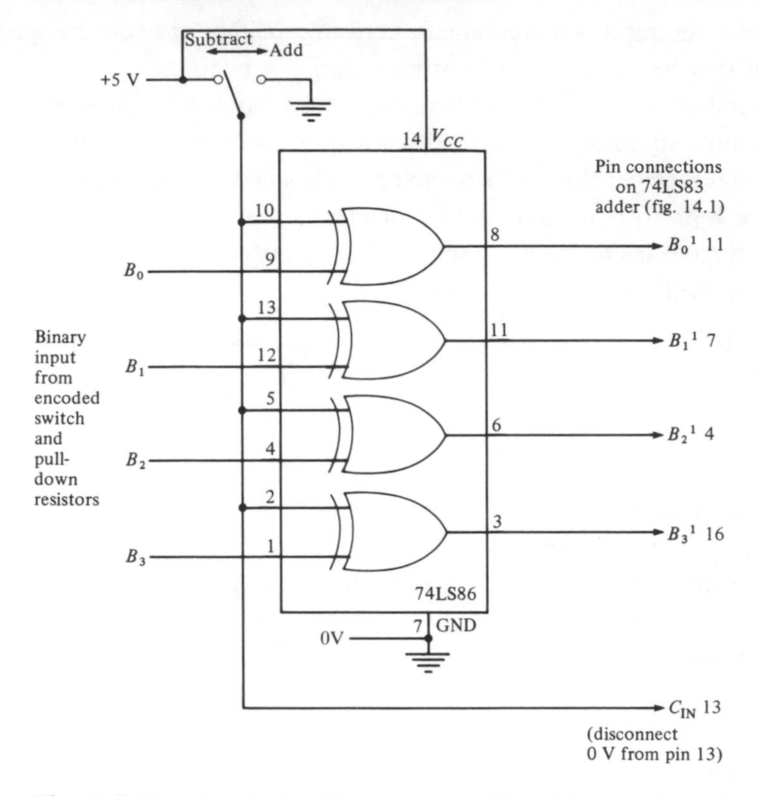

Fig. 14.5. Use of exclusive ORs to convert adder of fig. 14.1 into subtractor by complementing each bit and adding 1 via carry input.

C_{IN} between low and high, we can change the circuit function between a 4-bit adder and a 4-bit subtractor.

14.2.3 Binary multiplication and division

Whilst most binary arithmetic may appear unfamiliar to our practised decimal brains, the multiplication tables at least are simplicity itself:

$$0 \times 0 = 0$$
$$0 \times 1 = 0$$
$$1 \times 0 = 0$$
$$1 \times 1 = 1$$

We see that this is an AND function. It is worth noting in passing that the choice of the multiplication symbol for the AND function in Boolean algebra is no mere coincidence: the two operations are synonymous in the binary system.

Multiple digit binary multiplication involves the conventional 'long multiplication' operations of shifting and adding as well as the digit multiplication itself. These examples show the decimal equivalent alongside each binary calculation.

```
    101        5
    ×11       ×3
    ---       --
    101       15
   101
   ----
   1111

   1001        9
   ×101       ×5
   ----       --
   1001       45
  0000
 1001
 ------
 101101

   1101       13
  ×1011      ×11
  -----      ---
   1101       13
  1101       13
 0000        ---
1101         143
--------
10001111
```

The circuit hardware required for multiplication is therefore a combination of shift register and logical AND followed by addition. Conversely, division requires subtractions followed by shifts in the opposite direction. Both multiplication and division as expressed here in fundamental form are sequential operations and are therefore inherently rather slow. Some microprocessors do not even include them amongst their standard repertoire of functions, but require the programmer to implement the necessary sequence of AND, add, shift etc. One very fast but costly approach to high-speed multiplication is the TRW MPY 16HJ chip, which is essentially a large array of AND gates and full adders which carry out the operations in parallel. The 32-bit product of two 16-bit numbers can be produced in under 50 ns.

14.3 Bits, bytes and nibbles

Most of the exploration of combinational and sequential logic principles in chapter 13 worked with just one or two bits at a time. The longer numbers used in microcomputer systems have their own terminology, with 8 bits being conventionally known as a *byte*, its value being conveniently expressed as a two-digit hex number. It is perhaps inevitable therefore that the 4 bits expressed by a single hex digit should be dubbed a *nibble*, commonly also spelt as *nybble*.

14.4 Data buses

A computer is essentially a collection of data storage registers, an ALU and a means of moving numbers between them as required. Data switching is therefore of key importance. The multiplexer (section 13.16) behaves as an electronic selector switch which can be used for switching data to its required destination. We could therefore conceive of a computer system with a multiplexer on the input of each register with the inputs of the multiplexer wired from the outputs of all other registers. Unfortunately, such a design becomes extremely complicated, as fig. 14.6 illustrates, even with only four registers

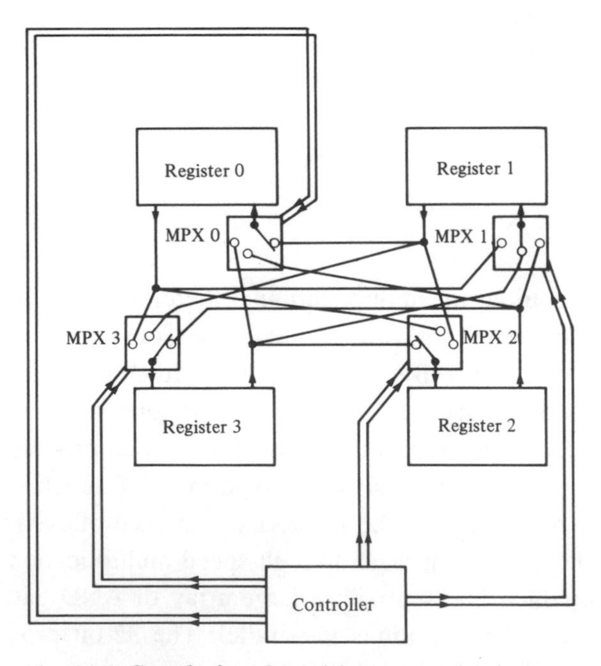

Fig. 14.6. Complexity of 'star' interconnection between registers for data transfer.

and only 1 bit per register. When we recognise that even a small computer memory will contain more than a million registers with 16 bits per register, the impracticality of such dedicated inter-connections becomes clear. It would be like building a railway system with a separate track and train linking every pair of stations that any passenger might conceivably wish to travel between.

Just as a real life railway uses a single line to interconnect many stations; so data communication uses a signal *bus* line to interconnect all the registers. The bus is appropriately named, being derived from the Latin *omnibus* (for all) because it provides a *data highway* available for transmission of numbers between any two elements of the digital system. The railway analogy cannot be pressed too far because, although several trains can run on one track at once, the electrical data bus can handle only one signal at a time. It is therefore a sequential system where the communications between registers happen in very fast succession. There are three essential features in a bus system:

(1) Any output not actually sending data must present a high output impedance so as not to interfere with signals on the bus.
(2) Every element connected to the bus has its own *address* number.
(3) In addition to the address and data lines, *control* lines are required. These may be as simple as the single read/write line shown or may also include additional device select lines where, as sometimes happens, different devices share common addresses.

The hardware simplification gained from the bus concept is illustrated in fig. 14.7. The four registers shown here require only 2 bits for the addresses: 00, 01, 10 and 11. Practical computer systems need more address capability; for instance even a tiny '64 K' (64 × 1024 bytes) memory needs 16 address lines for its 65 536 locations.

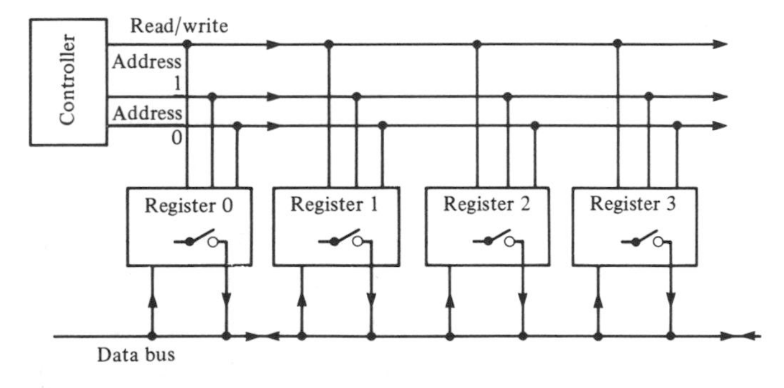

Fig. 14.7. Simplicity of data transfer using buses.

The data bus is shown here as one line for clarity, but would in practice be 8 or 16 lines for 1 or 2 byte numbers. Data inputs can all be paralleled safely on the bus so long as the outputs have sufficient drive capability (fan-out). The outputs themselves, as emphasised above, must not load the bus unless they are instructed to send data by being called on the address lines and sent a 'write' pulse. The normal TTL 'totem pole' output is unsuitable for direct bus driving because it is specifically designed to present a low output impedance at all times. One way of driving a bus is to use the *open collector* type of TTL chip where the upper totem pole drive transistor is replaced by an external resistor, a circuit type discussed in connection with driving numeric displays (section 13.13). A 2.2 kΩ load resistor is used with most LSTTL circuits such as the 74LS01 NAND gate chip and the 74LS05 inverter package (open collector versions of 74LS00 and 74LS04, respectively). When such outputs are at logic 1, no current is drawn by the output transistor so that several can be paralleled on a bus – provided only one output goes low at a time. This is sometimes known as a 'wired AND' arrangement because the bus is only high when all the outputs themselves are high. Useful in fairly complex bus systems is the 74LS38 powerful open collector NAND buffer with a fan-out of 30 and minimum load of 680 Ω, capable of driving up to 30 LSTTL inputs.

As was indicated in section 10.6, a switching circuit without active pull-up has the disadvantage of relying totally on the load resistor to charge stray capacitances, a much slower process than the 'brute force' available from the collector current of a pull-up transistor. Most bus driving circuits in both CMOS and TTL therefore use a more advanced design than open collector, called the *tri-state* output, which combines a totem pole driver with an isolating switch. Its three states are logic high, logic low and high impedance (disabled). A separate *output enable* (OE) input switches on the output circuit when required; when several tri-state outputs are paralleled on a single bus, it is important that only one OE input is activated at a time. Fig. 14.8 shows representative internal circuits for bus driving TTL gates, component values being typical of 'standard' TTL. Comparison with the conventional TTL gate in fig. 13.14 is interesting. The open collector gate (*a*) simply omits the pull-up transistor on the output. The tri-state gate (*b*) uses a simple and elegant way of decoupling the output via enable input OE. When OE is high everything works just like a normal gate, diode D_2 being cut off. When OE is taken low, the NAND action of the input emitters on T_1 ensures that T_2 and T_3 are cut off. Normally under these conditions T_4 would be conducting to pull output Y high, but in this case D_2 is conducting and thus cutting off T_4. With neither output transistor conducting, output Y has a high impedance condition, effec-

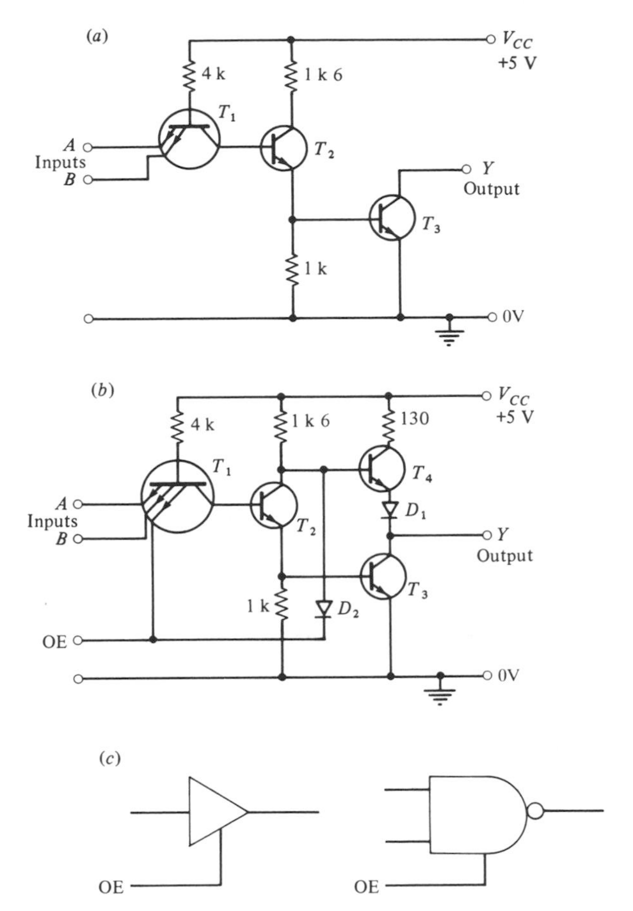

Fig. 14.8. TTL gates for bus driving. (*a*) Open collector, (*b*) Tri-state, (*c*) Tri-state symbols.

tively taking its 'hands off' the bus. Fig. 14.8(*c*) shows typical circuit symbols for tri-state gates. A wide range of tri-state devices is available, typical examples being 74LS244 containing eight buffers and 74LS374 positive edge-triggered transparent latch (8-bit register). A particularly useful chip for bus interfacing is the 74LS245 octal (8-line) bidirectional *transceiver*, which can be switched to either send or receive bus data as required; in its send direction, it has the high fanout of 30, whilst as a receiver its built-in hysteresis minimises spurious pickup of noise and reflected pulses from the bus. All are available as 74HC tri-state CMOS chips with a similar 'hands-off' capability.

The practical implementation of data buses requires careful observance of the design rules listed in section 13.17, particularly those associated with driving long lines. The large number of gate inputs hanging on a bus line increases its capacitance and makes termination resistors at the far end desirable if spurious pulses due to reflections are to be avoided. A 330 Ω resistor to 0 V and 220 Ω to V_{CC} make an effective termination for a bus driven by a 74LS245 transceiver.

14.5 Storage and memories

14.5.1 Magnetic and optical data storage

Storage and retrieval of numbers and characters is a vital part of every computer system. Long-term back-up storage of programs and data on magnetic tape, either reel or cassette, is familiar to professional and amateur user alike. Disk storage provides the advantage of random access, enabling fast retrieval of information in a different order from the way it was recorded. Magnetic recording is used for the universally popular *floppy disks*, giving capacities from 360 kbytes to 2 Mbytes, and for *Winchester (hard) disks* ranging up to 3000 Mbytes (3 Gbytes). Optical recording offers capacities up to several gigabytes and the advantage of a removable medium compared with the hard disk which is normally fixed in the computer system. A familiar application of optical data recording is, of course, the digital audio *Compact Disc* (CD), carrying over 500 Mbytes of digital data and widely used as the *CD-ROM* for PC programs.

14.5.2 Random access memory (RAM)

In sections 13.10 and 13.11, we examined the way that flip-flops are used to store bits and how they can be combined in the form of a register to store complete bytes. The shift register, with its ability to push data along a whole chain of flip-flops, is one type of memory, but in its usual form is essentially a serial, first in first out (FIFO), memory. Usually, data is used in a different order from the way it was stored: *random access memory* (RAM) is therefore required.

Millions of flip-flops can be fabricated in an array on one chip whilst allowing immediate access to every 'bit' or cell we are writing to or reading from. This is achieved by giving each cell a unique address as just discussed in connection with data buses. Fig. 14.9 shows how a *D*-type flip-flop is used as a RAM memory cell. The common data output bus is fed by the individual memory cells via open collector or tri-state outputs. Each cell is enabled by

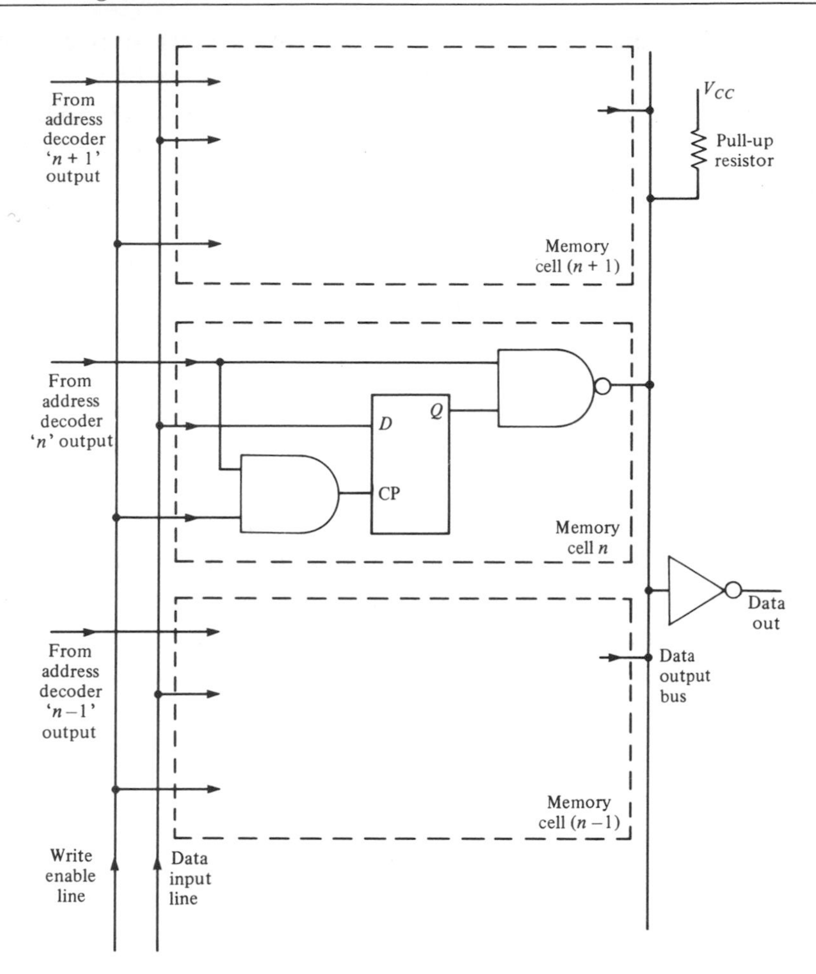

Fig. 14.9. Static RAM single bit cell.

a logic 1 from its own address decoder output, the AND and NAND acting to gate the write command and flip-flop output, respectively.

The most common type of memory is even simpler than the *static RAM* (SRAM) using the flip-flop cell just described. In *dynamic RAM* (DRAM) a binary number is temporarily stored as charge on the gate of a MOSFET, so that a very compact circuit indeed results. Fig. 14.10 shows such a memory cell; T_1, T_2 and T_3 are enhancement type MOSFETs. When the appropriate address line is activated together with the write input, a logic high from the AND gate makes T_1 conduct, thus charging or discharging C according to whether the data input is high or low (C is normally just the stray capacitance of the gate of T_2). If C is charged, T_2 conducts, taking its drain low. Whilst the cell is addressed, T_3 also conducts, thus connecting the 'wired AND' output bus to the drain of T_2 and taking it low if C is charged (data high). Correct polarity is restored by the inverter at the output. The charge on C cannot last indefinitely, so a *refresh* cycle is required, normally every 2 ms, in which each cell is read and rewritten. This operation is normally carried out by the central processor, but is sometimes incorporated in the memory chip itself. Despite this added complexity, DRAM has cost and size advantages over SRAM, its packing density extending to 256 M bits per chip. The main application for SRAM is its use for non-volatile memories, normally using CMOS circuits for low quiescent power consumption and battery supported power supplies to retain data when mains power is switched off.

A typical microcomputer memory may hold from 3 Mbytes to 64 Mbytes or even more. If each bit were to have an individual address line brought out, that would demand a massive address decoder with over half a million outputs. For convenience therefore, a memory is usually arranged as a matrix of bytes, using one address decoder for the rows and one for the columns of the matrix, thus drastically reducing the number of address decoder outputs required.

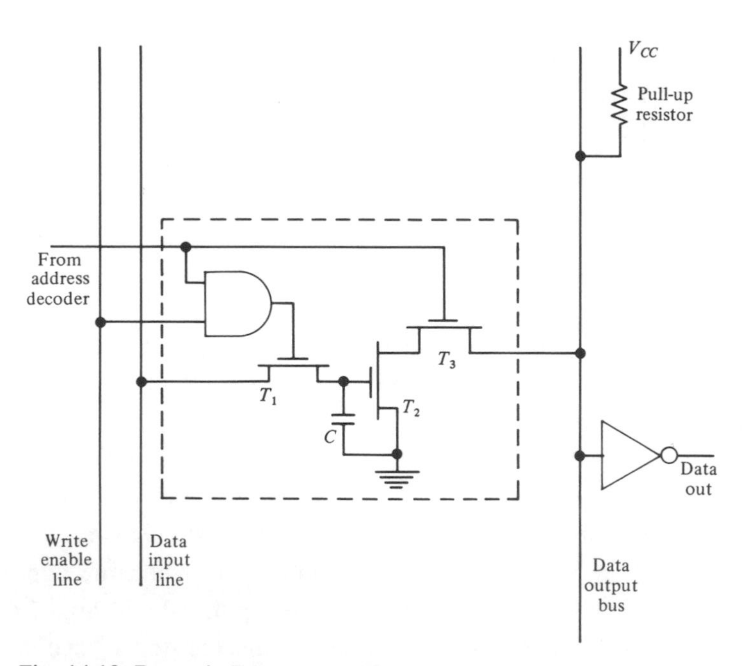

Fig. 14.10. Dynamic RAM single bit cell.

In a computer it is usual to represent the allocation of memory for different purposes such as program, data, display etc. in the form of an index of addresses known as the *memory map*, which guides the programmer to the appropriate addresses.

14.5.3 ROM, EPROM & E^{2}PROM

Most microcomputer applications use fixed data or computer programs. In such cases, the *read-only memory* (ROM) is the ideal storage medium, especially where many thousands of an identical product are produced. Rather like DRAM the ROM is basically an array of transistors (normally *N*-channel MOSFETs) but with the data pattern determined during manufacture. Where a memory cell contains a logic 1, a transistor is connected, and for logic 0 there is no transistor.

The programmable ROM (PROM) is manufactured with a transistor connected in every cell via a fusible link which can be burned out to an open circuit by passing a sufficiently large current through it. The user burns out the required links to produce his own data pattern using a PROM programmer which selects the addresses in sequence and supplies the necessary fusing current.

The erasable PROM (EPROM) is particularly valuable for prototype work and small production runs because it enables the data contents to be changed without replacing the chip. An EPROM uses *n*-channel enhancement MOSFETs as memory cells, but with special oxide-insulated floating gates. The MOSFETs prior to programming are pinched off and appear as a logic 1. Programming is achieved by applying a carefully controlled relatively high voltage (25 V) between drain and an auxiliary gate, causing temporary avalanche breakdown to occur in the oxide insulation between channel and floating gate. After removal of the voltage, breakdown ceases and charge is trapped on the floating gate, turning the transistor on and creating a logic 0 in the cell. The trapped charge is removed by exposing the chip to high-intensity ultra-violet light for about 30 minutes, thus knocking the charge off the floating gate and erasing data ready for reprogramming. Electrically erasable ROMs (EEPROMS) or E^{2}PROMS) have their gate charges removed by high voltages instead of ultra-violet light. Such devices are quicker to reprogram than EPROMs. In the popular *flash* E^{2}PROM, individual bit locations can be erased and rewritten from the normal +5 V supply; flash E^{2}PROM can therefore be treated as a non-volatile RAM.

The PROM, EPROM and E^{2}PROM can be usefully viewed by the designer not only as a data memory but also as a very compact and versatile combinational logic system. After all, a combinational logic circuit can always be expressed as a truth table producing a certain pattern of output bits for a given pattern of input bits. If the input bits are used as the address inputs to a PROM, the required output bits can be stored at the relevant addresses and very complex combinational logic functions achieved with only one or two chips. The addition of a counter with its outputs connected to the address bus turns the EPROM into a state machine which outputs a regular sequence of different bit patterns as the counter is incremented by the clock. Such circuits can be valuable in industrial control systems.

14.6 The microcomputer

Whether a computer is employed to move graphics across the screen of a video game, play music or control a chemical plant, the principles of its operation are the same – it implements a sequence of actions to send the right data to the right place at the right time. That sequence is precisely defined by the *program*, which is just a list of coded instructions executed in strict order by the control logic of the computer system. During execution the program is held as a series of binary numbers in memory, which is also used for data. Clearly, the areas of the memory map used for the program must be defined and not confused with the data areas: if the system were to try to interpret data as though it were program instructions, the resulting nonsense would immediately create chaos.

Fig. 14.11 illustrates the three elements of a microcomputer system: memory, input/output data interfaces and the *central processor unit* (CPU). The CPU is usually fabricated on one chip as a *microprocessor* containing several data registers (from 8 bits to 64 bits in width), the ALU for arithmetic and logical operations, together with the 'heart' of the system, which is the system control logic. This logic, together with the instruction register and decoder, fetches and executes instructions by switching the appropriate addresses on to the address bus and controlling the required registers, ALU and memory locations. The 'heartbeat' comes from the clock generator, a crystal-controlled oscillator running at a fixed frequency, usually within the range 10–200 MHz. Intimately connected to the system control logic is the program counter, which steps through the program memory locations as the instructions are executed, thus keeping the whole system running through the sequence specified in the program. The sequence of program execution is very rarely in strict memory location order, but often includes jumps forwards

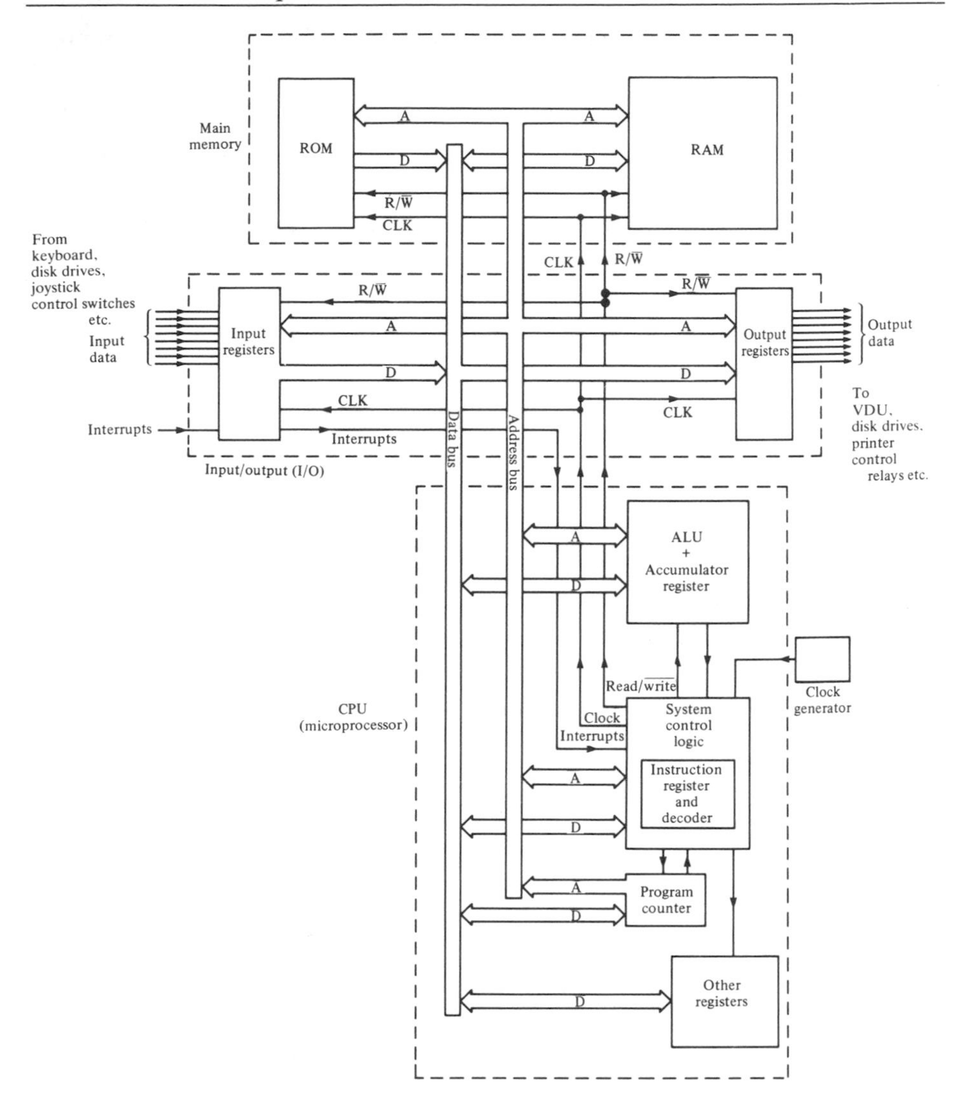

Fig. 14.11. Block diagram of typical microcomputer.

or backwards to different instructions depending on the results of individual operations (e.g. 'GOTO' instructions and jumps to subroutines and procedures). A further distraction to be handled by the system control logic is created by the *interrupt* lines associated with the input interface. These are specific hardware input connections to the processor and, when one is activated, the program stops after the instruction currently being executed and the processor is directed to run a program routine held at another defined memory address.

The instruction decoder and system control logic are most commonly implemented from interconnected logic on the CPU chip, though some of the most sophisticated CPUs employ a small subprocessor for the job which is internally preprogrammed with its own *microprogram*, giving greater flexibility in the chip design.

Microprocessor chips such as the 6502, Z80, 80486 and Pentium contain only the CPU, and need to be supplemented with I/O interface chips, RAM and ROM memory plus an ample number of latches and gates to 'glue' the microcomputer system together. For applications not requiring long programs, however, single chip microcontrollers such as the 8051 and 6805 are very useful because they include memory, I/O interfaces and even the clock generator on a single chip which can cost even less than the printed-circuit board it occupies.

Fig. 14.12 shows the pin connections for the popular 6502 microprocessor chip, the heart of the venerable Acorn/BBC Microcomputer. The 8-bit data bus and the 16-bit address bus dominate the picture. Note that the CPU chip is the source of the address bus, whilst data on the data bus can originate in any part of the system and hence it is shown as bidirectional. The read/write line and clock outputs complete the bus systems. The 16-bit address bus limits

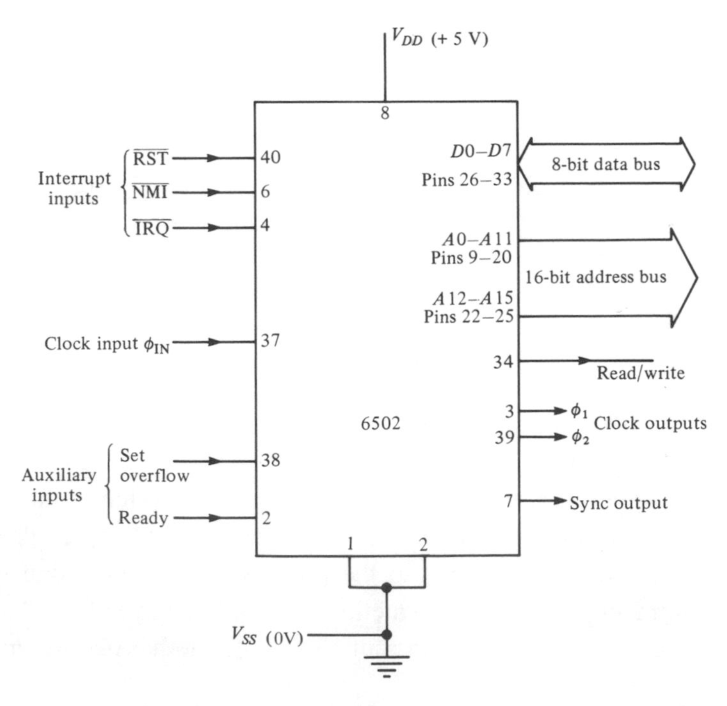

Fig. 14.12. Connections to the 6502 microprocessor IC (40-pin DIL package).

the maximum directly addressed memory to 64 kbytes. Although small by today's standards, where a 32-bit address bus is normal, the 6502 avoids unwieldy numbers of pin-out connections and is typical of the 8-bit processors at the heart of many microcontrollers. The BBC Micro or the BBC Micro emulator on today's 'Archimedes' can give the reader direct experience of 6502 programming, as we shall see shortly.

On the input side of the 6502, the interrupt lines are $\overline{\mathrm{RST}}$, $\overline{\mathrm{NMI}}$, $\overline{\mathrm{IRQ}}$. The clock input is fed from a simple external crystal oscillator such as the circuit of fig. 12.7 (typical clock frequency 2 MHz for the 6502A chip). The two auxiliary inputs are only used under special circumstances. The 'ready' input, together with the 'sync' output, allows modification of instruction execution timing. 'Set overflow' provides a simple external input able to set one bit in the internal 'flag' register to logic 1, which can be used to change the sequence of events during program execution.

14.7 Software

14.7.1 Instruction sequence

The microcomputer is a powerful and flexible machine, but its very flexibility means that without a list of instructions it can do nothing. Every computer user quickly appreciates that the hardware itself is only the starting point for a useful system – without its program it is as useless as an aircraft without a pilot and flight plan. It is to recognize the equality of its importance to hardware that the computer program is usually called the *software.*

A full study of software lies outside the scope of this book, but a brief introduction will help to guide the reader through the programs used in the computer experiments later in this chapter.

14.7.2 Machine code

Software programs come in many forms, but, like all types of communication, must use a defined language. Having now studied the microprocessor chip, it is appropriate for us to start by looking at the language actually handled by the CPU. This language is *machine code* and, as would be expected, it uses binary numbers which we usually group together into hexadecimal form for convenience in writing down. A machine code program simply consists of a sequence of coded instructions to the CPU interleaved with the relevant memory addresses or data. The CPU is designed to recognize that certain instructions are always followed by data or a memory address. For example,

Table 14.2

Memory location	Code	Meaning
70	A5	*op code* load accumulator with contents of memory location to be specified next
71	80	*operand* memory location of first number of the sum
72	18	*op code* set carry to zero
73	65	*op code* add to accumulator the contents of memory location to be specified next
74	81	*operand* memory location whose contents we want added to first number
75	85	*op code* store contents of accumulator in memory location to be specified next
76	82	*operand* memory location in which answer is to be stored
77	60	*op code* end of this part of program

the 6502 recognizes code &A5 as an instruction to load the accumulator register with the data present in the memory address which immediately follows. Furthermore, &A5 indicates that the location will be in the 'zero page' of memory (addresses &00–&FF), and therefore the CPU knows to expect only a single-byte address following the instruction. On the other hand, &AD also instructs the CPU to load the accumulator, but indicates that the memory address will be two bytes to allow the use of any location in the 64 K memory. Some instruction codes like &18 (set 'carry' bit to zero) need no memory location to be specified after them, and the CPU recognizes that it must treat the next byte as the next instruction. In this way, the CPU traces its path through the program, differentiating between those numbers which are meant to be treated as instructions and those which are data or memory addresses. The instruction numbers are known as *op codes* and the data or address following an op code is called the *operand*.

The machine code program in table 14.2 is a simple example in 6502 language which adds two numbers together. All digits are in hexadecimal. The left-hand column shows the memory locations which we have chosen for our program code; each location is then followed by the op code or operand.

14.7.3 Starting the program running

If we were dealing with a 'naked' microprocessor chip, we should enter the machine code into its memory locations directly by manually setting up addresses and data in binary using switches to inject the appropriate numbers onto the buses. The program is then started running by sending a *reset* pulse (usually 0 V) directly to the reset pin (RST) of the microcomputer chip. The internal logic of the CPU is designed to disable the system whilst RST is held low. When RST returns to +5 V, the CPU automatically loads the contents of specific memory locations into the program counter (&FFFC and &FFFD in the case of the 6502). The program counter then begins automatically to step through subsequent addresses to execute the program. Therefore, if programming a 6502, the designer must ensure that the starting address of his program is entered into locations &FFFC (low-order byte) and &FFFD (high-order byte) to act as the signpost so that the CPU can find his program. It is usual for RST to be triggered when the system is powered up, thus automatically starting the program, though a manual *RESET* key is sometimes also included on the computer. Memory locations &FFFC and &FFFD are usually ROM or EPROM memory for the obvious reason that the program start address would otherwise vanish on power down, removing the vital pointer to the first program line.

Whilst it is useful to know how the CPU itself opens for business, few readers will actually need to work with a naked microprocessor chip. That is because the quickest way of exploring this newest realm of electronic circuits is through a personal computer (PC) which comes complete with a software *operating system* already built into ROM. The operating system is a sophisticated software program which ensures that the CPU carries out all the routine tasks required of the system. For example, after a reset all registers must be cleared of random data. Then, if for example the machine is normally running BASIC, the CPU must address the first vacant memory location allocated to BASIC programs, prepare to read the keyboard and display the entered BASIC instructions on the screen. Without an operating system, all these facilities which we take for granted would be impossible and program entry would be a tedious sequence of operating many switches to present binary numbers for all codes, addresses and data. Most of the experiments which follow are designed to use the Acorn/BBC machine which incorporates the 6502 processor already discussed. Although obsolete by today's PC and 'Archimedes' standards the BBC Micro is still available in most UK colleges and schools. For our purpose here it has the advantage of providing easy access to its parallel input and output ports. Its straightforward assembly language facility is ideal for our experiments and is also available on the current Archimedes via the BBC Micro emulator.

The bus decoding examples can be applied to provide useful I/O ports for the IBM PC and compatibles (clones).

Its reliance on the operating system means that a computer does not normally start to execute the user's program in memory immediately after a reset, but prepares itself for a keyboard entry which instructs it to start. In BASIC, the word 'RUN' is all that is required.

Starting a machine code program is also easy: in common with many computers the BBC Micro requires the instruction CALL *start address* which then begins to execute machine code from that address. Interestingly enough, if the start address is wrong and the machine is asked to start executing something else, such as data, the CPU has no way of telling that this is not a program and it will plough in assuming that the numbers are op codes with chaotic results (usually a blank screen and no response from the keyboard). Such a *system crash* often requires power-down to retrieve the situation and, whilst no damage can result to the machine, much time-consuming work can be lost. The wise programmer always 'backs-up' the entered program on disk before attempting to run it.

14.7.4 Entering and running machine code directly

Our simple machine code addition program from Table 14.2 can be entered directly into memory using BASIC 'POKE' operations (?*location* = in the case of the BBC Micro). Although tedious, this technique gives a good feel for the operation of the microprocessor contained in the computer. To try out the machine code addition program, enter the following lines into the BBC Micro; no line numbers are needed:

```
?&70 =&A5
?&71 =&80
?&72 =&18
?&73 =&65
?&74 =&81
?&75 =&85
?&76 =&82
?&77 =&60
```

Having entered the program, we must poke the two numbers which we want to add together into the locations &80 and &81, respectively, which we have reserved for that purpose. Numbers may range from 0 to 255 (&00 to &FF) and can be in decimal or hex. For example, if our numbers are 13 and 19, enter

```
?&80 = 13     or in hex     ?&80 = &D
?&81 = 19     or in hex     ?&81 = &13
```

To run our simple program, we use the operating system, which responds to the BASIC instruction

```
CALL & 70
```

This entry sets the program counter to location &70 and begins execution from there onwards. The execution of the program is completed in a few microseconds, but we see no answer on the screen yet. This is because the result is left in memory location &82 where we instructed the processor to put it. The BASIC PEEK *location* instruction (?*location* on the BBC Micro) gives us easy access so we can 'open the oven door' to see our result by entering

```
PRINT? & 82
```

The computer responds after ⟨return⟩ by displaying the contents of location &82, which in this case is the sum of 13 and 19, *viz* 32. For an answer in hex, enter

```
PRINT ~ ?& 82
```

The screen will then display the answer 20.

14.7.5 Assembly code

The machine code program just entered, when shorn of the explanatory notes given earlier, is virtually impossible to understand. The binary codes used by the CPU are very remote from human language and make it difficult and tedious for any programmer to work in machine code. This problem is overcome by *assembly code*, which translates the machine code itself into alphabetical abbreviations chosen to remind the programmer of the function of each

Table 14.3

Op code	Operand	Comments
.ADDER		/label defining start of routine
LDA	BOXONE	/load first number into accumulator from memory location BOXONE
CLC		/set carry bit to zero
ADC	BOXTWO	/add second number from memory location BOXTWO into accumulator
STA	SUMBOX	/store this sum of numbers in location SUMBOX
RTS		/end of this part of program

instruction. For example, our instruction 'load accumulator', which in machine code is &A5, becomes LDA, a sensible abbreviation for the function. Because of their reminding function, assembly code instructions are called *mnemonics*.

Memory locations are also more easily handled by assembly code in that the programmer can name them by means of *labels* instead of handling hexadecimal numbers. Provided the relative location of each label is defined at the start of the program, the precise addresses in memory occupied by the program are no longer important and the program is said to be *relocatable* – an important feature for routines which may be used in a variety of different programs. As well as being used to identify operands, labels can also head up particular sections of program. This labelling is useful not only for the programmer to remember how the program works, but also for the clear signposting of branches and jumps in the program itself. For example, the conditional branch instruction BCC SMALL will divert a program to a routine labelled SMALL if the 'carry' digit in the status register is found to be zero.

A third useful feature of assembly code is that it allows the programmer's notes and explanatory comments to be included after each line of code. The inclusion of this kind of *documentation* is vital for every program. Without it, even the programmer himself may forget the reasoning behind the various routines and structures.

The value of assembly code is illustrated in the program shown in table 14.3, which is our 6502 machine code addition program expressed using mnemonics and labels.

To run an assembly code program, a 'decoder' called the *assembler* must first be run to convert the mnemonics and labels into binary machine code and strip off the comment field. The assembled machine code is then run using the usual CALL instruction, which can conveniently refer to the start label of the program, e.g. CALL ADDER.

The BBC Micro includes a useful assembler as part of the internal operating system. The assembler is particularly well integrated with the BASIC language facility so that sections of assembly code can be mixed in with BASIC program as long as they are identified with a pair of square brackets: [ ]. One or two lines of BASIC normally precede the assembly code. First we tell the program counter where to begin by assigning a memory location number to the variable *P%*, then we define memory locations for the labels used. The BASIC REMark facility can also be used for further documentation. Here is our assembly code addition program ready for direct entry into the BBC Micro.

```
 10  MODE 6: REM sets appropriate screen display
 20  P% = &70: REM sets program counter to &70
 30  BOXONE = &80:  BOXTWO = &81:  SUMBOX = &82:  REM
     defines label locations
 40  [
 50  .ADDER            /label defining start of routine
 60  LDA BOXONE        /load first number into accumulator from
                       memory location BOXONE
 70  CLC               /set carry bit to zero
 80  ADC BOXTWO        /add second number from memory location
                       BOXTWO into accumulator
 90  STA SUMBOX        /store this sum of numbers in location
                       SUMBOX
100  RTS               /end of assembly code; go back to BASIC
110  ]
```

Having entered the program, type RUN to run it in the usual way. This merely assembles the program: the machine code version will now appear on the screen with op codes and operands neatly set out in their respective columns. We must now poke our numbers to be added into their respective memory locations, remembering that we now have the convenience of referring to them by labels, e.g.

```
? BOXONE = 13
? BOXTWO = 19
```

To run the machine code, type CALL ADDER and look for the result in SUMBOX, i.e.

```
PRINT ? SUMBOX
```

and the answer will be displayed:

```
32
```

14.7.6 High-level languages

Whilst assembly code is a good deal more human than machine code, we have just seen that it is still tedious when it comes to a simple arithmetic operation. In particular, it is not well-suited to taking in data from the keyboard and displaying it on the screen, and in fact does emphasize the sophistication of these simple operations which we tend to take for granted.

Most programming today uses high-level languages, of which BASIC is the most widely used for home computers, other languages such as C and C++ being popular in professional applications. With high-level language, the programmer needs little or no knowledge of the detailed working of the processor, all the necessary register operations and carry bit handling being automatically implemented for him. Instructions are a series of meaningful English or algebraic statements, and remarks or comments can be freely included by preceding them with REM. Data can be entered readily from the keyboard without the need to poke it into specific memory locations, whilst results are displayed on the VDU or printed out.

Using BASIC, our addition sum becomes almost trivial in its simplicity. Enter on the keyboard:

```
PRINT 13+19
```

The computer responds by displaying the answer:

```
32
```

A BASIC program sequence easily adds further refinement to prompt the program user to enter his own numbers to be added. The PRINT statement is also enhanced to display 'SUM =' in addition to the answer.

```
10 REM PROGRAM TO ADD TWO NUMBERS
20 INPUT "ENTER FIRST NUMBER"; A
30 INPUT "ENTER SECOND NUMBER"; B
40 PRINT "SUM ="; A + B
50 END
```

Typing RUN will illustrate how this simple program has converted the computer hardware into an adding machine.

We can see how, instead of forcing the programmer to think in terms of memory locations and op codes, BASIC manages the detail behind the scenes and allows the use of ordinary algebraic variables such as A and B. BASIC, in common with other high-level languages can immediately understand an equation such as

$$\mathrm{SUM} = \mathrm{A} + \mathrm{B}$$

hence making mathematical operations very straightforward. In contrast, assembly code is convenient for quickly moving and sorting data and for reading and writing data to input and output ports where the use of defined addresses is essential. Most computer and video games are written in assembly code because they use a great deal of data shuffling but only simple calculations and logical decisions. On the other hand, evaluating an equation such as

$$f_0 = \frac{1}{2\pi}\left(\frac{1}{C_1 C_2 R_1 R_2}\right)^{1/2}$$

is very much quicker to program in a high-level language than in assembly code.

The translation of a high-level language program into the machine code to which the CPU responds is, understandably, a relatively complicated process which requires its own special program. This may be done line by line by an *interpreter* as the program is being executed, or the whole program may be translated into machine code by a *compiler* prior to execution. In general, languages using an interpreter such as BASIC run more slowly than compiled languages such as C because of the time spent in interpretation during execution. Alterations to interpreted programs are, however, quicker to try out because there is no time spent in recompiling before execution.

Although a compiled high-level program can run relatively fast, the machine code from an equivalent assembly program will normally run even faster because it uses the CPU more efficiently in carrying out the required operations. However, high-level programming is being used increasingly because of its clarity of expression and logical structure. Both these features contribute to getting complex programs working sooner, also helping to ensure that they are reliable in service and easy to modify should later changes be needed.

14.8 Input and output to microcomputers

14.8.1 Address decoding

In fig. 14.11, the all-important input and output ports are shown sitting on the address and data buses with the rest of the system. Many computers do not, however, include direct input/output facilities for logic signals, such access being restricted to the keyboard, printer and disk drives. The input/output port designs in this section can expand the use of a PC. At the same time they provide experience in address decoding and the use of tri-state gates on the data bus.

Some microprocessors such as the Intel 80486 have special provision for inputs and outputs which allows them to share bus addresses with memory. This is achieved by an additional control input on the bus called *I/O request.* When this is active (usually 0 V) it means that the data bus is required by an input or output device and any memory sharing those addresses is temporarily disconnected to avoid data corruption caused by bus *contention.* An advantage of dedicated I/O is that it avoids the need to decode the full processor address bus, which may be 32 bits long, at the I/O ports. 16 bits are usually sufficient and it is sometimes unnecessary to decode more than 8 bits.

An alternative approach is called *memory mapped I/O* which needs no I/O request line and simply dedicates a group of memory addresses to I/O. To avoid bus contention, no memory is normally fitted in these locations. Any processor chip can use memory mapped I/O; the 6502 and 68000 series can in fact use only this technique because they have no I/O request line.

Addresses used for I/O are normally near one end of the memory map in order to avoid clashes with the program and other data. The starting point in the design of an input or output port is to decode the address.

Fig. 14.13 shows a practical *address decoder* which gives a logic 1 output only when the inputs $A0$ to $A15$ are set to &H0300. The binary form of our chosen address is 0000 0011 0000 0000, and a few moments spent working through the logic will indicate how the decoder works.

Some applications need a more versatile address decoder which can be set to respond to different inputs, and here the universal decoder of fig. 14.14 is useful. This uses the 74LS85 *magnitude comparator*, which gives a logic 1 from pin 6 when the 4-bit number on its '*A*' inputs is identical to the 4-bit number on the '*B*' inputs. Here, the '*B*' inputs are connected to switches, the small dual-in-line or hexadecimal rotary type being ideal for the job. A switch

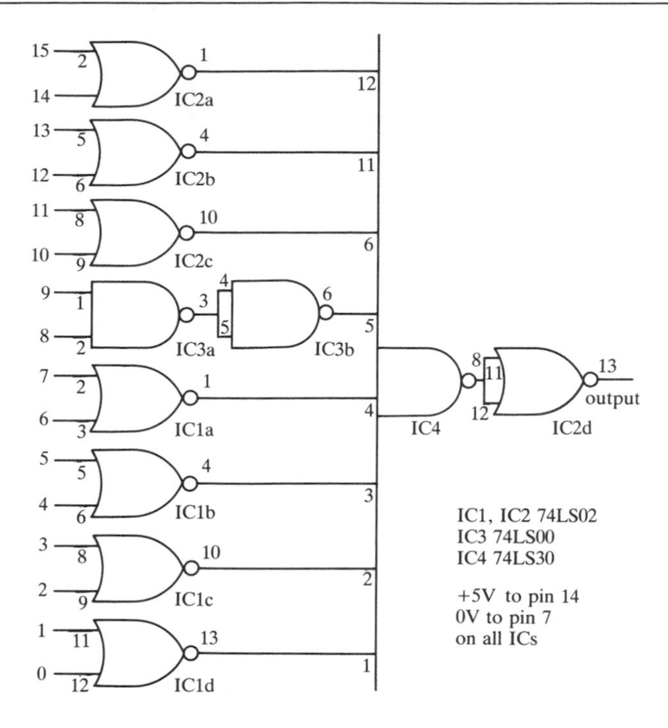

Fig. 14.13. Address decoder for PC and compatibles. Output is high only for input 0000 0011 0000 0000 (&H0300).

closure presents logic 1, with the 1 kΩ pull-down resistors ensuring a logic 0 when a switch is open. The four comparators are 'ANDed' together in the 74LS21 to give the decoded output. Any required 16-bit address code can be set up on the switches, and the output will only go high when that exact address is present on *A*0–*A*15.

As already mentioned, it is common for address decoders to save chips and PCB area by only partially decoding the address bus. For example, if in fig. 14.13 inputs A_0 to A_3 are ignored, the decoder will still respond to address & H0300. If in addition A_{12} to A_{15} are ignored, the decoder still works but it will respond again at &H1300, through to &HF300. This is immaterial as long as no other I/O ports reside at those addresses. Useful simplification and cost saving can be obtained by partial decoding.

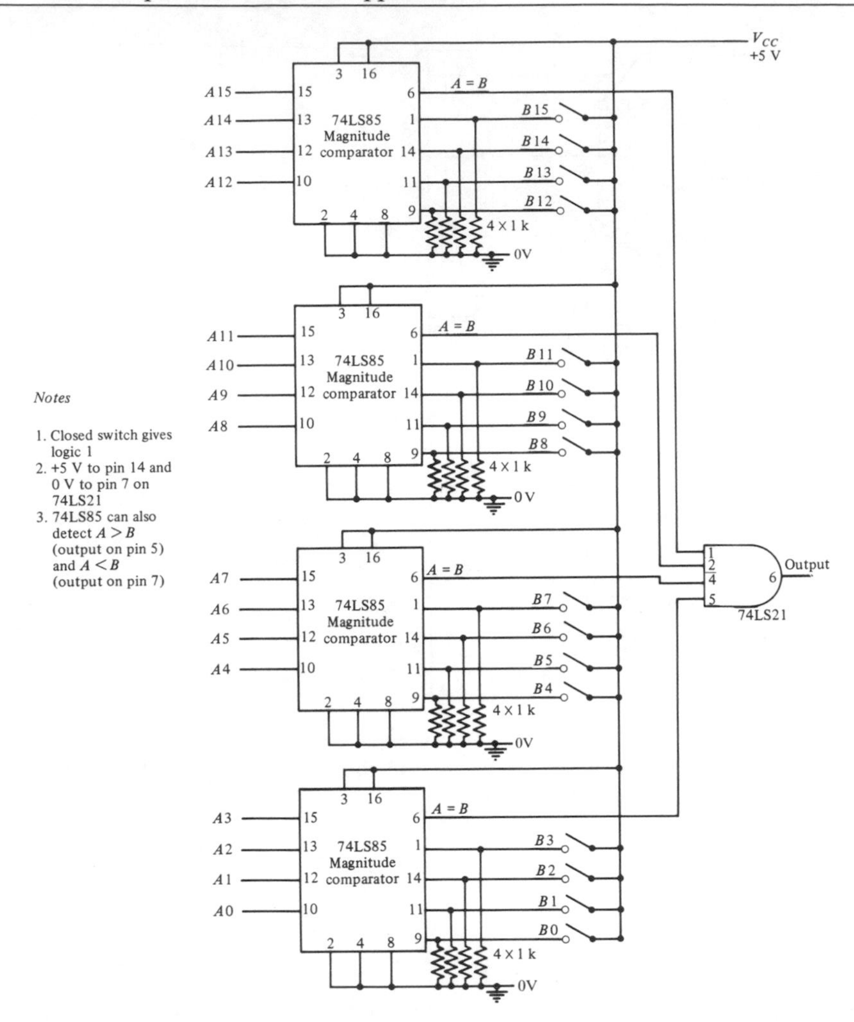

Fig. 14.14. Universal address decoder using 74LS85 magnitude comparator chips. Output gives logic 1 when input binary number (*A*0–*A*15) equals number set on switches (*B*0–*B*15).

14.8.2 Output port

Fig. 14.15 shows how one of our address decoders can be used with the 74LS273 latch to make an output port. The data present on data bus lines *D*0 to *D*7 is latched when the chosen address (&H0300 for fig. 14.13) is recognized at the same time that the write line $\overline{\text{WR}}$ is active (0 V). Having latched

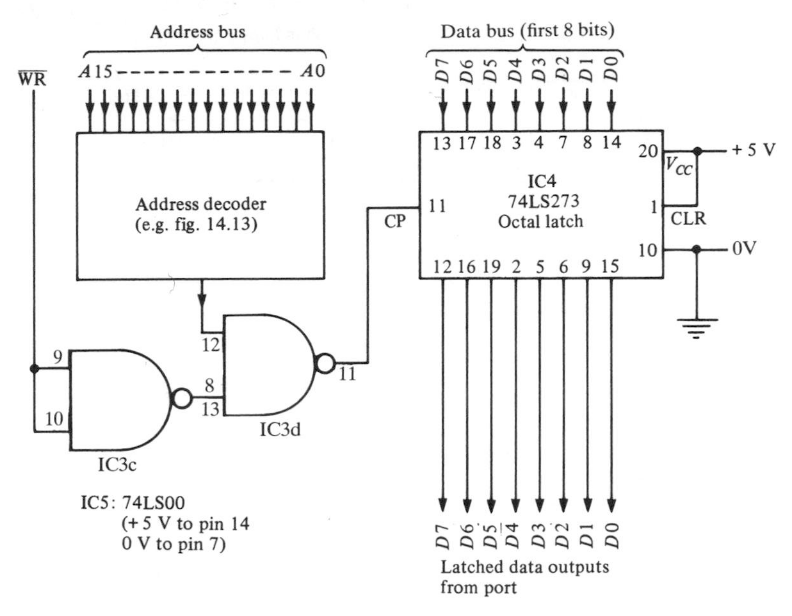

Fig. 14.15. Computer output port using 8-bit *D*-type latch clocked by address decoder and $\overline{\mathrm{WR}}$ pulse.

the data, LEDs may be used as binary data indicators using the driver circuit of fig. 13.35, logic 1 being shown by an illuminated LED. Alternatively, transistor relay drivers as discussed in chapter 1 (fig. 1.4(*b*)) extend the capability of the port. Use a 1 kΩ current limiting resistor in series with the base of each transistor. The circuit can mark a dramatic step forward for many PC users because the computer can now be used to control external devices such as motors, heaters and lamps. Naturally any mains-operated devices should be carefully isolated from the computer circuitry by a separate relay of proper mains voltage rating.

To assist in the practical construction of port circuits, fig. 14.16 shows the main connection data on the PC-compatible ISA 62-way expansion bus connector. This is used on most machines (except PS/2) from the PC XT upwards: there are normally one or two spare expansion slots available inside the PC. Most machines today are 16 bit and also include a 36-way connector to access data lines 9–15 and address lines 20–23. Appendix 5 gives full details.

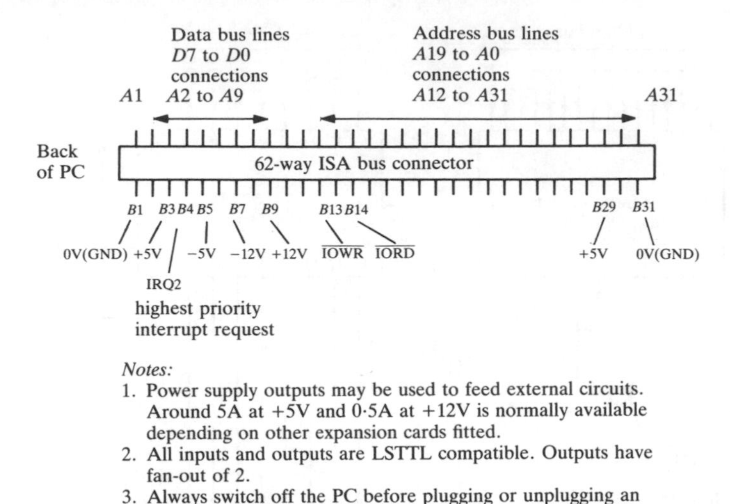

Fig. 14.16. Bus connections on expansion slots of PC and compatibles.

14.8.3 Input port

In fig. 14.17, our computer interface circuitry is extended to include an input port which uses the 74LS244 octal tri-state buffer to place data on the bus when enabled by the address decoder. The EXOR gate IC7*a* responds to the processor $\overline{\text{RD}}$ pulse to add one to the port address when the port is in its input mode. For example, using the decoder of fig. 14.13, we have an 8-bit output port at &H0300 and an 8-bit input port at &H0301. Alternatively, the universal address decoder (fig. 14.14) can be used in the same way.

14.8.4 Practical ports for the PC

The combined input and output port shown in fig. 14.17 can be constructed on a PC prototyping card designed to plug into a vacant expansion slot on the host PC's motherboard. Power, address, data, read, write and interrupt signals are all then available from the computer expansion bus. Suitable cards are available from Maplin Electronics.

Care must be taken to ensure proper isolation between the PC circuitry and any external voltages, otherwise expensive damage could occur to the PC.

The input and output ports can be readily addressed using suitable lines in a GWBASIC or QBASIC program. The key commands are INP, to read a

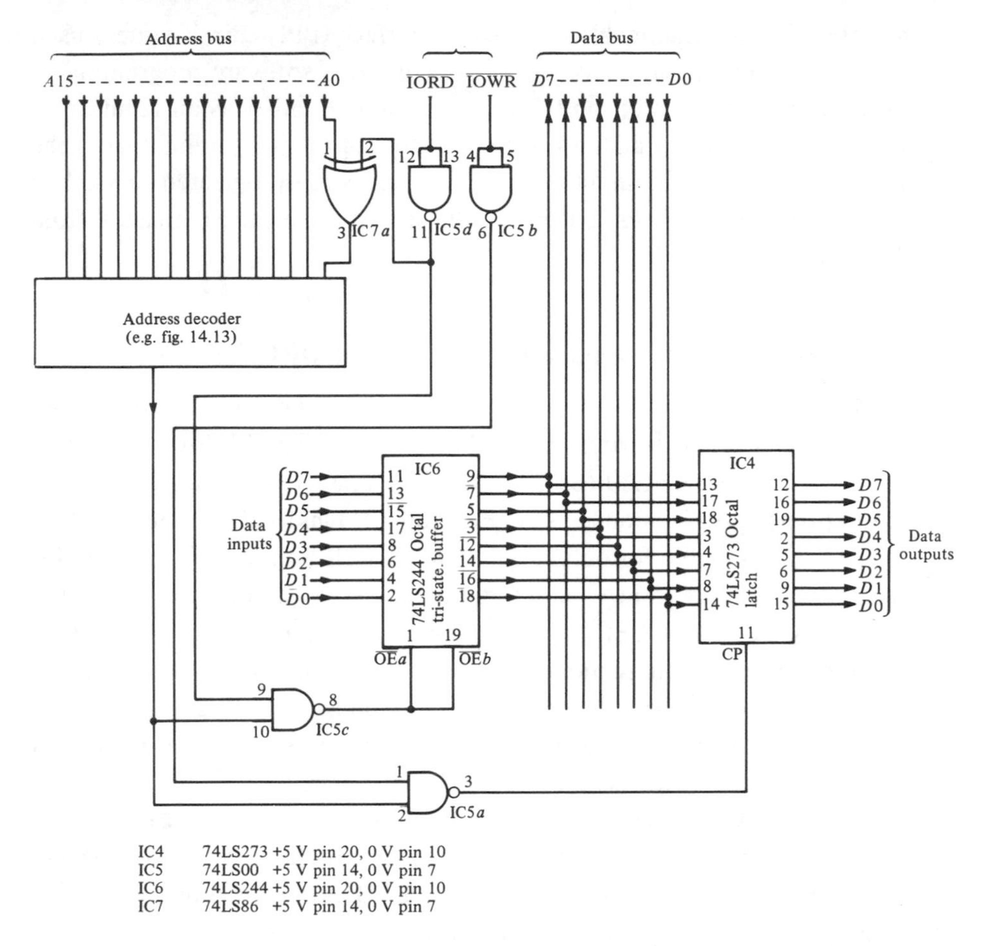

Fig. 14.17. Combined input and output port. Address decoder of fig. 14.13 gives output address &H0300 and input address &H0301.

hex number from a specified address and OUT to output a hex number to a specified address. Typical lines could be:-

To turn on all output lines to 1:

```
10 OUT &H0300, &HFF
```

To read a number from the input lines:

```
10 N = INP (&H0301)
20 PRINT N
```

To read a number from the input lines and send it to the output lines:

```
10 N = INP (&H0301)
20 OUT &H0300, N
30 GOTO 10
```

The Intel 8255 programmable peripheral interface (PPI) chip is widely used as a versatile input/output device in PC systems. It is software programmable so that each of its three 8-bit ports can be configured either as an input or an output. It is used with an address decoder, and can be programmed from the main processor so that its inputs and outputs are simply buffered or latched or act as bi-directional buses. Further useful information can be gleaned from the 8255 data sheets.

14.9 I/O experiments using the Acorn BBC Micro

The BBC Micro is particularly well-suited to input and output experiments because it includes two separate parallel 8-bit ports. The bi-directional user I/O port is at address &FE60 (65 120) and is accessible on connector *PL*10, whilst the printer port is at address &FE61 (65 121) and is connected to *PL*9. The latter port is suitable for output only, and is well-buffered by a 74LS244. This chip is capable of 15 mA output and hence is able to operate small reed relays directly as well as LEDs, a useful facility for experiments. The input/output direction of the bidirectional port may be set bit by bit by setting the corresponding bit of the data direction register at address &FE62 to 0 for input and 1 for output. In our experiments, we shall use the user I/O port for inputs only and therefore set address &FE62 to 0 at the start.

Fig. 14.18 shows the way that input switches and indicator LEDs can be connected to the ports for experimental use. The switches set the logic levels presented to the 8 bits of the user port; an LED is linked directly to each switch to give clear indication of the logic level. On the output side, the printer port feeds LEDs, an illuminated LED indicating logic 1. The extra switch connected to the *CB*1 input is used for interrupts in a later experiment.

The ports may be tested by a simple BASIC program, which is entered and RUN in the usual way:

```
10  ?&FE62 = 0 : REM set user I/O port to input mode
20  ?&FE61 = ?&FE60 : REM read I/O port and send that data to
    printer port output
30  GOTO 20
```

This routine simply scans or *polls* the input port continuously and transfers the data read there straight to the output port: whatever pattern is illuminated on the input LEDs is transferred to the output LEDs as soon as the program is set running. Line 10 *initialises* the user port to its input condition by setting the data direction register at &FE62 to zero.

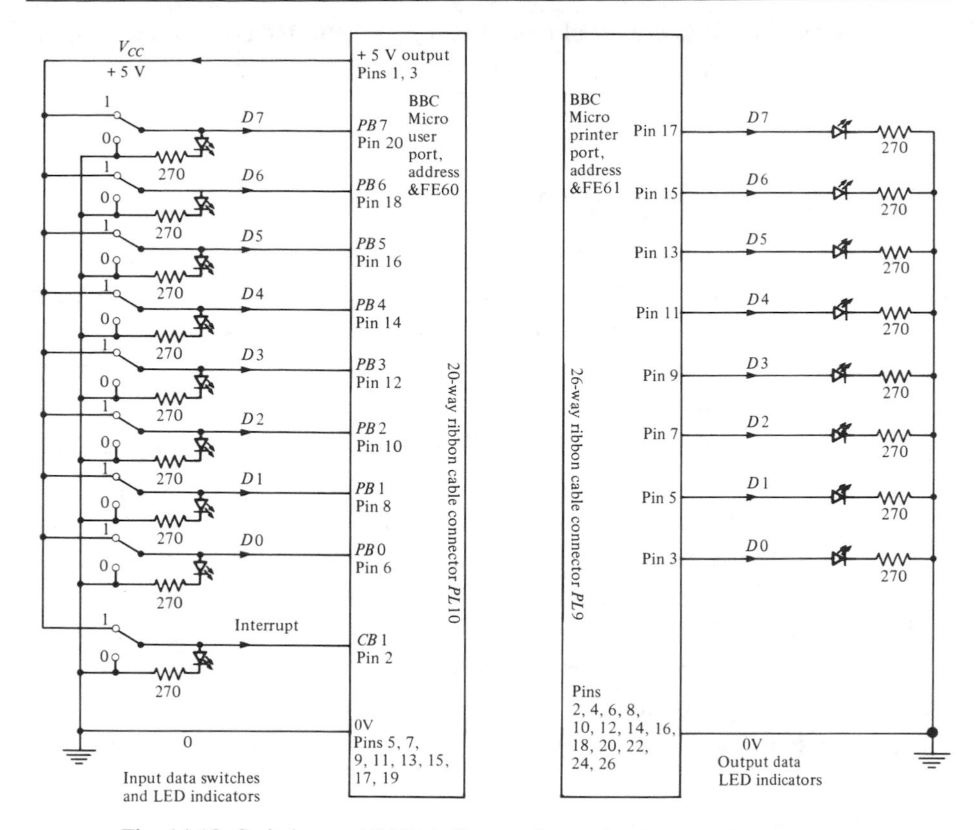

Fig. 14.18. Switches and LED indicators for testing input and output ports. Connections to BBC Micro ports are shown. The circuit can also be used with the PC ports in fig. 14.17.

There is plenty of scope for experiments with data handling programs. For example, a simple program can be written to transfer the data to the output only if its value lies within a certain range. Much more complex operations such as machine control can be invented by treating each of the eight input bits as a separate input from thermostat, photo-detector, pressure switch, etc. and similarly allocating the output bit lines to individual operations. Logical decisions can then be made in the program to implement a variety of output conditions depending on what inputs are present at the time.

We now see the microcomputer in its important role as a general-purpose element in an electronic system, able to accept inputs and generate outputs. Any amount of arithmetic and logical decision making can be introduced between input and output by appropriate software. This means enormous flexibility in application – for example, the functions of a machine control unit can be changed completely by plugging in a new EPROM containing a different software program.

14.10 Exploring the CPU

We are now in a position to delve deeper into the functioning of the CPU using our new input and output capability. The assembly code program 'EXPLORE' is written for the BBC Micro and demonstrates the basic arithmetic and logical functions of the computer CPU upon which all the more complex operations are built. The heart of the program is a series of assembly code subroutines between lines 40 and 520. Each begins with a label (e.g. .and) and finishes with the command RTS (return to BASIC from subroutine). Each subroutine illustrates an aspect of the functioning of the processor by using data set up manually on the input port switches and displaying the result on the output port LEDs. The comments associated with each line after the '\' explain its purpose. For convenience, the CALLing of each routine is assigned to a separate programmable function key in lines 560–650. The symbol ¦M at the end of each key programming line is equivalent to the ⟨return⟩ key so that a single stroke of the function key is sufficient to CALL the required routine. The auto-repeat facility on the keys allows a subroutine to be repeated rapidly just by holding down the relevant function key, thus repeatedly polling the input port and cycling through the selected operation.

After the program has been typed into the computer, it should immediately be saved on disk in case of any errors which might cause a frustrating 'crash'. Typing RUN ⟨return⟩ will then assemble the program into machine code. The function keys can now be used to CALL the routines as required. The various logical, arithmetic and counting functions will be observed in action. The experimenter will be impressed with the ease with which the microcomputer can be used to implement these functions compared with the lengthy business of building up the equivalent hardware logic circuits.

```
10 MODE 6
20 REM EXPLORE – an assembly code program to demonstrate
   directly the main logic/arithmetic operations in the computer's CPU.
30 P% = &1800 : REM sets program counter to &1800 where there is
   space for program.
40 [
50 .input
60 LDA#&00    \start by loading zero into accumulator
```

```
 70 STA&FE62  \copy accumulator (zero) into location &FE62 to
              set &FE60 port to 'input' mode
 80 RTS       \back to BASIC
 90 .data
100 LDA&FE60  \load accumulator with data from input port at &FE60
110 STA&FE61  \copy accumulator to output port at &FE61
120 RTS       \back to BASIC
130 .and
140 LDA&FE60  \load accumulator with data from input port
150 AND&FE61  \AND function with data at output port
160 STA&FE61  \copy result to output port
170 RTS       \back to BASIC
180 .or
190 LDA&FE60  \load accumulator with data from input port
200 ORA&FE61  \OR function with data at output port
210 STA&FE61  \copy result to output port
220 RTS       \back to BASIC
230 .exor
240 LDA&FE60  \load accumulator with data from input port
250 EOR&FE61  \EXOR function with data at output port
260 STA&FE61  \copy result to output port
270 RTS       \back to BASIC
280 .countup
290 INC&FE61  \increment data at output port
300 RTS       \back to BASIC
310 .countdown
320 DEC&FE61  \decrement data at output port
330 RTS       \back to BASIC
340 .shiftright
350 LSR&FE61  \shift data at output port one place to right
360 RTS       \back to BASIC
370 .shiftleft
380 ASL&FE61  \shift data at output port one place to left
390 RTS       \back to BASIC
400 .add
410 LDA&FE60  \load accumulator with data from input port
420 CLC       \clear any carry digit
430 ADC&FE61  \add data at output port to accumulator
440 STA&FE61  \copy result to output port
450 RTS       \back to BASIC
460 .subtract
470 LDA&FE61  \load accumulator with data at output port
```

```
480 SEC          \set carry digit to 1
490 SBC&FE60     \subtract data from input port from accumulator
500 STA&FE61     \copy result to output port
510 RTS          \back to BASIC
520 ]
530 REM program execution starts by running 'input' routine to
    initialize input port.
540 CALL input
550 REM function keys are programmed to call logic routines as
    follows:
560 *KEY0 CALL data |M
570 *KEY1 CALL and |M
580 *KEY2 CALL or |M
590 *KEY3 CALL exor |M
600 *KEY4 CALL countup |M
610 *KEY5 CALL countdown |M
620 *KEY6 CALL shiftright |M
630 *KEY7 CALL shiftleft |M
640 *KEY8 CALL add |M
650 *KEY9 CALL subtract |M
```

The main importance of this program is that it illustrates in an entertaining way the fact that the basic logic, counting and arithmetic functions which were discussed in chapter 13 in terms of hardware gates also constitute the basis of the CPU. From AND and OR to addition, subtraction and shift registers, we can now see that the awesome power of the computer is derived from the efficient marshalling of simple logical functions.

Whilst the assembly code used here in 'EXPLORE' is specific to the 6502 processor and BBC Micro machine, it can be adapted for the IBM PC and compatibles. The reader who has a PC equipped with input and output ports and an assembly language facility can experiment by rewriting 'EXPLORE' in Intel '80X86' assembler. Care should be taken to use the relevant input and output port addresses (&HO301 and &HO300 respectively in the circuit of fig. 14.17). As on the BBC Micro, the parallel printer port on the PC is a useful data output interface; it is located in the address range &HO378–O37F.

14.11 Processor digressions

14.11.1 Interrupts

So far our input/output experiments have illustrated the way that an input port can be polled by the processor and any input data handled as required by the program. In real applications, however, a processor is often required to take special action in response to an infrequent but important event where it is very wasteful of processing time to keep polling an input port repeatedly. For example, a processor might be normally sorting and evaluating data at a telemetry outstation when occasionally an external pulse arrives telling it to output its results back to base. That requires the processor to leave its routine job and jump into its special output procedure in response to the external command. This is accomplished by a special hardware *interrupt* input direct on the CPU chip which diverts the processor to a specific memory location. There it picks up a signpost or *vector* to a special part of the program.

Processors usually have more than one interrupt input, each having a different *priority level.* If several interrupts arrive together they are handled in strict priority order. The 6502 chip has two such input lines: the *non-maskable interrupt* (NMI), the top-priority input often used for emergency operations such as saving data when a power failure is detected, and the *interrupt request* (IRQ) used for general-purpose interrupts. The programmer can choose to make the processor ignore an IRQ by setting the *interrupt mask* bit in the processor status register. A NMI cannot, however, be ignored. More sophisticated processor systems have numerous interrupt hardware inputs at different priority levels. These are organised by a *programmable interrupt controller* (PIC) chip such as the 8259A used in the PC which manages eight levels of request and can be expanded by using additional 8259A chips. Even the two inputs in the basic 6502, however, can be used in conjunction with appropriate operating system software to prioritise a large number of different interrupt inputs. On the detection of an interrupt, the software can find out its origin by polling the various inputs such as keyboard, interval timer etc. Depending on the priority allocated to its input, the interrupt can either be executed immediately or it might be held temporarily until a higher priority interrupt has completed execution. The interrupt mask is used to prevent such an unwanted interruption of an interrupt.

The way that the processor handles an interrupt is rather similar to a reset operation except that after completing the required interrupt subroutine, the main program execution is resumed. This means that, before actually executing the special interrupt subroutine, the program must save the values of any registers which will be used by the interrupt subroutine. Those values may well be required by the main program again later, and failure to save and restore registers may result in their data being changed during the interrupt program with a consequential error when the main program is resumed. Registers are normally saved by pushing their values into that special area of memory allocated as the 'stack', from where the values can be retrieved in reverse order (LIFO) after completion of the interrupt routine. The CPU itself automatically saves the program counter and status registers in response to NMI or IRQ, these being restored later by the 'return from interrupt' (RTI) instruction.

The occurrence of an interrupt is like the telephone ringing on a clerk's desk – he immediately leaves his routine work to answer the call, carefully setting aside his partially completed items to make space for his telephone message pad, but is ready to return to them immediately after the call. It is interesting to reflect that a processor without an interrupt capability is like an office with no bell on the telephone. Someone would need to pick up the receiver every few seconds just to check whether a telephone call was coming in – the equivalent of polling an input port.

Interrupts are vital to the operation of multi-tasking operating systems, important software programs such as 'Windows' which enable a processor to appear to carry out a number of operations all at once. In reality, everything is implemented in sequence, but carefully organised interrupt priorities make them appear effectively simultaneous to the user.

A simple example on the BBC Micro will illustrate the power of interrupts. The way that the operating system handles the keyboard input employs an interrupt to load keyboard entries into their own reserved section of memory called the *keyboard buffer* even though the processor may be busy executing a program at the time. To illustrate this property, RUN the following program:

```
10  FOR I = 1 to 5000
20  PRINT ''BUSY''
30  NEXT
```

This will occupy the processor, apparently fulltime, filling the screen with the word ''BUSY'. Now, whilst the program is running, type in LIST followed by the ⟨return⟩ key. Nothing seems to happen immediately, but, after a few seconds, when the 'BUSY' printing has finished, the processor will respond by listing the program. The keyboard interrupted the processor during program execution for a sufficient time to put the LIST entry in the keyboard buffer in RAM memory before resuming its screen printing. To the operator, the computer was effectively handling two jobs at once.

The hardware interrupt input on the BBC Micro is accessible through the *CB*1 input on the user port, and the following assembly language program gives a clear illustration of interrupt operation. The program is designed to be entered after EXPLORE. Its main section is in the lines 1010 to 1070 called 'interruptroutine', which contains the actual program which the processor executes after an external interrupt on input *CB*1. Having saved the accumulator contents on the stack, 'interruptroutine' then has as its main operation a data transfer from the input port to the output port and thus gives a clear indication that the interrupt has happened, irrespective of anything else going on at the time. A good test is to run 'countdown' by holding down function key 5. Without releasing function key 5, operating the 'interrupt' switch to momentarily put logic 1 on the *CB*1 input will trigger an interrupt. At that instant, no matter where the countdown has got to, the output port will be 'refreshed' with whatever new data is set on the input port switches.

```
1000  REM experiments with interrupts from CB1 pin of user port.
      Enter this program after EXPLORE
1010  [
1020  .interruptroutine
1030  PHA          \save accumulator value on stack
1040  LDA&FE60     \load accumulator with data from input port
1050  STA&FE61     \copy accumulator to output port
1060  PLA          \retrieve accumulator value from stack
1070  RTI          \return to main program from interrupt
1080  .enableinterrupt
1090  LDA#0
1100  STA&FE6C     \taking location &FE6C temporarily to zero instructs
                   interface chip to respond to hi-lo transition on CB1
                   input
1110  LDA#&6F
1120  STA&FE6C     \setting location &FE6C to &6F disables all inter-
                   rupts apart from CB1
1130  LDA#&90
1140  STA&FE6E     \setting location &FE6E to &90 enables interrupts
                   from transition on CB1 to reach CPU
1150  .setvector
1160  LDA#interruptroutine MOD 256
                   \load accumulator with least significant byte of
                   location of 'interruptroutine'
```

```
1170  STA&0206   \change least significant byte of interrupt vector to
                 point to 'interruptroutine'
1180  LDA#interruptroutine DIV 256
                 \load accumulator with most significant byte of
                 location of 'interruptroutine'
1190  STA&0207   \change most significant byte of interrupt vector to
                 point to 'interruptroutine'
1200  RTS        \back to BASIC
1210  ]
1220  REM program execution starts by running 'enable interrupt' fol-
      lowed immediately by 'setvector'. Computer is thereby initialised
      to run 'interruptroutine' whenever interrupted by a pulse on CB1
1230  CALL enableinterrupt
```

The comments in the program explain the functions of the various statements, and it will be clear that the correct handling of an interrupt involves not only saving and restoring registers (only the accumulator is used in this case) but also storing the address of our 'interruptroutine' program in the memory locations &206 and &207 which the BBC Micro operating system allocates for the 2-byte 'vector' address pointing to the interrupt subroutine. When the interrupt is activated on *CB*1, the I/O chip pulses the interrupt request (IRQ) input directly on the CPU. Immediately the CPU branches to read the address contained in locations &FFFE, &FFFF (this operation is built into the 6502 chip). The BBC Micro operating system ROM has the address &DC1C in those locations, so the CPU jumps to execute the small routine starting at &DC1C. This code first checks by polling whether any internal interrupts require service; if it finds none have been activated, then it vectors (jumps) through to &206 and &207 where we have arranged that it finds the address of 'interruptroutine'. It is important to note that this interrupt vectoring is only included in BBC Micro operating systems of issue 1.0 or higher.

As usual, typing RUN initiates program assembly, followed immediately afterwards by the execution of line 1230 CALL enableinterrupt. This runs 'enableinterrupt', followed by 'setvector'. The first of these routines sets up the I/O interface to handle *CB*1 interrupts, whilst the second places the address of 'interruptroutine' as the interrupt vector in &206 and &207 ready for action.

So we see that programming to handle interrupts requires care, the basically simple principle needing quite a number of program instructions to ensure that everything happens as intended. However, the actual running of this sample program illustrates in a rewarding way the power of a microprocessor handling more than one task at once. Once more, we have an example which can trigger the reader's imagination in applying the computer in practical electronic systems. 'Interruptroutine' can be readily adapted to carry out any required action such as warning of low water pressure whilst, for example, the main program proceeds with running a washing machine sequence of operations.

14.11.2 Direct memory access (DMA)

For some peripherals, such as disk drives, where large quantities of data must be transferred in a short time, even interrupts are too slow. In such cases *direct memory access (DMA)* is used. Here the peripheral, such as the hard disk drive, gains direct communication with memory, avoiding the bottleneck of the CPU.

A DMA transfer begins with a peripheral requesting access to the bus by signalling on a special DMA line. Then, for the period of the DMA data transfer, the necessary memory addresses are generated (*asserted*) by the peripheral interface itself instead of by the CPU. The end of the DMA transfer is signalled by an interrupt from the peripheral interface to tell the CPU that normal processing can be resumed.

14.12 Digital signal processing (DSP)

In chapter 11 we saw how ordinary analogue electronic processes can be viewed as mathematical functions. For example amplification can be described as multiplication by a constant, signal mixing is addition and attenuation is division by a constant. We have also looked at more complicated functions such as multiplication of one signal by another (modulation). A digital computer is, of course, capable of carrying out all this arithmetic, and can therefore turn its hand to process signals once those signals have been converted into a sequence of binary digits like music on a CD.

The future of most signal processing, transmission and recording actually lies in digital techniques. Digital signal processing has the merit of total repeatability, freed from random variations in component values (tolerances) and offers the facility of digital remote control and memory of all functions. The ability to manipulate signals freely in time also provides processing functions such as comb filters which are costly or impossible in the analogue world.

In transmission and recording, the CD is an illustration of the powerful ability of a digital system to detect and correct signal degradations or errors due to interference (fingerprints and scratches). When these techniques are combined with processing intelligence, illustrated by the ability of a FAX machine to recognize blank areas of paper and not waste data capacity on them, some remarkable capabilities emerge. An example is the video telephone which transmits a television picture on a fraction of the bandwidth of an ordinary telephone voice circuit. It is also through DSP that a 1250 line high definition television picture with teletext and multichannel digital audio is comfortably fitted into the 6 MHz bandwidth previously found necessary for a 625 line analogue picture and mono sound.

This book would therefore be incomplete without a brief glimpse at DSP and a look at methods of converting signals between the analogue and digital domains.

DSP is a branch of digital design which is comprehensively covered in many other books and it is neither possible nor appropriate to discuss it in depth here. The ever-increasing number of dedicated DSP chips do all use the basic digital hardware and software functions discussed in these last two chapters. The specific configurations are, however, initially unfamiliar because they are optimized for fast sequences of arithmetic-intensive operations, often with processes happening in parallel.

As an illustration of DSP, the next section introduces an example of one of the most intriguing and important functions: the digital filter.

14.13 Digital filters

We saw in section 11.22 how low-pass, high-pass and band-pass filters can be constructed out of amplifiers and integrators in the state-variable (biquad) filter. In the switched capacitor filter of section 11.23 we saw the integration process itself implemented by the clocked successive addition of charge to a capacitor. It is therefore not difficult to see how integrators can be implemented in the digital domain by clocking successive addition into an accumulator register.

In section 11.21 we in fact started with an analogue computer and recognised its capability as a filter for analogue signals in section 11.22. Likewise we can take a digital computer and program it to act as a filter for digital signals.

If a digital signal, for example straight off a CD, is to be filtered, the stream of numbers can be subjected to fast software-implemented ‘circuitry’. Fig. 14.19 illustrates how a state-variable filter might be implemented as a sequence of events in software. Of course the right numbers have to be ready for processing at the right time and internal registers are used to hold numbers temporarily until the arithmetic processor is ready for them.

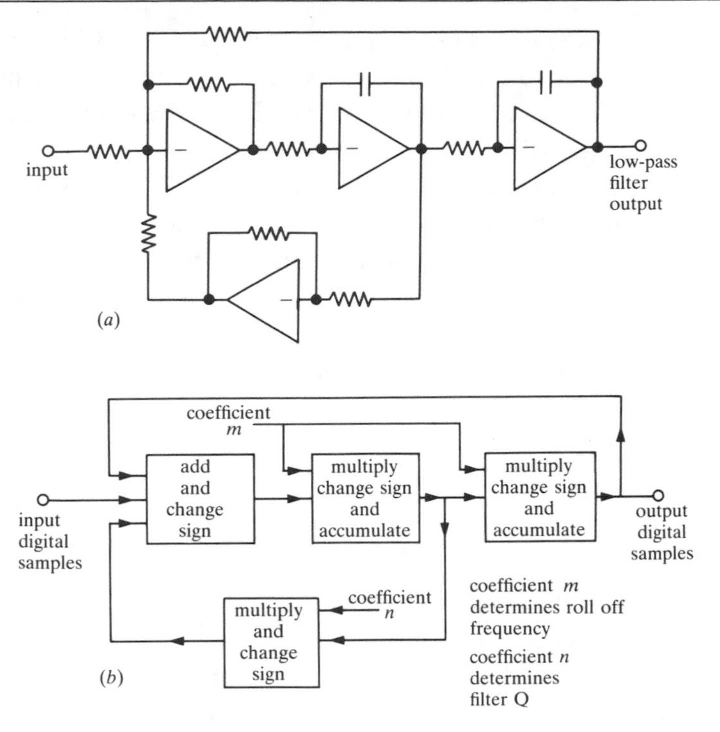

Fig. 14.19. Second order state-variable low-pass filter (*a*) and an outline of its form as a digital (DSP) process (*b*).

A full treatment of digital filters quickly begins to need mathematical language which lies outside the scope of this book, but this qualitative introduction to one approach emphasises the elegance of the process. Digital filters of great precision and fast roll-off with minimal phase-distortion can be achieved by repeating successively the kind of processing that has just been described. Today's digital processing speeds allow a remarkable amount of processing in the 22 μs interval between two audio samples. The precision and repeatability of digital signal processing (DSP) is leading to the all-digital recording studio where the microphone and instrument outputs are digitised at source and remain digital right through mixing, equalisation, compression and recording on to CD. The analogue decoding then finally happens in the listener's CD player.

14.14 Digital to analogue conversion

Digital to analogue (D/A) conversion is basically simple: each bit is given its appropriate analogue value and the outputs summed in an analogue adder.

Fig. 14.20 shows a practical 8-bit D/A converter (DAC) suitable for direct connection to the output port of the BBC Micro. It is designed to be powered by ±5 V supplies which are available from the BBC Micro auxiliary power socket. The two 74HC4016 CMOS chips provide eight analogue switches which connect the appropriate input resistors to the virtual earth adder according to the logic state of the input bits. Calibration is determined by the reference voltage V_{REF}, adjusted by $RV1$. Each input resistor value is determined by the inverse of the 'weight' of the bit which switches it. Thus the 'LSB' resistor (1M28 Ω) is 128 times the value of the 'MSB' resistor (10 kΩ). The feedback resistor is chosen here so that the analogue output voltage V_O is given by the simple equation

$$V_O = \frac{-V_{\text{REF}}}{100} \times D_{\text{IN}}$$

where D_{IN} is the decimal equivalent of the digital input value (0 to 255).

With this circuit connected to an output port, a binary countup or countdown produces a linear ramp analogue output. Sine and cosine functions can also be readily displayed on an oscilloscope by using the computer to generate the function in the first place.

It is useful to note that the analogue output V_O is proportional to the *product* of the digital input and the reference voltage. Most DACs have therefore an

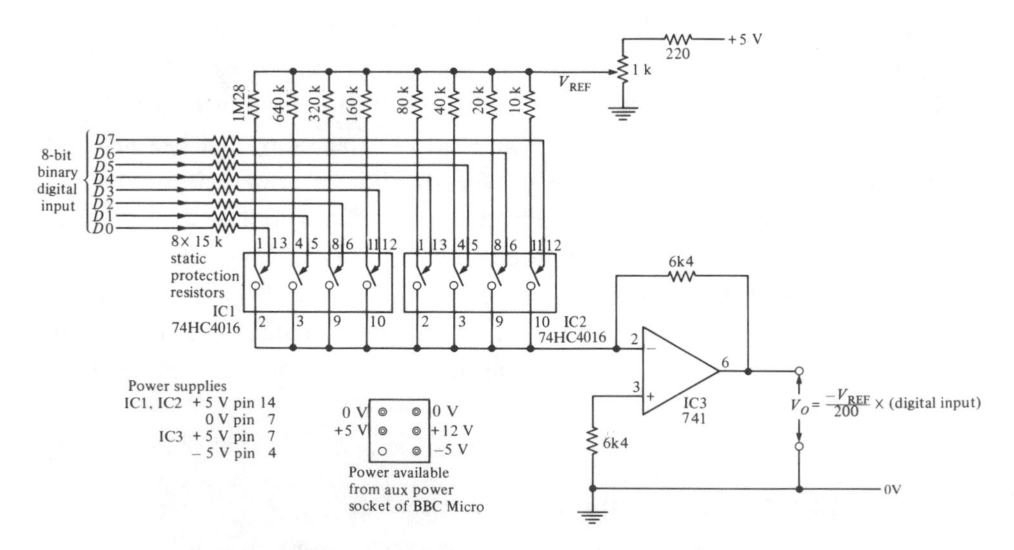

Fig. 14.20. Experimental 8-bit digital to analogue converter.

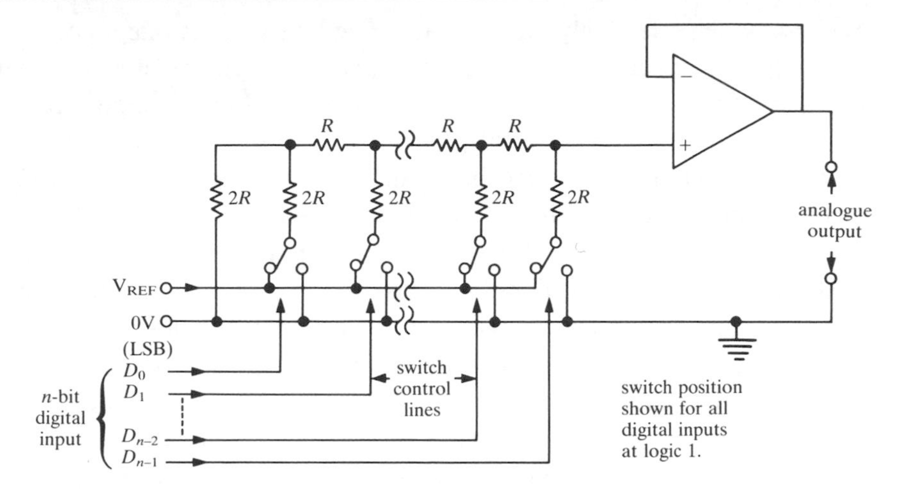

Fig. 14.21. Digital to analogue converter using R–2R ladder network.

inherent ability to multiply an analogue voltage by a digital input. Some converters are of the type optimized for this application, the multiplying DAC (MDAC), useful for the accurate digital control of analogue gain in a variety of applications.

Accuracy of the DAC summing resistors is of course of prime importance. In an IC DAC it is difficult to fabricate resistors of precise absolute value, but it is easy to make a lot of identically matched ones in the fabrication process. This has led to the 'R-2R' DAC shown in fig. 14.21. Potential divider calculations quickly suggest how it achieves the correct sequence of analogue outputs.

14.15 Analogue to digital conversion

14.15.1 Some basic ADC hardware

A/D conversion can be carried out in a number of ways. The very fast *flash* converters used for digital video signals use many parallel comparators to measure the analogue signal, one for each possible digital value. The comparator outputs then feed a binary encoder which generates a digital output corresponding to the highest comparator number activated by the input voltage. Conversion times are of the order of nanoseconds, but even an 8-bit converter needs 256 comparators and a 16-bit version becomes very unwieldy. A slower method, but one giving high resolution, is the single or dual slope integration method, which effectively measures the time taken for an integrator capacitor fed with constant current to charge to the same voltage level as the analogue input. A constant-frequency clock and counter then give a digital readout of the charge time, which is directly proportional to the analogue input voltage.

Probably the most popular technique is the *successive approximation* method, illustrated in its simplest version, known as a 'tracking' ADC, in fig. 14.22. This technique, which can readily be tried out experimentally, uses a DAC as the basis of its calibration. This DAC is used with a single analogue comparator to continuously compare the actual analogue input with the digital output presented by the up/down counter. If the digital output is too low, the counter is switched by the comparator to count up, if too high it counts down. This tight feedback quickly produces an accurate conversion.

Faster results may be obtained by using the true successive approximation technique which replaces the up/down counter with a logic system which sets each output bit to 1 in turn, starting with the MSB. If the resulting D/A output is less than the input voltage, the bit is left at 1, otherwise it is set back to 0 and the next bit tried. The converter quickly reaches the nearest digital output equivalent to the input voltage. It is essential that the input should remain constant during this approximation process, and it is normal to use a *sample/hold* circuit immediately prior to a successive approximation converter.

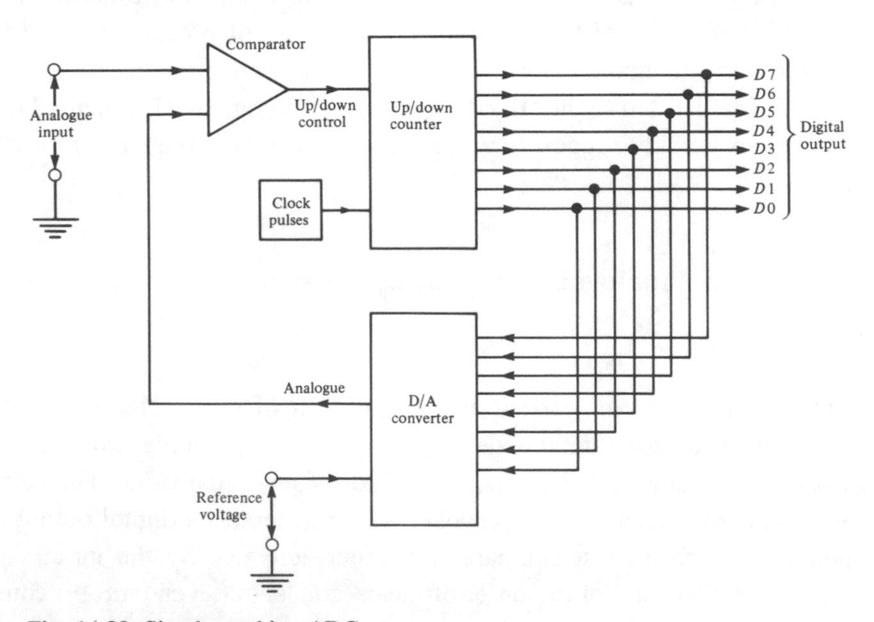

Fig. 14.22. Simple tracking ADC.

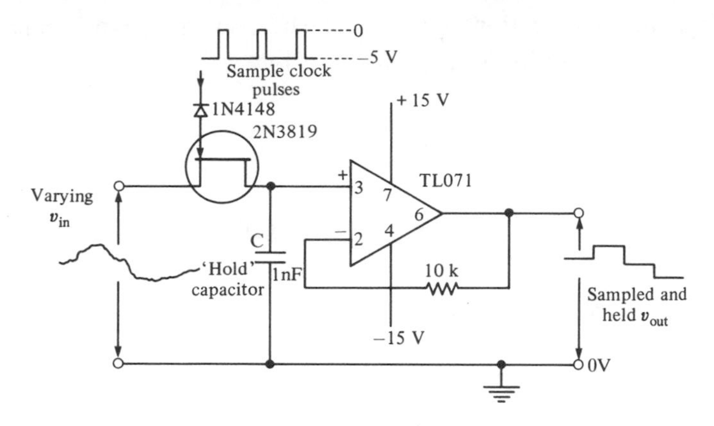

Fig. 14.23. Sample hold circuit for use prior to ADC.

14.15.2 Sample/hold circuit

A sample/hold circuit is shown in simple form in fig. 14.23. The FET allows the capacitor C to charge to the instantaneous value of the input voltage at every sample pulse. Between sample pulses, the FET switch is open and the capacitor holds its charge because of the high input resistance of the FET input op amp. Commercial sample/hold circuits use special low on-resistance MOSFETs to achieve fast charging and require a low offset high impedance amplifier to avoid 'droop' in capacitor voltage during the hold period.

14.15.3 Quantization noise and dither

However accurate an ADC may be, it still has the fundamental limitation that it cannot by itself respond to analogue signals less than $\frac{1}{2}$LSB in amplitude and that its response thereafter is in a series of steps each equal to one LSB: the image on a fax machine suffers from an analogous problem. With a smoothly changing clean analogue input, these steps create a form of distortion, known as *quantization noise*, which gets superimposed on the digital output. Because the steps are of constant height, as shown in fig. 14.24, the quantization noise as a percentage of the signal is worst at low signal levels, which in audio applications is perceived as a particularly objectionable distortion. This distortion problem can be turned into plain background noise by adding a precise level of noise, termed *dither*, to the analogue input signal to act as 'random bias'. Fig. 14.25 shows dither in action randomising, and therefore smoothing, the quantization steps. It can be shown that the maximum available signal to noise ratio, SNR_{max} is given by

$$SNR_{max} = \left(\frac{3}{2}\right) 2^{2N} \qquad (14.1)$$

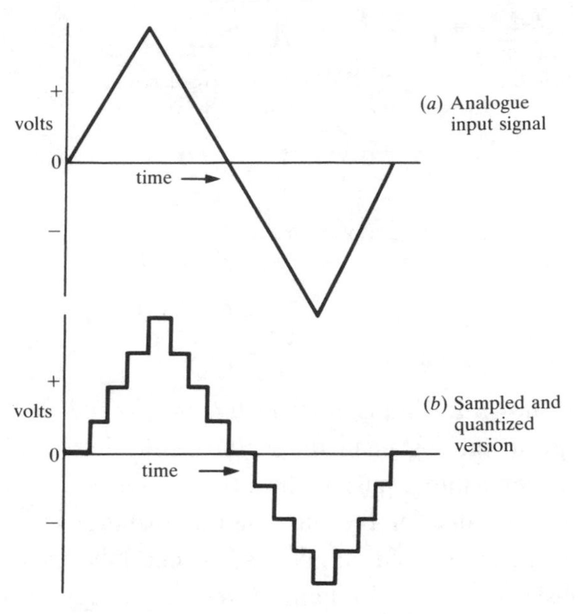

Fig. 14.24. Effect of quantization steps on an analogue signal.

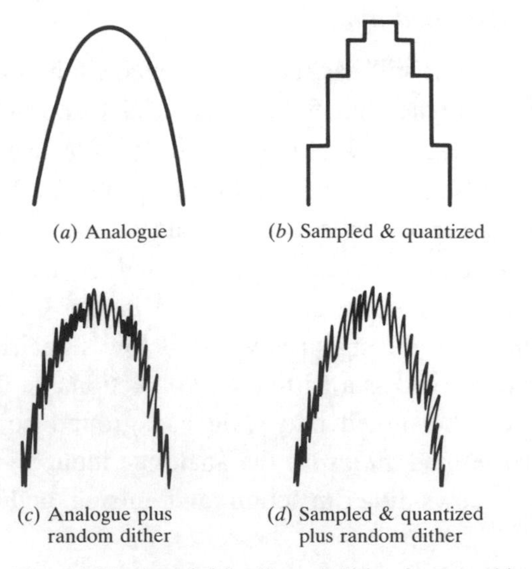

Fig. 14.25. Beneficial effect of dither in smoothing out quantization steps.

where N is the number of bits in the ADC and the noise is assumed to be random.

In decibel form, assuming a sinusoidal signal ($\sqrt{2}$ peak/rms ratio)

$$\mathrm{SNR}_{\mathrm{max}} \approx 6.02N + 1.76 \text{ dB}. \tag{14.2}$$

Thus, for example, an 'ideal' 16-bit conventional ADC theoretically gives a peak sinusoidal signal SNR of nearly 98 dB. With the usual method of measuring SNR relative to r.m.s signal this reduces to 95 dB. However, clever dither techniques are now used in digital audio recording whereby the spectrum of the dither noise is shaped so that more of the noise power lies towards the high frequency end of the audio band where the ear is less sensitive. In this way very clean results are obtained with low quantization distortion. This noise shaping is usually applied in conjunction with the technique of oversampling conversion described in section 14.15.5. It is also possible in some applications to arrange for the dither subsequently to be subtracted from the signal after it has carried out its task in the conversion processes, giving further SNR improvement.

14.15.4 Sampling frequency

Digitizing a signal consists of taking a regular series of rapid snapshots. How rapidly must these snapshots or samples be taken in order to represent the signal fully? Fig. 14.26 shows a commonsense approach to exploring the correct sampling frequency. Fig. 14.26(*a*), where the sample frequency is comparable to the signal frequency clearly gives the wrong result, whilst the much higher sample frequency in fig. 14.26(*b*) gives an output much closer to the original signal.

This problem was studied by Nyquist in connection with telecoms, which resulted in the Nyquist sampling theorem:

> In order not to lose any signal information, then the criterion for sampling frequency f_s is:
>
> $$f_s \geqslant 2f_{\mathrm{max}} \tag{14.3}$$
>
> where f_{max} is the maximum frequency component of the signal being sampled.

At first sight, this requirement for just one sample every half cycle looks surprisingly low, but the theorem rightly assumes that the sampled signal will be reconstituted through a steep-cut filter of precisely f_{max} bandwidth. This neatly removes the 'square corners' from the waveform since these represent 'rubbish' beyond the band of interest.

Fig. 14.27 shows how the sampling theorem can be understood in terms of sampling as a modulation process. The sampling frequency f_s is being multiplied by all the frequencies in the input signal spectrum. A bit of trigonometrical expansion with sine or cosine functions soon shows that this yields intermodulation products filling up the spectrum either side of f_s. If these intermodulation products get mixed in with the original signal in the spectrum from 0 to f_{max} then they interfere and mean that the original signal can never be recovered. Such interference is known as *aliasing distortion*. For this reason, sampling frequency f_s must be at least twice f_{max} in order to avoid this clash.

For example, the CD sampling frequency of 44.1 kHz is chosen to be comfortably twice the audio bandwidth of 20 kHz.

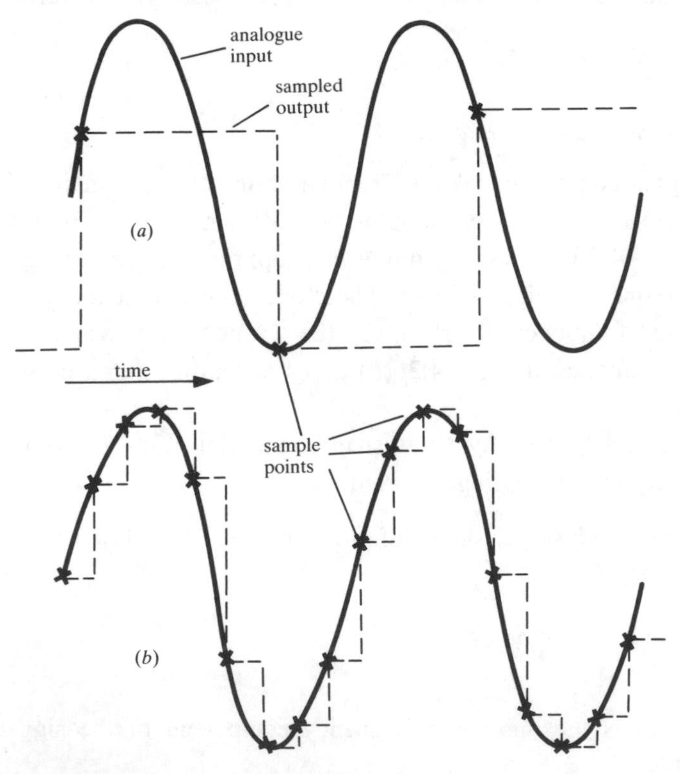

Fig. 14.26. Sampling of analogue signals prior to A/D conversion. (*a*) Inadequate sample rate. (*b*) Adequate sample rate.

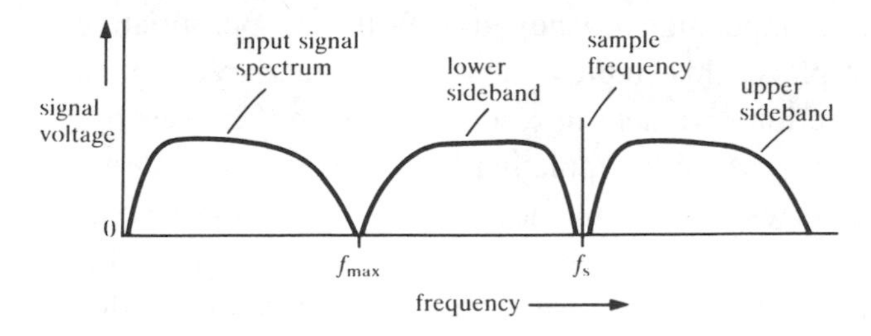

Notes
1. Upper and lower sidebands are the modulation products formed when input signal is multiplied by sample frequency f_s.
2. Nyquist Sampling Theorem recognizes that, in order to avoid lower sideband overlapping input signal spectrum, $f_s \geqslant 2f_{max}$, where f_{max} is maximum frequency in input signal.

Fig. 14.27. Spectrum of sampled signal.

Aliasing problems become clearer if, for example, we imagine that CDs were sampled at only 22 kHz. If then, say, a 17 kHz audio signal were to arrive at the sampling ADC, it would intermodulate with the 22 kHz sample frequency to produce a spurious 5 kHz signal. This spurious *alias* product, being within the audio band could never then be removed by the subsequent filtering. It is therefore vital that analogue signals are subject to *anti-alias filtering* prior to sampling to ensure that no frequency components higher than $f_s/2$ are present.

14.15.5 Delta–sigma (oversampling) data conversion

(*a*) *Oversampling.* Where better than 16-bit resolution is required at high speed, conversion techniques such as successive approximation discussed so far become progressively less effective, especially at the lower signal levels. The requirement for an accurately calibrated multi-bit DAC as an integral part of the converter becomes very demanding. Minor discrepancies between DAC switching levels due to component tolerances and variations in timing of the bit switches can give rise to notches in the transfer characteristic and even to missing codes in the output.

Today's high speed digital circuits have made possible converters using *oversampling* techniques which work at sampling frequencies much higher than the theoretical minimum for the signal bandwidth. Oversampling has the benefit of spreading the quantization noise over a wider bandwidth. This enables much of the noise to be 'left behind' when the digital signal is filtered back to the original input signal bandwidth. With the appropriate dither at the input in the first place, this improvement in the measured signal/noise ratio becomes a genuine improvement in resolution which increases the number of bits in the ADC process by interpolating between converter steps. Each 6 dB increase in SNR corresponds to the addition of 1 bit to the resolution so that, if the sampling is fast enough, even a 1-bit converter can give high resolution.

Known as *delta–sigma conversion*, the oversampling method typically employs oversampling factors of 256 times and more. Using circuitry which provides optimum noise shaping, it can provide over 18-bits of resolution from a 1-bit converter. The principle is also applied in the D/A converters of most CD players where a number of trade names are used. PDM (pulse density modulation) and MASH (multi-stage noise shaping) are two examples of the technique also known as *Bitstream* conversion.

(*b*) *1-bit A/D conversion.* The essential feature of a 1-bit ADC compared with multibit converters is that of repeating the use of the same analogue components over and over again in the time interval between output samples. The analogue input voltage is converted into digital bits by the repetitions instead of by the different components of different values used in a multibit converter. A fast clock rate for the repetitions can then achieve great precision independent of component tolerances.

Essentially, if we were to look at the 1-bit digitized signal at the over-sampled clock rate, we would see a higher concentration of binary ones when the analogue input is high and a higher concentration of zeros at a lower input voltage.

(*c*) *Delta–sigma operation.* Fig. 14.28 shows the outline of a delta–sigma

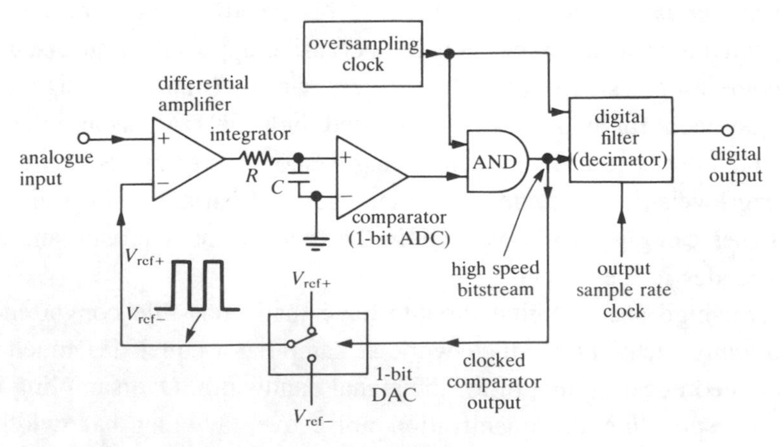

Fig. 14.28. Delta–sigma oversampling ADC.

oversampling ADC. The differential amplifier at the input is continuously comparing the analogue input signal with the output of the internal 1-bit DAC which is typically working at 256 times the required sample frequency of the final digital output. For example if the final sample rate were 44.1 kHz the internal oversampling clock would be 11.2896 MHz. The output of the differential amplifier is integrated, then fed to a comparator whose output is clocked out at the oversampling rate. If the integrator output is greater than 0 V the comparator outputs a stream of binary ones and if it is less than 0 V a stream of zeros is the result. The comparator is in fact a 1-bit ADC and generates a serial stream of 1s and 0s according to the value of the integrated output of the differential amplifier.

The loop is closed by feeding the clocked output of the comparator into the 1-bit DAC input.

This results in the differential amplifier producing the difference between the instantaneous voltage on the analogue input and the average value of the immediately preceding analogue samples, hence the word *delta* (small difference) in the name. Another name for this technique is the *charge balancing* converter because the whole aim of the DAC – differential amplifier–comparator, loop is to maintain zero charge on the integrator capacitor. The clocked comparator produces just sufficient logic 1 pulses at any one time to cancel (via the differential amplifier) the charge due to the analogue input flowing into the integrator. In other words the output of the AND gate carries a high speed bitstream clocked at typically 11.2896 MHz where the density of 'ones' is proportional to the analogue input voltage.

Converting this bitstream into, say, a 16-bit parallel digital output clocked at 44.1 kHz can just be carried out with a counter and latch. In practice it is normally carried out with a low-pass digital filter whose output is resampled at the output sample rate of 44.1 kHz. The low-pass filter removes fast changes in the digital signal and hence time-averages it ready for resampling at any chosen rate. This process is known as *decimation* and is much used for *sample rate conversion* where, for example, signals at the studio sampling rate of 48 kHz can be resampled to produce CD masters at 44.1 kHz or at 32 kHz for telecom transmission to broadcast transmitters and subsequent D/A conversion in NICAM decoders in TV sets.

(*d*) *Noise shaping*. The reader may be puzzled by an inconsistency implied in the description so far. The oversampling described is a factor of 256 times and yet by this means a 1-bit converter is claimed to produce a resolution of 16-bits or better (around 96 dB signal/noise ratio). On its own, 256 times oversampling only spreads out the quantization noise bandwidth a factor of

256, to give a 24 dB improvement above the 7.7 dB signal/noise ratio of the 1-bit converter, ending up with only about 5-bits of equivalent resolution.

The answer to finding the other bits of resolution lies in the integrator in the oversampling converter. Because it is effectively within a 'feedback' loop, the integrator's low-pass character actually exhibits a high-pass influence on the noise, pushing most of the quantization noise arising in the comparator up into the high-frequency part of the spectrum. This is a remarkable example of a circuit exhibiting synergy. A side-effect of a necessary feature like the integrator produces further advantage for the designer by pushing the noise out of the way beyond the band of interest. The integrator–comparator combination produces a quantization noise spectrum which rises at 6 dB per octave. In other words, there is much more noise power per Hz up in the megahertz region than in the input frequency band. Therefore the final decimating filter at the output removes much more noise than we would expect from a white-noise spectrum.

It can be shown that for an oversampling factor D and a first-order (single RC) integrator:

$$\text{SNR enhancement} \approx \frac{3D^3}{\pi^2}. \tag{14.4}$$

In practice, two or more stages of integration are used, often consisting of switched capacitor filters. For a second-order system the improvement becomes:

$$\text{SNR enhancement} \approx \frac{5D^5}{\pi^4}. \tag{14.5}$$

For $D = 256$ in such a second order system,

$$\begin{aligned}\text{SNR enhancement} &\approx 10 \log_{10}\left[\frac{5\times(256)^5}{\pi^4}\right] \text{dB}\\ &\approx 107.5 \text{ dB}.\end{aligned}$$

Added to the theoretical 7.7 dB SNR of the 1-bit converter, we get over 114 dB of SNR, amply exceeding the theoretical 109 dB given by 18-bits. Given special attention to loop stability, higher order noise shaping is possible, giving even better resolution.

Today's high speed processing means that oversampling converters are not only more accurate than multi-bit circuits, they are also cheaper because difficult precision components are replaced by precision timing which is much more straightforward. Oversampling is much more tolerant of hardware imperfections. The need for a sample and hold circuit is also generally eliminated because the frequency of the input processing is so much higher than that of the analogue input.

(*e*) *Oversampling D/A converters.* The delta–sigma oversampling technique is just as relevant in a DAC as in an ADC. For example in a CD player the input 44.1 kHz sample rate 16-bit digital signal has its sample rate increased by typically 256 times. This is carried out in a digital filter which is a decimator in reverse, known as an *interpolator* which in this 256 times example will produce a 1-bit output clocked at 11.2896 MHz. Noise shaping is used as in the ADC case to maximize SNR. Once again we have a bitstream where the density of binary ones is proportional to the required analogue amplitude. Only analogue integration (low pass filtering) of the resulting pulse train is then required to produce a high resolution analogue output (like the charge pump ratemeter in section 10.12).

(*f*) *Anti-alias filtering.* A bonus associated with oversampling is the elimination of complex analogue anti-alias filters. The Nyquist sampling theorem is just as relevant at every point in a bitstream system as in multi-bit conversion, but a digital anti-alias filter is an integral part of the decimator or interpolator.

For example in a 256 times oversampled 16-bit audio ADC, the input audio signal is initially sampled not at 44.1 kHz but at 11.2896 MHz. Therefore our input anti-alias filter needs only to remove components higher than 5.6 MHz. A simple RC low-pass filter generally suffices, the main concern being to remove any RF pickup which could beat with the oversampling clock to produce heterodyne whistles.

This elimination of 'brick wall' steep cut anti-alias filters at both the ADC input and DAC output reduces cost and also generally improves phase performance. It is extremely difficult to design a steep cut analogue filter without introducing phase distortion near cut-off (i.e. the high frequencies emerging at a different time from the rest of the band). The digital filters used for decimation and interpolation are much more straightforward in this respect and can avoid significant phase distortion.

With the kind of developments in ADCs and DACs just described, the boundary between analogue signals and digital signals is becoming almost seamless. Of course, care needs to be taken to avoid unwanted crosstalk between the analogue and digital parts of a circuit, but increasingly the conversion process is just another chip on the PCB, facilitating the trend towards DSP for more and more functions in electronics.

Appendix 1

Component identification

Resistor codes

Resistors are normally coded with coloured bands to enable quick identification of the value.

The colour code specifies:

(1) The value of the resistor, in ohms.
(2) The maximum deviation from the stated value (tolerance).

Preferred values

Most resistors are manufactured in a limited number of values which are internationally standard and are termed 'preferred values'. The following values, and decade multiples thereof, are preferred:

E12 range	10, 12, 15, 18, 22, 27, 33, 39, 47, 56, 68, 82,
E24 range	10, 11, 12, 13, 15, 16, 18, 20, 22, 24, 27, 30, 33, 36, 39, 43, 47, 51, 56, 62, 68, 75, 82, 91.

Tolerance

Most resistors have a manufacturer's tolerance of ±5%, which is adequate for most electronic applications. For more accurate work, 2% and 1% tolerance are readily available.

Interpreting the colour code

The four colour code bands are at one end of the component (fig. A1.1). Counting from the end, the first three bands give the value and the fourth the tolerance. The significance of the colours is shown in table A1.1.

Table A1.1

First 3 bands	Tolerance band
Black 0	Brown ±1%
Brown 1	Red ±2%
Red 2	Gold ±5%
Orange 3	Silver ±10%
Yellow 4	No band ±20%
Green 5	A fifth band of salmon pink, indicates a high-stability resistor.
Blue 6	
Violet 7	
Grey 8	
White 9	

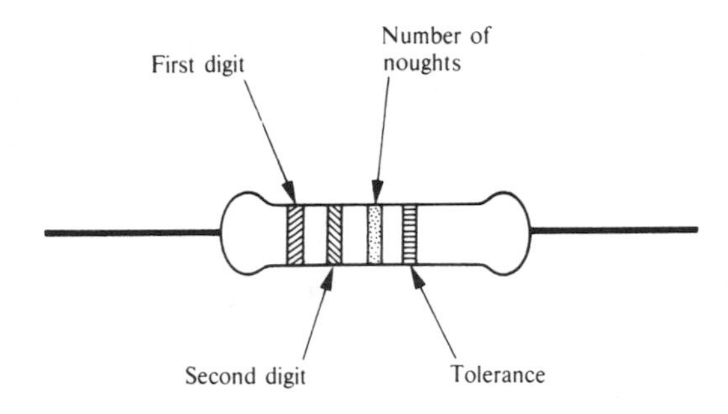

Fig. A1.1.

Reading the value

Preferred values are specified to two significant figures; hence the first two bands are used to specify those two figures and the third band specifies the number of noughts to be added:

first band – first digit,
second band – second digit,
third band – number of noughts.

Thus a 4700 Ω (4.7 kΩ) resistor ±10% tolerance has the following colours:

yellow, violet, red, silver.

Surface mount technology (SMT)

Surface mount components do not normally use colour coding. Instead, the significant digits and the number of noughts are specified directly in figures. For example, a 4.7 kΩ resistor is labelled 472 and a 10 kΩ resistor 103.

Table A1.2

Value	Code	Value	Code
0.47 Ω	R47	100 Ω	100R
1.0 Ω	1R0	1 kΩ	1K0
4.7 Ω	4R7	10 kΩ	10K
10 Ω	10R	100 kΩ	100K
47 Ω	47R	1 MΩ	1M0
		10 MΩ	10M

Power rating

The permissible maximum power dissipation in a resistor is not specified in the colour coding, but is normally $\frac{1}{4}$ W for the most common size (approximately 6 mm in length). Resistors of 1 W dissipation are also used, these being between 1.5 cm and 2 cm in length and of a correspondingly larger diameter. If still higher wattages are required, wirewound resistors having dissipations of 5 W, 10 W and 20 W are also readily available.

B.S. 1852 Resistor Code

This code, a combination of figures and letters to give value and tolerance, is used on circuit diagrams and on some components which are not colour coded. It is best explained by the examples in table A1.2.

After the code can be added a letter to indicate tolerance.

F ± 1%
G ± 2%
J ± 5%
K ± 10%
M ± 20%

Hence 4K7K indicates 4.7 kΩ ± 10%.

Capacitor coding

Most capacitors have the value and working voltage printed on them, however the resistor colour code is often used, the value being specified in picofarads (pF). Tolerance bands do not use gold and silver, and are coded as follows:

black ±20%	orange ± 2.5%
brown ±1%	green ± 5%
white ±10%	red ± 2%

A fifth band is sometimes added to specify a particular characteristic, such as temperature coefficient or working voltage. For example, a red fifth band is often used on polyester capacitors to indicate 250 V wkg, yellow for 400 V, and blue 630 V.

1 picofarad (pF) = 10^{-12} F
1 nanofarad (nF) = 10^{-9} F = 1000 pF
1 microfarad (μF) = 10^{-6} F = 1000 nF

Appendix 2

Transistor selection

Anyone studying the product lists of a semiconductor manufacturer might feel somewhat daunted at the array of different type numbers. In practice, things are very much more simple than they might appear. Very few circuits indeed require exotic transistors, and there is enormous flexibility in substituting one type of number for another. Most transistor circuits in this book specify the versatile BC107, though any general-purpose npn transistor will be satisfactory. For instance, the 2N2222 can be directly substituted for the BC107, whilst plastic-encapsulated transistors such as the 2N3704 can be used, noting the different connections.

When studying the manufacturer's data books to choose a transistor for a particular circuit, the chief parameters to look for are as follows:

maximum collector–emitter voltage, $V_{CE(\mathrm{max})}$
maximum collect current $I_{C(\mathrm{max})}$
maximum power dissipation at 25 °C air temperature P_{max}
minimum transition frequency f_T
typical current gain h_{FE}

Table A2.1 lists parameters for some common npn transistors. Also included are the all-important connection configurations, which refer to the diagrams in Fig. A2.1. For each of the npn transistors quoted, a pnp device can be selected with basically similar characteristics, which is suitable for use in complementary circuits; these are included at the foot of table A2.1.

Table A2.1

	BC107	2N3704	2N3053	BFY50	2N3055
$V_{CE(\text{max})}$ V	45	30	40	35	60
$I_{C(\text{max})}$ mA	300	500	700	1 A	15 A
P_{max} mW	360	360	1 W	800	115 W (case at 25 °C)
f_T MHz	150	100	100	60	1
h_{FE}	200	200	150	30	50
Connection diagram	(*a*)	(*b*)	(*a*)	(*a*)	(*c*)
pnp complement	BC177	2N3702	2N4037	2N4036	PNP3055 or MJ2955

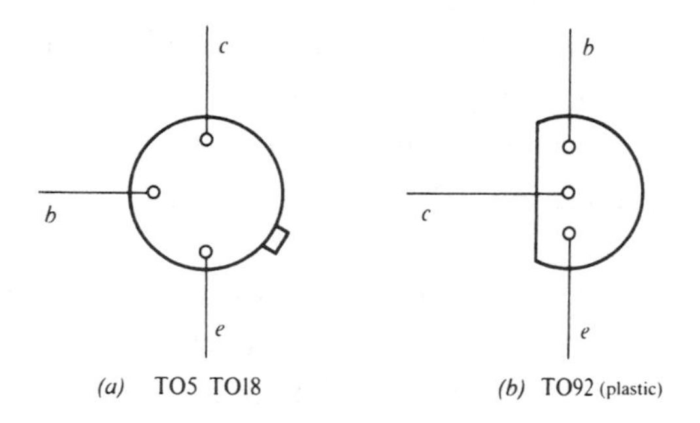

(*a*) TO5 TO18 (*b*) TO92 (plastic)

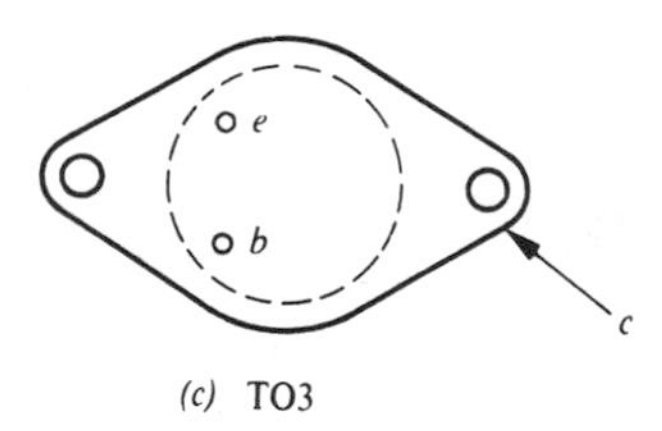

(*c*) TO3

Fig. A2.1. Transistor connections viewed from wire side.

Appendix 3

Op amp data

General

Most of the op amp circuits in this book are designed round the popular 'industry standard' 741 IC. Some applications might, however, require a different type of op amp, such as one with very low input bias current or high slew rate. This appendix gives the major electrical characteristics of the 741 and 748, together with those of three other op amps of different specifications.

The LM308 is basically similar to the 741, but with lower input bias current. The NE5534 offers a very high slew rate and low-noise audio performance, maintaining full-output up to 90 kHz, whereas the 741 has restricted output swing above 10 kHz. The TL081 is typical of IC op amps with a FET input. Input bias and offset currents are exceedingly low, and input resistance very high. Its high slew rate also makes it useful for high quality audio work, especially in its low noise version, TL071. Dual (072/082) and quad (074/084) versions are useful for compact layouts.

Electrical characteristics

In table A3.1, symbols have the following meanings:

A_{VOL}	Open-loop d.c. voltage gain
R_{in}	Input resistance
$\pm V_{CC(max)}$	Maximum supply voltage
CMRR	Common-mode rejection ratio
V_{io}	Input offset voltage
I_b	Input bias current
I_{io}	Input offset current
B	Bandwidth (unity gain)
S	Slew rate

Table A3.1

	741	748	LM308	NE5534	TL071/TL081
A_{VOL} dB	106	106	102	100	106
R_{in} Ω	2 M	2 M	40 M	100 k	10^{12}
$\pm V_{CC(\max)}$	±18	±18	±18	±20	±18
CMRR dB	90	90	100	100	76
V_{io} mV	1	2	10	0.5	3
I_b	80 nA	80 nA	7 nA	500 nA	30 pA
I_{io}	20 nA	20 nA	1.5 nA	20 nA	5 pA
B Hz	1 M	1 M	1 M	10 M	3 M
S V/μs	0.5	0.5	0.4	13	13
C_c pF	—	30	30	22	—
Conn	(*a*)	(*b*)	(*c*)	(*d*)	(*a*)

Table A3.2

Pin	(*a*)	(*b*)	(*c*)	(*d*)
1	Offset null	Offset null/C_c	Offset null	Offset null
2	Inverting input	Inverting input	Inverting input	Inverting input
3	Non-inverting input	Non-inverting input	Non-inverting input	Non-inverting input
4	V_{CC-}	V_{CC-}	V_{CC-}	V_{CC-}
5	Offset null	Offset null	Offset null	C_c
6	Output	Output	Output/C_c	Output
7	V_{CC+}	V_{CC+}	V_{CC+}	V_{CC+}
8	No connection	C_c	C_c	Offset null/C_c

C_c Recommended value of external compensation capacitor (unity gain)

Conn Pin connections (see fig. A3.1 and table A3.2)

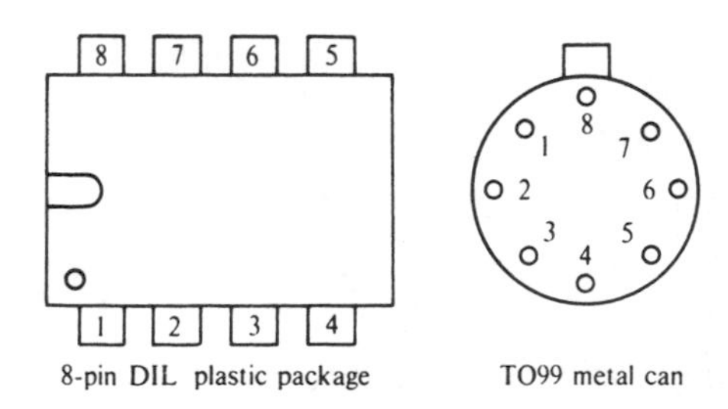

Fig. A3.1

Appendix 4
Digital IC connections

This appendix gives pin connection information for some of the most popular TTL and CMOS digital ICs.

The 7400 series diagrams also apply to the 74LS00 TTL chips and 74AC00, 74HCT00, 74LV00, 74LVT00 CMOS chips of the same number.

TTL circuits (7400 series) use $V_{CC} = +5\text{ V} \pm 0.25\text{ V}$. Most CMOS circuits will operate with V_{DD} from +2 V to +6 V (except LV/LVT limited to 3.6 V max), earth (0 V or ground) is connected to the pin marked V_{SS}.

Most TTL outputs have a fan-out of ten and most TTL inputs a load factor of one. Input load factors higher than one are indicated by the appropriate number opposite the input on the pin diagram.

A selection of '7400' series TTL and CMOS logic

7400	quad 2-input NAND
7402	quad 2-input NOR
7404	hex inverter
7410	triple 3-input NAND
7413	dual 4-input NAND Schmitt trigger
7420	dual 4-inputer NAND
7430	8-input NAND
7442	BCD–decimal decoder
7447A/74247	BCD–7-segment indicator decoder
7473	dual *J–K* flip-flop with clear
7474	dual *D* flip-flop
7475	quad latch
7476	dual *J–K* flip-flops with preset and clear
7483A	4-bit binary full adder

7486	quad exclusive OR
7490A	decade counter
7493A	4-bit binary counter
74121	monostable
74122	retriggerable monostable
74123	dual retriggerable monostable
74141	BCD–decimal decoder and Nixie driver
74LS138	1 of 8 demultiplexer
74151	8 into 1 multiplexer
74154	1 of 16 decoder

A selection of 4000 series CMOS logic

This original CMOS logic series, largely obsolete for new designs because of its slow speed, still has appeal in electrically noisy environments where its slow propagation times (>50 ns) and high logic thresholds give good noise immunity. Supply voltage can be from +3 V to +15 V. Many 4000 series chips are now produced with HC characteristics as 74HC4000 etc. They retain the 4000 series pin connections.

4000	dual 3-input NOR plus inverter
4001	quad 2-input NOR
4002	dual 4-input NOR
4008	4-bit binary full adder
4011	quad 2-input NAND
4012	dual 3-input NAND
4013	dual *D* flip-flop
4016/4066	quad bilateral switch
4017	decade counter (decimal output)
4024	7-bit binary counter
4030	quad exclusive OR
4049	hex inverter
4068	8-input NAND
4528	dual retriggerable monostable

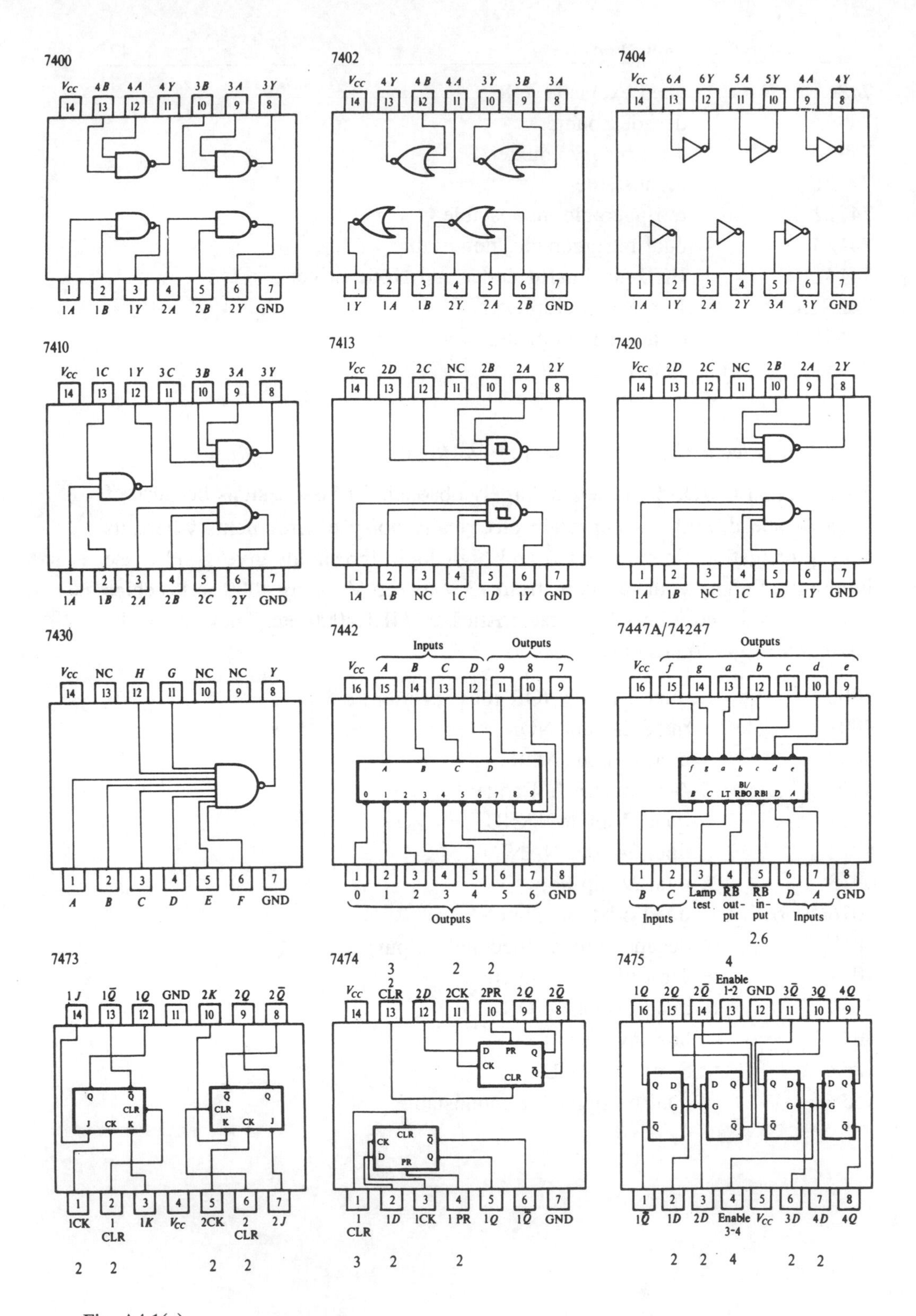
7400
Vcc 4B 4A 4Y 3B 3A 3Y
14 13 12 11 10 9 8
1 2 3 4 5 6 7
1A 1B 1Y 2A 2B 2Y GND
7402
Vcc 4Y 4B 4A 3Y 3B 3A
14 13 12 11 10 9 8
1 2 3 4 5 6 7
1Y 1A 1B 2Y 2A 2B GND
7404
Vcc 6A 6Y 5A 5Y 4A 4Y
14 13 12 11 10 9 8
1 2 3 4 5 6 7
1A 1Y 2A 2Y 3A 3Y GND
7410
Vcc 1C 1Y 3C 3B 3A 3Y
14 13 12 11 10 9 8
1 2 3 4 5 6 7
1A 1B 2A 2B 2C 2Y GND
7413
Vcc 2D 2C NC 2B 2A 2Y
14 13 12 11 10 9 8
1 2 3 4 5 6 7
1A 1B NC 1C 1D 1Y GND
7420
Vcc 2D 2C NC 2B 2A 2Y
14 13 12 11 10 9 8
1 2 3 4 5 6 7
1A 1B NC 1C 1D 1Y GND
7430
Vcc NC H G NC NC Y
14 13 12 11 10 9 8
1 2 3 4 5 6 7
A B C D E F GND
7442
Inputs
Outputs
Vcc A B C D 9 8 7
16 15 14 13 12 11 10 9
A B C D
0 1 2 3 4 5 6 7 8 9
1 2 3 4 5 6 7 8
0 1 2 3 4 5 6 GND
Outputs
7447A/74247
Outputs
Vcc f g a b c d e
16 15 14 13 12 11 10 9
f g a b c d e
BI/
B C LT RBO RBI D A
1 2 3 4 5 6 7 8
B C Lamp test RB output RB input D A GND
Inputs
Inputs
2.6
7473
1J 1Q̄ 1Q GND 2K 2Q 2Q̄
14 13 12 11 10 9 8
Q Q̄ CLR J CK K
Q̄ Q CLR K CK J
1 2 3 4 5 6 7
1CK 1 CLR 1K Vcc 2CK 2 CLR 2J
2 2 2 2
7474
Vcc 2 CLR 2D 2CK 2PR 2Q 2Q̄
3 2 2
14 13 12 11 10 9 8
D PR Q CK CLR Q̄
CK CLR Q̄ D PR Q
1 2 3 4 5 6 7
1 CLR 1D 1CK 1 PR 1Q 1Q̄ GND
3 2 2
7475
4
1Q 2Q 2Q̄ Enable 1-2 GND 3Q̄ 3Q 4Q
16 15 14 13 12 11 10 9
Q D G Q̄
D Q G Q̄
Q D G Q̄
D Q G Q̄
1 2 3 4 5 6 7 8
1Q̄ 1D 2D Enable 3-4 Vcc 3D 4D 4Q
2 2 4 2 2

Fig. A4.1(*a*)

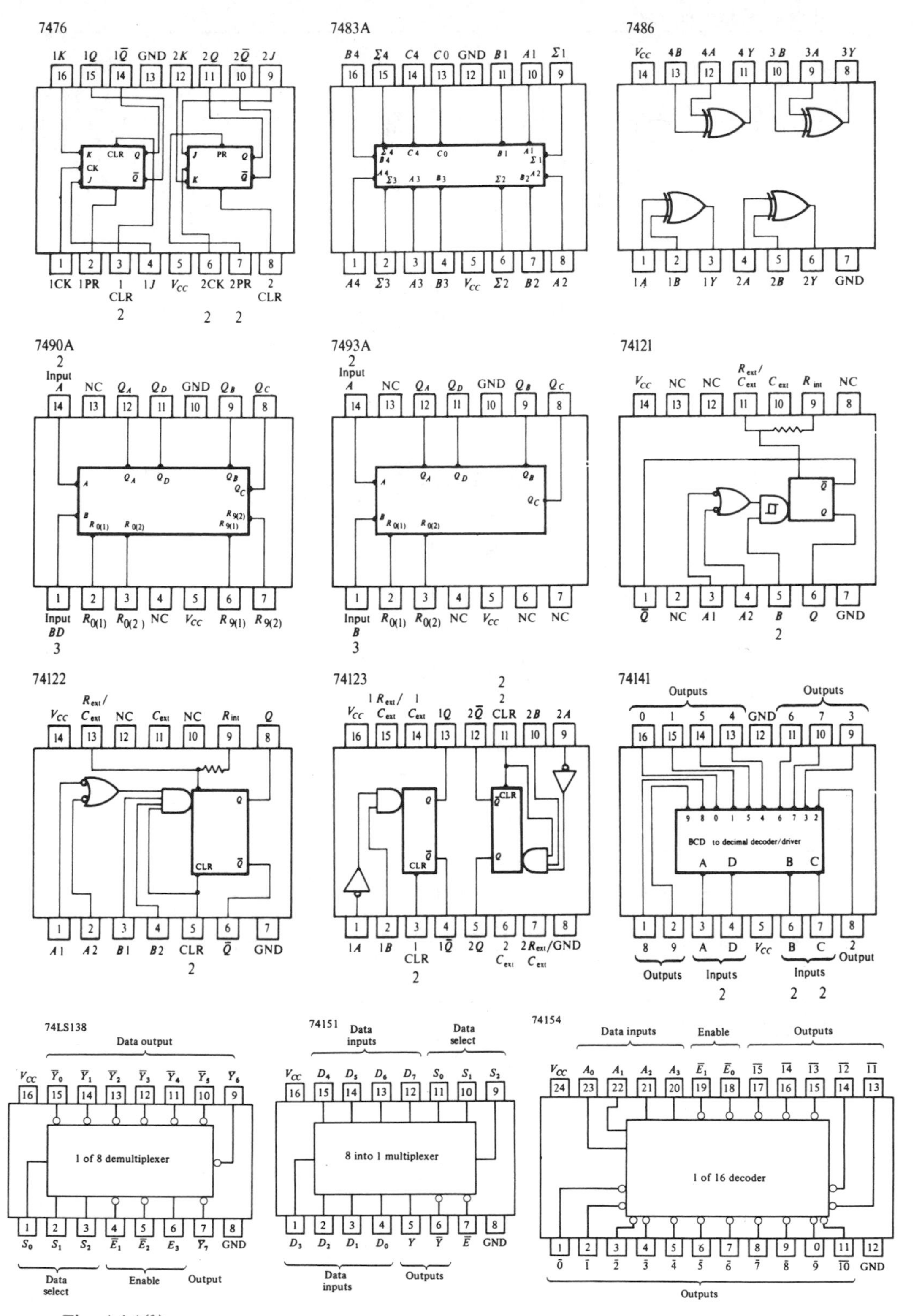
7476
7483A
7486
7490A
7493A
74121
74122
74123
74141
74LS138
74151
74154
1 of 8 demultiplexer
8 into 1 multiplexer
1 of 16 decoder
BCD to decimal decoder/driver
Data output
Data inputs
Data select
Enable
Outputs
Inputs
Output

Fig. A4.1(*b*)

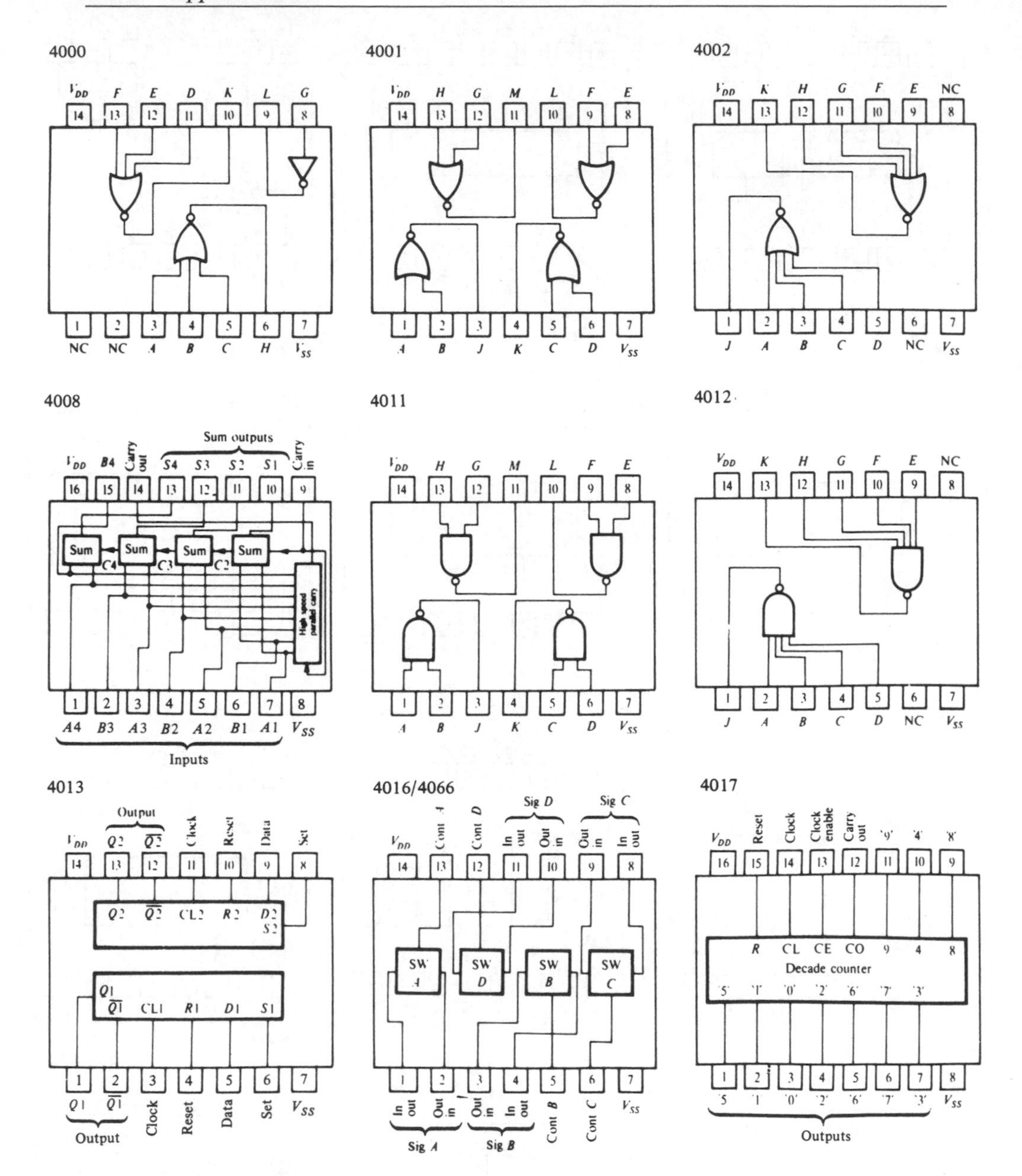
4000
V_{DD} F E D K L G
14 13 12 11 10 9 8
1 2 3 4 5 6 7
NC NC A B C H V_{SS}
4001
V_{DD} H G M L F E
A B J K C D V_{SS}
4002
V_{DD} K H G F E NC
J A B C D NC V_{SS}
4008
Sum outputs
V_{DD} B4 Carry out S4 S3 S2 S1 Carry in
Sum
C4 C3 C2
High speed parallel carry
A4 B3 A3 B2 A2 B1 A1 V_{SS}
Inputs
4011
V_{DD} H G M L F E
A B J K C D V_{SS}
4012
V_{DD} K H G F E NC
J A B C D NC V_{SS}
4013
Output
V_{DD} Q2 $\overline{Q2}$ Clock Reset Data Set
Q2 $\overline{Q2}$ CL2 R2 D2 S2
Q1 $\overline{Q1}$ CL1 R1 D1 S1
Q1 $\overline{Q1}$ Clock Reset Data Set V_{SS}
Output
4016/4066
Sig D Sig C
V_{DD} Cont A Cont D In out Out in Out in In out
SW A SW D SW B SW C
In out Out in Out in In out Cont B Cont C V_{SS}
Sig A Sig B
4017
V_{DD} Reset Clock Clock enable Carry out '9' '4' '8'
R CL CE CO 9 4 8
Decade counter
'5' '1' '0' '2' '6' '7' '3'
'5' '1' '0' '2' '6' '7' '3' V_{SS}
Outputs

Fig. A4.1(*c*)

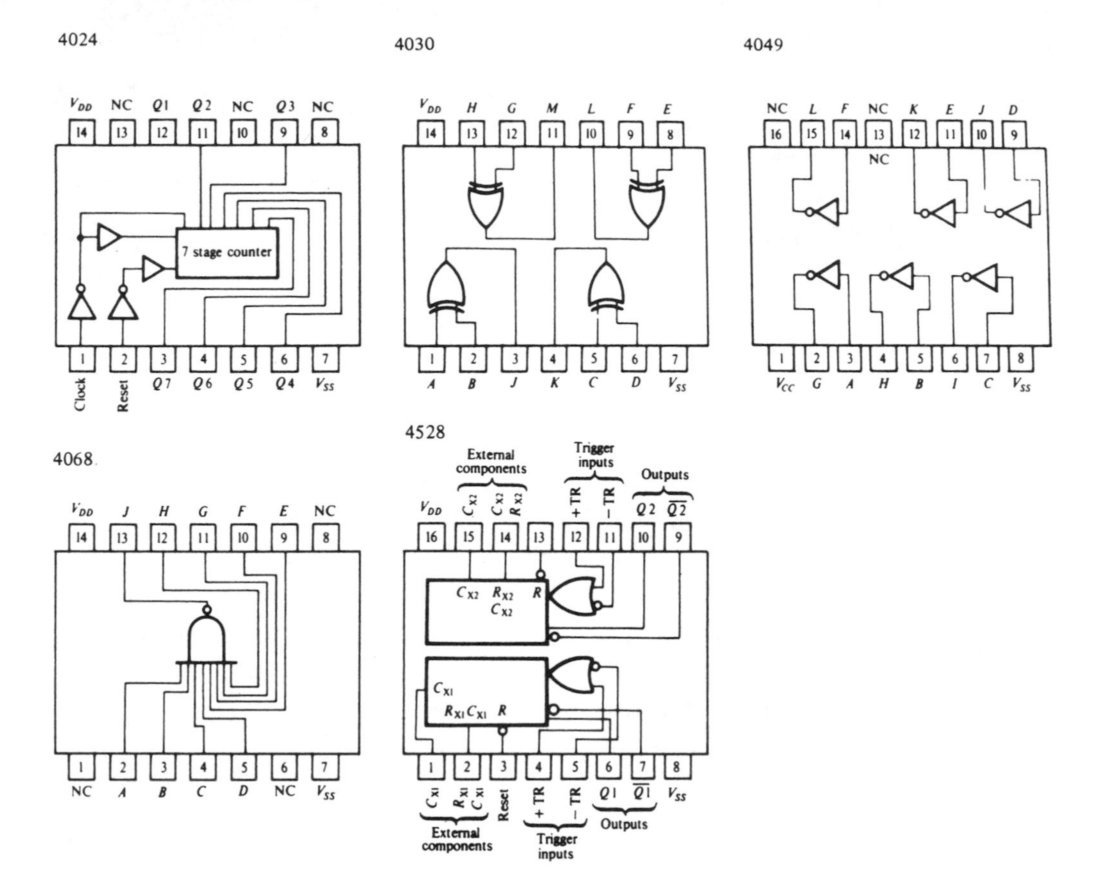
4024
V_{DD}
NC
Q1
Q2
NC
Q3
NC
7 stage counter
Clock
Reset
Q7
Q6
Q5
Q4
V_{SS}
4030
V_{DD}
H
G
M
L
F
E
A
B
J
K
C
D
V_{SS}
4049
NC
L
F
NC
K
E
J
D
V_{CC}
G
A
H
B
I
C
V_{SS}
4068
V_{DD}
J
H
G
F
E
NC
NC
A
B
C
D
NC
V_{SS}
4528
External components
Trigger inputs
Outputs
V_{DD}
C_{X2}
C_{X2} R_{X2}
+TR
−TR
Q2
$\overline{Q2}$
R
C_{X1}
R_{X1} C_{X1}
C_{X1}
R_{X1} C_{X1}
Reset
+TR
−TR
Q1
$\overline{Q1}$
V_{SS}
External components
Trigger inputs
Outputs

Fig. A4.1(*d*)

Appendix 5

Interfacing to the PC

Most PCs are fitted with a number of expansion bus sockets, each consisting of a 62-way connector in line with a 36-way connector. The connectors are of the direct edge type fitted to the motherboard. The 62-way connector carries the original *Industry Standard Architecture* (ISA) connections, whilst the addition of the 36-way connector caters for the more recent *Extended Industry Standard Architecture* (EISA) with its 32-bit address bus and 16-bit data bus.

Fig. A5.1 shows the EISA connectors with their pin numbering configuration. Tables A5.1 and A5.2 list the pin connections and functions.

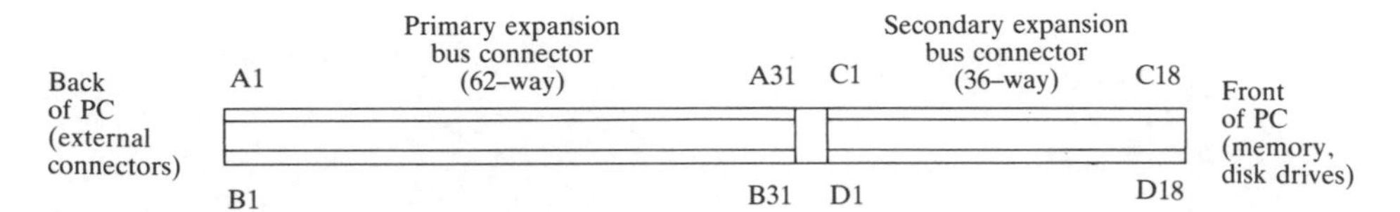

Fig. A5.1 ISA/EISA expansion bus connectors (inside the PC) viewed from above.

Table A5.1 *Signals present on the 62-way ISA bus connector*

Pin No.	Abbrev.	Direction	Signal	Function
A1	$\overline{\text{IOCHK}}$	I	I/O Channel check	Take low to indicate a parity error in a memory or I/O device.
A2	D7	I/O	Data line 7	Data bus line
A3	D6	I/O	Data line 6	Data bus line
A4	D5	I/O	Data line 5	Data bus line
A5	D4	I/O	Data line 4	Data bus line
A6	D3	I/O	Data line 3	Data bus line
A7	D2	I/O	Data line 2	Data bus line
A8	D1	I/O	Data line 1	Data bus line
A9	D0	I/O	Data line 0	Data bus line
A10	$\overline{\text{IOCHRDY}}$	I	I/O channel ready	Pulsed low by a slow memory or I/O device to signal that it is not ready for data transfer.
A11	AEN	O	Address enable	Issued by the DMA controller to indicate that a DMA cycle is in progress. Disables port I/O during a DMA operation in which IORD and IOWR may be asserted.
A12	A19	I/O	Address line 19	Address bus line
A13	A18	I/O	Address line 18	Address bus line
A14	A17	I/O	Address line 17	Address bus line
A15	A16	I/O	Address line 16	Address bus line
A16	A15	I/O	Address line 15	Address bus line
A17	A14	I/O	Address line 14	Address bus line
A18	A13	I/O	Address line 13	Address bus line
A19	A12	I/O	Address line 12	Address bus line
A20	A11	I/O	Address line 11	Address bus line
A21	A10	I/O	Address line 10	Address bus line
A22	A9	I/O	Address line 9	Address bus line
A23	A8	I/O	Address line 8	Address bus line
A24	A7	I/O	Address line 7	Address bus line
A25	A6	I/O	Address line 6	Address bus line
A26	A5	I/O	Address line 5	Address bus line
A27	A4	I/O	Address line 4	Address bus line

Table A5.1 (*cont.*)

Pin No.	Abbrev.	Direction	Signal	Function
A28	A3	I/O	Address line 3	Address bus line
A29	A2	I/O	Address line 2	Address bus line
A30	A1	I/O	Address line 1	Address bus line
A31	A0	I/O	Address line 0	Address bus line
B1	GND	n.a.	Ground	Ground/common 0 V
B2	RESET	O	Reset	When taken high this signal resets all expansion cards
B3	+5 V	n.a.	+5 V d.c.	Supply voltage rail
B4	IRQ2	I	Interrupt request level 2	Interrupt request (highest priority, active high).
B5	–5 V	n.a.	–5 V d.c.	Supply voltage rail
B6	DRQ2	I	Direct memory access request level 2	Taken high when a DMA transfer is required. The signal remains high until the corresponding DACK line goes low.
B7	–12 V	n.a.	–12 V d.c.	Supply voltage rail
B8	0 WS	I	Zero wait state	Indicates to the microprocessor that the present bus cycle can be completed without inserting any additional wait cycles.
B9	+12 V	n.a.	+12 V d.c.	Supply voltage rail
B10	GND	n.a.	Ground	Ground/common 0 V
B11	$\overline{\text{MEMW}}$	O	Memory write	Taken low to signal a memory write operation.
B12	$\overline{\text{MEMR}}$	O	Memory read	Taken low to signal a memory read operation
B13	$\overline{\text{IOWR}}$	O	I/O write	Taken low to signal an I/O write operation.
B14	$\overline{\text{IORD}}$	O	I/O read	Taken low to signal an I/O read operation.

Table A5.1 (*cont.*)

Pin No.	Abbrev.	Direction	Signal	Function
B15	$\overline{\text{DACK3}}$	O	Direct memory access acknowledge level 3	Taken low to acknowledge a DMA request on the corresponding level
B16	DRQ3	I	Direct memory access request level 3	Taken high when a DMA transfer is required. The signal remains high until the corresponding DACK line goes low.
B17	$\overline{\text{DACK1}}$	O	Direct memory access acknowlege level 1	Taken low to acknowledge a DMA request on the corresonding level
B18	DRQ1	I	Direct memory access request level 1	Taken high when a DMA transfer is required. The signal remains high until the corresponding DACK line goes low.
B19	$\overline{\text{DACK0}}$	O	Direct memory access acknowledge level 0	Taken low to acknowledge a DMA request on the corresponding level
B20	CLK4	O	4.77 MHz clock (typical)	CPU clock divided by 3, 210 ns period, 33% duty cycle.
B21	IRQ7	I	Interrupt request level 7	Taken high by an I/O device when it requires service
B22	IRQ6	I	Interrupt request level 6	Taken high by an I/O device when it requires service
B23	IRQ5	I	Interrupt request level 5	Taken high by an I/O device when it requires service
B24	IRQ4	I	Interrupt request level 4	Taken high by an I/O device when it requires service
B25	IRQ3	I	Interrupt request level 3	Taken high by an I/O device when it requires service

Table A5.1 (*cont.*)

Pin No.	Abbrev.	Direction	Signal	Function
B26	$\overline{\text{DACK2}}$	O	Direct memory access acknowledge level 2	Taken low to acknowledge a DMA request on the corresponding level
B27	TC	O	Terminal count	Pulsed high to indicate that a DMA transfer terminal count has been reached.
B28	$\overline{\text{ALE}}$	O	Address latch enable	A falling edge indicates that the address latch is to be enabled. The signal is taken high during DMA transfers.
B29	+5 V	n.a.	+5 V d.c.	Supply voltage rail
B30	OSC	O	14.31818 MHz clock (typical)	Fast clock with 790 ns period, 50% duty cycle.
B31	GND	n.a.	Ground	Ground/common 0 V

Notes:

(a) Signal directions are quoted relative to the system motherboard; I represents input, O represents output, and I/O represents a bidirectional signal used both for input and also for output (n.a. indicates not applicable).

(b) IRQ4, IRQ6 and IRQ7 are generated by the motherboard serial, disk, and parallel interfaces respectively.

(c) DACK0 (sometimes labelled REFRESH) is used to refresh dynamic memory whilst DACK1 to DACK3 are used to acknowledge other DMA requests.

Table A5.2 *Signals present on the 36-way additional connector (EISA)*

Pin No.	Abbrev.	Direction	Signal	Function
C1	SBHE	I/O	System bus high enable	When asserted this signal indicates that the high byte (D8 to D15) is present on the DATA bus.
C2	LA23	I/O	Latched address line 23	Address bus line
C3	LA22	I/O	Latched address line 22	Address bus line
C4	LA21	I/O	Latched address line 21	Address bus line
C5	LA20	I/O	Latched address line 20	Address bus line
C6	LA23	I/O	Latched address line 19	Address bus line
C7	LA22	I/O	Latched address line 18	Address bus line
C8	LA23	I/O	Latched address line 17	Address bus line
C9	$\overline{\text{MEMWR}}$	I/O	Memory write	Taken low to signal a memory write operation.
C10	$\overline{\text{MEMRD}}$	I/O	Memory read	Taken low to signal a memory read operation.
C11	D8	I/O	Data line 8	Data bus line
C12	D9	I/O	Data line 9	Data bus line
C13	D10	I/O	Data line 10	Data bus line
C14	D11	I/O	Data line 11	Data bus line
C15	D12	I/O	Data line 12	Data bus line
C16	D13	I/O	Data line 13	Data bus line
C17	D14	I/O	Data line 14	Data bus line
C18	D15	I/O	Data line 15	Data bus line
D1	$\overline{\text{MEMCS16}}$	I	Memory chip-select 16	Taken low to indicate that the current data transfer is a 16-bit (single wait state) memory operation.

Table A5.2 (*cont.*)

Pin No.	Abbrev.	Direction	Signal	Function
D2	$\overline{\text{IOCS16}}$	I	I/O chip-select 16	Taken low to indicate that the current data transfer is a 16-bit (single wait state) I/O operation.
D3	IRQ10	I	Interrupt request level 10	Taken high by an I/O device when it requires service.
D4	IRQ11	I	Interrupt request level 11	Taken high by an I/O device when it requires service.
D5	IRQ12	I	Interrupt request level 12	Taken high by an I/O device when it requires service.
D6	IRQ13	I	Interrupt request level 13	Taken high by an I/O device when it requires service.
D7	IRQ14	I	Interrupt request level 14	Taken high by an I/O device when it requires service.
D8	$\overline{\text{DACK0}}$	O	Direct memory access acknowledge level 0	Taken low to acknowledge a DMA request on the corresponding level.
D9	DRQ0	I	Direct memory access request level 0	Taken high when a DMA transfer is required. The signal remains high until the corresponding DACK line goes low.
D10	$\overline{\text{DACK5}}$	O	Direct memory access acknowledge level 5	Taken low to acknowledge a DMA request on the corresponding level.
D11	DRQ5	I	Direct memory access request level 5	Taken high when a DMA transfer is required. The signal remains high until the corresponding DACK line goes low.

Table A5.2 (*cont.*)

Pin No.	Abbrev.	Direction	Signal	Function
D12	$\overline{\text{DACK6}}$	O	Direct memory access acknowledge level 6	Taken low to acknowledge a DMA request on the corresponding level.
D13	DRQ6	I	Direct memory access request level 6	Taken high when a DMA transfer is required. The signal remains high until the corresponding DACK line goes low.
D14	$\overline{\text{DACK7}}$	O	Direct memory access acknowledge level 7	Taken low to acknowledge a DMA request on the corresponding level.
D15	DRQ7	I	Direct memory access request level 7	Taken high when a DMA transfer is required. The signal remains high until the corresponding DACK line goes low.
D16	+5 V	n.a.	+5 V d.c.	Supply voltage rail
D17	$\overline{\text{MASTER}}$	I	Master	Taken low by the I/O processor when controlling the system address, data and control bus lines.
D18	GND	n.a.	Ground	Ground/common 0 V

BIBLIOGRAPHY

Introductory texts

J.M. Ivison. *Electric Circuit Theory*. Van Nostrand Reinhold. 1977.
J.E. Lackey, M.D. Hehn and J.L. Massey. *Fundamentals of Electricity and Electronics*. Holt, Rinehart and Winston, 1983.
I. McKenzie Smith. *Hughes Electrical Technology*, 6th edn. Longman, 1987.

Chapter 1

D.V. Morgan and K. Board. *An Introduction to Semiconductor Microtechnology*. Wiley, 1983.
J. Seymour. *Electronic Devices and Components*. Pitman, 1981.
S.M. Sze. *Physics of Semiconductor Devices*. Wiley, 1981.

Chapter 2

Siliconix Inc Engineering Staff. *Designing with Field-Effect Transistors*, 2nd edn. McGraw-Hill, 1990.
T.D. Towers and N.S. Towers. *Towers' International Mospower and other Fet Selector*. Foulsham, 1983.

Chapter 3

M.G. Scroggie and S.W. Amos. *Foundations of Wireless and Electronics*, 10th edn. Iliffe, 1984.

Chapter 5

M.J. Buckingham. *Noise in Electronic Devices and Systems*. Ellis Horwood, 1983.

R.A. Penfold. *Preamplifier and Filter Circuits*. Babani, 1991.
R.A. Penfold. *High Power Audio Amplifier Construction*. Babani, 1991.

Chapter 7

A.W. Scott. *Understanding Microwaves*. Wiley, 1993.
J.B. Scott. *Analog Electronic Design*. Prentice Hall, 1991.

Chapter 9

M. Brown (Motorola). *Practical Switching Power Supply Design*. Academic Press, 1990.
C.W. Lander. *Power Electronics*. 3rd edn. McGraw-Hill, 1993.
V. Martin. *IC Voltage Regulator Sourcebook with Experiments*. Tab, 1983.

Chapter 11

G.B. Clayton. *Operational Amplifiers*, 2nd edn. Butterworths, 1979.
W.G. Jung. *IC Op-Amp Cookbook*, 3rd edn. Howard Sams, 1986.

Chapter 12

R.J. Matthys. *Crystal Oscillator Circuits*. Wiley, 1983.

Chapter 13

J.R. Gibson. *Electronic Logic Circuits*, 3rd edn. Arnold, 1992.
High Speed CMOS Logic Data Book. Texas Instruments Ltd, 1991.
LVT Low Voltage Technology. Texas Instruments Ltd, 1992.
The TTL Data Book for Design Engineers (volumes 1 and 2). Texas Instruments Ltd, 1989.

Chapter 14

P. Abel. *IBM PC Assembly Language and Programming*. Prentice Hall, 1991.
J. Locha and P. Norton. *Assembly Language for the PC*. Brady, 1989.
M. Minasi. *The Complete PC Upgrade and Maintenance Guide*, 3rd edn. Sybex, 1994.
C.G. Morgan. *The Microcomputer in the Laboratory*. Sigma, 1983.
P. Norton. *Inside the IBM PC and PS/2*, 4th edn. 1991.
R.A. Penfold. *Interfacing P.C.s and Compatibles*. Babani, 1992.
I. Robertson. *PC Pocket Book*. Newnes, 1994.

A.P. Stephenson and D. J. Stephenson. *Advanced Machine Code Techniques for the BBC Micro*. Granada, 1984.

M. Tooley. *P.C. Troubleshooting Pocket Book*. Newnes, 1994.

General

A.M. Ball. *Semiconductor Data Book*, 11th edn. Newnes-Butterworths, 1981.

A. Basak. *Analogue Electronic Circuits and Systems*. Cambridge University Press, 1991.

K. Blair Benson and D.G. Fink. *Advanced Television for the 1990s*. McGraw-Hill, 1991.

D. Boswell and M. Wickham. *Surface Mount Guidelines*. McGraw-Hill, 1992.

P. Horowitz and W. Hill. *The Art of Electronics*, 2nd edn. Cambridge University Press, 1989.

R.M. Marston. *Electronic Circuits Pocket Book*. Newnes, 1991.

R.A. Pease. *Troubleshooting Analog Circuits*. Newnes, 1991.

R.A. Penfold. *How to Design and Make Your Own PCBs*. Babani, 1983.

T.D. Towers. *Towers' International Transistor Selector (Update four)*. Foulsham, 1990.

E. Trundle. *Newnes Guide to TV & Video Technology*, 2nd edn. Newnes, 1992.

Index